Fractional Calculus and the Future of Science

Fractional Calculus and the Future of Science

Editor

Bruce J. West

MDPI • Basel • Beijing • Wuhan • Barcelona • Belgrade • Manchester • Tokyo • Cluj • Tianjin

Editor
Bruce J. West
Center for Nonlinear Science
University of North Texas
Denton, TX
USA

Editorial Office
MDPI
St. Alban-Anlage 66
4052 Basel, Switzerland

This is a reprint of articles from the Special Issue published online in the open access journal *Entropy* (ISSN 1099-4300) (available at: https://www.mdpi.com/journal/entropy/special_issues/ fract_future).

For citation purposes, cite each article independently as indicated on the article page online and as indicated below:

LastName, A.A.; LastName, B.B.; LastName, C.C. Article Title. *Journal Name* **Year**, *Volume Number*, Page Range.

ISBN 978-3-0365-2826-7 (Hbk)
ISBN 978-3-0365-2827-4 (PDF)

Contents

About the Editor

Bruce J. West Prof. Dr. Bruce J. West was Chief Scientist Mathematics, US Army Research Office, 1999–2021 (retired). He has worked on the development of the fractional calculus for the modeling of complex phenomena and his book Physics of Fractal Operators (with Bologna and Grigolini, Springer, 2003) received the Army Research Laboratory Award for Publication in 2003; Fractional Calculus View of Complexity: Tomorrow's Science (CRC Press, 2016) and Nature's Patterns and the Fractional Calculus (Walter de Gruyter, 2017). Among many awards he received: Meritorious Senior Professional Presidential Rank Award 2012 and Distinguished Senior Professional Presidential Rank Award 2018. Before coming to ARO, Dr. West was Professor of Physics, University of North Texas, 1989-1999; Chair of the Department of Physics 1989-1993. During his time at the university, he did research into the quantum manifestations of chaos (energy level repulsion, ionization rate enhancement, breakdown of the Correspondence Principle); the foundations of statistical mechanics (getting random fluctuations without statistics, failure of the Green-Kubo relation, Lévy statistics); and nonlinear processing techniques applied to biomedical phenomena. He was elected a Fellow of the American Physical Society in 1992; he received the Decker Scholar Award in 1993 and the UNT President's Award for research in 1994. Prior to joining the university, Dr. West was Director, Division of Applied Nonlinear Science, La Jolla Institute, 1983-1989. During this period, he worked on the development of nonlinear dynamical models of biomedical phenomena, physical oceanography, and the statistical mechanical foundations of thermodynamics. Specifically, he helped developed ways to use renormalization group concepts to extract pattern information from biomedical time series. This latter research eventually led to the development of a new discipline, Fractal Physiology (with Bassingthwaighte and Leibovitch, Oxford University Press, 1994), explaining how fractals have revolutionized the way we think about dynamics in the human body. The prequel to this book was his Fractal Physiology and Chaos in Medicine (World Scientific, 1990). Dr. West was Associate Director, Center for the Studies of Nonlinear Dynamics, La Jolla Institute, 1979-1983. He applied some of the newly emerging concepts in nonlinear dynamics systems theory to nonlinear water wave fields and turbulence. He also explained how the branching structure of the lung and other such physiological structures could be described by scaling, this led to a totally new fractal model for the architecture of the mammalian lung.

MDPI

Editorial

Fractional Calculus and the Future of Science

Bruce J. West

Center for Nonlinear Science, University of North Texas, Denton, TX 76203, USA; brucejwest213@gmail.com

Citation: West, B.J. Fractional Calculus and the Future of Science. *Entropy* **2021**, *23*, 1566. https://doi.org/10.3390/e23121566

Received: 15 November 2021
Accepted: 19 November 2021
Published: 25 November 2021

The invitation to contribute to this anthology of articles on the fractional calculus (FC) encouraged submissions in which the authors look behind the mathematics and examine what must be true about the phenomenon to justify the replacement of an integer-order derivative with a non-integer-order (fractional) derivative (FD) before discussing ways to solve the new equations. The desired articles are intended to provide the reader with a window into the future of specific science disciplines by peering through the lens of the fractional calculus (FC) and suggesting how what is seen entails a difference in thinking about that area of science. Thus, a perfect submission would be more about the implications and utility of the FC than about its formal structure in chemistry, epidemiology, sociology, psychology, physics, or any of the other scientific disciplines. Imaginative articles that implement FC in new and interesting ways that reveal its transformational nature, including, but not limited to, such things as: how a fractional derivative in time incorporates memory into the solution of the dynamic description of an earthquake, a brain quake, or a crash in the stock market; how the fractional derivative in space incorporates spatial non-locality into the solution of the complex dynamical descriptions of a riot, the collective intelligence of social groups, or the neuronal activity of the brain. Finally, we are interested in how the combined fractional derivatives in time and space of functional measures of uncertainty incorporate both memory and nonlocality into the phase space solution to capture the limited uncertainty of an ensemble of fractal trajectories, or the scaling behavior of complex dynamical networks.

As West points out [1], Sir Isaac Newton transformed *Natural Philosophy* into today's *Science* by focusing on the fundamental nature of motion, and he did so by using geometry in a way that resonated with the scientific community of his day. What Newton accomplished was to reveal what was entailed by fluxions (the differential calculus) without explicitly referencing them and in so doing he convinced generations of scientists of the value of their analyzing how physical phenomena change in space and over time. Whether by conscious plan or by serendipity, Newton cleared a path for less talented investigators to follow and contribute to the nascent discipline of mechanics and thereby determined the direction of quantitative reasoning in physics for the next three centuries.

This was the starting point for a discussion of how science, through its intermittent turning of its investigative tools on itself, has reached another epoch of transition. However, unlike the paradigm shifts within a scientific discipline introduced by Thomas Kuhn in 1962 the present shift addresses the whole of science. West argues that the dominance of Newton's world view is drawing to a close and he weaves the threads of chaos theory, fractals, non-ergodic behavior of dynamic systems and fractional kinetic theory into a fractal tapestry of the physical, social, and life sciences. It is the complexity of this tapestry that is shown to be fundamentally incompatible with Newton's notions of space and time.

Angstmann and Henry (AH) [2] address the discipline of chemistry, specifically the equations governing the time evolution of population densities of chemical species taking into account the spatial movement and their interactions using reaction–diffusion equations. They discuss the failure of classical techniques to properly describe diffusion and review fractional subdiffusion resulting from particles being trapped for arbitrarily long times, which are modeled using the continuous time random walk (CTRW) model. This fractional subdiffusion is characterized by the mean square displacement of a chemical population

1

spreading as a sublinear power law in time and is described using a Caputo fractional derivative in time in a reactive–subdiffusion equation.

AH emphasize that the main lessons learned from their analysis: (i) The governing equations are different depending on whether newborn particles inherit the waiting times of their parents. (ii) Birth and death terms must be treated differently. (iii) In the case where particles are removed, but not instantaneously at the start of the waiting times between jumps, the reaction and subdiffusive terms are not additive. They go on to explore the analytic solution to several exemplars, including a mass-conserving tempered time fractional diffusion equation that is subdiffusive for short times but manifests standard diffusion at long times.

Machado [3] chose to explore the fractal nature of financial time series using the Dow Jones industrial average (DJIA) but avoided using standard time series analysis. Instead, he uses multidimensional scaling (MDS) together with the concepts of distance, entropy, fractal dimension, and the FC. Introducing ten distinct definitions of distance, most of which I had not previously encountered, he was able to generalize MDS as an extension of the traditional metric formulation to construct a smooth non-Euclidean space. In this space, the fractal dimension and entropy measures for analyzing the three-dimensional portraits produced by the generalized MDS are interpreted.

Several known relations between the fractal dimension, using the box counting definition, and the idea that a random variable's entropy is its average level of "information" of the corresponding PDF, are used to define the information as an entropy of non-integer order α. This parameter α gives an extra degree of freedom to adapt the sensitivity of the entropy calculation to each specific dataset. Machado points out that time is viewed as a continuous and linear flow so that any perturbation is automatically assigned to the variable under analysis. Stated differently, since people are entities immersed in the time flow, apparently, we are incapable of distinguishing between perturbations in the time or the measured variable. Machado's analysis explores an alternative strategy to that of time series analysis for reading the relationship between the variables.

In the reaction–diffusion equation discussed by AH the chemical reaction term was modeled using the Verhulst equation for population growth. In 1798, Malthus argued that the integer-order rate equation produced an exponential population growth and 40 years later Verhulst replaced the constant growth rate of Malthus with one which decreased linearly with the growing population to mitigate the dire predictions of Malthus. The Verhulst (logistic) equation has the advantage of being one of the few nonlinear equations that has an analytic solution depicting a sigmoidal growth to a finite maximum population. Izadi and Srivastava (IS) [4] recount this bit of history and note that the integer-order derivative has been replaced by a fractional derivative in applications in numerous disciplines of science and engineering. They go on to point out that for most fractional differential equations (FDEs) there is to date no possibility of finding an exact analytic solution.

IS provide a brief review of the analytic and numerical methods that have been developed and applied for the FDEs which are based upon various loosely related models of real-world problems. Given the popularity of the fractional logistic equation (FLE) in the modeling of such phenomena as the growth of tumors in medicine, IS use it as a prototype on which to utilize the local discontinuous Galerkin (LDG) discretization approach for numerically solving the FLE. Given that the FLE has a second-order non-linearity IS rewrite it as two linear first order FDEs to apply the LDG scheme. Consequently, LDG is employed to discretize the resulting system, as well as the fractional operator. The mathematical details of the technique are presented along with comparison with alternative numerical and approximation approaches.

The transport of particles through continuous media is described by transport equations often based on general principles, such as conservation of energy and momentum. As pointed out by Masoliver [5], in general these transport equations are unsolvable non-linear integrodifferential equations. He elected to discuss transport processes using the telegrapher's equation (TE) and its generalization to the fractional case, emphasizing that random walk (RW) models are fundamental in describing TEs because they try to reproduce the microscopic mechanisms of transport. In this way he accounts for "diffusion with finite velocity".

The integer-order TE (IOTE) at early times behaves like a wave front and at late times like ordinary diffusion, it is generalized to a fractional-order TE (FOTE) in transport through highly disordered media, for instance, random media or fractal structures. Masoliver shows how this is done using fractional RWs resulting in a three-dimensional FOTE that is fractional is both space and time. The two different dynamics governing transport are fractional wave behavior at early times and fractional diffusion behavior as late times. Masoliver also shows analytic solutions for various combinations of IO-time, FO-space, FO-time and IO-space derivatives in one, two and three dimensions.

There are some investigators whose papers I always anticipate reading because I know that in addition to my gaining technical knowledge the author will put their work into a context that I would have missed left to my own resources. Professor Mainardi [6] falls into that category and his review paper falls into another category as well, that being: "Everything you wanted to know about _?_ but were afraid to ask.", here the blank is filled in with the Mittag–Leffler function (MLF). Just as the exponential function is the workhorse of linear IO differential equations, the MLF is the workhorse of linear FO differential equations, and the latter reduces to the former when the FO index goes to unity. His paper is written with the skill and insight that only one who has worked with the leaders in the field and has himself made lasting contributions to our understanding could manage.

His survey interleaves and draws connections among stochastic processes, such as the fractional Poisson process, the thinning of renewal processes, using the MLF. The CTRW is used to generalize the classical Kolmogorov–Feller equation (KFE) to the fractional KFE (FKFE) where the MLF is the waiting-time PDF. The fractional diffusion-wave equation has solutions to boundary value problems in terms of Wright functions that are inverse Laplace transforms of two parameter MLFs and are Greens functions in the solutions to specific boundary value problems.

Big Data (BD) and Machine Learning (ML) are two of the more visible areas of research in which investigators are working to span the gap separating the understanding based on modeling in social and life sciences from the more quantitative models of physics and engineering. Niu et al. (NCW) [7] maintain that the future success of these research activities is tied to the successful application of the FC and fractional order thinking (FOT) to the understanding of complex systems, to improving the processing and control of those systems and even to extending the enabling of creativity itself. The heart of the matter is that BD and ML seek to characterize complexity and of the ten characteristics used to describe BD *variability* is selected by NCW as the most important.

The complexity observed in most BD is almost invariably manifest through inverse power law (IPL) resulting from the processed data. The heavy-tailed nature of multiple PDFs is discussed along with its connection to the FC through fractional diffusion equations that are fractional in space, in time, or both. A variety of fractional discrete data processing techniques are sketched out to model the variability of BD along with a discussion of the CTRW. The key for the learning process is the optimization method and NCW inquire into how to use the FC to improve on existing methods of optimization in ML.

A Skellam distribution is generated by taking the difference between two independent Poisson random variables, which results in an integer valued Lévy process. Gupta et al. (GKL) [8] discuss a time fractional Skellam process that describes the inter-arrival times between positive and negative jumps as a MLF distribution rather than an exponential distribution and this formulation has been applied to financial and competitive games datasets. GKL show how the formalism is extended to a Skellam process of order k. A Skellam process is used to model the difference between the number of goals between two teams in a soccer match. Similarly, a Skellam process of order k can be used to model the difference in the number of points scored by two competing teams in a basketball match where $k = 3$, meaning there are three distinct ways to score points. Elsewhere, the authors show that a fractional Skellam process is better at modeling a tick-by-tick financial dataset than the Skellam process, or equivalently that the MLF is superior to the exponential in describing the inter-arrival times between successive ticks.

The FC has the potential to improve the performance of control systems as demonstrated by Zheng et al. (ZLCW) [9]. They argue that the improvement is due to the greater flexibility in modeling of systems and in the controller design methodology using fractional-order proportional-integral-derivative (FOPID) which is a generalization of the classical PID controller. ZLCW point out that although the FOPID controller provides better performance it is also more difficult to implement. Consequently, rather than presenting a general theory they present a case study to demonstrate the advantages of the proposed method. The classical frequency-domain method is the analytic design method for the FOPID controller used in the case study.

Digital watermarking is a form of embedding a signal (a watermark) within another signal known as the cover, which might be a digital media, such as an image, audio, video, or other digital media, and has become popular as a copyright enforcement tool in the last few decades. Gonzalez-Lee et al. (GVMNPL) [10] explore the advantages of a FC watermarking system for detecting Gaussian watermarks. They briefly critique multiple FC strategies that have been adopted to replace the linear additive rule more commonly used to watermark a signal. Watermarking includes using fractional derivatives, since there is a relationship between the order of the derivative and the resulting function; fractional Fourier transforms (random, continuous, and discrete), since there is a strong dependency between the orders and the resulting coefficient set; as well as, fractional Wavelet transforms.

CVMNPL emphasize that all the techniques discussed that use the FC have the same starting point and the overall difference among them is the use of some transform coefficients set for watermarking. In a previous work, the authors investigated the case of Gaussian watermarks and their results suggested that the FC reduces the false positives percentages (FPP). In that earlier work, however, they had limited testing and a deeper study of the fractional scheme for detecting Gaussian watermarks was called for. The present work accomplishes this task and confirms that the fractional scheme is reliable for the unambiguous (error-free) detection of Gaussian watermarks.

Song and Karniadakis (SK) [11] open their contribution to the anthology with the assertion that the modeling of wall-bounded turbulent flows is presently an unsolved problem in classical physics. They propose a fundamentally new approach for modeling the entire average velocity profile from the wall to the centerline of the pipe based on the FC. They were surprised to find that representing the Reynolds stresses with a non-local variable-order fractional derivative that decays with distance from the wall results in a universal form for all Reynolds numbers for channel flow, pipe flow, and Couette flow.

A remarkable feature of this paper is the exhaustive numerical testing of the new theoretical results against existing datasets from direct numerical simulation of the equations, as well as from experimental measurements. Taken together these results support the hypothesis that the rate of turbulent diffusion changes continuously with distance from the wall and the strong non-locality of turbulent interactions intensify away from the wall.

The final paper in this anthology presents an overview of the rapidly expanding area of Distributed-Order Fractional Calculus (DOFC) by Ding et al. (DPSS) [12]. DOFC generalizes the intrinsic multiscale nature of constant-order and variable-order fractional operators, which provides new ways to think about and model systems whose behavior emerges from the complex interplay and superposition of non-local and memory effects across a multitude of scales. They discuss the various ways the fractional order in space and/or time can be distributed and review the multiple ways these equations can be numerically integrated. The areas of application on which they focus are engineering and the physical sciences, with applications to viscoelasticity, transport processes and control theory taking center stage.

Mechanisms, such as multiple relaxation time in viscoelasticity, multiple temporal and spatial effects in transport processes, and mixtures of time delays in control theory, have all illustrated the significance of DOFC over more traditional integer-order methods. This review provides a glimpse into the various ways the DOFC has established its utility in the modeling of previously unsolved or partially solved complex problems. Hopefully, the attentive reader will see a way in which the DOFC may provide insight into a problem they have put on the backburner because they could not see a way forward.

Conflicts of Interest: The author declare no conflict of interest.

References

1. West, B.J. Sir Isaac Newton Stranger in a Strange Land. *Entropy* **2020**, *22*, 1204. [CrossRef] [PubMed]
2. Angstmann, C.N.; Henry, B.I. Time Fractional Fisher–KPP and Fitzhugh–Nagumo Equations. *Entropy* **2020**, *22*, 1035. [CrossRef] [PubMed]
3. Machado, J.A.T. Fractal and Entropy Analysis of the Dow Jones Index Using Multidimensional Scaling. *Entropy* **2020**, *22*, 1138. [CrossRef] [PubMed]
4. Izadi, M.; Srivastava, H.M. A Discretization Approach for the Nonlinear Fractional Logistic Equation. *Entropy* **2020**, *22*, 1328. [CrossRef] [PubMed]
5. Masoliver, J. Telegraphic Transport Processes and Their Fractional Generalization: A Review and Some Extensions. *Entropy* **2021**, *23*, 364. [CrossRef] [PubMed]
6. Mainardi, F. Why the Mittag-Leffler Function Can Be Considered the Queen Function of the Fractional Calculus? *Entropy* **2020**, *22*, 1359. [CrossRef] [PubMed]
7. Niu, H.; Chen, Y.; West, B.J. Why Do Big Data and Machine Learning Entail the Fractional Dynamics? *Entropy* **2021**, *23*, 297. [CrossRef] [PubMed]
8. Gupta, N.; Kumar, A.; Leonenko, N. Skellam Type Processes of Order k and Beyond. *Entropy* **2020**, *22*, 1193. [CrossRef] [PubMed]
9. Zheng, W.; Luo, Y.; Chen, Y.; Wang, X. A Simplified Fractional Order PID Controller's Optimal Tuning: A Case Study on a PMSM Speed Servo. *Entropy* **2021**, *23*, 130. [CrossRef] [PubMed]
10. Gonzalez-Lee, M.; Vazquez-Leal, H.; Morales-Mendoza, L.J.; Nakano-Miyatake, M.; Perez-Meana, H.; Laguna-Camacho, J.R. Statistical Assessment of Discrimination Capabilities of a Fractional Calculus Based Image Watermarking System for Gaussian Watermarks. *Entropy* **2021**, *23*, 255. [CrossRef] [PubMed]
11. Song, F.; Karniadakis, G.E. Variable-Order Fractional Models for Wall-Bounded Turbulent Flows. *Entropy* **2021**, *23*, 782. [CrossRef] [PubMed]
12. Ding, W.; Patnaik, S.; Sidhardh, S.; Semperlotti, F. Applications of Distributed-Order Fractional Operators: A Review. *Entropy* **2021**, *23*, 110. [CrossRef] [PubMed]

Article

Sir Isaac Newton Stranger in a Strange Land

Bruce J. West

Office of the Director Army Research, Research Triangle Park, NC 27709, USA; brucejwest213@gmail.com

Received: 27 September 2020; Accepted: 19 October 2020; Published: 25 October 2020

Abstract: The theme of this essay is that the time of dominance of Newton's world view in science is drawing to a close. The harbinger of its demise was the work of Poincaré on the three-body problem and its culmination into what is now called chaos theory. The signature of chaos is the sensitive dependence on initial conditions resulting in the unpredictability of single particle trajectories. Classical determinism has become increasingly rare with the advent of chaos, being replaced by erratic stochastic processes. However, even the probability calculus could not withstand the non-Newtonian assault from the social and life sciences. The ordinary partial differential equations that traditionally determined the evolution of probability density functions (PDFs) in phase space are replaced with their fractional counterparts. Allometry relation is proven to result from a system's complexity using exact solutions for the PDF of the Fractional Kinetic Theory (FKT). Complexity theory is shown to be incompatible with Newton's unquestioning reliance on an absolute space and time upon which he built his discrete calculus.

Keywords: complexity; chaos; fractional calculus; subordination

1. Introduction

Three centuries ago, Newton transformed Natural Philosophy into today's Science by focusing on change and mathematical quantification and he did so in a way that resonated with the scientific community of his day. His arguments appeared to be geometric in character, and nowhere in the *Principia* do you find explicit reference to fluxions, or to differentials. What Newton did was reveal the entailments of the calculus and convince generations of scientists of the value of their focusing on how physical objects change their location in time. Some contemporary mathematicians of his generation recognized what he had done, but their number can be counted on one hand, and their comments are primarily of historical interest.

Fast forward to today, where modern science, from Anatomy to Zoology, is seen to have absorbed the transformational effect of Newton's contribution to how we quantitatively and qualitatively understand the world, the fundamental importance of motion. However, it has occurred to a number of the more philosophically attuned contemporary scientists that we are now at another point of transition, where the implications of complexity, memory, and uncertainty have revealed themselves to be barriers to our future understanding of our technological society. The fractional calculus (FC) has emerged from the shadows as a way of taming these three disrupters with a methodology capable of analytically smoothing their singular natures.

If Sir Isaac Newton were reincarnated into the modern world would he again achieve scientific greatness using his prodigious intellect? Of course we cannot know the answer to this counterfactual, but what we can determine is whether his fundamental assumptions upon which the physical laws of analytic mechanics are based remain valid in the today's world of complexity science. Whether or not Newton would remain a stranger in this strange land of today's science is the question we seek to answer in this essay. Not literally, of course, but more to the point whether the fundamental assumptions on which his mechanics is based can be sufficiently modified to be compatible with the mathematics found necessary to describe today's complex phenomena, without being distorted to the point of being abandoned. Can Newton's view of the world be made compatible with the FC?

The FC moldered in the mathematical backwaters for over 300 years. Since the time of Newton it was mostly ignored by the social, physical, and life scientists, intermittently emerging from the shadows of formalism with an application. Historically, the community of international physical scientists saw no need for a new calculus, or if occasionally seeing the need thought it not worthy of acknowledgment. The community agreed that the ordinary differential calculus of Newton and Leibniz, along with the analytic functions entailed by solving the equations resulting from Newton's force law, are all that is required to provide a scientific description of the macroscopic physical world.

In his *Mathematical Principles of Natural Philosophy* [1], Newton introduced mathematics into the study of *Natural Philosophy*. He argued the need for quantification of scientific knowledge through the introduction of mathematics in the form of *fluxions* and thereby changed the historical goal of natural philosophy from that of wisdom to that of knowledge. This new term fluxion does not appear anywhere in the *Principles*, but scholars have found numerous geometric arguments, which, in fact, were in all probability based on limits in which Newton, no doubt, had differentials in the back of his mind. The Marquis de l'Hôpital commented that Newton's *magnum opus* was "a book dense with the theory and application of the infinitesimal calculus"; an observation also made in modern times by Whiteside [2].

Along with mathematics, Newton also introduced a number of definitions that determined how scientists were to understand his vision of the physical world for the next few hundred years. We do not quote his definitions of such well-known things as inertia and force here, but instead we record the notions of space and time that he believed were the accepted understanding of their meanings as explained in his first scholium (A scholium is a marginal note or explanatory comment made by a scholar), which are [1] as follows.

I   Absolute, time, and mathematical time, of itself, and from its own nature, flows equably without relation to anything external, and by another name is called duration: relative, apparent, and common time, is some sensible and external (whether accurate or unequable) measure of duration by the means of motion, which is commonly used instead of true time; such as an hour, a day, a month, a year.

II   Absolute space, in its own nature, without relation to anything external, remains always similar and immovable. Relative space is some movable dimension or measure of the absolute space; which our senses determine by its position to bodies; and which is commonly taken for immovable space; such is the dimension of subterraneous, an aerial, or celestial space, determined by its position in respect of the earth. Absolute and relative space are the same in figure and magnitude; but they do not remain always numerically the same. For if the earth, for instance, moves, a space of our air, which relatively and in respect of the earth remains always the same, will at one time be one part of the absolute space into which the air passes; at another time it will be another part of the same, and so, absolutely understood, it will be continually changed.

Newton's understanding of these two notions of the absolute are what enabled him to invent fluxions and introduce motion as the basis for his new physics. Of course, the mathematically awkward discrete notation of fluxions was subsequently elbowed out of history by the user-friendly notation of Leibniz, which became known as the differential calculus. The differential calculus enabled subsequent generations of scientists to describe the motion of particles in terms of continuous single particle trajectories in space and time. The differential calculus fills literally thousands of mathematics/physics text books; all assuming that I and II codify the real world and are taught to eager students and novitiate scientists throughout the world. Herein, we argue for a mathematics that provides a logical framework for understanding the more complex world of the Information Age, in which I and II must be applied with extreme caution, if at all.

The increase in sensitivity of diagnostic tools, advances in data processing techniques, and expanding computational capabilities have all contributed to the broadening of science in ways that have brought many phenomena from borderline interest to center stage. These curious complex processes are now cataloged under the heading of non-integer scaling phenomena. An understanding of the fundamental dynamics underlying such scaling requires a new mathematical perspective, such as that obtained using the dynamics described by non-integer (fractional) operators and such descriptions ushered in the sunset for much of what remains of Newton's world view.

Much of what is written in this Introduction will be familiar to those with a background in physics, even if the organization of the material is not. However the reasons why classical physics fails to explain a given complex phenomena remains a mystery to those without such a background as well as to many who do. Therefore, we express the purpose of this paper in the form of a hypothesis and present arguments in support of the Complexity Hypothesis (CH):

> Complex phenomena entailing description by chaos theory, fractional Kinetic Theory, or the fractional calculus in general, are incompatible with the ordinary calculus and consequently are incompatible with Newtonian Physics.

1.1. The Demise of Newton's World View?

The evidence is all around us that the domain of application of Newton's view of the physical world is contracting dramatically. His view was reluctantly contracted with the introduction of quantum mechanics along with relativity over a century ago. However, physicists took consolation in the fact that the dynamic predictions of the very fast, the very large, and the very small, all reduce to those of Newton in the appropriate limits. For special relativity, the dramatic changes in time occur as the speed of light is approached [3]; for general relativity, space curves in the neighborhood of a large mass [4]; and for quantum phenomena, the correspondence principle associated with the size of Planck's constant insures the quantized nature of energy is lost at large quantum numbers and energy is continuous on the scale we live our lives [5]. However, the more recent constrictions produced by chaotic dynamics is different; so much so that once made, there is no limit in which the view of Newton can reemerge. This requires more explanation, as the inappropriate application of the differential calculus to describe the dynamics of strongly nonlinear phenomena often yields misleading results. In the author's view, one such misinterpretation arose in support of the political interpretation of climate change.

It should be evident that the rubric *climate change* provides an example of such a misapplication of the nonlinear hydrodynamic partial differential equations that purport to describe the internal motion of the earth's atmosphere involving the multiple interactions with the earth's temperature field, solar radiation, cloud cover, and all the rest. Climate change is not just a problem in Newtonian physics, because if it were we would have the answer to the problem in hand, which some few scientists believe we do. I say this with full appreciation for the criticism such a statement will draw, from both the believers in climate change and the sceptics who do not. Let me be absolutely clear in stating that I believe in climate change, but belief is the wrong word. Climate change is a scientific fact not a matter of faith or belief. What I am skeptical about concerns the quasi-scientific arguments used in the political arena that assign causality of that change to human activity followed by the assertion that climate change can be significantly influenced by political action.

I came to this conclusion, not through a "eureka" moment, or flash of insight, but more through the weight of evidence drawn from my own scientific research. I even coauthored a book about it [6] with a colleague who was then a post-doctoral researcher of mine. Our book addressed climate change as a problem in physics and was greeted with a yawn from the scientific community. It was the last scientific contribution I made to that debate and the science has not moved significantly since its publication. My epiphany was that those who successfully communicate technically difficult ideas tell a story. Thus, I have decided to populate this essay with a sequence of technically-based stories.

Each one lending additional support to the CH. The first story concerns chaos theory and some of what that entails.

1.2. Chaos Theory

The chaos story begins in the middle nineteenth century with Oscar II, the King of Sweden and Norway, and his concern over how long the Earth will survive. More pointedly, he wondered whether the solar system was stable. Could one expect the moon to spiral out of its orbit and crash into the Earth? Would the Earth break from its timeless trajectory and collide with the Sun? Let me stop here and say this is the beginning of the somewhat romanticized historical account of how chaos came into being that I learned when I was first introduced to the "three-body problem" as a freshly mined minted physics PhD in 1970. The actual historical account is a bit more banal, but not much.

Oscar II had done well in mathematics while a university student and had grown into an active patron of the subject [7], so his sponsorship of a prize in mathematics, unrelated to any particular institution was not surprising. Mittag-Leffler, who was then the editor of the Swedish journal *Acta*, made the original announcement of the King's mathematics competition, in the science magazine *Nature*. In that announcement Mittag-Leffler listed four categories to which international scientists could submit contributions. The category concerning the stability of the solar system was written in the following arcane way [7].

(1) A system being given of a number whatever of particles attracting one another mutually according to Newton's law, it is proposed, on the assumption that there never takes place an impact of two particles to expand the coordinates of each particle in a series proceeding according to some known functions of time and converging uniformly for any space of time.

The committee that evaluated the submissions to the competition consisted of, along with Mittag-Leffler, two other giants of nineteenth century mathematics, Hermite and Weierstrass. To avoid any possibility of bias the entrants and their submissions remained anonymous until the winner was selected, at which time the name was to be published in *Acta*. Out of a field of 12 entrants, the committee selected Henri Poincaré, who had responded to question (1). He extended the analysis of the solvable two-body problem to the addition of one additional body, which was much less massive than the other two. Poincaré proved that the solution to Newton's dynamic equations for his restricted three-body problem could not have a simple analytic form. His published proof entailed the invention of new mathematics, the implications of which have kept the best mathematician in the world actively engaged for over a century.

In reviewing the prize-winning memoire for publication in *Acta*, a referee pointed out an error in the manuscript. Part of the drama associated with publishing the final version of the paper concerned the secrecy surrounding that error. Correcting this error entailed a major rewrite, which took Poincaré nearly a year to complete. In composing the revision, he conceived of and implemented in the manuscript the idea of a homoclinic point [7], which is the basis of our understanding of what today goes by the popular name of chaos theory. In short, he introduced the *Three-Body Problem* to the scientific community as being of fundamental importance and proved that the elliptic orbits of the two-body problem were replaced by orbits in the restricted three-body problem that resembled nothing so much as a plate of spaghetti. A single strand of entangled spaghetti was the convoluted trajectory of the third body and the asymptotic position of the body along that trajectory at any time was unpredictable. Today we call such trajectories fractals [8].

Sir James Lighthill, on the three-hundred-year anniversary of the communication of Newton's *Principia* to the Royal Society, and while he was president of the *International Union of Theoretical and Applied Mechanics*, published the paper *The recently recognized failure of predictability in Newtonian dynamics* [9]. In this paper, Lighthill traces the history of mechanics from Tycho Brahe collecting astronomical data as a court astronomer, through Poincaré's proof of the limited predictability horizon of Newton's law of the dynamics of mechanical systems. To put this in a proper perspective let us use Lighthill's words:

> We are all deeply conscious today that the enthusiasm of our forebears for the marvelous achievements of Newtonian mechanics led them to make generalizations in this area of predictability which, indeed, we may have generally tended to believe before 1960, but which we now recognize were false. We collectively wish to apologize for having misled the general educated public by spreading ideas about determinism of systems satisfying Newton's laws of motion that, after 1960, were to be proved incorrect…

This reluctant indictment of the Newtonian system of nonlinear partial differential equations that describe how the radiation from the sun is absorbed by the earth's atmosphere and redistributed around the globe has to the best of my knowledge never been explicitly refuted. This is not unexpected as Sir James was the scientific leader in the area of applied mathematics involving those same equations for over thirty years. If the unpredictability of coupled systems of nonlinear differential equation were expressed as a theorem, then one can draw a corollary regarding the nature of the computer simulations based on those same equations. The reader is free to infer from these remarks if Newton's view is truly dead or whether it is just confined to an ever decreasing domain of analytic application.

What we can conclude with certainty is that Newton's force law typically breaks down when the system being analyzed is not linear and the equations of motion are nonlinear. Such equations typically do not have analytic solutions, their solutions are generically chaotic [10,11]. As scientists, this loss of predictability, which is the foundation of the physical sciences, ought to be our greatest concern, or at least the mathematical foundation of all our physical models, the differential calculus, ought to be the focus of our concern.

It is worth mentioning that in his philosophical writings Poincaré recognized that his mathematical analysis entailed the loss of predictability and the existence of a new kind of chance [12]:

> A very slight cause, which escapes us, determines a considerable effect which we can not help seeing, and then we say this effect is due to chance. If we could know exactly the laws of nature and the situation of the universe at the initial instant, we should be able to predict exactly the situation of this same universe at a subsequent instant. But even when the natural laws should have no further secret for us, we could know the initial situation only *approximately*. If that permits us to foresee the subsequent situation *with the same degree of approximation*, this is all we require, we say the phenomenon has been predicted, that it is ruled by laws. But this is not always the case: it may happen that slight differences in the initial conditions produce very great differences in the final phenomena: a slight error in the former would make an enormous error in the latter. Predication become impossible and we have the fortuitous phenomenon.

For over a century, some of the world's leading mathematicians have been working on what might be a proper replacement for, or extension of, Newton's physics. They typically begin with the notion that a conservative nonlinear dynamical system with three or more degrees-of-freedom is chaotic [13], which means that its dependence on initial conditions is so sensitive that an infinitesimal change in the initial state will produce a trajectory that exponentially diverges from the trajectory predicted by the original state. Such an exponential separation of trajectories means that the perturbed state is unstable in the sense that its asymptotic location cannot be predicted from the initial state.

The work that Lighthill was alluding to in his remarks quoted earlier were those of the meteorologist Ed Lorenz, whose ground breaking paper opened the world of fluid dynamics to the importance of chaos [14], and ended dreams of long-term weather forecasting. Those that have considered chaos as a possible obstacle to climate forecasting as well, treat it in much the same way that the nineteen century physicists Maxwell and Boltzmann treated many-body effects to produce Kinetic Theory. Only now the modern climate physicist examines large-scale computer simulations of the earth's atmosphere as having random fluctuations around the average dynamical behavior of the atmosphere's velocity field and temperature. The established procedure is to carry out a large number of computer simulations, all starting from the "same state", and from them construct an ensemble of atmospheres with which to calculate the average dynamics of the interesting physical quantities.

The general impression in the meteorology community is that such ensemble averages ought to be sufficient to smooth out the influence of chaotic trajectories and thereby provide the appropriate phase space probability density function in the kinetic theory sense. The problem with the approach is when one actually attempts to average over an ensemble of chaotic trajectories the integer moments diverge leaving the coefficients ill-defined in the kinetic theory of Maxwell and Boltzmann. Here again we find a need for a new kind of mathematics and the fractional calculus comes to the rescue, providing a fractional Kinetic Theory (FKT).

In Section 2, we generalize the traditional phase space partial differential equations for the probability density function (PDF) to the fractional calculus. This is done by averaging over an ensemble of chaotic trajectories, and following the mathematical arguments of Zaslavsky [15] create a FKT. The solution to a simple fractional diffusion equation is shown to have a generic analytic form.

1.3. Allometry Relations

Scientists believed that phenomena whose dynamic description is the result of using non-integer operators, such as fractional derivatives, were interesting curiosities, but lay outside the mainstream of science. Even such empirical laws as allometry relations (ARs), in which the functionality of a system is related to a non-integer power of the system's size, were thought to have causal relations, with traditional differential dynamic descriptions [16–18]. Perhaps the most famous allometry relation is that between the average metabolic rates of mammals and their average total body masses (TBMs) as depicted by the "mouse-to-elephant" curve in Figure 1. In this figure, the solid curve is a fit to data by a power-law relation of the form

$$\langle Y \rangle = a \langle X \rangle^{b}, \tag{1}$$

which is a straight line on log-log graph paper with slope b :

$$\log \langle Y \rangle = \log a + b \log \langle X \rangle. \tag{2}$$

The functionality of the system Y, here the average metabolic rate is denoted by $\langle Y \rangle$ and the size of the system X, here the average TBM is denoted by $\langle X \rangle$. Note that the brackets here denote the empirical averaging process.

Historically such ARs were explained using biophysical arguments, for example, Sarrus and Rameaux [17,18] used simple geometrical arguments for heat transfer. They assumed the heat generated by a body is proportional to its volume and the heat is lost at the body's surface and is proportional to surface area. The balance between the two suggested that the allometry parameter is given by the ratio of dimensions to be $2/3$, which does not fit the data very well. The empirical value of the allometry parameter is $b \approx 0.74$, which was subsequently accounted for by using fractal scaling arguments [19]. A statistical technique based on the fractional calculus was developed in [20] to explain the averaging brackets in Equation (1), which in due course we use herein as an exemplar of complexity in the fractal statistics of physiological phenomena.

In Section 3, selected applications of the FC are presented with the intent of persuading the reader that as systems become more complex the value of the ordinary differential calculus to describe their behavior increasingly diminishes, until it is eventually nearly lost altogether. The analytic PDF that solves the simple FKT problem is shown to explain the empirical AR using a complexity-based arguments.

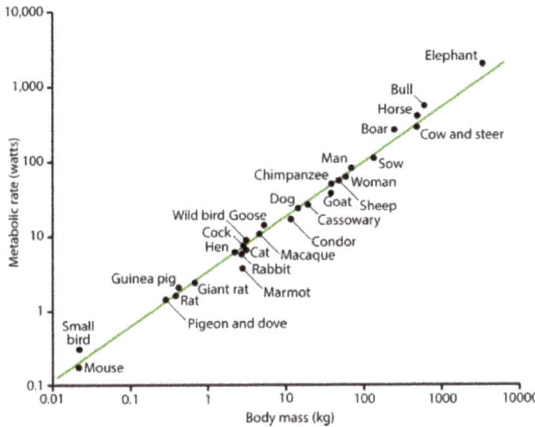

Figure 1. The mouse-to-elephant curve. The average metabolic rates of mammals and birds are plotted versus the average body weight (TBM) on log-log graph paper covering five orders of magnitude in size. The solid line segment is the best linear regression to the data from Schmidt-Neilson [18] with permission.

1.4. Another Time

The willingness of his contemporaries to accept Newton's view of time flowing as an uninterrupted featureless stream is understandable. However, the reluctance of physicists to directly challenge Newton's view of time outside extreme conditions in the physical sciences is unclear. This reluctance is not evident in psychology where everything we see, smell, taste and otherwise experience is in a continuous state of change. Consequently, the changes in the physical world are not experienced uniformly, which is another way of saying that there is an objective time associated with the physical and a subjective time associated with the psychological world. The physical scientists dismissed subjective time out of hand, prior to Einstein, but even after relatively the experiential time they accepted was considered to be a local physical time.

Here, we follow the discussion of Turalska and West [21]. The idea of different clocks telling different times arises naturally in physics; the linear transformation of Lorentz in relativistic physics being a familiar example. However, we are interested in the notion of multiple clocks in the biological and social sciences wherein they have begun distinguishing between cell-specific and organ-specific clocks in biology and person-specific and group-specific clocks in sociology [22]. Of course, the distinction between subjective and objective time dates back to the empirical Weber–Fechner Law [23] in the latter half of the nineteenth century.

While the global behavior of an organ, say the heart, might be characterized by apparently periodic cycles, the activity of single neurons demonstrate burstiness and noise. In a similar way people in a social group operate according to their individual schedules, not always performing particular actions in the same global time frame. Consequently, because of the stochastic behavior of one or both clocks, a probabilistic transformation between times is often necessary. An example of such a transformation is given by the subordination procedure.

Insight into the subordination procedure is provided if we begin by defining two clocks that operationalize time in two distinct ways. The ticking of the first clock records a subjective or operational discrete time n, which measures an individual's time $T(n)$. The ticking of the second clock records the objective or chronological time t, which measures the social time $T(t)$ upon which a society of individuals agree. If each tick of the discrete clock n is considered to be an event, the relation between operational and chronological time is given by the waiting time PDF of those events in chronological time $\psi(t)$. Assuming a renewal property for events, as given by a chain condition (convolution) from renewal theory in Section 2.1, one can relate operational to chronological time [21]:

$$\langle T(t) \rangle = \sum_{n=0}^{\infty} \int_0^t \Psi\left(t - t'\right) \psi_n\left(t'\right) T(n) dt' \tag{3}$$

Every tick of the operational clock is an event, which in the chronological time occurs at time intervals drawn from the renewal waiting-time PDF. This randomness entails the sum over all events and the result is an average over many realizations of the transformation. The last of the n events occurs at time t' and the survival probability $\Psi\left(t - t'\right)$ insures that no further event occurs before the time t.

For example, consider the behavior of a two-state operational clock, whose evolution is depicted in Figure 2, where the clock switches back and forth (tick tock) between its two states at equal time intervals. However, in chronological time this regular behavior is significantly distorted as seen in the figure. The time transformation was taken to be an inverse power law (IPL) waiting time PDF $\psi(t)$. Thus, a single time step in the operational time corresponds to a random time interval being drawn from $\psi(t)$ in chronological time. The tail of the IPL PDF leads to especially strong distortions of the operational time trajectory, as there exist a non-zero probability of drawing very large time intervals between events. However, as the transformation between the operational and chronological time scales involves a random process, one needs to consider infinitely many trajectories in the chronological time, which leads to the average behavior of the clock in the chronological time denoted in Equation (3) by brackets.

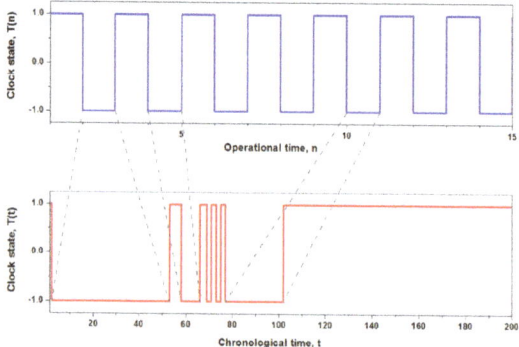

Figure 2. The upper curve is the regular transition between the two states of the individual in operational time. The lower curve is the subordination of the transition times to an IPL PDF to obtain chronological time.

Newton's view of homogeneous isotropic time is shown to be incompatible with multiple phenomena in the social and life sciences in Section 3.2 using subordination theory. In that section the disciplines of biophysics, psychophysics, and sociophysics, to the degree they have adopted the Newtonian viewpoint, are shown to be misleading. The complexity of these disciplines require a new calculus to describe their dynamics.

In Section 3.2, we establish a direct link between subordination theory and the FC. This has been done in the literature in a number of different ways. In Section 2, we show how the probability calculus can be generalized to the FC in order to include temporal memory and spatial heterogeneity with probability theory.

What is entailed by the results presented herein is discussed in Section 5 and some conclusions are drawn.

2. Fractional Kinetic Theory

Zaslavsky [24] considered chaotic dynamics, as a physical phenomenon, to be a bridge spanning the gap between deterministic and stochastic dynamic systems. The dynamic states in the first case are described by regular functions and in the second by kinetic or other probabilistic equations. He developed the mathematics for the fractional kinetics corresponding to chaotic dynamics that is intermediate between completely regular (integrable) and completely random cases. The kinetics become "strange" because some moments of the PDF are infinite and the Onsager Principle is violated in that it takes infinitely long for fluctuations to relax back to the equilibrium state. An alternative to the derivation of the fractional kinetic equation (FKE) given by Zaslavsky [24] is presented by West and Grigolini [25]. In this section we present the overlapping highlights of these two derivations in schematic form, emphasizing the physical interpretation.

2.1. Generalizing Kinetic Theory

We sketch Zaslavsky's arguments leading to the FKT resulting from the underlying dynamics being chaotic and consequently the dynamic trajectories being fractal. We begin with the chain condition of Bachelier, Smoluchowsky, Chapman, and Kolmogorov (BSCK) [26]:

$$P(x,t|x_0,t_0) = \int P(x,t|x',t')P(x',t'|x_0,t_0)dy, \tag{4}$$

where $P(x,t|x',t')$ is the probability density of having a particle at position x at time t if at time $t' \leq t$ the particle was at the point x'. We make the assumption that the PDF is stationary such that

$$P(x,t|x_0,t_0) = P(x,x_0;t-t_0), \tag{5}$$

corresponding to the regular scheme for the kinetic derivation [26] and with $\Delta t \equiv t - t_0$ we have for the initial condition

$$\lim_{\Delta t \to 0} P(x,x_0;\Delta t) = \delta(x-x_0). \tag{6}$$

The first generalization of the historical kinetic theory argument is made by taking into account the fractal nature of the set generated by the ensemble of chaotic trajectories initiated by an underlying non-integrable Hamiltonian. Inserting the time limit for a fractional time differential into the BSCK chain condition enables us to write

$$\partial_t^\alpha [P(x,t)] = \lim_{\Delta t \to 0} \frac{1}{\Delta t^\alpha} \int dy [P(x,y;\Delta t) - \delta(x-y)]P(y;t). \tag{7}$$

This expression can be simplified using a second generalization, that being introducing the generalized Taylor expansion

$$P(x,y;\Delta t) = \delta(x-y) + A_1(y;\Delta t)\delta^{(\beta)}(x-y) + A_2(y;\Delta t)\delta^{(\beta+1)}(x-y), \tag{8}$$

for a set characterized by the fractal dimension $0 < \beta \leq 1$. Inserting this expansion into Equation (7) simplifies the generalized BSCK chain condition by introducing the quantities

$$\mathcal{A}(x) \equiv \lim_{\Delta t \to 0} \frac{A_1(x;\Delta t)}{\Delta t^\alpha} = \lim_{\Delta t \to 0} \int dy \frac{|x-y|^\beta}{\Delta t^\alpha} P(x,y;\Delta t), \tag{9}$$

$$\mathcal{B}(x) \equiv \lim_{\Delta t \to 0} \frac{A_2(x;\Delta t)}{\Delta t^\alpha} = \lim_{\Delta t \to 0} \int dy \frac{|x-y|^{\beta+1}}{\Delta t^\alpha} P(x,y;\Delta t). \tag{10}$$

Zaslavsky [15] explained that the limit in these two expressions are the result of the fractal dimensionality of the space-time set along which the state of the system is meandering in the $\Delta t \to 0$ limit.

We do not reproduce the mathematical details from the open literature and instead jump to the result for the one-dimensional Fractional Kinetic equation (FKE) [15,25] and write the fractional Fokker–Planck equation (FFPE):

$$\partial_t^\alpha \left[P(x,t) \right] = \partial_{|x|}^\beta \left[\mathcal{A}(x) P(x,t) \right] + \partial_{|x|}^{\beta+1} \left[\mathcal{B}(x) P(x,t) \right]. \tag{11}$$

The FFPE has fractional indices in the domain $0 < \alpha, \beta \leq 1$, the fractional time derivative is of the Caputo form, and the fractional spatial derivative is of the symmetric Reisz–Feller form.

So how different are the solutions to the above FFPE from those of the ordinary FPE even when $\beta = 1$?

2.2. Solution to a Simple FKE

One of the simplest dynamical processes described by the FFPE having far-reaching implications has a constant fractional diffusion coefficient and a vanishing fractional velocity:

$$\mathcal{A}(x) = 0 \text{ and } \mathcal{B}(x) = K_\beta, \tag{12}$$

thereby reducing Equation (11) to

$$\partial_t^\alpha \left[P(x,t) \right] = K_\beta \partial_{|x|}^{\beta+1} \left[P(x,t) \right]. \tag{13}$$

This is one of the simplest form of anomalous diffusion, first discussed in terms of the continuous time random walk (CTRW) by Montroll and Scher [27].

The solution to this fractional diffusion equation is readily obtained by taking its combined Fourier–Laplace transform and introducing the notation

$$\mathcal{F}\left\{ \partial_{|x|}^{\beta+1} \left[f(x) \right]; k \right\} = -|k|^{\beta+1} \widetilde{f}(k), \tag{14}$$

where $\widetilde{f}(k)$ is the Fourier transform of $f(x)$ and correspondingly

$$\mathcal{L}\left\{ \partial_t^\alpha \left[g(t) \right]; u \right\} = u^\alpha \widehat{g}(u) - u^{\alpha-1} g(0) \tag{15}$$

where $\widehat{g}(u)$ is the Laplace transform of $g(t)$. Note that in Equation (14) we used the Fourier transform of the Reisz–Feller derivative in space and in Equation (15) we used the Laplace transform of the Caputo derivative in time. Consequently we obtain from the Fourier-Laplace transform of the FFPE:

$$u^\alpha P^*(k,u) - u^{\alpha-1} \widetilde{P}(k, t = 0) = -K_\beta |k|^{\beta+1} P^*(k,u), \tag{16}$$

where the asterisk denotes the double transform of the PDF and the indices lie in the interval $0 < \alpha, \beta \leq 1$. This equation is simplified for the initial value problem:

$$P(x, t = 0) = \delta(x) \implies \widetilde{P}(k, t = 0) = 1, \tag{17}$$

to the form

$$P^*(k,u) = \frac{u^{\alpha-1}}{u^\alpha + K_\beta |k|^{\beta+1}}. \tag{18}$$

The inverse Fourier–Laplace transform of this expression yields the solution to the initial value problem for the PDF.

Metzler and Klafter [28] derived the FFPE using the CTRW formalism of Montroll and Weiss [29] and reviewed the potential functions for various combinations of indices. It has also been derived using subordination theory by West [30]. The inverse Laplace transform of $P^*(k,u)$ yields the characteristic function

$$\widetilde{P}(k,t) = E_\alpha \left(-K_\beta \, |k|^{\beta+1} \, t^\alpha \right) \tag{19}$$

expressed in terms of the Mittag–Leffler function (MLF):

$$E_\alpha \, (z) = \sum_{n=0}^{\infty} \frac{z^{n\alpha}}{\Gamma \, (n\alpha + 1)}. \tag{20}$$

The inverse Fourier transform of the characteristic function yields the PDF solution

$$P(x,t) = \mathcal{F}^{-1} \left[E_\alpha \left(-K_\beta \, |k|^{\beta+1} \, t^\alpha \right) ; x \right]. \tag{21}$$

The simple substitution $k' = kt^\delta$ into Equation (21), with $\delta = \frac{\alpha}{\beta+1}$, after some algebra reduces the formal solution to

$$P(x,t) = \frac{1}{t^\delta} \mathcal{F}^{-1} \left[E_\alpha \left(-K_\beta \, |k'|^{\beta+1} \right) ; \frac{x}{t^\delta} \right], \tag{22}$$

or in a more familiar scaling form:

$$P(x,t) = \frac{1}{t^\delta} F \left(\frac{x}{t^\delta} \right), \tag{23}$$

where the new function is defined:

$$F \left(\frac{x}{t^\delta} \right) \equiv P \left(\frac{x}{t^\delta}, 1 \right). \tag{24}$$

The function $F(\cdot)$ is analytic in the scaled variable x/t^δ, is properly normalized and can therefore be treated as a PDF. For a standard diffusion process, $\alpha = 1$, in which case the MLF becomes an exponential so that for $\beta = 1$ the Fourier transform can be carried out and this function becomes a Gaussian with $\delta = 1/2$. When $\alpha = 1 \neq \beta$ the result is a stable Lévy process [26,31] with the Lévy index given by $0 < 1/\delta \leq 2$. However, for general chaotic systems there is a broad class of distributions for which the functional form is neither Gaussian nor Lévy.

Mainardi et al. [32] obtained a variety of other solutions to the FKE in terms of the properties of the MLF for $0 < \alpha < 1$. The inverse Fourier transform of the scaled PDF solution for $\beta = 1$ asymptotically relaxes as the IPL $t^{-\alpha/2}$.

2.3. Self-Similar Random Walks

Zaslavsky et al. [33] worked to visualize the underlying landscape produced by averaging over chaotic trajectories and to describe the formal structure uncovered by extensive numerical calculations. They discuss the notion of a "stochastic web" to characterize the chaotic dynamics generated by Hamiltionian systems in which "weak" chaotic orbits are concentrated on small measure domains of phase space thereby constituting a "web". They note that transport through stochastic webs could produce non-Gaussian, i.e., intrinsically anomalous, diffusion.

The nexus points of the web constitute traps were homoclinic points have dissolved into a spray of local points that locally entrap trajectories for IPL lengths of time. Exiting a trap the orbit undergoes a long–range flight having self-similar properties. The process can be realized as passing through the turnstiles of "cantori" [34]. This argument is realized by replacing the complete simulation of the Hamiltonian dynamics with a random walk (RW) containing the appropriate qualitative features. They do this by way of example whereby they construct a RW determined by a Weierstrass (W) function [35]. Consider the discrete probability described by the stepping PDF for the Weierstrass random walk (WRW) on a one-dimensional lattice with sites indexed by x [35]:

$$p(x) = \frac{a-1}{2a} \sum_{n=0}^{\infty} \frac{1}{a^n} \left[\delta_{x,b^n} + \delta_{x,-b^n} \right], \tag{25}$$

where a and b are dimensionless constants greater than one. and δ_{ij} is the Kronecker delta function: $\delta_{ij} = 1$ for $i = j$ and $\delta_{ij} = 0$ for $i \neq j$. We follow the analysis of this discrete process given by West and Grigolini [6]. The first notable property of the PDF generated by the WRW is that the second moment of this RW process diverges:

$$\left\langle x^2 \right\rangle = \frac{a-1}{a} \sum_{n=0}^{\infty} \left(\frac{b^2}{a} \right)^n , \tag{26}$$

for $b^2 > a$ as the series is infinite. The discrete Fourier transform of the PDF given by Equation (25) yields the discrete characteristic function

$$\widehat{p}(k) = \frac{a-1}{a} \sum_{n=0}^{\infty} \frac{1}{a^n} \cos\left[b^n k\right] . \tag{27}$$

This series was introduced by Weierstrass in 1872 in response to Cantor, a former student and subsequent colleague, who challenged him to construct an analytic function that is continuous everywhere but is nowhere differentiable. Thanks to Mandelbrot [8] we now know that this was the first consciously constructed fractal function and the divergence of the second moment is a consequence of its non-analytic properties.

As the WRW process unfolds the set of sites visited mimics the influence of localized chaotic islands, interspersed by gaps, nested within clusters of clumps over ever-larger spatial scales. The WRW generates a hierarchy of traps that are statistically self-similar, as suggested by Figure 3. The parameter a determines the number of subclusters within a cluster and the parameter b determines the scale size between clusters.

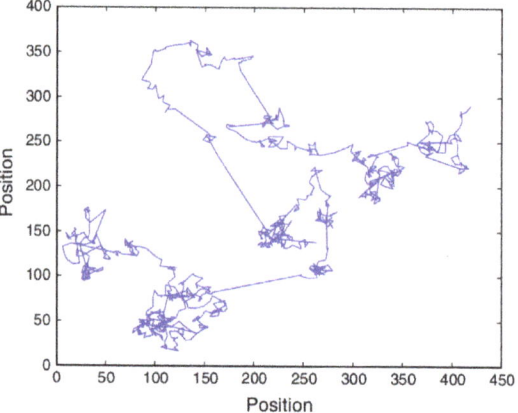

Figure 3. The landing sites for the WRW are depicted and the islands of clusters discussed in the text are readily seen.

The Weierstrass form of the characteristic function allows for a renormalization group (RG) solution [36] from which we can determine the scaling properties of the WRW. Scaling the argument of the characteristic function by b and reordering terms in the series allows us to write [33,36]

$$\widehat{p}(bk) = a\widehat{p}(k) - \frac{a-1}{a} \cos k. \tag{28}$$

The RG solution to Equation (28) can be separated into a homogeneous part and a singular part:

$$\widehat{p}(k) = \widehat{p}_s(k) + \widehat{p}_h(k) , \tag{29}$$

where $\widehat{p}_h(k)$ is analytic in the neighborhood $k = 0$ and $\widehat{p}_s(k)$ is singular in this neighborhood. The singular part $\widehat{p}_s(k)$ is obtained by solving the scaling equation:

$$\widehat{p}_s(bk) = a\widehat{p}_s(k), \tag{30}$$

where we assume the formal solution:

$$\widehat{p}_s(k) = A(k)k^{\delta}. \tag{31}$$

Inserting this form of the singular solution into Equation (30) yields

$$A(bk)b^{\delta}k^{\delta} = aA(k)k^{\delta}, \tag{32}$$

providing the distinct equalities

$$b^{\delta} = a, \tag{33}$$
$$A(bk) = A(k). \tag{34}$$

The first equality yields for the power index in terms of the series parameters $\delta = \ln a / \ln b$. The second equality implies that $A(k)$ is periodic in the logarithm of k with period $\ln b$. Consequently, the singular part of the RG solution is written

$$\widehat{p}_s(k) = \sum_{n=-\infty}^{\infty} A_n |k|^{H_n}, \tag{35}$$

with the complex power–law index:

$$H_n = \delta + in\frac{2\pi}{\ln b} = \frac{\ln a}{\ln b} + in\frac{2\pi}{\ln b}. \tag{36}$$

The analytic forms of the Fourier coefficients in Equation (35) are given in [35].

Hughes et al. [35] prove that the dominant behavior of the WRW is determined by the lowest-order term in the singular part of the solution for the discrete characteristic function, but we do not show that here. Instead we assume that the dominant behavior is given by the $n = 0$ term in the series:

$$\widehat{p}_s(k) \approx A_0 |k|^{\delta}, \tag{37}$$

whose inverse Fourier transform is determined by a Tauberian theorem to be the IPL:

$$p(x) = \frac{K(\delta)}{|x|^{\delta+1}}, \tag{38}$$

and $K(\delta)$ is a known function of δ. Thus, the singular part of the WRW has an IPL stepping PDF and this dominant behavior intuitively justifies ignoring all the other terms in the series.

We now write for the asymptotic time-dependent form of the discrete PDF resulting from the WRW:

$$\begin{aligned} P(x, n+1) &= \sum_{x'} p(x - x')P(x', n) \\ &= \sum_{x'} \frac{K(\delta)}{|x - x'|^{\delta+1}} P(x', n), \end{aligned} \tag{39}$$

where we assume that each step n in WRW process occurs at equal time intervals. Equation (39) was analyzed in 1970 by Gillis and Weiss [37], who determined that its solution is a Lévy PDF, thereby connecting the RG solution of the WRW to our discussion of the fractional diffusion equation given earlier.

Stable Lévy processes can therefore arise from the "weak" chaotic nature of the phase space trajectories. This is, in part, a consequence of the asymptotic behavior $k \to 0$ corresponding to the asymptotic $x \to \infty$, which is of significance in determining the transport behavior of the anomalous diffusion process.

3. Patterns and Complexity

In the Introduction we identified one of those patterns that is not restricted to a particular discipline, but pops up in every discipline from anatomy to zoology, and that pattern is an allometry relation (AR). However, what distinguishes such patterns from, for example, simple periodic motion? Of course, the existence of such regularity, the pattern of reproducibility in space and time, is what motivated the first investigators to seek common causes to associate with those patterns. Periodic motions, such as vibrations, motivated Hook to introduce his law using Newton's mechanical force for its explanation. The amazing success of such laws reinforced the idea that other phenomena including the beating of the heart, walking, and the propagation of light could all be described by adopting a similar modeling strategy. However, the *luminiferous aether* is now a quaint historical myth concerning the assumed need for a medium with remarkable properties to support the propagation of electromagnetic waves. In addition, the *normal sinus rhythm* of the heart is a medical myth as heartbeats are not sinusoidal. The more complex the phenomenon being considered the less well the patterns are reproduced using Newton's view of science.

Much of the present discussion stems from the need to replace Newton's atavistic characterization of space and time, because they fail to capture the rich structure of the complexity of the modern world. The failure to systematically reexamine these fundamental assumptions have restricted the utility of the modeling techniques of modern physics in the study of the psychology, sociology and the life sciences. The experience of space and time differs between those of the claustrophobic or agoraphobic, from the performer on the stage or the surgeon operating on the brain, from the warrior on the battlefield to the physician on the critical care ward. We require a mathematics that can capture all of this and so much more. The conclusions drawn herein were anticipated a couple of years ago [38]:

> What is becoming increasingly clear ... is that the technical intensity of the world has become so dense that the mathematical language initiated by Newton is no longer adequate for its understanding. In fact we now find that we have been looking at the world through a lens that often suppresses the most important aspects of phenomena, most of which are not "simple". These are characteristics of the phenomena that cannot be described using differential equations and we refer to them as complex.

3.1. Allometry through Complexity

We have argued elsewhere [20,39] that the empirical AR given by Equation (1) is a consequence of the imbalance between the complexity associated with the system functionality and the complexity associated with the system size, both being measured by Shannon information. We refer to this as the allometry/information hypothesis (A/I–H) [40] and postulate that in a complex network, composed of two or more interacting subnetworks, the flow of information is driven by the complexity gradient between the subnetworks, transported from that with the greater to that with the lesser complexity.

Implicit in the A/I–H is the assumed existence of dependencies of both system size and system functionality on complexity. Such dependencies have been observed in the positive feedback between social complexity and the size of human social groups [41,42], as well as in ant colony size [43], and the increase in biological complexity with ecosystem size [44]. Other relations have been observed in multiple disciplines, including the increase of prey refuge from predators with habitat complexity [45], computational complexity increasing with program size [46], and gene functionality depending on system complexity [47]. We abstract from these observations that the complexity of a phenomenon increases with system size and that the system functionality increases with system complexity.

The argument presented in this section follows that given recently by West et al. [48] in their discussion of the evolution of military technology over the past millennium. It is intuitively understood, but not often explicitly stated, that size and complexity grow together and are inextricably intertwined through criticality. Moreover, although tied together, their changes are not in direct proportion to one another. A similar connection exists between complexity and system functionality [38]. These interconnections are represented through homogeneous scaling relations, as shown below. West argued that as a system increases in size it provides increasing opportunity for variability, which is necessary in order to maintain stability. Scaling provides a measure of complexity in dynamic systems, indicating that the system's observables can simultaneously fluctuate over many time and/or space scales. An observable $Z(t)$ scales if for a constant λ it satisfies the homogeneous relation

$$Z(\lambda t) = \lambda^{\mu_z} Z(t) \tag{40}$$

with the scaling index given by μ_z. Note that if we consider the AR given by Equation (1), but without the averaging brackets, the size and functionality depend on a parameter t, and scale in the manner indicated by Equation (40), each with a distinct power law index, then $b = \mu_Y / \mu_X$ in order for the AR to be satisfied.

The hallmarks of fractal statistics are spatial (z) inhomogeneity and temporal (t) intermittency and the phase space trajectory $(z; t)$ replaces the dynamic variable $Z(t)$. In phase space, the scaling of the dynamic variable is replaced by a scaling of the PDF $P(z; t)$:

$$P(z; t) = \frac{1}{t^{\mu_z}} F_z \left(\frac{z}{t^{\mu_z}} \right) \tag{41}$$

as given by Equation (23) for general complex phenomena. There is a broad class of PDFs for which the functional form of $F_z(\cdot)$ is left unspecified.

It is straightforward to calculate the average value of $Z(t)$ using the PDF given by Equation (41):

$$\langle Z(t) \rangle = \int z P(z, t) dz = \bar{q}_z t^{\mu_z}, \tag{42}$$

and the overall constant is determined by the scaling variable $q = z/t^{\mu_z}$ averaged over the PDF $F(q)$:

$$\bar{q}_z \equiv \int q F_z(q) \, dq. \tag{43}$$

Interpreting $Z(t)$ as the system's TBM $X(t)$ Equation (42) describes the growth in the overall average size of a complex system with the time t, due to the intrinsic dynamics generating increasing complexity. A similar observation can be made interpreting the dynamic variable with a functionality of the system $Y(t)$. Consequently, the same functional form results for both $Y(t)$ and $X(t)$, each with its own index. This is not entirely unexpected since both the functionality and size of the system grow with complexity, but at different rates.

Notice that using the scaling PDF that the average of the dynamic variable now has the scaling property:

$$\langle Z(\lambda t) \rangle = \lambda^{\mu_z} \langle Z(t) \rangle. \tag{44}$$

If both the size and functionality of the system can be characterized in terms of the system's complexity by the same form of scaling PDF we obtain two equations in t for the averages. Setting the scaling parameter to $\lambda = 1/t$, after some algebra we obtain the equalities

$$t = \left(\frac{\langle Y(t) \rangle}{\langle Y(1) \rangle} \right)^{\frac{1}{\mu_Y}} = \left(\frac{\langle X(t) \rangle}{\langle X(1) \rangle} \right)^{\frac{1}{\mu_X}}, \tag{45}$$

which can rewritten in the form of the empirical AR given by Equation (1):

$$\langle Y \rangle = a \langle X \rangle^{b}, \tag{46}$$

with the allometry parameters:

$$a = \frac{\langle Y(1) \rangle}{\langle X(1) \rangle^{b}} = \frac{\bar{q}_{Y}}{\bar{q}_{X}^{b}} \text{ and } b = \frac{\mu_{Y}}{\mu_{X}}. \tag{47}$$

Here, we have used Equation (42) to obtain the second equality for the allometry coefficient. Thus, demonstrating that the empirical AR is the result of the self-similar behavior of the PDF.

Note that the allometry index b is expressed as the ratio of $b = \alpha / (\beta + 1)$ for the system functionality to that for the system size. In general, this ratio is less than one for both the system size and functionality. It is also the case that for physiological systems $b < 1$. The more the index for the fractional time derivative deviates downward from one, the greater influence the complexity history has on the present behavior of the independent variable, whether functionality or size. The more the index of the fractional variate derivative deviates downward from two, the greater is the nonlocal coupling of the independent variables (functionality or size) across scales. However, these two mechanisms do not independently determine the scaled PDF. It is their ratio that determines the balancing of effects in the functionality and size separately, and then through their ratio to obtain b.

It is this coupling across scales in size as well as in physiologic time that entails the temporal AR with $b < 1$, as well as, the positive growth of entropy in approaching the steady state asymptotically. The results of these brief arguments are encapsulated in the Principle of Complexity Management (PCM), which establishes that in the interaction between two complex networks, information flows from the more complex to the less complex network. Information transfer is maximally efficient when the complexities of the two networks are matched [38]. In the time-size application of this section, the PCM takes the form *The origin of natural patterns manifest by temporal ARs is the imbalance between the complexity associated with a system's measure of time and the complexity associated with a system's size. In both networks the complexity is measured by the Wiener/Shannon entropy.*

3.2. Its about Time

The fundamental question addressed in this section is whether time outside the physical sciences, say the time for a scurrying mouse at the lower left of Figure 1 is the same as that of the lumbering elephant at the upper right of the metabolic AR curve. Newton would assert that they are identical and we would agree that the time shared by the two animals is the same when referenced to an external mechanical clock. However, are the two times the same when referenced to their individual physiological clocks? This question arises because the lifespans of the two creatures are essentially the same when their lifetimes are measured using the product of the number of heartbeats times the average time interval between beats. This is very different from the comparison of their separate lifespans when referenced to an external clock in which case the two differ by years. This change of reference of time measures, from the ticking of a clock to the beating of a heart, suggests that physiological time may be a monotonically decreasing function of physical time [49].

This difference in the meaning of time has lead to such concepts as biological time [50], physiologic time [51], and metabolic time [52], all in an effort to highlight the distinction between time in living and in inanimate systems. The intrinsic time in a living process was first called biological time by Hill [53], who reasoned that since so many properties of an organism change with size that time itself ought to scale with TBM. Natural scientists have subsequently hypothesized that physiologic time differs from the time measured by the ticking of a mechanical clock, or Newtonian time, in that the former changes with the size of the animal [17,18], whereas the latter does not [54].

Lindstedt and Calder [55] developed the concept of biological time further and determined experimentally that biological time, such as species longevity, satisfies a temporal AR with the functionality of the system being the physiologic time $Y = \tau$ and X the TBM M [56]:

$$\langle \tau \rangle = a \langle M \rangle^b \tag{48}$$

which describes the average duration of biological events. In Figure 4, we record the average heart rate $R = 1/\langle \tau \rangle$ for sixteen animals [57] covering six orders of magnitude in average TBM. The solid line segment is the fit to the data with empirical values to the allometry parameters given by $a = 205$ and $b = 0.248$, with a quality of fit measured by $r^2 = 0.96$. Other, more exhaustive, fits to larger data sets, made by other investigators, support the notion that physiologic time is extensive and may be found in many other places [17,18], but the results are equivalent.

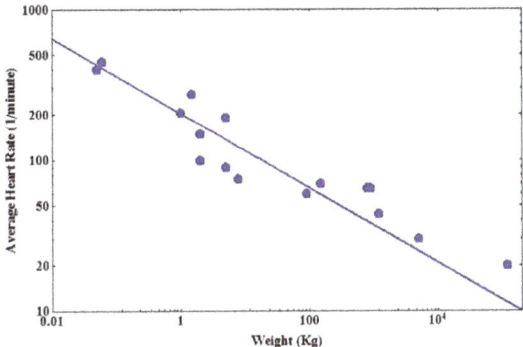

Figure 4. The average heart rate in beats per minute for 16 animals from the fastest, hamsters, to the slowest, large whales, with humans being in the middle of a fitting curve. The data were obtained from [57] and the solid line segment is fitted to the temporal AR. From the work in [49] with permission.

In an allometry context, one version of the FKE, would be given by Equation (13) where the phase space variables (z, t) are here given by (m, τ) [30] and $P(m, \tau)dm$ is the probability that the dynamic mass variable $M(\tau)$ lies in the interval $(m, m + dm)$ at time τ. $M(\tau)$ represents the TBM of a mature individual species member, within an ensemble of realizations, at the physiological time τ. The exact solution to the FKE has been obtained as the inverse Fourier transform of the characteristic function, expressed in terms of the Mittag–Leffler function given by Equation (21) with the variables properly defined. The allometry coefficient in this temporal AR has a theoretical value expressed in terms of the average of the scaled variable $q = m/\tau^\delta$. Consequently, the complexity of the underlying physiology of an animal entails the physiologic time through the scaling statistics.

The dependence of the empirical AR on the overall state of the system is captured by the entropy. The Wiener/Shannon information entropy associated with the system manifesting temporal allometry has the value

$$S(\tau) = -\int P(m, \tau) \log_2 P(m, \tau) dm \tag{49}$$

which when the scaled PDF given by Equation (41) is inserted into the integral yields

$$S(\tau) = S_0 + \delta \log_2 \tau \tag{50}$$

where S_0 is the entropy referenced to the PDF $F(\cdot)$.

Consequently, as we mentioned earlier, given a monotonic function relating physical and physiologic time $t = g(\tau)$, such that

$$\frac{dg(\tau)}{d\tau} \equiv \dot{g} \geq 0 \qquad (51)$$

we have for the physical time derivative of the entropy Equation (50):

$$\frac{dS(\tau)}{dt} = \frac{\delta}{\tau}\frac{1}{\dot{g}} \geq 0 \qquad (52)$$

Consequently, the entropy generation in physical time for the physiologic process entailing the temporal AR is positive semidefinite. Thus, the rate of entropy generation in Newtonian time is consistent with the dynamics of living systems having their own physiological time.

It is worth pointing out that empirical ARs are not necessarily restricted to living systems, but also arise in social systems as well. This is not entirely unexpected, as the average mass in an empirical AR is actually a surrogate for the living system's complexity. Proceeding by analogy, one might anticipate that such an AR should appear in a social context, where the average TBM is replaced with the average population or population density. This does, in fact, occur in the form of ARs where the functionality is expressed in terms of the rate at which an event occurs. An exemplar is Farr's Law, which dates back to the nineteenth century, and quantifies the "evil effects of crowding", relating a population's mortality rate to an institution's patient population density in the form of a rate AR [38,58]. Other examples of social ARs include an increasing urban crime rates, the more rapid spread of infectious diseases, and a speedup in pedestrian walking, all with increasing city size, as quantitatively confirmed by Bettencourt et al. [59]. Unlike the biological case, in the social rate ARs the allometry index has a value greater than one, $b > 1$, confirming that cities have, at all times and in all places, throughout history, entailed increased rates in human activity, for good or ill.

4. Subordination

The Montroll–Weiss (MW) perspective of CTRW [29] has been used to support the assumption that there are at least two distinct, but related, interpretations of time associated with a system's dynamics. As noted in the Introduction, the first is the external time associated with an objective observer who records the behavior of the system. This is Newton's assumption of what constitutes time: it is experimental or clock time. The second kind of time is the local time associated with the internal dynamics of the system, called subjective or operational time. In a psychological experiment the latter time is what is experienced by the participant. The experimental observation, carried out in the clock time t, is subordinated to a process occurring in the operational time n. For simplicity, we assume the operational time n to be an integer number so large as to become indistinguishable from a continuous variable. In the operational time n the evolution of the PDF describing the process is described by the ordinary diffusion equation

$$\frac{\partial P(x,n)}{\partial n} = D\frac{\partial^2 P(x,n)}{\partial x^2} = \mathcal{L}P(x,n), \qquad (53)$$

where $\mathcal{L} \equiv D\frac{\partial^2}{\partial x^2}$ is the diffusion operator.

The dynamics generating the diffusion process is the simple Langevin equation

$$\frac{dX(n)}{dn} = \eta(n), \qquad (54)$$

where $X(n)$ is the space coordinate at time n and $\eta(n)$ is the fluctuating velocity. If the velocity is a stochastic process with delta correlated fluctuations, this equation yields a diffusion process with scaling index $\delta = 1/2$. If $\delta \neq 1/2$ the diffusion is anomalous and is the result of memory influencing the fluctuations. In the present representation $\eta(n)$ of Equation (54) is totally random, i.e., it has no memory. However, in the clock time, the event $\eta(n)$ occurs at time $t(n)$ and the independent event

$\eta(n+1)$ at time $t(n+1)$ with the time distance $\tau(n) = t(n+1) - t(n)$ derived from a waiting time PDF $\psi(\tau)$. We are interested in the case where the waiting time PDF has the hyperbolic form:

$$\psi(\tau) = (\mu - 1) \frac{T^{\mu-1}}{(T+\tau)^{\mu}} \tag{55}$$

We use this hyperbolic form to define the concept of crucial event.

Crucial events are defined by the time interval separating the occurrence of consecutive events. The time intervals between crucial events are determined by a waiting time PDF given by Equation (55), with the condition $1 < \mu < 3$. In clock time we use the theoretical MW prescription [29] to obtain

$$P(x,t) = \sum_{n=0}^{\infty} \int_0^t dt' \, \psi_n(t') \, \Psi(t - t') \, e^{n\mathcal{L}} P(x,0). \tag{56}$$

Note that $\psi_n(t')$ is the PDF that n events have occurred and that the last event took place at time t'.

For the formula given by Equation (56) to hold with n going to ∞, we must assume that for the random walker to travel the distance x in a time t a virtual infinitely large number of events may occur, thereby implying the diffusion coefficient D is extremely small. In the case $\mu < 2$, the mean waiting time $\langle \tau \rangle$ diverges, thereby providing an additional reason for the experimental observation time t to be large.

It is possible to prove, using the arguments developed by Allegrini et al. [60] with a minor notational change, that Equation (56) is equivalent to the integro-differential phase space equation:

$$\frac{\partial P(x,t)}{\partial t} = \int_0^t dt' \Phi(t - t') \, \mathcal{L} P(x,t'), \tag{57}$$

where $\Phi(t)$ is the MW memory kernel related to the waiting–time PDF and $\psi(t) = \psi_{n=1}(t)$. In the Laplace transform representation where $\widehat{f}(u)$ denotes the Laplace transform of $f(t)$, this latter relation is

$$\widehat{\Phi}(u) = \frac{u\widehat{\psi}(u)}{1 - \widehat{\psi}(u)}. \tag{58}$$

In the case where the index for the hyperbolic PDF, which asymptotically is the IPL index, is in the interval $1 < \mu < 2$, using Equation (58) it is shown [61] that asymptotically $u \to 0$:

$$\widehat{\Phi}(u) \approx u^{1-\alpha}. \tag{59}$$

Inserting this asymptotic expression into the Laplace transform of Equation (57) and taking the inverse Laplace transform yields the fractional diffusion equation (FDE):

$$\frac{\partial^{\alpha} P(x,t)}{\partial t^{\alpha}} = \mathcal{L} P(x,t) \tag{60}$$

Here, the fractional time derivative is of the Caputo form with $\alpha = \mu - 1 < 1$. We note here that the analytic solution to Equation (60) is given by the scaling PDF Equation (23) when $\beta = 1$ and $\delta = \alpha/2$.

Culbreth et al. [62] stress certain subtleties of these formal results to provide a context with which to appreciate their contribution to the field of cognition and to the fractional calculus. First, they notice that we can use psychological arguments to interpret the connection between operational time and clock time, as done in [63]. The operational time is subjective in this psychological context with a logarithmic connection with the clock time t, which changes an exponential waiting time PDF into the hyperbolic structure of Equation (55). This property provides the rationale for why they [62] consider the CTRW formalism to be closely connected to the issue of cognition. As they point out, earlier work [60] analyzed a series of events using the hyperbolic waiting time PDF using the Kolmogorov–Sinai definition of complexity and determined that the signal becomes computationally compressible for $2 < \mu < 3$. This is equivalent to assessing that the time series hosts messages that can be decoded.

On the other hand, the Kolmogorov–Sinai entropy vanishes for $\mu < 2$ and has been recently generalized to take into account the rare crucial events [64] of this region. These crucial events are conjectured to be the signal of swarm intelligence [65], while the observation of the dynamics of the brain leads to the conclusion that $\mu = 2$ is a proper signature of the brain of an awake subject [66]. In summary, the events characterized by the inter-event or hyperbolic waiting time PDF are considered to be a signature of cognition and are known to be responsible for the transport of information from one intelligent system to another [67,68]. The term crucial events is a proper nomenclature to acknowledge the importance of these rare events.

5. Discussion and Conclusions

We began this essay with the stated intent of supporting the Complexity Hypothesis by demonstrating to the reader why Newton's dynamic view of physical objects is not just inappropriate for living and social systems but its domain of application within the physical sciences is shrinking dramatically as well. The unexamined assumptions regarding the nature of space and time, with which Newton opened his *Principles,* make his force law invalid for the study of complex phenomena. Yet, these are the phenomena of interest to scientists in the 21st century, whether such phenomena reside in the physical, social, or life sciences.

As mentioned, Newton's equations have been shown to require changes when particles are moving very fast (approaching the speed of light), when the spatial scales are very large (cosmological) and when they are very small (quantum mechanical). In each of these domains the dynamic laws follow a correspondence principle in that they converge on Newton's laws by changing a parameter value to replicate the world of our five senses. Herein we have shown that in this world of experience we continually encounter deviations from Newton's laws at normal speeds and spatial scales, due to chaos. Chaotic dynamics led to replacement of the probability calculus of Kinetic Theory with that of FKT, as well as to operational time. One way to measure the degree of complexity generated by chaotic attractors is by using the entropy of the behavior.

Crutchfield et al. [69] interpreted the entropy of a dynamic process as the average rate of information generation by a chaotic process in that the more precisely an initial state of a system is specified, the more information one has available. The amount of information contained in the initial state is inversely proportional to the state space volume V_i localized by measurement. Trajectories initiated in a local volume of a regular attractor remain close to one another as the system evolves, and therefore no new information is generated, while the initial information is preserved in time. Consequently, the initial information can be used to predict the system's final state.

On the other hand, on a chaotic attractor the initial volume gets smeared out, consequently, as the system evolves the initial information is destroyed and replaced by newly created information. Thus, the volume in the specification of the initial system is eventually spread over the entire attractor and all predictive power is lost since the probability of being anywhere on the attractor is the same. All causal connection between the present state of the system and its future or final state is lost. This is referred to as the sensitive dependence on initial conditions.

Let us denote the final region of phase space the system occupies by V_f so that the change in the observable information ΔI is determined by the volume change from the initial to final state [70,71]:

$$\Delta I = \log_2 \left(\frac{V_f}{V_i} \right). \tag{61}$$

The time rate of information change (creation or dissipation) is therefore

$$\frac{dI}{dt} = \frac{1}{V}\frac{dV}{dt},\tag{62}$$

where the time-dependent volume V over which the initial conditions are spread determines the ultimate fate of the initial information. In regular, which is to say non-chaotic, systems the sensitivity of the flow in the initial conditions grows with time no more rapidly than a polynomial. Let $\Omega(t)$ be the number of states at time t that can be distinguished such that if the greatest polynomial index is n such that $\Omega(t) \propto t^n$. The ratio of the final to initial volume in such a system is equal to the relative number of states independently of the time $\frac{V_f}{V_i} = \frac{\Omega_f}{\Omega_i}$, so that for the rate at which information changes [71]:

$$\frac{dI}{dt} \sim \frac{n}{t}.\tag{63}$$

Thus, the rate of generation of new information decreases with time and converges to zero as $t \to \infty$. As in Poincaré's quote in the Introduction, the final state is approximately predictable from the approximate initial information.

On the other hand, in chaotic systems two trajectories separate exponentially and therefore the number of distinguishable states grows exponentially with time $\Omega(t) \propto \exp(\lambda t)$, where λ is the Liapunov coefficient. In this case, the rate at which information is generated is constant:

$$\frac{dI}{dt} \sim \lambda.\tag{64}$$

In this latter system, information is continuously generated by the attractor independently of the initial state. Nicolis and Tsuda [70] used this property of chaotic dynamic systems in the early modeling of cognitive systems using nonlinear dynamics and subsequently for information processing in neurophysiology, cognitive psychology, and perception [72].

Thus, Newton's statements about the absolute nature of space is contradicted by the chaotic trajectories entailed by his own force law when applied to complex systems. Subsequently, even Kinetic Theory and the introduction of stochastic differential equations, which were early attempts to make the differential calculus and complex phenomena compatible, could only be salvaged by means of the FC. In a similar way, Newton's statements regarding the absolute nature of time have been shown to have little place, if any, outside restricted domains of the physical sciences.

Funding: This research received no external funding.

Conflicts of Interest: The author declares no conflicts of interest.

References

1. Newton, I. *Mathematical Principles of Natural Philosophy*; Cambridge, UK, 1686; *Great Books of the Western World*; Encyclopedia Britannica Inc.: Chicago, IL, USA, 1952; Volume 34.
2. Whiteside, D.T. The mathematical principles underlying Newton's Principia Mathematica. *J. Hist. Astron.* **1970**, *1*, 116–138. [CrossRef]
3. Mills, R. *Sapac, Tiem, and Quanta*; W.H. Freeman and Co.: New York, NY, USA, 1994.
4. Einstein, A. *Relativity, the Special and General Theory*; Philosophical Library: New York, NY, USA, 1961.
5. Jammer, M. *The Philosophy of Quantum Mechanics*; Wiley–Interscience Publication: New York, NY, USA, 1974.
6. West, B.J.; Scafetta, N. *Disrupted Networks, from Physics to Climate Change, Studies of Nonlinear Phenomena in Life Science*; World Scientific: Hackensack, NJ, USA, 2010; Volume 13.
7. Barrow-Green, J. *Poincaré and the Three Body Problem*; American Mathematical Society, London Mathematical Society: Provodince, RI, USA, 1997; Volume 11.
8. Mandelbrot, B.B. *Fractals: Form, Chance and Dimension*; W.H. Freeman and Company: San Francisco, CA, USA, 1977.

9. Lighthill, J. The recently recognized failure of predictability in Newtonian dynamics. *Proc. R. Soc. Lond. A* **1986**, *407*, 35–50.
10. Siegel, C.L.; Moser, J.K. *Lectures on Celestial Mechanics*; Springer: Berlin, Germany, 1971.
11. Ott, E. *Chaos in Dynamical Systems*; Cambridge University Press: New York, NY, USA, 1993.
12. Poincaré, H. *The Foundations of Science*; Halsted, G.B., Translator; The Science Press: New York, NY, USA, 1913.
13. Li, T.Y.; Yorke, J.A. Period Three Implies Chaos. *Am. Math. Mon.* **1975**, *82*, 985. [CrossRef]
14. Lorenz, E.N. Deterministic Nonperiodic Flow. *J. Atmos. Sci.* **1963**, *20*, 130–141. [CrossRef]
15. Zaslavsky, G.M. Chaos, fractional kinetics, and anomalous transport. *Phys. Rep.* **2002**, *371*, 461–580. [CrossRef]
16. Huxley, J.S. *Problems of Relative Growth*; Dial Press: New York, NY, USA, 1931.
17. Calder, W.W., III. *Size, Function and Life History*; Harvard University Press: Cambridge, MA, USA, 1984.
18. Schmidt-Nielsen, K. *Scaling, Why Is Animal Size so Important?* Cambridge University Press: Cambridge, UK, 1984.
19. West, G.B.; Brown, J.H.; Enquist, B.J. A general model for the origin of allometric scaling laws in biology. *Science* **1997**, *276*, 122–124. [CrossRef]
20. West, B.J.; West, D. Fractional dynamics of allometry. *Fract. Calc. Appl. Anal.* **2012**, *15*, 2012. [CrossRef]
21. Turalska, M.; West, B.J. Fractional Dynamics of Individuals in Complex Networks. *Front. Phys.* **2018**, *6*, 110. [CrossRef]
22. Roberts, F.S. *Measurement Theory with Applcations to Decisionmaking, Utility, and the Social Sciences*; Encyclopedia of Mathematics and Its Applications Vol. 7; Addison-Wesley: Reading, MA, USA, 1979.
23. Fechner, G.T. *Elemente der Psychophysik*; Breitkopf and Härtel: Leipzig, Germany, 1860.
24. Zaslavsky, G.M. Fractional Kinetics of Hamiltonian Chaotic sytems. In *Applications of Fractioanl Calculus in Physics*; Hilfer, R., Ed.; World Scientific: River Edge, NJ, USA, 2000; pp. 203–240.
25. West, B.J.; Grigolini, P. Frational Differences, Derivatives and Fractal Time Series. In *Applications of Fractioanl Calculus in Physics*; Hilfer, R., Ed.; World Scientific: River Edge, NJ, USA, 2000; pp. 171–202.
26. Montroll, E.W.; West, B.J. On an enriched collection of stochatic processes. In *Fluctuation Phenomena*; Montroll, E.W., Lebowitz, J.L., Eds.; North-Holand Personal Library: New York, NY, USA, 1987.
27. Montroll, E.W.; Scher, H. Random walks on lattices. IV. Continuous-time walks and influence of absorbing boundaries. *J. Stat. Phys.* **1973**, *9*, 101–135. [CrossRef]
28. Metzler, R.; Klafter, J. The random walk's guide to anomalous diffusion: A fractional dynamics approach. *Phys. Rep.* **2000**, *339*, 1–77. [CrossRef]
29. Montroll, E.W.; Weiss, G.H. Random walks on lattices. II. *J. Math. Phys.* **1965**, *6*, 167–181. [CrossRef]
30. West, B.J. *Fractional View of Complexity, Tomorrow's Science*; CRC Press: Boca Raton, FL, USA, 2016.
31. Zolotarev, V.M. *One-Dimensional Stable Distributions*; American Mathematical Soc.: Providence, RI, USA, 1986.
32. Mainardi, F.; Goreflo, R.; Li, B.-L. A fractional generalization of the Poisson process. *Vietnam J. Math.* **2004**, *32*, 53–64.
33. Zaslavsky, F.M.; Stevens, D.; Weitzner, H. Self–similar transport in omplete chaos. *Phys. Rev. E* **1993**, *48*, 1683. [CrossRef] [PubMed]
34. Meiss, J.D. Class renormalization: Islands around islands. *Phys. Rev. A* **1986**, *34*, 2375. [CrossRef]
35. Hughes, B.; Montroll, E.; Shlesinger, M. Fractal random walks. *J. Stat. Phys.* **1982**, *28*, 111–126, doi:10.1007/BF01011626. [CrossRef]
36. Montroll, E.W.; Shlesinger, M. Wonderful World of Random Walks. In *Studies in Statistical Mechanics*; Leibowitz, J., Montroll, E.W., Eds.; North-Holland: Amstrdam, The Netherlands, 1984; Volume II, pp. 1–121.
37. Gillis, J.E.; Weiss, G.H. Expected number of distinct sites visited by a random walk with an infinite variance. *J. Math. Phys.* **1970**, *11*, 1307–1312. [CrossRef]
38. West, B.J. *Nature's Patterns and the Fractional Calculus*; Fractional Calculus in Applied Science and Engineering 2; Walter de Gruther GmbH: Berlin, Germany; Boston, MA, USA, 2017.
39. West D.; West, B.J. On allometry relations. *Int. J. Mod. Phys. B* **2012**, *26*, 1230013. [CrossRef]
40. West, B.J. Information forces. *J. Theor. Comput. Sci.* **2016**, *3*, 144. [CrossRef]
41. Collard, M.; Ruttle, A.; Buchanan, B.; O'Brien, M.J. Population size and cultural evolution in nonindustrial food-producing societies. *PLoS ONE* **2013**, *8*, e72628. [CrossRef]

42. West, B.J.; Massari, G.F.; Culbreth, G.; Failla, R.; Bologna, M.; Dunbard, R.I.M.; Grigolini, P. Relating size and functionality in human social networks through complexity. *Proc. Natl. Acad. Sci. USA* **2020**, *117*, 18355–18358. [CrossRef]

43. Ferguson-Gow, H.; Sumner, S.; Bourke, A.F.G.; Jones, K.E. Colony size predicts division of labour in attine ants. *Proc. R. Soc. B* **2014**, *281*, 20141411. [CrossRef] [PubMed]

44. Cadenasso, M.L.; Pickett, S.T.A.; Grove, J.M. Dimensions of ecosystem complexity: Heterogeneity, connectivity, and history. *Ecol. Complex.* **2006**, *3*, 1–12. [CrossRef]

45. Gotceitas, V.V.; Colgan, P. Predator foraging success and habitat complexity: Quantitative test of the threshold hypothesis. *Oecologia* **1989**, *80*, 158–166. [CrossRef] [PubMed]

46. Joosten, J.J.; Soler-Toscano, F.; Zenil, H. Program-size versus time complexity. *Int. J. Unconv. Comput.* **2011**, *7*, 353–387.

47. Jain, R.; Rivera, M.C.; Lake, J.A. Horizontal gene transfer among genomes: The complexity hypothesis. *Proc. Natl. Acad. Sci. USA* **1999**, *96*, 3801–3806. [CrossRef]

48. West, B.J.; West, D.; Kott, A. Allometry Relation of Technology Systems. *J. Def. Model. Simul. Appl. Methodol. Technol.* **2020**, 1–6. [CrossRef]

49. West, D.; West, B.J. Physiological time: A hypothesis. *Phys. Life Rev.* **2013**, *10*, 210–224. [CrossRef]

50. Winfree, A.T. *Timing of Biological Clocks*; Princeton University Press: Princeton, NJ, USA, 1987.

51. Brody, S. *Bioenergetics and Growth*; Reinhold: New York, NY, USA, 1945.

52. Schmidt-Nielsen, K. *Animal Physiology*; Cambridge University Press: Cambridge, UK, 1997.

53. Hill, A.V. The dimensions of animals and their muscular dynamics. *Sci. Prog.* **1950**, *38*, 209–230.

54. Prigogine, I.; Stengers, I. *Order out of Chaos: Man's New Dialogue with Nature*; Bantam Books: Toronto, CA, USA, 1984.

55. Lindstedt, S.L.; Calder, W.A., III. Body size and longevity in birds. *Condor* **1976**, *78*, 91–94. [CrossRef]

56. Lindstedt, S.L.; Miller, B.J.; Buskirk, S.W. Home range, time and body size in mammals. *Ecology* **1986**, *67*, 413–418. [CrossRef]

57. Al-Dabaan, B.B. Scaling Laws in Biology. Available online: http://www.math-physics-tutor.com/web_documents/bader (accessed on 18 October 2020).

58. Humphreys, N.S. (Ed.) *Vital Statistics: A Memorial Volume of Selections from the Reports and Writings of William Farr*; The Sanitory Institute of Great Britian: London, UK, 1885.

59. Bettencourt, L.M.A. The origins of scaling in cities. *Science* **2013**, *340*, 1438–1441. [CrossRef]

60. Allegrini, P.; Aquino, G.; Grigolini, P.; Palatella, L.; Rosa, A. Generalized master equation via aging continuous-time random walks. *Phys. Rev. E* **2003**, *68*, 056123. [CrossRef] [PubMed]

61. Pramukkul, P.; Svenkeson, A.; Grigolini, P.; Bologna, M.; West, B. complexity and the Fractional Calculus. *Adv. Math. Phys.* **2013**, *2013*, 1–7.. [CrossRef]

62. Culbreth, G.; Bologna, M.; West, B.J.; Grigolini, P. Caputo Fractional Derivative verus Quantum Coherence. *Entropy* , under review.

63. Grigolini, P.; Aquino, G.; Bologna, M.; Lukovic, M.; West, B.J. A theory of 1/f noise in human cognition. *Phys. A Stat. Mech. Appl.* **2009**, *388*, 4192–4204. [CrossRef]

64. Korabel, N.; Barkai, E. Pesin-type identity for intermittent dynamics with a zero Lyaponov exponent. *Phys. Rev. Lett.* **2009**, *102*, 050601. [CrossRef]

65. Vanni, F.; Lukovic, M.; Grigolini, P. Criticality and transmission of information in a swarm of cooperative units. *Phys. Rev. Lett.* **2011**, *107*, 078103. [CrossRef]

66. Allegrini, P.; Menicucci, D.; Bedini, R.; Fronzoni, L.; Gemignani, A.; Grigolini, P.; West, B.J.; Paradisi, P. Spontaneous brain activity as a source of ideal 1/f noise. *Phys. Rev. E* **2009**, *80*, 061914. [CrossRef]

67. Mahmoodi, K.; West, B.J.; Grigolini, P. Selfish Algorithm and Emergence of Collective Intelligence. *Front. Physiol.* **2020**, in press. [CrossRef]

68. West, B.J.; Geneston, E.L.; Grigolini, P. Maximizing information exchange between complex networks. *Phys. Rep.* **2008**, *468*, 1–99. [CrossRef]

69. Crutchfield, J.P.; Farmer, J.D.; Packard, N.H.; Shaw, R.S. Chaos. *Sci. Am.***1987**, *255*, 46–57. [CrossRef]

70. Nicolis, J.S.; Tsuda, I. Chaotic dynamics of information processing: The magic number of seven plus two revisited. *Bull. Math. Biol.* **1985**, *47*, 343–365. [PubMed]

71. Shaw, R. Strange attractors, chaotic behavior, and information flow. *Z. Naturforsch A* **1981**, *36*, 80–112. [CrossRef]

72. Nicolis, J.S. *Chaos and Information Processing*; World Scientific: Singapore, 1991.

Publisher's Note: MDPI stays neutral with regard to jurisdictional claims in published maps and institutional affiliations.

Article

Time Fractional Fisher–KPP and Fitzhugh–Nagumo Equations

Christopher N. Angstmann * and Bruce I. Henry *

School of Mathematics and Statistics, UNSW, Sydney 2052 NSW, Australia
* Correspondence: c.angstmann@unsw.edu.au (C.N.A.); b.henry@unsw.edu.au (B.I.H.)

Received: 28 August 2020; Accepted: 12 September 2020; Published: 16 September 2020

Abstract: A standard reaction–diffusion equation consists of two additive terms, a diffusion term and a reaction rate term. The latter term is obtained directly from a reaction rate equation which is itself derived from known reaction kinetics, together with modelling assumptions such as the law of mass action for well-mixed systems. In formulating a reaction–subdiffusion equation, it is not sufficient to know the reaction rate equation. It is also necessary to know details of the reaction kinetics, even in well-mixed systems where reactions are not diffusion limited. This is because, at a fundamental level, birth and death processes need to be dealt with differently in subdiffusive environments. While there has been some discussion of this in the published literature, few examples have been provided, and there are still very many papers being published with Caputo fractional time derivatives simply replacing first order time derivatives in reaction–diffusion equations. In this paper, we formulate clear examples of reaction–subdiffusion systems, based on; equal birth and death rate dynamics, Fisher–Kolmogorov, Petrovsky and Piskunov (Fisher–KPP) equation dynamics, and Fitzhugh–Nagumo equation dynamics. These examples illustrate how to incorporate considerations of reaction kinetics into fractional reaction–diffusion equations. We also show how the dynamics of a system with birth rates and death rates cancelling, in an otherwise subdiffusive environment, are governed by a mass-conserving tempered time fractional diffusion equation that is subdiffusive for short times but standard diffusion for long times.

Keywords: fractional diffusion; continuous time random walks; reaction–diffusion equations; reaction kinetics

1. Introduction

Reaction–diffusion partial differential equations are among the most widely used equations in applied mathematics modelling. These equations govern the time evolution of concentrations, or population densities, of species, at different spatial locations, that are diffusing and reacting. Applications include the spatio-temporal spread of epidemics, the spatial spread of invasive species and the development of animal coat patterns [1–3]. In these modelling equations, diffusion is represented by a spatial Laplacian operating on the population densities, and reactions are included as additive terms representing changes per unit time in population densities through reaction rates. In well-mixed systems the reaction rate equations can often be derived from the law of mass-action [4]. A famous example of a reaction–diffusion equation is the Fisher–KPP equation named after Fisher [5] and Kolmogorov, Petrovsky and Piskunov [6]. The standard reaction–diffusion representation of this equation is

$$\frac{\partial u(x,t)}{\partial t} = D\frac{\partial^2 u(x,t)}{\partial x^2} + ru(x,t)(1 - u(x,t)), \quad D > 0, r > 0. \tag{1}$$

Here, $u(x,t)$ represents the population density of a species, $D\frac{\partial^2 u(x,t)}{\partial x^2}$ represents the diffusion of the species and $ru(x,t)(1 - u(x,t))$ represents the reactions of the species. In the absence of diffusion, the time rate of change in the population density is the same at all points in space and is given by

$$\frac{\partial u(x,t)}{\partial t} = ru(x,t)(1 - u(x,t)). \tag{2}$$

In this example and in the following, for simplicity, we have considered systems in one spatial dimension. Extensions to higher spatial dimensions are possible.

Over the past two decades, there has been a growing awareness of fractional diffusion, where diffusion cannot be modelled using a standard Laplacian and the mean square displacement of diffusing species does not grow linearly in time, as anticipated by Einstein's famous modelling of Brownian motion [7]. In particular, following widespread observations in biological systems, there has been a great deal of attention focussed on fractional subdiffusion, characterized by the mean square displacement of a population spreading as a sublinear power law in time. It is now generally accepted that if subdiffusion arises from particles being trapped for arbitrarily long periods of time, the appropriate equation to model subdiffusion is the time fractional diffusion equation [8]

$$\frac{\partial u(x,t)}{\partial t} = {}_0D_t^{1-\gamma}\frac{\partial^2 u(x,t)}{\partial x^2}, \quad 0 < \gamma < 1, \tag{3}$$

which can be derived [9,10] from a continuous time random walk (CTRW) [11] with a power law waiting time density. In this equation,

$${}_0D_t^{1-\gamma}y(x,t) = \frac{1}{\Gamma(\gamma)}\frac{\partial}{\partial t}\int_0^t \frac{y(x,t')}{(t-t')^{1-\gamma}}\,dt' \tag{4}$$

is the Riemann–Liouville fractional derivative of order $1 - \gamma$, see, for example, reference [12]. It might be anticipated that the appropriate evolution equation to model subdiffusion, with reactions governed by the reaction rate equation,

$$\frac{\partial u(x,t)}{\partial t} = f(u(x,t)), \tag{5}$$

would be

$$\frac{\partial u(x,t)}{\partial t} = {}_0D_t^{1-\gamma}\frac{\partial^2 u(x,t)}{\partial x^2} + f(u(x,t)). \tag{6}$$

Indeed, such an equation had been derived from an underyling CTRW model, under certain assumptions, [13], however it is not valid in general. For example, the simple model equation

$$\frac{\partial u(x,t)}{\partial t} = {}_0D_t^{1-\gamma}\frac{\partial^2 u(x,t)}{\partial x^2} - u(x,t), \tag{7}$$

can have unphysical negative solutions [14].

The time fractional subdiffusion equation is also often written as [15]

$$\frac{\partial^\gamma u(x,t)}{\partial t^\gamma} = \frac{\partial^2 u(x,t)}{\partial x^2}, \quad 0 < \gamma < 1, \tag{8}$$

where

$$\frac{\partial^\gamma}{\partial t^\gamma}y(x,t) = \frac{1}{\Gamma(1-\gamma)}\int_0^t \frac{\frac{\partial}{\partial t'}y(x,t')}{(t-t')^\gamma}\,dt' \tag{9}$$

denotes a Caputo fractional derivative, see, for example, reference [12]. There has been quite a bit written in the published literature on the greater physical practicality of the Caputo derivative over the Riemann–Liouville derivative, but this is largely unfounded [12]. Note, however, that if one takes

Equation (8) as the starting evolution equation for subdiffusion then this is suggestive of the following reaction–subdiffusion equation,

$$\frac{\partial^\gamma u(x,t)}{\partial t^\gamma} = \frac{\partial^2 u(x,t)}{\partial x^2} + f(u(x,t)). \tag{10}$$

Equations along the lines of Equation (10) are particularly widespread in the literature with the motivation that fractional derivatives incorporate a history dependence, and solutions of Equation (10) remain positive. Equation (10) can be derived from a CTRW where particles are being removed or added instantaneously at the start of the waiting times between jumps, but only under the contrived constraint that $\frac{\partial^{1-\gamma} f(u(x,t))}{\partial t^{1-\gamma}}$ represents the cumulative total of additions and removals to the arrival density of particles at position x and time t [14].

The derivation of reaction–subdiffusion equations from physically consistent CTRWs has been carried out in a series of papers [14,16–26]. The main lessons from this body of work are: (i) The governing equations are different depending on whether or not new born particles inherit the waiting times of their parents. (ii) Birth terms and death terms must be treated differently. (iii) In the case where particles are removed, but not instantaneously at the start of the waiting time between jumps, the reaction and subdiffusion terms are not additive. The following equation [21,24],

$$\begin{aligned}
\frac{\partial u(x,t)}{\partial t} = {} & D_\gamma \frac{\partial^2}{\partial x^2} \left[e^{-\int_0^t a(u(x,t'),x,t')\,dt'} \, {}_0\mathcal{D}_t^{1-\gamma} \left(e^{\int_0^t a(u(x,t'),x,t')\,dt'} u(x,t) \right) \right] \\
& + c(u(x,t),x,t) - a(u(x,t),x,t)u(x,t),
\end{aligned} \tag{11}$$

which was derived from a continuous time random walk model, provides the evolution equation for particles undergoing subdiffusion with particles annihilated at a per capita rate, $a(u(x,t),x,t)$ and created at a rate $c(u(x,t),x,t)$. In the derivation of this equation it was assumed that newborn particles do not inherit the waiting times of their parents.

In the remainder of this paper we explore examples related to Equation (11). These examples have been selected to emphasize the importance of considering the details of the reaction kinetics when dealing with reaction–subdiffusion problems. Whilst there have been many papers published on various methods of solution for variants of Equation (10) (see, for example, [27–31]), there have been very few papers published considering algebraic or numerical solution methods for variants of Equation (11). We hope that the examples below will stimulate further activity in this area, where the physical motivation for the modelling equation is stronger.

2. Examples

2.1. Birth and Death Balance

As a first example, we consider a population density of $u(x,t)$ particles per unit volume that are diffusing with a per capita death rate α and a birth rate $\alpha u(x,t)$. The reaction rate equation reflecting this balance between births and deaths, in a well-mixed population, at a location x is

$$\frac{\partial u(x,t)}{dt} = 0, \tag{12}$$

and thus the standard reaction–diffusion equation describing this system is

$$\frac{\partial u(x,t)}{dt} = D\frac{\partial^2 u(x,t)}{\partial x^2}. \tag{13}$$

The simple generalization of this equation for subdiffusive transport is

$$\frac{\partial u(x,t)}{dt} = D_\gamma\, {}_0\mathcal{D}_t^{1-\gamma}\frac{\partial^2 u(x,t)}{\partial x^2}$$

$$= D_\gamma \frac{\partial^2}{\partial x^2} \left[{}_0\mathcal{D}_t^{1-\gamma} u(x,t) \right]. \tag{14}$$

Indeed, if there were no births or deaths then the reaction rate equation would still be given by Equation (12); and Equation (14) is the appropriate equation to describe subdiffusion without births or deaths. However, the reaction–subdiffusion equation, following Equation (11), and using the reaction rate kinetics $a(u(x,t),x,t) = \alpha$ and $c(u(x,t),x,t) = \alpha u(x,t)$, which are also consistent with the rate equation, Equation (12), is remarkably different;

$$\frac{\partial u(x,t)}{\partial t} = D_\gamma \frac{\partial^2}{\partial x^2} \left[e^{-\alpha t} {}_0\mathcal{D}_t^{1-\gamma} \left(e^{\alpha t} u(x,t) \right) \right], \quad \alpha > 0. \tag{15}$$

The fundamental difference between Equations (14) and (15) is that in the former equation the Laplacian operates on a time fractional derivative and in the latter the Laplacian operates on a tempered time fractional derivative [32,33]. In the more general time fractional reaction–diffusion equation, Equation (11), the term in brackets following the Laplacian defines a generalized tempered time fractional derivative. The physical interpretation of the tempering is that if particles are being annihilated at a given rate while they wait then they cannot wait an arbitrarily long time at a given location. Note that both Equations (14) and (15) are mass conserving and thus Equation (15) then defines a mass conserving, tempered, time fractional diffusion equation.

The mean square displacement of the diffusing particles, $\langle x^2(t) \rangle$, provides a clear measurable difference between particles following Equation (14) or Equation (15). In the former case, identified as $\langle x_I^2(t) \rangle$, we have [8],

$$\langle x_I^2(t) \rangle = \frac{2D_\gamma}{\Gamma(1+\gamma)} t^\gamma, \tag{16}$$

and in the latter case, identified as $\langle x_{II}^2(t) \rangle$, we have (Appendix A)

$$\langle x_{II}^2(t) \rangle = 2D_\gamma e^{-\alpha t} t^\gamma E_{1,\gamma}^{(1)}(\alpha t), \tag{17}$$

where

$$E_{1,\gamma}^{(1)}(z) = \frac{d}{dz} \sum_{k=0}^{\infty} \frac{z^k}{\Gamma(\gamma+k)} \tag{18}$$

is the derivative of a generalized Mittag–Leffler function [34]. Note that at short times,

$$\langle x_{II}^2(t) \rangle \sim \frac{2D_\gamma}{\Gamma(1+\gamma)} t^\gamma, \tag{19}$$

but at large times, using the asymptotic expansion of the generalized Mittag–Leffler function (Equation (6) in [35]),

$$\langle x_{II}^2(t) \rangle \sim 2D_\gamma \alpha^{1-\gamma} t. \tag{20}$$

Thus, mass conserving tempered time fractional diffusion is not anomalous at long times.

We can also write down explicit expressions for solutions to Equations (14) and (15), labelled as $u_I(x,t)$ and $u_{II}(x,t)$, respectively. For simplicity we consider the infinite domain Greens function solutions with initial condition $u(x,0) = \delta(x)$.

The Greens function solution of the fractional diffusion equation Equation (14) can be written as [8]

$$u_I(x,t) = \frac{1}{\sqrt{4\pi D_\gamma t^\gamma}} H_{1,2}^{2,0} \left[\frac{x^2}{4D_\gamma t^\gamma} \middle| \begin{array}{c} (1-\frac{\gamma}{2},\gamma) \\ (0,1) \qquad (\frac{1}{2},1) \end{array} \right], \tag{21}$$

where H denotes a Fox H-function [36], see Equation (A11).

To find the Greens function solution $u_{II}(x,t)$ we first note that Equation (15) can be re-written as

$$\frac{\partial v(x,t)}{\partial t} = D_\gamma \frac{\partial^2}{\partial x^2}\, _0\mathcal{D}_t^{1-\gamma} v(x,t) + \alpha v(x,t), \tag{22}$$

where

$$v(x,t) = e^{\alpha t} u_{II}(x,t). \tag{23}$$

The Greens function solution of Equation (22) can be obtained as a special case of the more general results in Appendix B of [14], yielding

$$v(x,t) = \frac{1}{\sqrt{4\pi D_\gamma t^\gamma}} \sum_{j=0}^{\infty} \frac{(\alpha t)^j}{j!} H_{1,2}^{2,0} \left[\frac{x^2}{4D_\gamma t^\gamma} \,\middle|\, \begin{array}{c} (1 - \frac{\gamma}{2} + j, \gamma) \\ (0,1) \end{array} \quad (\tfrac{1}{2} + j, 1) \right], \tag{24}$$

and then using Equation (23) we have

$$u_{II}(x,t) = e^{-\alpha t} \frac{1}{\sqrt{4\pi D_\gamma t^\gamma}} \sum_{j=0}^{\infty} \frac{(\alpha t)^j}{j!} H_{1,2}^{2,0} \left[\frac{x^2}{4D_\gamma t^\gamma} \,\middle|\, \begin{array}{c} (1 - \frac{\gamma}{2} + j, \gamma) \\ (0,1) \end{array} \quad (\tfrac{1}{2} + j, 1) \right]. \tag{25}$$

In Appendix B, we show that the Fox functions in Equations (21) and (25) can be simplified for $\gamma = \frac{1}{2}$ in terms of Miejer G-Functions [37], see Equation (A12), which have the advantage that they can readily be evaluated using computer algebra packages such as MATHEMATICA and MAPLE. Using the result of Equation (A19) from the Appendix B, we can write (see also [8] in the case of $u_I(x,t)$)

$$u_I(x,t) = \frac{1}{\sqrt{8\pi^3 D t^{\frac{1}{2}}}} G_{0,3}^{3,0} \left[\left(\frac{x^2}{16D t^{\frac{1}{2}}} \right)^2 \,\middle|\, \begin{array}{c} - \\ 0, \frac{1}{4}, \frac{1}{2} \end{array} \right], \tag{26}$$

and

$$u_{II}(x,t) = e^{-\alpha t} \frac{1}{\sqrt{8\pi^3 D t^{\frac{1}{2}}}} \sum_{j=0}^{\infty} \frac{(2\alpha t)^j}{j!} G_{1,4}^{4,0} \left[\left(\frac{x^2}{16D t^{\frac{1}{2}}} \right)^2 \,\middle|\, \begin{array}{c} \frac{3}{4} + j \\ 0, \frac{1}{2}, \frac{1}{4} + \frac{j}{2}, \frac{3}{4} + \frac{j}{2} \end{array} \right]. \tag{27}$$

Note that the expression for $u_{II}(x,t)$ simplifies to the expression for $u_I(x,t)$ if $\alpha = 0$. If $|x| \gg 4\sqrt{D t^{\frac{1}{2}}}$ then we can use asymptotic expansions for $G_{0,3}^{3,0}(z)$ and $G_{1,4}^{4,0}(z)$ with $z \gg 1$ (see Appendix B) to write

$$u_I(x,t) \sim \frac{1}{\sqrt{8\pi^3 D t^{\frac{1}{2}}}} \exp\left(-3\left(\frac{x^2}{16D t^{\frac{1}{2}}}\right)^{\frac{2}{3}}\right) \left(\frac{x^2}{16D t^{\frac{1}{2}}}\right)^{\frac{1}{2}} \frac{M_0}{\left(\frac{x^2}{16D t^{\frac{1}{2}}}\right)^{\frac{2}{3}} - 1}, \tag{28}$$

and

$$u_{II}(x,t) \sim M e^{\alpha t} u_I(x,t), \tag{29}$$

where M_0 and M are constant terms. The solutions $u_I(t)$, Equation (26), and $u_{II}(t)$, Equation (27) are plotted in Figure 1, with $\alpha = 1$ and $D = 1$, at times $t = 0.1$, $t = 1.0$ and $t = 10.0$. The solutions are very similar at early times but the corner at the origin, which is characteristic of subdiffusion, is less sharp at longer times in the solution of Equation (27).

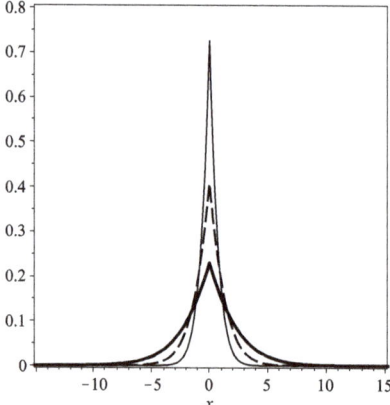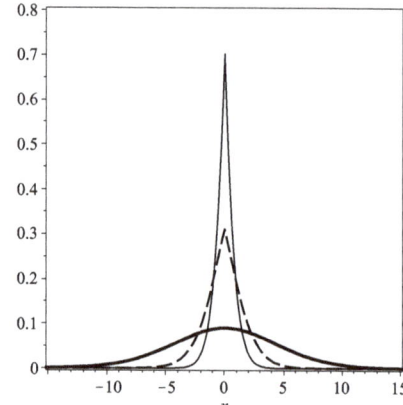

Figure 1. Plots of Equation (26), the algebraic solution to Equation (14), (**left**), and Equation (27), the algebraic solution to Equation (15), (**right**), at times $t = 0.1$ (solid line), $t = 1.0$ (dashed line) and $t = 10.0$ (bold solid line). The reaction parameter $\alpha = 1$, and the fractional order derivative is taken to be $\gamma = 0.5$ in each of these plots.

The lesson from this simple example is that reaction dynamics equations do not contain sufficient information on their own to provide model equations for reaction–subdiffusion systems even in well-mixed systems. In the case of standard diffusion, the evolution of the population density is only affected by the overall reaction rates, in a well-mixed system, but not the details of the reaction kinetics. In a standard reaction–diffusion system, the dynamics with no births and no deaths is the same as if there were births and deaths but the rates cancelled out. The reaction–diffusion equation with these reaction kinetics has no memory of the birth and death processes. This is very different in the case of subdiffusion where the details of the reaction kinetics are important to the overall dynamics of the system. The subdiffusive system retains a memory that there were particles that were created and annihilated. Moreover, the particle deaths temper the fractional diffusion. The example in the next section further highlights the significance of the reaction kinetics in reaction–subdiffusion systems.

2.2. Fractional Fisher–KPP Equation

The reaction rate equation for the Fisher–KPP Equation (1) is given in Equation (2). There are many different reaction kinetics that could be considered that are consistent with Equation (2). For example, the term $(1 - u(x, t))$ in its entirety could represent a per capita birth rate if is is strictly positive, or a per capita death rate if it is strictly negative. This term could also be regarded as being composed of two terms, a constant per capita birth term and a linear per capita death term. These three possibilities are highlighted for illustrative purposes below to show how different subdiffusion–reaction equations apply depending on the reaction kinetics.

(i) Constant per capita birth rate, $c(u(x, t), x, t) = ru(x, t)$, linear per capita death rate, $a(u(x, t), x, t) = ru(x, t)$,

$$
\begin{aligned}
\frac{\partial u(x, t)}{\partial t} &= D_\gamma \frac{\partial^2}{\partial x^2} \left[e^{-\int_0^t ru(x,t')\,dt'} {}_0\mathcal{D}_t^{1-\gamma} \left(e^{\int_0^t ru(x,t')\,dt'} u(x, t) \right) \right] \\
&\quad + ru(x, t)(1 - u(x, t)), \quad u(x, t) \geq 0
\end{aligned}
\tag{30}
$$

(ii) No births, $c(u(x,t),x,t) = 0$, linear per capita death rate, $a(u(x,t),x,t) = r(1 - u(x,t))$,

$$
\begin{aligned}
\frac{\partial u(x,t)}{\partial t} &= D_\gamma \frac{\partial^2}{\partial x^2} \left[e^{-\int_0^t r(1-u(x,t'))\,dt'} \, {}_0\mathcal{D}_t^{1-\gamma} \left(e^{\int_0^t r(1-u(x,t'))\,dt'} u(x,t) \right) \right] \\
&\quad + r(1 - u(x,t))u(x,t), \quad u(x,t) \geq 1.
\end{aligned} \tag{31}
$$

(iii) Linear per capita birth rate, $c(u(x,t),x,t) = ru(x,t)(1 - u(x,t))$, no deaths, $a(u(x,t),x,t) = 0$,

$$
\frac{\partial u(x,t)}{\partial t} = D_\gamma \frac{\partial^2}{\partial x^2} \left[{}_0\mathcal{D}_t^{1-\gamma} u(x,t) \right] + ru(x,t)(1 - u(x,t)), \quad 0 \leq u(x,t) \ll 1. \tag{32}
$$

Note that none of the factional Fisher–KPP reaction–diffusion equations can be expressed in the form

$$
\frac{\partial^\gamma u(x,t)}{\partial t^\gamma} = D_\gamma \frac{\partial^2 u(x,t)}{\partial x^2} + ru(x,t)(1 - u(x,t)), \tag{33}
$$

which results from simply replacing the integer order time derivative with a fractional order Caputo derivative. As noted above, an equation of this form could only be obtained from a CTRW if $\frac{\partial^{1-\gamma}}{\partial t^{1-\gamma}} (ru(x,t)(1 - u(x,t)))$ is contrived as the cumulative instantaneous creation and annihilation of particles at the start of the waiting time between particle jumps at position x and time t [14].

The Greens function solutions for the nonlinear fractional reaction–diffusion equations, Equations (30)–(32), cannot be obtained simply using Fourier–Laplace transform methods. However, it is possible to find numerical solutions using the discrete time random walk methods described in [38].

The Fisher–KPP reaction rate equation, Equation (2) can be motivated by different chemical reactions consistent with the law of mass action [4]. One possibility is that of a single species A which undergoes coalescence reactions $A + A \xrightarrow{r} A$, and degradation reactions $A \xrightarrow{r} A + A$; also referrred to as reversible coagulation dynamics [39]. In this scenario the creation term, $ru(x,t)$, arises from degradation and the annihilation term, $-r(u^2(x,t))$, arises from coalescence. Another possibility is a branching–coalescence scheme [17], $B + X \rightleftharpoons X + X$, with the concentration of B maintained at a constant level. Equation (30) is a fractional Fisher–KPP reaction–diffusion equation consistent with each of the reaction schemes described here and it was obtained earlier for the branching–coalescence reaction scheme in [17].

2.3. Fractional Fitzhugh–Nagumo Equation

A widely studied reaction–diffusion system used to model wave propagation and pattern formation in excitable media is the Fitzhugh–Nagumo system of equations [40,41]

$$
\frac{\partial v(x,t)}{\partial t} = D_v \frac{\partial^2 v(x,t)}{\partial x^2} + v(x,t)(v(x,t) - a)(1 - v(x,t)) - w(x,t), \quad D_v \geq 0, a \geq 0 \tag{34}
$$

$$
\frac{\partial w(x,t)}{\partial t} = D_w \frac{\partial^2 w(x,t)}{\partial x^2} + \epsilon (v(x,t) - bw(x,t)) \quad D_w \geq 0, \epsilon \geq 0, b \geq 0, \tag{35}
$$

named after Fizthugh [42] and Nagumo [43]. In recent years, the single component fractional equation

$$
\frac{\partial^\alpha u(x,t)}{\partial t^\alpha} = D_u \frac{\partial^2 u(x,t)}{\partial x^2} + u(x,t)(u(x,t) - a)(1 - u(x,t)) \tag{36}
$$

has been studied as a test equation for various methods of solution of time fractional reaction–diffusion equations (see, for example, [27,30] and references there-in).

A time fractional Fitzhugh–Nagumo system of equations consistent with Equation (11), derived from a CTRW formalism, can be obtained by identifying per capita annihilation rates, a_v and a_w, and creation rates, c_v and c_w, as follows:

$$a_v(v(x,t), w(x,t)) = a + v^2(x,t) + \frac{w(x,t)}{v(x,t)}, \tag{37}$$

$$c_v(v(x,t), w(x,t)) = (1+a)v^2(x,t), \tag{38}$$

$$a_w(v(x,t), w(x,t)) = \epsilon b, \tag{39}$$

$$c_w(v(x,t), w(x,t)) = \epsilon v(x,t). \tag{40}$$

The corresponding time fractional Fitzhugh–Nagumo system is given by

$$\frac{\partial v(x,t)}{\partial t} = D_{v,\gamma} \frac{\partial^2}{\partial x^2} \left[e^{-\int_0^t (v^2(x,t')+a+w(x,t'))\,dt'} {}_0\mathcal{D}_t^{1-\gamma} \left(e^{\int_0^t (v^2(x,t')+a+w(x,t'))\,dt'} v(x,t) \right) \right]$$
$$+ v(x,t)(v(x,t)-a)(1-v(x,t)-w(x,t)), \tag{41}$$

$$\frac{\partial w(x,t)}{\partial t} = D_{w,\gamma} \frac{\partial^2}{\partial x^2} \left[e^{-\epsilon b t} {}_0\mathcal{D}_t^{1-\gamma} \left(e^{\epsilon b t} w(x,t) \right) \right] + \epsilon v(x,t) - \epsilon b w(x,t). \tag{42}$$

If $w(x,t) = 0$ this identifies a single component time fractional equation

$$\frac{\partial u(x,t)}{\partial t} = D_\gamma \frac{\partial^2}{\partial x^2} \left[e^{-\int_0^t (u^2(x,t')+a)\,dt'} {}_0\mathcal{D}_t^{1-\gamma} \left(e^{\int_0^t (u^2(x,t')+a)\,dt'} u(x,t) \right) \right]$$
$$+ u(x,t)(u(x,t)-a)(1-u(x,t)), \tag{43}$$

which could be called a time fractional Fitzhugh–Nagumo equation, although the nomenclature could be misleading because a single component equation, without external sources or sinks, could not display Fitzhugh–Nagumo dynamics. Equation (43) is, however, well posed as a nonlinear time fractional reaction–diffusion equation that can be derived from a physically consistent CTRW, and thus it should be preferred for testing numerical methods of solution over the single component model equation, Equation (36), obtained by replacing an integer order time derivative with a Caputo fractional order derivative.

3. Discussion

Over the past two decades there have been large numbers of papers published on numerical methods for nonlinear fractional reaction–diffusion equations. The original motivation for including time fractional derivatives in reaction–diffusion equations was based on a CTRW description of diffusion with traps and reactions [13]. This description was refined and improved in a series of papers [14,16–24], leading to the formulation of time fractional reaction–diffusion equations along the lines of Equation (11). However, many investigations of time fractional reaction–diffusion equations have been carried out on systems obtained by simply replacing integer order time derivatives with Caputo fractional order derivatives. These studies may be interesting from a mathematical analysis point of view but they may not be directly relevant to mathematical modelling applications.

In this paper we have illustrated, through examples, how different time fractional reaction–diffusion equations can be formulated, consistent with an underlying CTRW formalism, taking into account the reaction kinetics. There are three points worth noting in this context: (i) The fractional reaction–diffusion systems considered in this approach are relevant to well-mixed reactions that are not diffusion limited. The reaction dynamics can often be formulated using the law of mass action in these systems. (ii) Different time fractional reaction–diffusion systems can be formulated that are consistent with the same equation for the reaction dynamics. It is important to know the reaction kinetics. (iii) Reaction–subdiffusion equations typically involve a spatial Laplacian operating on a generalized tempered time fractional derivative. The solution of these types of equations would typically require very different numerical approaches than those proposed for reaction–diffusion systems with a fractional order Caputo time derivative replacing the integer order time derivative.

It is hoped that the physically motivated time fractional reaction–diffusion equations, such as Equations (30) and (43), will become more widely used, replacing the simpler ad-hoc equations, such as Equations (33) and (36), as a test for different methods of solution of nonlinear fractional reaction–diffusion systems. Beyond this, there is a real need for physical experiments to be devised and carried out to validate and calibrate time fractional reaction–diffusion models.

Author Contributions: The authors have contributed equally to all aspects of this work including; conceptualization, methodology, formal analysis, writing, project administration, funding acquisition. All authors have read and agreed to the published version of the manuscript.

Funding: This research was funded by the Australian Commonwealth Government ARC DP200100345.

Conflicts of Interest: The authors declare no conflict of interest.

Appendix A. Mean Square Displacements

The mean square displacement of particles evolving according to the fractional diffusion equation

$$\frac{\partial u(x,t)}{\partial t} = D_\gamma \frac{\partial^2}{\partial x^2} \left[e^{-\alpha t} \, {}_0\mathcal{D}_t^{1-\gamma} \left(e^{\alpha t} u(x,t) \right) \right], \quad \alpha > 0. \tag{A1}$$

can simply be obtained from the infinite domain Greens function solution $G(x,t)$ with initial condition $G(x,0) = \delta(x)$, via

$$\langle x^2(t) \rangle = \lim_{q \to 0} -\frac{d^2}{dq^2} \hat{G}(q,t) \tag{A2}$$

where $\hat{G}(q,t)$ denotes the Fourier transform w.r.t. x. We begin by taking the Fourier transform of Equation (A2) and re-arranging terms to write

$$\frac{\partial}{\partial t} \left(e^{\alpha t} \hat{G}(q,t) \right) = -q^2 D_\gamma \, {}_0\mathcal{D}_t^{1-\gamma} \left(e^{\alpha t} \hat{G}(q,t) \right) + \alpha e^{\alpha t} \hat{G}(q,t). \tag{A3}$$

We now introduce

$$\hat{F}(q,t) = e^{\alpha t} \hat{G}(q,t), \tag{A4}$$

noting that $\hat{F}(q,0) = \hat{G}(q,0) = 1$, and then

$$\langle x^2(t) \rangle = e^{-\alpha t} \lim_{q \to 0} -\frac{d^2}{dq^2} \hat{F}(q,t). \tag{A5}$$

Starting with the differential equation for $\hat{F}(q,t)$,

$$\frac{\partial}{\partial t} \left(\hat{F}(q,t) \right) = -q^2 D_\gamma \, {}_0\mathcal{D}_t^{1-\gamma} \left(\hat{F}(q,t) \right) + \alpha \hat{F}(q,t), \tag{A6}$$

we take the Laplace transform w.r.t. time and rearrange terms to write

$$\hat{F}(q,s) = \frac{1}{(s + D_\gamma q^2 s^{1-\gamma} - \alpha)} \tag{A7}$$

From this we have

$$\lim_{q \to 0} -\frac{d^2}{dq^2} \hat{F}(q,s) = \frac{2D_\gamma}{s^{\gamma-1}\alpha^2 - 2s^\gamma \alpha + s^{\gamma+1}},$$

$$= 2D_\gamma \frac{s^{1-\gamma}}{(s-\alpha)^2}. \tag{A8}$$

We now take the inverse Laplace transform using Equation (2.3.26) of [34] to write

$$\lim_{q \to 0} -\frac{d^2}{dq^2}\hat{\tilde{F}}(q,t) = 2D_\gamma t^\gamma E_{1,\gamma}^{(1)}(\alpha t),\tag{A9}$$

and then

$$\langle x^2(t)\rangle = 2D_\gamma e^{-\alpha t}t^\gamma E_{1,\gamma}^{(1)}(\alpha t).\tag{A10}$$

Appendix B. Fox H-Function and Meijer G-Function Solutions

The Fox H-function and the Meijer G-function are defined as path integrals [36]

$$H_{p,q}^{m,n}\left[z \left| \begin{array}{c}(a_1,A_1)(a_2,A_2)\ldots(a_p,A_p)\\(b_1,B_1)(b_2,B_2)\ldots(b_q,B_q)\end{array}\right.\right] = \frac{1}{2\pi i}\int_L \frac{\prod_{j=1}^m \Gamma(b_j+B_js)\prod_{j=1}^n \Gamma(1-a_j-A_js)}{\prod_{j=m+1}^q \Gamma(1-b_j-B_js)\prod_{j=n+1}^p \Gamma(a_j+A_js)}z^{-s}\,ds,\tag{A11}$$

and

$$G_{p,q}^{m,n}\left[z \left| \begin{array}{c}a_1,a_2,\ldots a_p\\b_1,b_2,\ldots b_q\end{array}\right.\right] = \frac{1}{2\pi i}\int_L \frac{\prod_{j=1}^m \Gamma(b_j-s)\prod_{j=1}^n \Gamma(1-a_j+s)}{\prod_{j=m+1}^q \Gamma(1-b_j+s)\prod_{j=n+1}^p \Gamma(a_j-s)}z^s\,ds,\tag{A12}$$

respectively, where $0 \le n \le p$, $1 \le m \le q$, $\{a_j,b_j\} \in \mathbb{C}$, $\{\alpha_j,\beta_j\} \in \mathbb{R}^+$, and L is a suitably chosen contour. With a simple change of variables it follows that if $A_j = C, j = 1..p$ and $B_j = C, j = 1..q$ then

$$H_{p,q}^{m,n}\left[z \left| \begin{array}{c}(a_1,C)(a_2,C)\ldots(a_p,C)\\(b_1,C)(b_2,C)\ldots(b_q,C)\end{array}\right.\right] = \frac{1}{C}G_{p,q}^{m,n}\left[z^{\frac{1}{C}} \left| \begin{array}{c}a_1,a_2,\ldots a_p\\b_1,b_2,\ldots b_q\end{array}\right.\right]\tag{A13}$$

The Legendre duplication formula

$$\Gamma(2z) = \frac{2^{2z-1}}{\sqrt{\pi}}\Gamma(z)\Gamma(z+\frac{1}{2})\tag{A14}$$

is useful for reducing Fox H-functions to Meijer G-functions in the expressions below.

The Fox-H function

$$H_{1,2}^{2,0}\left[z \left| \begin{array}{c}(1-\frac{\gamma}{2}+j,\gamma)\\(0,1)(\frac{1}{2}+j+1)\end{array}\right.\right] = \frac{1}{2\pi i}\int_L \frac{\Gamma(s)\Gamma(\frac{1}{2}+j+s)}{\Gamma(1-\frac{\gamma}{2}+j+\gamma s)}z^{-s}\,ds\tag{A15}$$

appears in the solutions, Equations (21) and (25). Here we show how, in the case $\gamma = \frac{1}{2}$, this can be represented as a Meijer G-function leading to the solutions in Equations (26) and (27). With $\gamma = \frac{1}{2}$ in Equation (A15) we have

$$H_{1,2}^{2,0}\left[z \left| \begin{array}{c}(\frac{3}{4}+j,\frac{1}{2})\\(0,1)(\frac{1}{2}+j+1)\end{array}\right.\right] = \frac{1}{2\pi i}\int_L \frac{\Gamma(s)\Gamma(\frac{1}{2}+j+s)}{\Gamma(\frac{3}{4}+j+\frac{s}{2})}z^{-s}\,ds.\tag{A16}$$

We now use the duplication formula, Equation (A14), to replace

$$\Gamma(s) = \frac{2^{s-1}}{\sqrt{\pi}}\Gamma(\frac{s}{2})\Gamma(\frac{s}{2}+\frac{1}{2}),\tag{A17}$$

and

$$\Gamma(\frac{1}{2}+j+s) = \frac{2^{\frac{1}{2}+j+s-1}}{\sqrt{\pi}}\Gamma(\frac{1}{4}+\frac{j}{2}+\frac{s}{2})\Gamma(\frac{3}{4}+\frac{j}{2}+\frac{s}{2}),\tag{A18}$$

so that

$$
\begin{aligned}
H^{2,0}_{1,2}\left[z \,\middle|\, \begin{matrix} (\tfrac{3}{4}+j,\tfrac{1}{2}) \\ (0,1)(\tfrac{1}{2}+j+1) \end{matrix}\right] &= \frac{1}{2\pi i}\int_{L}\frac{2^{\frac{1}{2}+j+2s-2}}{\pi}\frac{\Gamma(\tfrac{s}{2})\Gamma(\tfrac{1}{2}+\tfrac{s}{2})\Gamma(\tfrac{1}{4}+\tfrac{j}{2}+\tfrac{s}{2})\Gamma(\tfrac{3}{4}+\tfrac{j}{2}+\tfrac{s}{2})}{\Gamma(\tfrac{3}{4}+j+\tfrac{s}{2})}z^{-s}\,ds, \\[2mm]
&= \frac{2^{\frac{1}{2}+j-2}}{\pi}H^{4,0}_{1,4}\left[\frac{z}{4}\,\middle|\,\begin{matrix}(\tfrac{3}{4}+j,\tfrac{1}{2}) \\ (0,\tfrac{1}{2})(\tfrac{1}{2},\tfrac{1}{2})(\tfrac{1}{4}+\tfrac{j}{2},\tfrac{1}{2})(\tfrac{3}{4}+\tfrac{j}{2},\tfrac{1}{2})\end{matrix}\right] \\[2mm]
&= \frac{2^{j}}{\sqrt{2\pi}}G^{4,0}_{1,4}\left[(\tfrac{z}{4})^{2}\,\middle|\,\begin{matrix}\tfrac{3}{4}+j \\ 0,\tfrac{1}{2},\tfrac{1}{4}+\tfrac{j}{2},\tfrac{3}{4}+\tfrac{j}{2}\end{matrix}\right].
\end{aligned}
\tag{A19}
$$

Note that if $j=0$ this simplifies further to

$$
\frac{1}{\sqrt{2\pi}}G^{4,0}_{1,4}\left[(\tfrac{z}{4})^{2}\,\middle|\,\begin{matrix}\tfrac{3}{4} \\ 0,\tfrac{1}{2},\tfrac{1}{4},\tfrac{3}{4}\end{matrix}\right] = \frac{1}{\sqrt{2\pi}}G^{3,0}_{0,3}\left[(\tfrac{z}{4})^{2}\,\middle|\,\begin{matrix}- \\ 0,\tfrac{1}{2},\tfrac{1}{4}\end{matrix}\right].
\tag{A20}
$$

The Meijer G-functions above are of the general form $G^{q,0}_{p,q}(z)$ where asymptotic expansions are known for $z \gg 1$ [37]. In particular, using Equation (22) in [37], we have

$$
G^{4,0}_{1,4}\left[z\,\middle|\,\begin{matrix}\tfrac{3}{4}+j \\ 0,\tfrac{1}{2},\tfrac{1}{4}+\tfrac{j}{2},\tfrac{3}{4}+\tfrac{j}{2}\end{matrix}\right] \sim \exp(-3z^{\frac{1}{3}})z^{-\frac{1}{12}}\sum_{k=0}^{\infty}\frac{M_{k}(j)}{z^{\frac{k}{3}}}
\tag{A21}
$$

for all $j \in \mathbb{N}$ and $z \gg 1$, where the $M_{k}(j)$ are functions of the parameters, including j, but not the variable z. If we let M denote the largest $M_{k}(j)$ then we can evaluate the sum as a geometric series to write

$$
G^{4,0}_{1,4}\left[z\,\middle|\,\begin{matrix}\tfrac{3}{4}+j \\ 0,\tfrac{1}{2},\tfrac{1}{4}+\tfrac{j}{2},\tfrac{3}{4}+\tfrac{j}{2}\end{matrix}\right] \sim M\frac{\exp(-3z^{\frac{1}{3}})z^{\frac{1}{4}}}{z^{\frac{1}{3}}-1}
\tag{A22}
$$

and similarly for $G^{3,0}_{0,3}$.

References

1. Okubo, A. Diffusion and ecological problems: Mathematical models. *Biomathematics* **1980**, *10*, 114.
2. Britton, N.F. *Reaction-Diffusion Equations and Their Applications to Biology*; Academic Press: London, UK, 1986.
3. Murray, J.D. *Mathematical Biology. II Spatial Models and Biomedical Applications*; Springer: New York, NY, USA, 2003.
4. Chellaboina, V.; Bhat, S.P.; Haddad, W.M.; Bernstein, D.S. Modeling and analysis of mass-action kinetics. *IEEE Control Syst.* **2009**, *29*, 60–78.
5. Fisher, R.A. The Wave of Advance of Advantageous Genes. *Ann. Eugen.* **1937**, *7*, 353–369. [CrossRef]
6. Kolmogorov, A.; Petrovskii, I.; Piskunov, N. A study of the diffusion equation with increase in the amount of substance, and its application to a biological problem. *Bull. Mosc. Univ. Math. Mech.* **1937**, *1*, 1–25.
7. Einstein, A. On the motion of small particles suspended in liquids at rest required by the molecular-kinetic theory of heat. *Ann. Der Phys.* **1905**, *17*, 549–560. [CrossRef]
8. Metzler, R.; Klafter, J. The random walk's guide to anomalous diffusion: A fractional dynamics approach. *Phys. Rep.* **2000**, *339*, 1–77. [CrossRef]
9. Hilfer, R.; Anton, L. Fractional master equations and fractal time random walks. *Phys. Rev. E* **1995**, *51*, R848. [CrossRef]
10. Compte, A. Stochastic foundations of fractional dynamics. *Phys. Rev. E* **1996**, *53*, 4191–4193. [CrossRef]
11. Montroll, E.; Weiss, G. Random walks on lattices II. *J. Math. Phys.* **1965**, *6*, 167 . [CrossRef]
12. Li, C.; Qiang, D.; Chen, Y.Q. On Riemann-Liouville and Caputo Derivatives. *Discret. Dyn. Nat. Soc.* **2011**, *2011*, 562494. [CrossRef]
13. Henry, B.I.; Wearne, S.L. Fractional reaction-diffusion. *Phys. A* **2000**, *276*, 448–455. [CrossRef]
14. Henry, B.I.; Langlands, T.A.M.; Wearne, S.L. Anomalous diffusion with linear reaction dynamics: From continuous time random walks to fractional reaction-diffusion equations. *Phys. Rev. E* **2006**, *74*, 031116. [CrossRef]

15. Gorenflo, R.; Luchko, Y.; Mainardi, F. Wright functions as scale-invariant solutions of the diffusion wave equation. *J. Comput. Appl. Math.* **2000**, *118*, 175–191. [CrossRef]

16. Sokolov, I.M.; Schmidt, M.G.W.; Sagués, F. Reaction-subdiffusion equations. *Phys. Rev. E* **2006**, *73*, 031102. [CrossRef]

17. Yadav, A.; Fedotov, S.; Méndez, V.; Horsthemke, W. Progagating fronts in reaction-transport systems with memory. *Phys. Letts. A* **2007**, *371*, 374–378. [CrossRef]

18. Langlands, T.A.M.; Henry, B.I.; Wearne, S.L. Anomalous subdiffusion with multispecies linear reaction dynamics. *Phys. Rev. E* **2008**, *77*, 021111. [CrossRef]

19. Campos, D.; Fedotov, S.; Mendez, V. Anomalous reaction-transport processes: The dynamics beyond the law of mass action. *Phys. Rev. E* **2008**, *77*, 061130. [CrossRef]

20. Froemberg, D.; Schmidt-Martens, H.; Sokolov, I.M.; Sagués, F. Front propagation in $A + B \rightarrow 2A$ reaction under subdiffusion. *Phys. Rev. E* **2008**, *78*, 011128. [CrossRef]

21. Fedotov, S. Non-Markovian random walks and nonlinear reactions: Subdiffusion and propagating fronts. *Phys. Rev. E* **2010**, *81*, 011117. [CrossRef]

22. Abad, E.; Yuste, S.B.; Lindenberg, K. Reaction-subdiffusion and reaction-superdiffusion equations for evanescent particles performing continuous-time random walks. *Phys. Rev. E* **2010**, *81*, 031115. [CrossRef]

23. Yuste, S.B.; Abad, E.; Lindenberg, K. Reaction-subdiffusion model of morphogen gradient formation. *Phys. Rev. E* **2010** *82*, 061123. [CrossRef]

24. Angstmann, C.N.; Donnelly, I.C.; Henry, B.I. Continuous time random walks with reactions forcing and trapping. *Math. Model. Nat. Phenom.* **2013**, *8*, 17–27. [CrossRef]

25. Nepomnyashchy, A.A. Mathematical modelling of sub-diffusion reaction systems. *Math. Model. Nat. Phenom.* **2016**, *11*, 26–36. [CrossRef]

26. Abad, E.; Angstmann, C.N.; Henry, B.I.; McGann, A.V.; Vot, F.L.; Yuste, S.B. Reaction-diffusion and reaction-subdiffusion equations on arbitrarily evolving domains. *Phys. Rev. E* **2020**, *102*, 032111. [CrossRef]

27. Rida, S.Z.; El-Sayed, A.M.A.; Arafa, A.A.M. On the solutions of time-fractional reaction-diffusion equations. *Commun. Nonlinear Sci. Numer. Simul.* **2010**, *15*, 3847–3854. [CrossRef]

28. Zhang, J.; Yang, X. A class of efficient difference method for time fractional reaction-dffusion equation. *Comput. Appl. Math.* **2018**, *37*, 4376–4396. [CrossRef]

29. Li, C.; Wang, Z. The local discontinuous Galerkin finite element methods for Caputo-type partial differential equations: Numerical analysis. *Appl. Numer. Math.* **2019**, *140*, 1–22. [CrossRef]

30. Prakash, A.; Kaur, H. A reliable numerical algorithm for a fractional model of Fitzhugh-Nagumo equation arising in the transmission of nerve impulses. *Nonlinear Eng.* **2019**, *8*, 719–727. [CrossRef]

31. Kanth, A.S.V.R.; Garg, N. A numerical approach for a class of time-fractional reaction-diffusion equation through exponential B-spline method. *Comput. Appl. Math.* **2020**, *39*, 1–24. [CrossRef]

32. Meerschaert, M.M.; Zhang, Y.; Baeumer, B. Tempered anomalous diffusion in heterogeneous systems. *Geophys. Res. Letts.* **2008**, *35*, L17403. [CrossRef]

33. Sabzikar, F.; Meerschaert, M.M.; Chen, J. Tempered fractional calculus. *J. Comput. Phys.* **2015**, *293*, 14–28. [CrossRef] [PubMed]

34. Mathai, A.M.; Haubold, H.J. Mittag-Leffler Functions and Fractional Calculus. In *Special Functions for Applied Scientists*; Springer: New York, NY, USA, 2008; pp. 79–134.

35. Gorenflo, R.; Loutchko, J.; Luchko, Y. Computation of the Mittag-Leffler function $E_{\alpha,\beta}(z)$ and its derivative. *Fract. Calc. Appl. Anal.* **2002**, *5*, 1–26.

36. Fox, C. The G and H functions as symmetrical Fourier kernels. *Trans. Am. Math. Soc.* **1961**, *98*, 395–429.

37. Meijer, C.S. On the G-function. *Mathematics* **1946**, *26*, 227–237.

38. Angstmann, C.N.; Donnelly, I.C.; Henry, B.I.; Jacobs, B.A.; Langlands, T.A.M.; Nichols, J.A. From stochastic processes to numerical methods: A new scheme for solving reaction subdiffusion fractional partial differential equations. *J. Comput. Phys.* **2016**, *307*, 508–534. [CrossRef]

39. ben-Avraham, D.; Burschka, M.A.; Doering, C.R. Statics and dynamics of a diffusion-limited reaction: Anomalous kinetics, non-equilibrium self-ordering, and a dynamic transition. *J. Stat. Phys.* **1990**, *60*, 695–728. [CrossRef]

40. Jones, C.K.R.T. Stability of the travelling wave solution of the Fitzhugh-Nagumo system. *Trans. Am. Math. Soc.* **1984**, *286*, 431–469. [CrossRef]

41. Zheng, Q.; Shen, J. Pattern formation in the FitzHugh-Nagumo model. *Comput. Math. Appl.* **2015**, *70*, 1082–1097. [CrossRef]

42. Fitzhugh, R. Impulse and physiological states in theoretical models of nerve membrane. *Biophys. J.* **1961**, *1*, 445–466. [CrossRef]

43. Nagumo, J.S.; Arimoto, S.; Yoshizawa, S. An active pulse transmission line stimulating nerve axon. *Proc. IRE* **1962**, *50*, 2061–2070. [CrossRef]

Article

Fractal and Entropy Analysis of the Dow Jones Index Using Multidimensional Scaling

José A. Tenreiro Machado

Department of Electrical Engineering, Institute of Engineering, Polytechnic Institute of Porto, 4249-015 Porto, Portugal; jtm@isep.ipp.pt; Tel.: +351-228340500

Received: 4 September 2020; Accepted: 30 September 2020; Published: 8 October 2020

Abstract: Financial time series have a fractal nature that poses challenges for their dynamical characterization. The Dow Jones Industrial Average (DJIA) is one of the most influential financial indices, and due to its importance, it is adopted as a test bed for this study. The paper explores an alternative strategy to the standard time analysis, by joining the multidimensional scaling (MDS) computational tool and the concepts of distance, entropy, fractal dimension, and fractional calculus. First, several distances are considered to measure the similarities between objects under study and to yield proper input information to the MDS. Then, the MDS constructs a representation based on the similarity of the objects, where time can be viewed as a parametric variable. The resulting plots show a complex structure that is further analyzed with the Shannon entropy and fractal dimension. In a final step, a deeper and more detailed assessment is achieved by associating the concepts of fractional calculus and entropy. Indeed, the fractional-order entropy highlights the results obtained by the other tools, namely that the DJIA fractal nature is visible at different time scales with a fractional order memory that permeates the time series.

Keywords: multidimensional scaling; fractals; fractional calculus; financial indices; entropy; Dow Jones; complex systems

1. Introduction

The Dow Jones Industrial Average (DJIA), or Dow Jones, is a stock market index that reflects the stock performance of 30 relevant companies included in the U.S. stock exchanges. The DJIA is the second-oldest among the U.S. market indices and started on 26 May 1896. The DJIA is the best-known index in finance and is considered a key benchmark for assessing the global business trend in the world.

The financial time series reflect intricate effects between a variety of agents coming from economic and social processes, geophysical phenomena, health crisis, and political strategies [1–4]. At present, we find all sorts of financial indices for capturing the dynamics of markets and stock exchange institutions. In general, all have a fractal nature with variations that are difficult to predict [5–13]. A number of techniques have been proposed to investigate the financial indices and to unravel the embedded complex dynamics [14–18]. Such studies adopt the underlying concept of linear time flow and consider that the fractal nature of the index is intrinsic to its own artificial nature.

This paper studies the interplay between the DJIA values and the time flow. The present day standard assumption is that time is a continuous linear succession of events often called the "arrow of time". We must clarify that (i) the nature of the time variable, either continuous or discrete, either with a constant rhythm of variation or not, is simply under the light of the financial index, so that we are independent of the classical laws of physics, (ii) merely the DJIA is adopted since other financial indices reveal the same type of behavior, but are limited to much shorter time series, and (iii) no financial foreseeing is intended. Therefore, the Gedankenexperiment in the follow-up addresses the controversy about the texture of time [19–22], but just in the limited scope of financial indices.

For this purpose, the concepts of multidimensional scaling (MDS), fractional dimension, entropy, and fractional calculus are brought up as useful tools to tackle complex systems. MDS is a computational tool for visualizing the level of similarity between items of a dataset. The MDS translates information regarding the pairwise distances among a set of items into a configuration of representative points of an abstract Cartesian space [23–29]. Mandelbrot coined the word "fractal" [30,31] for complex objects that are self-similar across different scales. Fractals can be characterized by the so-called fractal dimension, which may be seen as quantifying complexity [32–34]. Information theory was introduced by Claude Shannon [35] and has as the primary concept the information content of a given event, which is a decreasing function of its probability [36–39]. The entropy of a random variable is the average value of information and has been proven to be a valuable tool for assessing complex phenomena [40–42]. Fractional calculus (FC) is the branch of mathematical analysis that generalizes differentiation and integration to real or complex orders [43–48]. The topic was raised by Gottfried Leibniz in 1695 and remained an exotic field until the Twentieth Century. In the last few decades, FC became a popular tool for analyzing phenomena with long-range memory and non-locality [49–57].

The association of these mathematical and computational tools yields relevant viewpoints when analyzing financial indices [7–9,11,58–61].

Bearing these ideas in mind, this paper is organized as follows. Section 2 introduces the dataset and methods and develops some initial experiments using MDS. Section 3 explores the use of fractal and entropy analysis of the MDS loci. Finally, Section 4 draws the main conclusions.

2. Dataset and Methods

2.1. The DJIA Dataset

The dataset consists of the daily close values of the DJIA from 28 December 1959, up to 1 September 2020, corresponding to a time series of $T = 15,832$ days, covering approximately half a century. Each week consists of 5 working days, and some missing data due to special events were estimated by means of linear interpolation between adjacent values.

We assess the dynamics of the DJIA by comparing its values $x(t)$ for a given time window of t_w days. Therefore, the ith vector of DJIA values consists of $\xi_i = [x(1), \ldots, x(t_w)]$, where days "1" and "$t_w$" denote the start and end time instants in the time window. Hereafter, for simplicity, we consider consecutive disjoint time windows, and a number of experiments with t_w having values multiples of 5 days. Therefore, the total number of time windows (and vectors) is $N_w = \left\lfloor \frac{T}{t_w} \right\rfloor$, where $\lfloor \cdot \rfloor$ denotes the floor function, which gives as the output the greatest integer less than or equal to the input value.

The evolution of the DJIA in time reveals a fractal nature as represented in Figure 1. If we calculate the histogram of the logarithm of the returns, that is of $lr = \ln\left(\frac{x(t+1)}{x(t)}\right)$, we verify a sustained noisy behavior and fat tails in the statistical distribution as depicted in Figure 2 for time windows of $t_w = 60$ days.

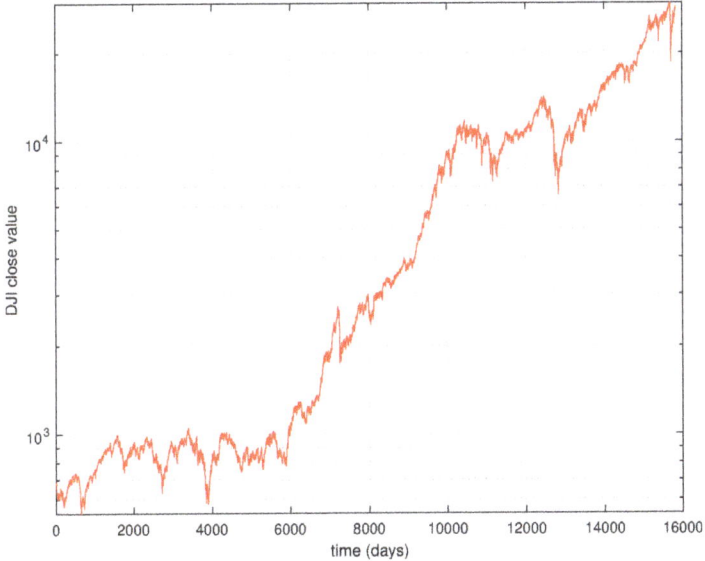

Figure 1. Daily close values of the DJIA from 28 December 1959, up to 1 September 2020.

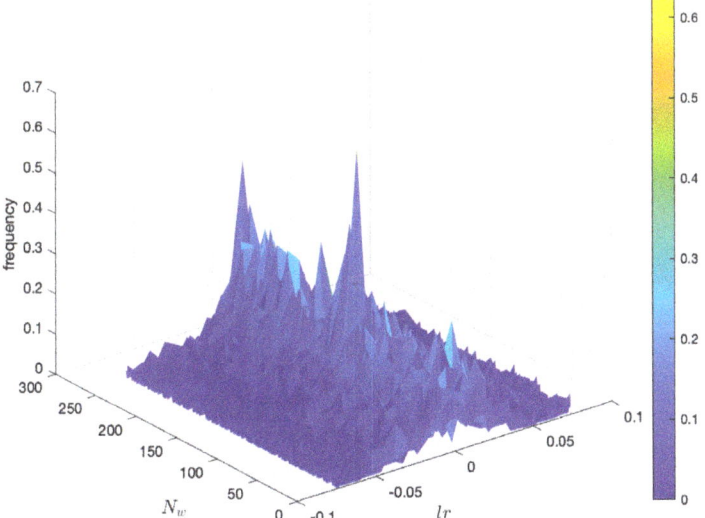

Figure 2. Histogram of the logarithm of the returns of the DJIA from 28 December 1959, up to 1 September 2020, for time windows of $t_w = 60$ days.

2.2. Distances

The DJIA dynamics is studied indirectly through the MDS by comparing the vectors $(\xi_i(1), \ldots, \xi_i(t_w)), i = 1, \ldots, N_w, t = 1, \ldots, t_w$, and analyzing the properties of the resulting plot in the perspective of entropy and fractal dimension. This approach requires the definition of an appropriate distance [62]. A function $d: \mathcal{A} \times \mathcal{A} \to \mathbb{R}$ on a set $\mathcal{A}$ is a "distance" when, for the items $\xi_i, \xi_j, \xi_k \in \mathcal{A}$, it satisfies the conditions (i) $d(\xi_i, \xi_j) \geq 0$ (non-negativity), (ii) $d(\xi_i, \xi_j) = 0$ (identity of indiscernibles) if and only if $\xi_i = \xi_j$, (iii) $d(\xi_i, \xi_j) = d(\xi_j, \xi_i)$ (symmetry), and (iv) $d(\xi_i, \xi_k) \leq d(\xi_i, \xi_j) + d(\xi_j, \xi_k)$ (triangle inequality). If the three conditions are followed, then the function is a "metric" and together with $\mathcal{A}$ yields a "metric space". Obviously, these conditions still allow a considerable freedom, and we find in the literature a plethora of possible metrics each with its own pros and cons. In practice, users adopt one or more distances if they capture adequately the characteristics of the items under assessment. Therefore, we start by considering a test bench of 10 distinct indices, namely the Manhattan, Euclidean, Tchebychev, Lorentzian, Sørensen, Canberra, Clark, divergence, angular, and Jaccard distances (denoted as {Ma, Eu, Tc, Lo, So, Ca, Cl, Dv, Ac, Ja}), given by [63]:

$$d_{i,j}^{Ma} = \sum_{t=1}^{t_w} \left| \xi_i(t) - \xi_j(t) \right|, \tag{1a}$$

$$d_{i,j}^{Eu} = \sqrt{\sum_{t=1}^{t_w} \left(\xi_i(t) - \xi_j(t) \right)^2}, \tag{1b}$$

$$d_{i,j}^{Tc} = \max_t \left(\xi_i(t) - \xi_j(t) \right), \tag{1c}$$

$$d_{i,j}^{Lo} = \sum_{t=1}^{t_w} \log \left(1 + \left| \xi_i(t) - \xi_j(t) \right| \right), \tag{1d}$$

$$d_{i,j}^{So} = \frac{\sum_{t=1}^{t_w} \left| \xi_i(t) - \xi_j(t) \right|}{\sum_{t=1}^{t_w} \left(\left| \xi_i(t) \right| + \left| \xi_j(t) \right| \right)}, \tag{1e}$$

$$d_{i,j}^{Ca} = \sum_{t=1}^{t_w} \frac{\left| \xi_i(t) - \xi_j(t) \right|}{\left| \xi_i(t) \right| + \left| \xi_j(t) \right|}, \tag{1f}$$

$$d_{i,j}^{Cl} = \sqrt{\sum_{t=1}^{t_w} \left(\frac{\left| \xi_i(t) - \xi_j(t) \right|}{\left| \xi_i(t) \right| + \left| \xi_j(t) \right|} \right)^2}, \tag{1g}$$

$$d_{i,j}^{Dv} = \sum_{t=1}^{t_w} \frac{\left(\xi_i(t) - \xi_j(t) \right)^2}{\left(\left| \xi_i(t) \right| + \left| \xi_j(t) \right| \right)^2}, \tag{1h}$$

$$d_{i,j}^{Ac} = \arccos\left(r_{ij} \right), \; r_{ij} = \frac{\sum_{t=1}^{t_w} \xi_i(t) \xi_j(t)}{\sqrt{\sum_{t=1}^{t_w} \xi_i^2(t) \sum_{t=1}^{t_w} \xi_j^2(t)}}, \tag{1i}$$

$$d_{i,j}^{Ja} = \frac{\sum_{t=1}^{t_w} \left(\xi_i(t) - \xi_j(t) \right)^2}{\sum_{t=1}^{t_w} \xi_i^2(t) + \sum_{t=1}^{t_w} \xi_j^2(t) - \sum_{t=1}^{t_w} \xi_i(t) \xi_j(t)}, \tag{1j}$$

where ξ_i and ξ_j, $i, j = 1, \ldots, N_w$, are the ith and jth vectors of the DJIA time series, each of dimension t_w. The Manhattan, Euclidean, and Tchebychev distances are particular cases of the Minkowski distance $d_{i,j}^{Mi} = \left(\sum_{t=1}^{t_w} \left| \xi_i(t) - \xi_j(t) \right|^q \right)^{\frac{1}{q}}$, namely for $q = 1$, $q = 2$ and $q \to \infty$, respectively. The Lorentzian distance applies the natural logarithm to the absolute difference with 1 added to guarantee the non-negativity property and to eschew the log of zero. We find in the literature several distinct versions of the Sørensen distance, eventually with other names, and representing a statistic used for comparing the similarity between two samples. The Canberra and Clark distances are weighted versions of the Manhattan and Euclidean distances. These expressions replace $\left| \xi_i(t) - \xi_j(t) \right|$ by $\left| \xi(t) - \xi_j(t) \right| / \left(\left| \xi_i(t) \right| + \left| \xi_j(t) \right| \right)$ and are sensitive to small changes near zero. The angular cosine distance follows the cosine similarity r_{ij} that comes from the inner product of two vectors, $\xi_i \cdot \xi_j$. The angular cosine distance $d_{i,j}^{Ac}$ gives the angle between the vectors ξ_i and ξ_j. The Jaccard distance is the ratio of the size of the symmetric difference to the union of two sets.

2.3. The MDS Loci

Once having defined the metric for comparing the vectors, the MDS requires the construction of a matrix $\mathbf{D} = [d_{i,j}]$ of item-to-item distances. In our case, "item" corresponds to a t_w-dim vectors. Therefore, the square matrix $\mathbf{D}$ is symmetric, with the main diagonal of zeros and dimension $N_w \times N_w$ equal to the number of items. The MDS computational algorithm tries to plot the items in a low-dimensional space so that users can easily analyze possible relationships that are difficult to unravel in a high number of dimensions. In other words, the MDS performs a dimension reduction and plots items in a $p < N_w$ dimensional space, by estimating a matrix $\hat{\mathbf{D}} = \left[\hat{d}_{i,j}\right]$, corresponding to the p-dim items $\hat{x}_i$, so that the distances, $\hat{d}_{i,j}$, mimic the original ones, $d_{i,j}$.

The classical MDS can perform the optimization procedure based on a variety of loss functions, often called "strain", that are a form of minimizing the residual sum of squares. The metric MDS generalizes the optimization procedure called "stress", S_D, such as:

$$S_D(\xi_1, \ldots, \xi) = \left[\sum_{i,j} \left(\hat{d}_{i,j} - d_{i,j} \right)^2 \right]^{\frac{1}{2}}, \tag{2}$$

or:

$$S_D(\xi_1, \ldots, \xi) = \left[\frac{\sum_{i,j} \left(\hat{d}_{i,j} - d_{i,j} \right)^2}{\sum_{i,j} d_{i,j}^2} \right]^{\frac{1}{2}}, \tag{3}$$

where $d_{i,j} = |\xi_i - \xi_j|$, $i,j = 1, \ldots, N_w$.

The generalized MDS is an extension of metric formulation, so that the target space is an arbitrary smooth non-Euclidean space.

Once having obtained the MDS estimate coordinates of the objects $\hat{x}_i$, the user can decide the dimension p for visualization. Usually, the values $p = 2$ and $p = 3$ are selected since they allow a direct representation. Moreover, the quality of the MDS approximation can be assessed by means of the Sheppard and stress charts. The Sheppard diagram plots $\hat{d}_{i,j}$ vs. $d_{i,j}$. If the points follow a straight/curved line, this means a linear/non-linear relationship, but in both cases, the smaller the scatter, the better the approximation is. A second assessment tool consists of the plot of S_D vs. p. Usually, the curve is monotonic decreasing with a large diminishing at first and a slow variation afterwards.

Since the MDS locus results from relative information (i.e., the distances), the coordinates usually do not have some physical meaning, and the user can rotate, shift, or magnify the representation to have a better view. Moreover, distinct distances lead to different plots that are correct from the mathematical and computational viewpoints, but that reflect distinct characteristics of the dataset. Therefore, it is up to the user to choose one or more distances that better highlight the aspects of the dataset under study.

Often, it is recommended to pre-process the data before calculating the distances in order to reduce the sensitivity to some details such as different units or a high variation of numerical values. In the case of the DJIA, two data pre-processing schemes (also called normalizing, or data transformation), $\mathcal{P}_1$ and $\mathcal{P}_2$, are considered: (i) subtracting the arithmetic average and dividing by the standard variation, that is by calculating $\left\{ \mathcal{P}_1 : x(t) \leftarrow \frac{x(t)-\mu}{\sigma} \right\}$, where $\mu = \frac{1}{T} \sum_{t=1}^{T} x(t)$ and $\sigma = \sqrt{\frac{1}{T-1} \sum_{t=1}^{T} (x(t) - \mu)^2}$, and (ii) by applying a logarithm so that $\{\mathcal{P}_2 : x(t) \leftarrow \lg(x(t))\}$. The linear transformation $\mathcal{P}_1$ is often adopted in statistics and signal processing [64–68], while the non-linear transformation $\mathcal{P}_2$ can be adopted with signals revealing an exponential-like evolution [69–73]. Of course, other data transformations could be envisaged, but these two are commonly adopted. Therefore, the main question concerning this issue is to understand to what extend the pre-processing influences the final results.

2.3.1. Data Pre-Processing Using $\mathcal{P}_1$

Figure 3 shows the MDS locus for $p = 3$ and $t_w = 60$ days, with pre-processing $\mathcal{P}_1$ and using the Lorentzian and Canberra distances, $d_{i,j}^{Lo}$ and $d_{i,j}^{Ca}$. The larger circle represents the first vector, and the lines connect two consecutive dots (representing the vectors from two consecutive time windows). The lines are included simply for auxiliary purposes and for highlighting the discontinuities. The MATLAB nonclassical multidimensional scaling algorithm mdscale and the Sammon's nonlinear mapping criterion sammon were used. Figure 4 illustrates the corresponding Sheppard and stress diagrams for the Canberra distance (1f). For the sake of parsimony, the other charts are not represented.

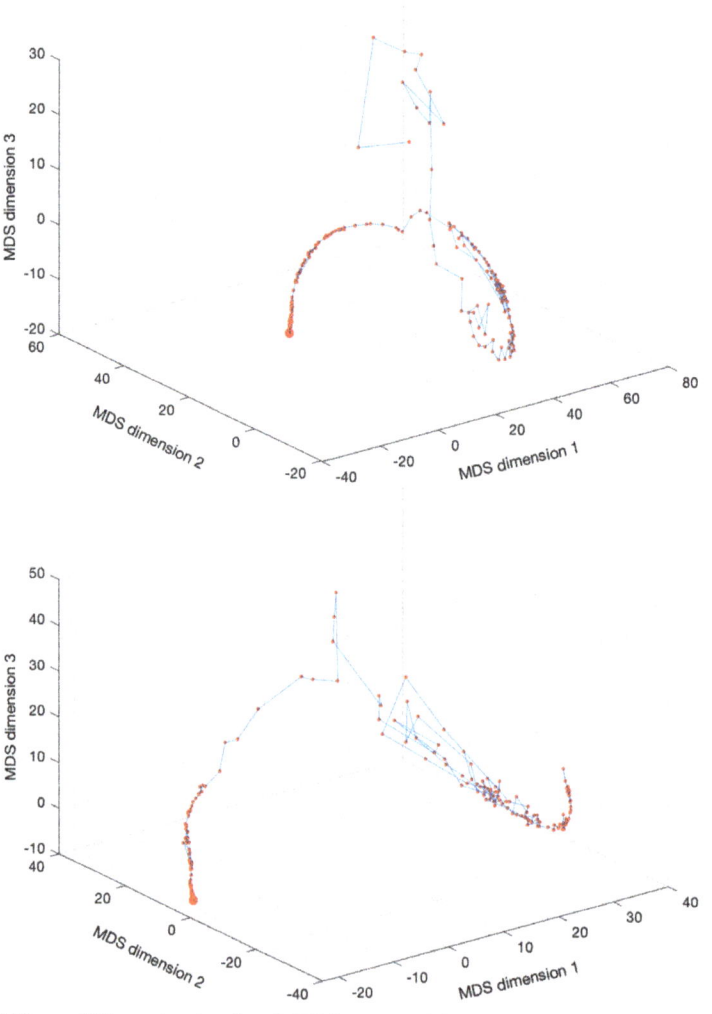

Figure 3. The multidimensional scaling (MDS) locus, $\hat{x}_i$, of the DJIA dataset for $p = 3$ and $t_w = 60$ days ($N_w = 263$), with pre-processing $\mathcal{P}_1$ and using the Lorentzian (1d) and Canberra (1f) distances.

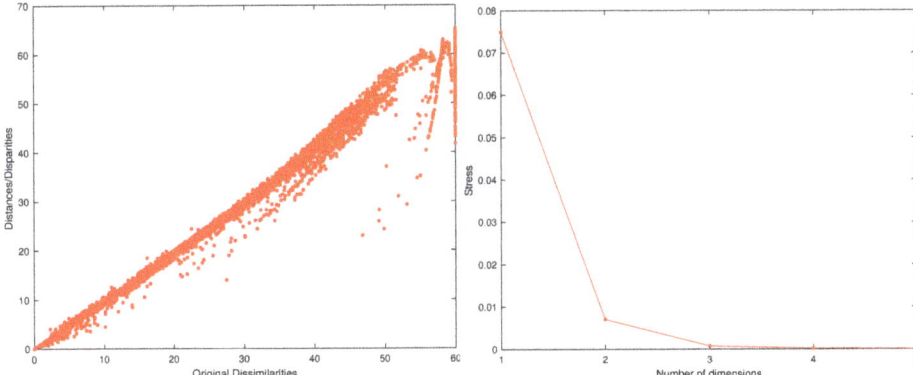

Figure 4. The Sheppard diagram, $\hat{d}_{i,j}$ vs. $d_{i,j}$, for $p = 3$, and stress plot, S_D vs. p, of the DJIA dataset with $t_w = 60$ days, with pre-processing $\mathcal{P}_1$ and using the Canberra distance (1f).

We verify that the MDS loci exhibit segments where we have an almost continuous evolution and others with strong discontinuities. The first segments portray relatively smooth dynamics, while the second ones represent dramatic variations, in the perspective of the adopted distance and visualization technique. These dynamical effects are not read in the same way as with the standard time representations. Moreover, their visualization varies according to the type of distance adopted to construct the matrix **D**. This should be expected, since it is well known that each distance highlights a specific set of properties embedded in the original time series and that the selection of one of more distances has to be performed on a case-by-case basis, before deciding those more adapted to the dataset.

Another relevant topic is the effect of the time window t_w on the results. In other words, we can ask how the dimension of the vector ξ_i, $i = 1, \ldots, N_w$, capturing the DJIA time dynamics, influences the MDS representation. For example, Figure 5 shows the MDS locus for $p = 3$, $t_w = 10$ days ($N_w = 1583$), and the Canberra distance (1e).

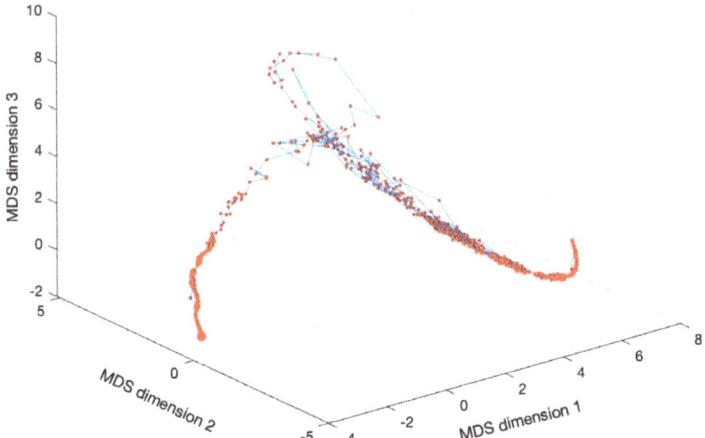

Figure 5. The MDS locus, $\hat{x}_i$, of the DJIA dataset for $p = 3$ and $t_w = 10$ days ($N_w = 1583$), with pre-processing $\mathcal{P}_1$ and using the Canberra distance (1e).

2.3.2. Data Pre-Processing Using $\mathcal{P}_2$

Figure 6 shows the MDS locus for $p = 3$ and $t_w = 60$ days, with pre-processing $\mathcal{P}_2$ and using the Lorentzian and Canberra distances, $d_{i,j}^{Lo}$ and $d_{i,j}^{Ca}$. Figure 7 depicts the Sheppard and stress diagrams for the Canberra distance (1f).

We can also check the effect of the time window t_w. Figure 8 shows the MDS locus for $p = 3$, $t_w = 10$ days ($N_w = 1583$), and the Canberra distance (1e) revealing, again, a slight diminishing of the volatility.

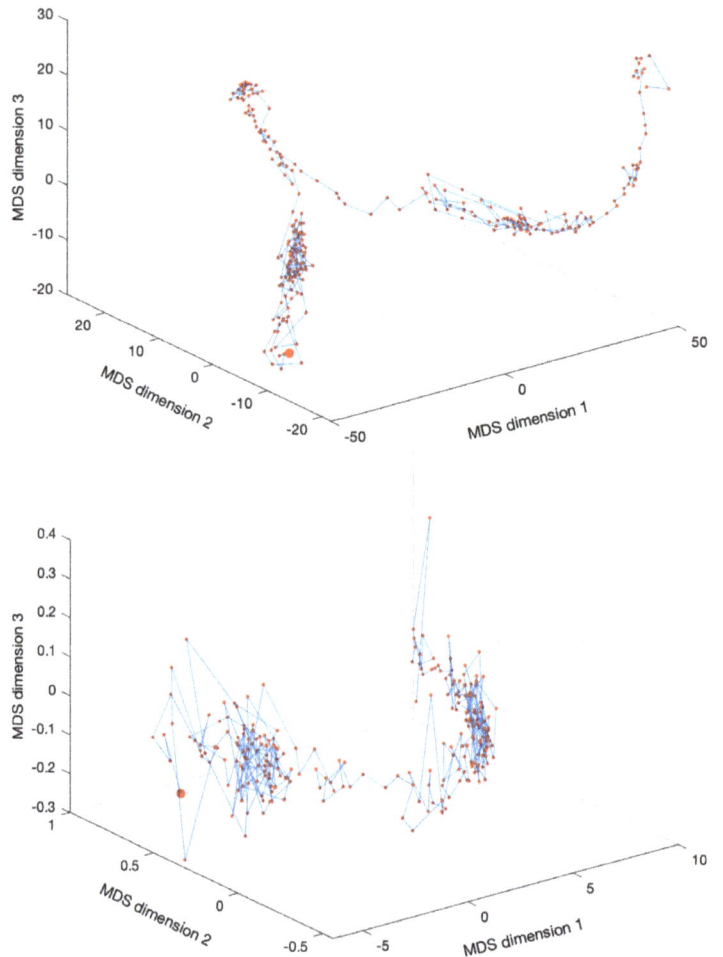

Figure 6. The MDS locus, $\hat{x}_i$, of the DJIA dataset for $p = 3$ and $t_w = 60$ days ($N_w = 263$), with pre-processing $\mathcal{P}_2$ and using the Lorentzian (1d) and Canberra (1f) distances.

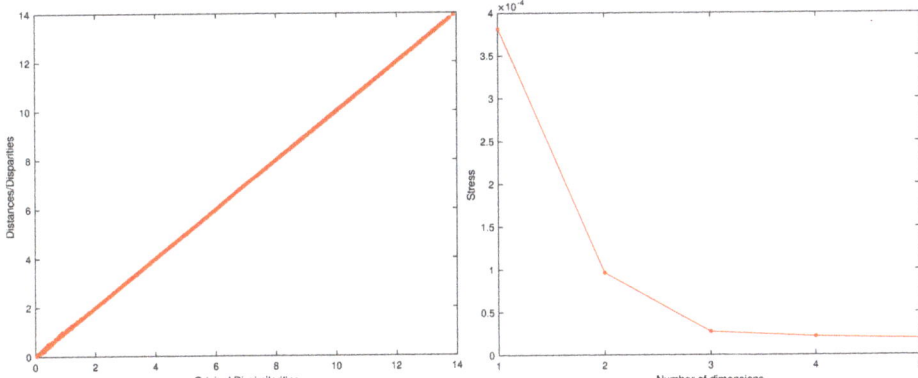

Figure 7. The Sheppard diagram, $\hat{d}_{i,j}$ vs. $d_{i,j}$, for $p = 3$, and the stress plot, S_D vs. p, of the DJIA dataset with $t_w = 60$ days, with pre-processing $\mathcal{P}_2$ and using the Canberra distance (1f).

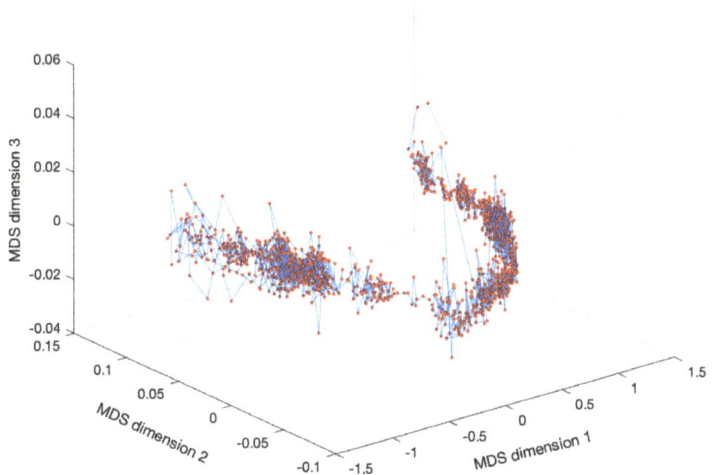

Figure 8. The MDS locus, $\hat{x}_i$, of the DJIA dataset for $p = 3$ and $t_w = 10$ days ($N_w = 1583$), with pre-processing $\mathcal{P}_2$ and using the Canberra distance (1e).

As in the previous sub-section, we observe that the MDS plots reveal some segments almost with a continuous evolution and some with discontinuities. Furthermore, as before, increasing t_w reduces the volatility in the MDS representations. These results, with regions of smooth variation, interspersed with abrupt changes, were already noticed since they reflect relativistic time effects [74,75]. Such dynamics was interpreted as a portrait of the fundamental non-smooth nature of the flow of the time variable underlying the DJIA evolution. Nonetheless, we are still far from a comprehensive understanding of the MDS loci, and we need to design additional tools to extract additional conclusions.

3. Fractal, Entropy, and Fractional Analysis

We consider the fractal dimension and entropy measures for analyzing the 3-dim portraits produced by the MDS.

The fractal dimension, f_d, characterizes the fractal pattern of a given object by quantifying the ratio of the change in detail to the change in scale. Several types of fractal dimension can be found in the literature. In our case, f_d is calculated by means of the box counting method as the exponent of a power law $N(\epsilon) = a\epsilon^{-f_d}$, where a is a parameter that depends on the shape and size of the object, and N and ϵ stand for the number of boxes required to capture the object and the size (or scale) of the box, respectively. Therefore, f_d can be estimated as:

$$f_d = -\lim_{\epsilon \to 0} \frac{\ln(N(\epsilon))}{\ln(\epsilon)}. \tag{4}$$

The entropy of a random variable is the average level of "information" of the corresponding probability distribution. The key cornerstone of the Shannon theory consists of the information content, which for an event having probability of occurrence p_i, is given by:

$$I(p_i) = -\ln p_i. \tag{5}$$

For a 3-dim random variable (X, Y, Z) with probability distribution p_{XYZ}, the Shannon entropy, H_{XYZ}, is given by:

$$H_{XYZ} = -\sum_X \sum_Y \sum_Z p_{XYZ} \ln(p_{XYZ}), \tag{6}$$

where $-\ln(p_{XYZ})$ is the information for the event with probability p_{XYZ}.

The concept of entropy can be generalized in the scope of fractional calculus [76–86]. This approach gives more freedom to adapt the entropy measure to the phenomenon under study by adjusting the fractional order. The information and entropy of order $\alpha \in \mathbb{R}$ are given by [77,87]:

$$I_\alpha(p_i) = D^\alpha I(p_i) = -\frac{p_i^{-\alpha}}{\Gamma(\alpha + 1)} [\ln p_i + \psi(1) - \psi(1 - \alpha)] \tag{7}$$

$$H_{XYZ}^\alpha = \sum_i \left\{ -\frac{p_i^{-\alpha}}{\Gamma(\alpha + 1)} [\ln p_i + \psi(1) - \psi(1 - \alpha)] \right\} p_i \tag{8}$$

where $\Gamma(\cdot)$ and $\psi(\cdot)$ represent the gamma and digamma functions.

The parameter α gives an extra degree of freedom to adapt the sensitivity of the entropy calculation of each specific data series.

In an algorithmic perspective, these measures require the adoption of some grid (or box) for capturing and counting the objects, the main difference being that the fractal dimension just considers a Boolean perspective of "1" and "0", that is the box is either full or empty, while the entropy considers the number of counts in each box.

In the follow-up, a 3-dim grid defined between the minimum and maximum values obtained for each axis of the MDS locus is considered. For the fractal dimension, we obtain f_d by the slope of $N(\epsilon)$ versus ϵ for 10 decreasing values of the box sizes. In the case of the entropy, we calculate H_{XYZ} when adopting 20 bins for each MDS axis. The auxiliary lines connecting the object (i.e., the points) are not considered for the calculations.

Figures 9 and 10 show the variation of f_d and H_{XYZ} with t_w, with pre-processing $\mathcal{P}_1$ and $\mathcal{P}_2$, respectively, when using the distances (1a)–(1j). For $t_w \in \{5, \ldots, 240\}$, we have correspondingly MDS with N_w ($t_w \in \{3166, \ldots, 65\}$) points.

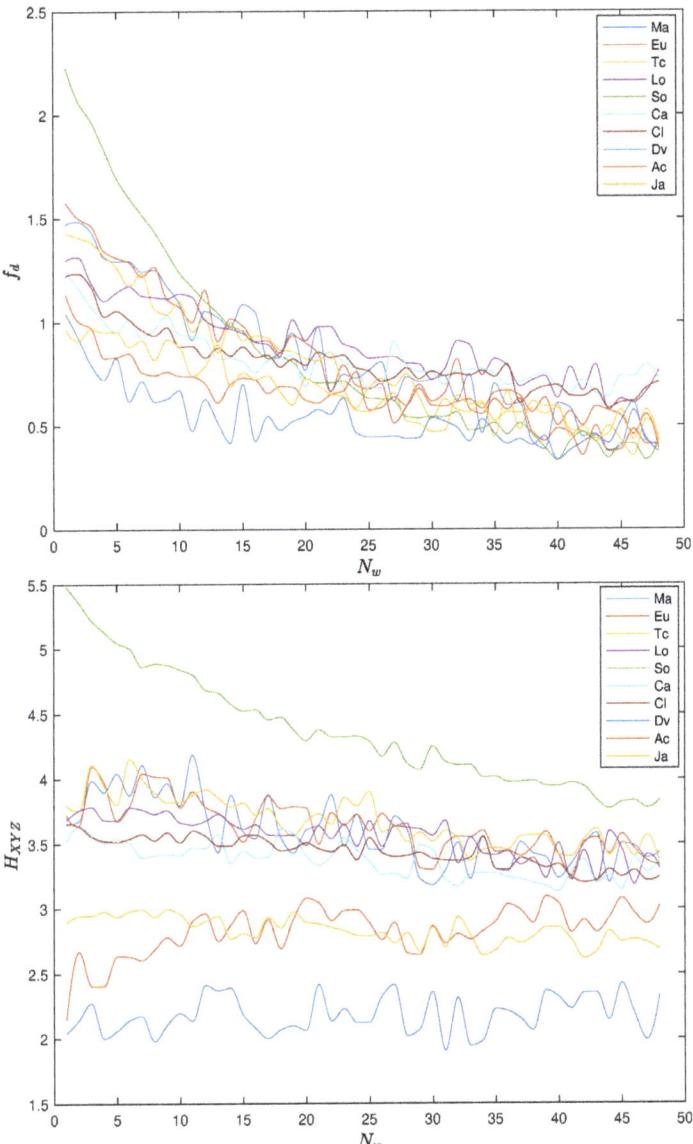

Figure 9. Plot of fractal dimension, f_d, and Shannon entropy, H_{XYZ}, versus N_w ($t_w \in \{5, \ldots, 240\}$), with pre-processing $\mathcal{P}_1$ and using the distances (1a)–(1j).

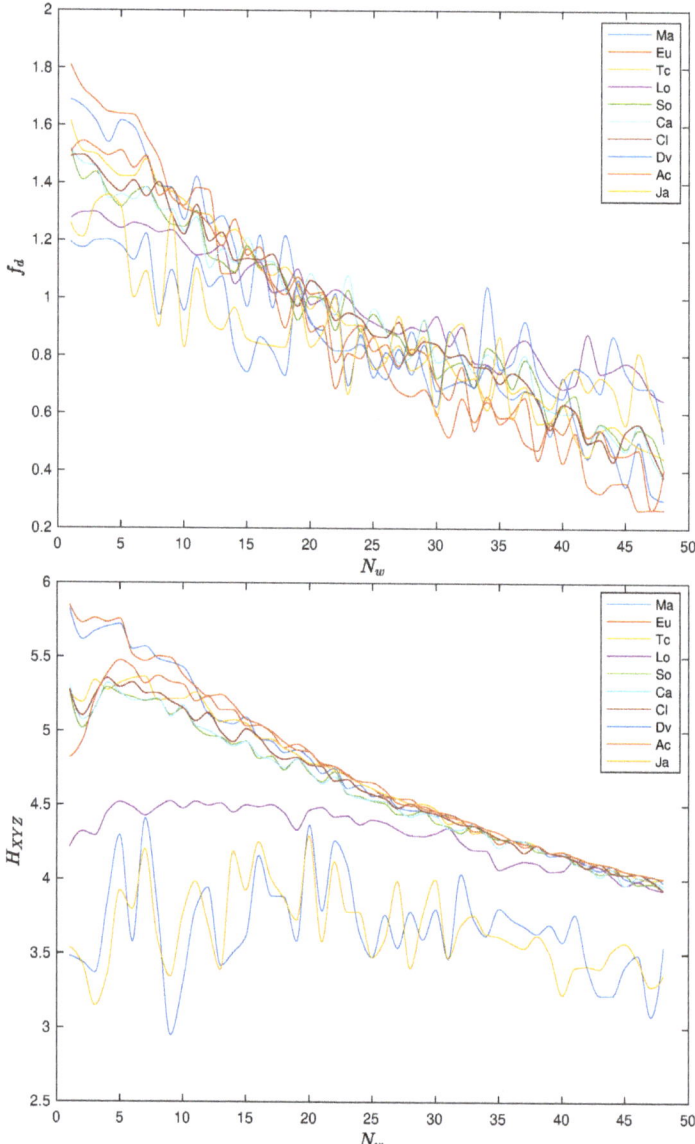

Figure 10. Plot of fractal dimension, f_d, and Shannon entropy, H_{XYZ}, versus N_w ($t_w \in \{5, \ldots, 240\}$), with pre-processing $\mathcal{P}_2$ and using the distances (1a)–(1j). The Manhattan, Euclidean, Tchebychev, Lorentzian, Sørensen, Canberra, Clark, divergence, angular, and Jaccard distances (denoted as {Ma, Eu, Tc, Lo, So, Ca, Cl, Dv, Ac, Ja}).

We note some "noise", but that should be expected due to the numerical nature of the experiments. In general, the two indices decrease with t_w, revealing, again, the "low pass filtering" effect of the dimension of the time window. We note a considerable difference of the values of f_d and H_{XYZ} for small values of t_w, but a stabilization and some convergence to closer values when t_w increases.

In the case of the fractional entropy, H_{XYZ}^α, we can tune the value of α to achieve a maximum sensitivity. In other words, we can select the value $\alpha_{max(H)}$ to obtain max $\left(H_{XYZ}^\alpha\right)$. Figures 11 and 12 depict $\max\left(H_{XYZ}^{\alpha}\right)$ vs. $\alpha_{max(H)}$ with $t_w \in \{5, 10, \ldots, 240\}$, with pre-processing $\mathcal{P}_1$ and $\mathcal{P}_2$, respectively, and using the distances (1a)–(1j).

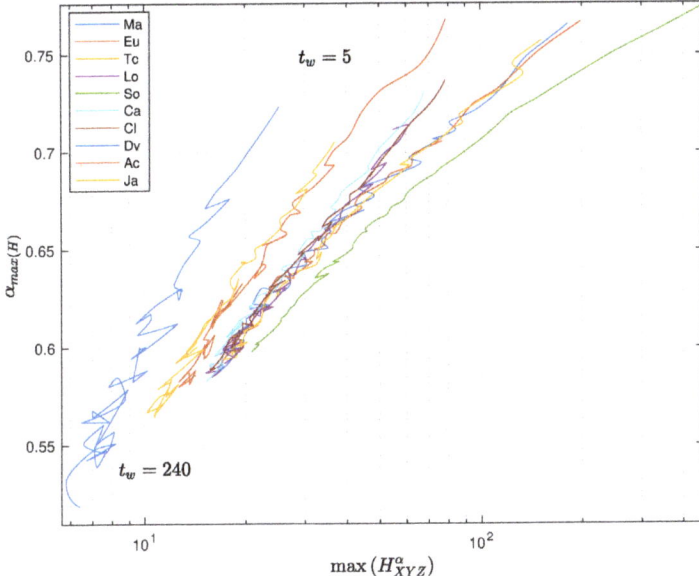

Figure 11. Plot of $\alpha_{max(H)}$ versus $\max\left(H_{XYZ}^{\alpha}\right)$, with $t_w \in \{5, 10, \ldots, 240\}$, with pre-processing $\mathcal{P}_1$ and using the distances (1a)–(1j).

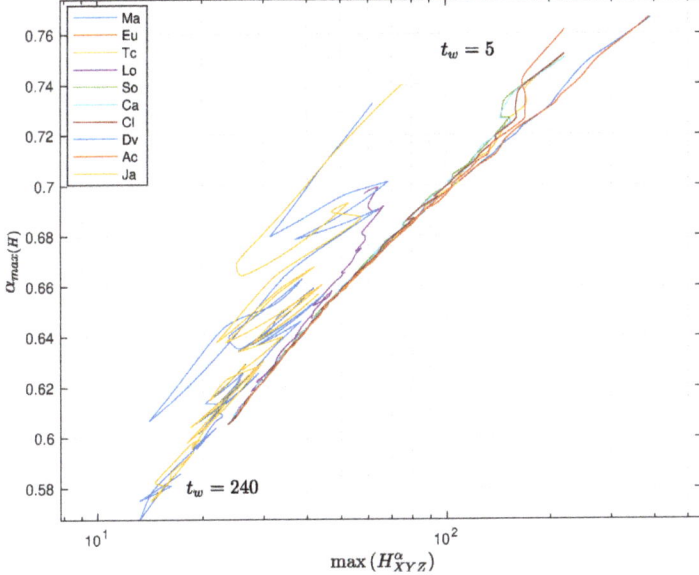

Figure 12. Plot of $\alpha_{max(H)}$ versus $\max\left(H_{XYZ}^{\alpha}\right)$, with $t_w \in \{5, 10, \ldots, 240\}$, with pre-processing $\mathcal{P}_2$ and using the distances (1a)–(1j).

We verify a strong correlation between the entropy and the value of the fractional order. Furthermore, we note that $0.55 \leq \alpha_{max(H)} \leq 0.75$ and $0.57 \leq \alpha_{max(H)} \leq 0.77$ for $\mathcal{P}_1$ and $\mathcal{P}_2$, respectively, far from integer values and clearly representative of fractional dynamics. For small time windows, each distance has a distinct behavior, but when the time window increases, all distances converge to almost similar points of $\alpha_{max(H)}$, both for $\mathcal{P}_1$ and $\mathcal{P}_2$. Obviously, with larger time windows, we have a smaller number of points in the MDS locus, and that influences the result. The convergence towards a common behavior for all distances is observed after the first values of t_w. This means that we are unraveling the fractional dynamics, that is a characteristic of long-range memory effects embedded in the time series.

For the pre-processing $\mathcal{P}_1$, the divergence distance produces a slightly separated plot to the left, while for $\mathcal{P}_2$, we see that position is occupied the divergence and Jaccard distances, but with a fuzzier behavior. As before, we note that the type of pre-processing does not yield any significant modification of the global conclusions.

4. Conclusions

Commonly, time is viewed as a continuous and linear flow so that any perturbation, such as noise and volatility, is automatically assigned to the variable under analysis. In other words, since we are entities immersed in the time flow, apparently, we are incapable of distinguishing between perturbations in the time and the measured variable. This paper explored an alternative strategy of reading the relationship between the variables. For that purpose, the DJIA, from 28 December 1959, up to 1 September 2020, was adopted as the vehicle for the numerical experiments. This dataset corresponds to a human-made phenomenon, and therefore, any conjecture about the nature of time is independent of the presently accepted conceptions about its flux. In the proposed approach, the time series was organized into vectors corresponding to specified time windows. Those vectors were then compared by means of a panoply of distances and the resulting information plotted in a three-dimensional space by means of MDS. Indeed, the MDS representation corresponds to a "customized projection" of high-dimensional data into a low-dimensional space. Loosely speaking, we can say "customized projection" since we do not pose any a priori requirements, the algorithm merely being based on the idea of minimizing the difference between the original measurements and the replicated (approximated) value. Therefore, the MDS does not automatically guarantee the success of such a "projection", but the quality results were assessed by the stress and Shepard diagrams. In the case of the DJIA and the adopted distances, the good quality of the MDS technique was confirmed.

The MDS loci have distinct shapes, according to the type of distance adopted to compare vectors. Therefore, additional tools were necessary to highlight the main characteristics of these representations where time is no longer the explicit variable. For that purpose, several mathematical tools were considered, namely the Shannon entropy and fractal dimension. In all cases, we observed some variability with the time window, which occurs naturally due to the numerical treatment of this type of data. The Shannon entropy and fractal dimension exhibited the same type of behavior, with a progressive variation with the time window and a stabilization toward a common value for large t_w. While these results can be read merely as the effect of a low pass filtering provided by the large time window, we can also foresee that another property inherent to the DJIA is their origin.

The fractional entropy was brought up to further analyze the MDS locus. This tool allows a better sensitivity to the dataset than the Shannon entropy, since users can tune the calculations by means of the fractional order. In the case of the DJIA, the tuning of α for achieving the maximum entropy revealed not only that such values are independent of the distance, but also that we clearly have orders far from integer values, characteristic of fractional dynamics with non-local effects.

Some concepts are debatable and do not follow the standard orthodoxy, but the set of experiments with an artificial time series allows thinking outside the box and provides a strategy for exploring the texture of time in the perspective of entropy and fractional calculus.

Entropy **2020**, *22*, 1138

Funding: This research received no external funding.

Conflicts of Interest: The author declares no conflict of interest.

References

1. Trippi, R. *Chaos & Nonlinear Dynamics in the Financial Markets*; Irwin Professional Publishing Company: Burr Ridge, IL, USA, 1995.
2. Vialar, T. *Complex and Chaotic Nonlinear Dynamics: Advances in Economics and Finance, Mathematics and Statistics*; Springer: Berlin/Heidelberg, Germany, 2009.
3. Bischi, G.I.; Chiarella, C.; Gardini, L. (Eds.) *Nonlinear Dynamics in Economics, Finance and the Social Sciences*; Springer: Berlin/Heidelberg, Germany, 2010.
4. Meyers, R.A. (Ed.) *Complex Systems in Finance and Econometrics*; Springer: New York, NY, USA, 2010.
5. Machado, J.A.T. Calculation of Fractional Derivatives of Noisy Data with Genetic Algorithms. *Nonlinear Dyn.* **2009**, *57*, 253–260. [CrossRef]
6. Duarte, F.B.; Machado, J.A.T.; Duarte, G.M. Dynamics of the Dow Jones and the NASDAQ Stock Indexes. *Nonlinear Dyn.* **2010**, *61*, 691–705. [CrossRef]
7. Machado, J.T.; Duarte, F.B.; Duarte, G.M. Analysis of stock market indices through multidimensional scaling. *Commun. Nonlinear Sci. Numer. Simul.* **2011**, *16*, 4610–4618. [CrossRef]
8. Machado, J.A.T.; Duarte, G.M.; Duarte, F.B. Analysis of Financial Data Series Using Fractional Fourier Transform and Multidimensional Scaling. *Nonlinear Dyn.* **2011**, *65*, 235–245. [CrossRef]
9. Machado, J.A.T.; Duarte, G.M.; Duarte, F.B. Identifying Economic Periods and Crisis with the Multidimensional Scaling. *Nonlinear Dyn.* **2011**, *63*, 611–622. [CrossRef]
10. Machado, J.T.; Duarte, F.B.; Duarte, G.M. Analysis of Financial Indices by Means of The Windowed Fourier Transform. *Signal Image Video Process.* **2012**, *6*, 487–494. [CrossRef]
11. Machado, J.T.; Duarte, G.M.; Duarte, F.B. Analysis of stock market indices with multidimensional scaling and wavelets. *Math. Probl. Eng.* **2012**, *2012*, 14.
12. Machado, J.A.T.; Duarte, G.M.; Duarte, F.B. Fractional dynamics in financial indexes. *Int. J. Bifurc. Chaos* **2012**, *22*, 1250249. [CrossRef]
13. Machado, J.T.; Duarte, F.B.; Duarte, G.M. Power Law Analysis of Financial Index Dynamics. *Discret. Dyn. Nat. Soc.* **2012**, *2012*, 12.
14. Da Silva, S.; Matsushita, R.; Gleria, I.; Figueiredo, A.; Rathie, P. International finance, Lévy distributions, and the econophysics of exchange rates. *Commun. Nonlinear Sci. Numer. Simul.* **2005**, *10*, 355–466. [CrossRef]
15. Chen, W.C. Nonlinear dynamics and chaos in a fractional-order financial system. *Chaos Solitons Fractals* **2008**, *36*, 1305–1314. [CrossRef]
16. Piqueira, J.R.C.; Mortoza, L.P.D. Complexity analysis research of financial and economic system under the condition of three parameters' change circumstances. *Commun. Nonlinear Sci. Numer. Simul.* **2012**, *17*, 1690–1695. [CrossRef]
17. Ma, J.; Bangura, H.I. Complexity analysis research of financial and economic system under the condition of three parameters' change circumstances. *Nonlinear Dyn.* **2012**, *70*, 2313–2326. [CrossRef]
18. Ngounda, E.; Patidar, K.C.; Pindza, E. Contour integral method for European options with jumps. *Commun. Nonlinear Sci. Numer. Simul.* **2013**, *18*, 478–492. [CrossRef]
19. Horwich, P. *Asymmetries in Time: Problems in the Philosophy of Science*; The MIT Press: Cambridge, MA, USA, 1987.
20. Reichenbach, H. *The Direction of Time*; University of California Press: New York, NY, USA, 1991.
21. Dainton, B. *Time and Space*, 2nd ed.; Acumen Publishing, Limited: Chesham, UK, 2001.
22. Callender, C. (Ed.) *The Oxford Handbook of Philosophy of Time*; Oxford University Press: New York, NY, USA, 2011.
23. Torgerson, W. *Theory and Methods of Scaling*; Wiley: New York, NY, USA, 1958.
24. Shepard, R.N. The analysis of proximities: Multidimensional scaling with an unknown distance function. *Psychometrika* **1962**, *27*, 219–246. [CrossRef]

25. Kruskal, J. Multidimensional scaling by optimizing goodness of fit to a nonmetric hypothesis. *Psychometrika* **1964**, *29*, 1–27. [CrossRef]

26. Sammon, J. A nonlinear mapping for data structure analysis. *IEEE Trans. Comput.* **1969**, *18*, 401–409. [CrossRef]

27. Kruskal, J.B.; Wish, M. *Multidimensional Scaling*; Sage Publications: Newbury Park, CA, USA, 1978.

28. Borg, I.; Groenen, P.J. *Modern Multidimensional Scaling-Theory and Applications*; Springer: New York, NY, USA, 2005.

29. Saeed, N.; Nam, H.; Imtiaz, M.; Saqib, D.B.M. A Survey on Multidimensional Scaling. *ACM Comput. Surv. CSUR* **2018**, *51*, 47. [CrossRef]

30. Mandelbrot, B.B.; Ness, J.W.V. The fractional Brownian motions, fractional noises and applications. *SIAM Rev.* **1968**, *10*, 422–437. [CrossRef]

31. Mandelbrot, B.B. *The Fractal Geometry of Nature*; W. H. Freeman: New York, NY, USA, 1983.

32. Berry, M.V. Diffractals. *J. Phys. A Math. Gen.* **1979**, *12*, 781–797. [CrossRef]

33. Lapidus, M.L.; Fleckinger-Pellé, J. Tambour fractal: Vers une résolution de la conjecture de Weyl-Berry pour les valeurs propres du Laplacien. *C. R. L'Académie Sci. Paris Sér. I Math.* **1988**, *306*, 171–175.

34. Schroeder, M. *Fractals, Chaos, Power Laws: Minutes from an Infinite Paradise*; W. H. Freeman: New York, NY, USA, 1991.

35. Shannon, C.E. A mathematical theory of communication. *Bell Syst. Tech. J.* **1948**, *27*, 379–423, 623–656. [CrossRef]

36. Khinchin, A.I. *Mathematical Foundations of Information Theory*; Dover: New York, NY, USA, 1957.

37. Jaynes, E.T. Information Theory and Statistical Mechanics. *Phys. Rev.* **1957**, *106*, 620–630. [CrossRef]

38. Rényi, A. On measures of information and entropy. In *Proceedings of the fourth Berkeley Symposium on Mathematics, Statistics and Probability, Volume 1: Contributions to the Theory of Statistics, Berkeley, CA, USA, 20 June–30 July 1960*; University of California Press: Berkeley, CA, USA, 1961; pp. 547–561. Available online: https://projecteuclid.org/euclid.bsmsp/1200512181 (accessed on 4 September 2020).

39. Brillouin, L. *Science and Information Theory*; Academic Press: London, UK, 1962.

40. Lin, J. Divergence measures based on the Shannon entropy. *IEEE Trans. Inf. Theory* **1991**, *37*, 145–151. [CrossRef]

41. Beck, C. Generalised information and entropy measures in physics. *Contemp. Phys.* **2009**, *50*, 495–510. [CrossRef]

42. Gray, R.M. *Entropy and Information Theory*; Springer: New York, NY, USA, 2011.

43. Oldham, K.; Spanier, J. *The Fractional Calculus: Theory and Application of Differentiation and Integration to Arbitrary Order*; Academic Press: New York, NY, USA, 1974.

44. Samko, S.; Kilbas, A.; Marichev, O. *Fractional Integrals and Derivatives: Theory and Applications*; Gordon and Breach Science Publishers: Amsterdam, The Netherlands, 1993.

45. Miller, K.; Ross, B. *An Introduction to the Fractional Calculus and Fractional Differential Equations*; John Wiley and Sons: New York, NY, USA, 1993.

46. Kilbas, A.; Srivastava, H.; Trujillo, J. Volume 204, North-Holland Mathematics Studies. In *Theory and Applications of Fractional Differential Equations*; Elsevier: Amsterdam, The Netherlands, 2006.

47. Kochubei, A.; Luchko, Y. (Eds.) Volume 1, De Gruyter Reference. In *Handbook of Fractional Calculus with Applications: Basic Theory*; De Gruyter: Berlin, Germany, 2019.

48. Kochubei, A.; Luchko, Y. (Eds.) Volume 2, De Gruyter Reference. In *Handbook of Fractional Calculus with Applications: Fractional Differential Equations*; De Gruyter: Berlin, Germany, 2019.

49. Westerlund, S.; Ekstam, L. Capacitor theory. *IEEE Trans. Dielectr. Electr. Insul.* **1994**, *1*, 826–839. [CrossRef]

50. Grigolini, P.; Aquino, G.; Bologna, M.; Luković, M.; West, B.J. A theory of $1/f$ noise in human cognition. *Phys. A Stat. Mech. Its Appl.* **2009**, *388*, 4192–4204. [CrossRef]

51. West, B.J.; Grigolini, P. *Complex Webs: Anticipating the Improbable*; Cambridge University Press: New York, NY, USA, 2010.

52. Tarasov, V. *Fractional Dynamics: Applications of Fractional Calculus to Dynamics of Particles, Fields and Media*; Springer: New York, NY, USA, 2010.

53. Mainardi, F. *Fractional Calculus and Waves in Linear Viscoelasticity: An Introduction to Mathematical Models*; Imperial College Press: London, UK, 2010.

54. Ortigueira, M.D. *Fractional Calculus for Scientists and Engineers*; Lecture Notes in Electrical Engineering; Springer: Dordrecht, The Netherlands, 2011.

55. West, B.J. *Fractional Calculus View of Complexity: Tomorrow's Science*; CRC Press: Boca Raton, FL, USA, 2015.

56. Tarasov, V.E. (Ed.) Volume 4, De Gruyter Reference. In *Handbook of Fractional Calculus with Applications: Applications in Physics, Part A*; De Gruyter: Berlin, Germany, 2019.

57. Tarasov, V.E. (Ed.) Volume 5, De Gruyter Reference. In *Handbook of Fractional Calculus with Applications: Applications in Physics, Part B*; De Gruyter: Berlin, Germany, 2019.

58. Machado, J.A.T.; Mata, M.E. A multidimensional scaling perspective of Rostow's forecasts with the track-record (1960s–2011) of pioneers and latecomers. In *Dynamical Systems: Theory, Proceedings of the 12th International Conference on Dynamical Systems—Theory and Applications*; Awrejcewicz, J., Kazmierczak, M., Olejnik, P., Mrozowski, J., Eds.; Łódź University of Technology: Łódź, Poland, 2013; pp. 361–378.

59. Machado, J.A.T.; Mata, M.E. Analysis of World Economic Variables Using Multidimensional Scaling. *PLoS ONE* **2013**, *10*, e0121277. [CrossRef]

60. Tenreiro Machado, J.A.; Lopes, A.M.; Galhano, A.M. Multidimensional scaling visualization using parametric similarity indices. *Entropy* **2015**, *17*, 1775–1794. [CrossRef]

61. Mata, M.; Machado, J. Entropy Analysis of Monetary Unions. *Entropy* **2017**, *19*, 245. [CrossRef]

62. Deza, M.M.; Deza, E. *Encyclopedia of Distances*; Springer: Berlin/Heidelberg, Germany, 2009.

63. Cha, S.H. Measures between Probability Density Functions. *Int. J. Math. Model. Methods Appl. Sci.* **2007**, *1*, 300–307.

64. Papoulis, A. *Signal Analysis*; McGraw-Hill: New York, NY, USA, 1977.

65. Oppenheim, A.V.; Schafer, R.W. *Digital Signal Processing*; Prentice Hall: Upper Saddle River, NJ, USA, 1989.

66. Parzen, E. *Modern Probability Theory and Its Applications*; Wiley-Interscience: New York, NY, USA, 1992.

67. Pollock, D.S.; Green, R.C.; Nguyen, T. (Eds.) *Handbook of Time Series Analysis, Signal Processing, and Dynamics (Signal Processing and Its Applications)*; Academic Press: London, UK, 1999.

68. Small, M. *Applied Nonlinear Time Series Analysis: Applications in Physics, Physiology and Finance*; World Scientific Publishing: Singapore, 2005.

69. Keene, O.N. The log transformation is special. *Stat. Med.* **1995**, *14*, 811–819. [CrossRef] [PubMed]

70. Leydesdorff, L.; Bensman, S. Classification and powerlaws: The logarithmic transformation. *J. Am. Soc. Inf. Sci. Technol.* **2006**, *57*, 1470–1486. [CrossRef]

71. Stango, V.; Zinman, J. Exponential Growth Bias and Household Finance. *J. Financ.* **2009**, *64*, 2807–2849. [CrossRef]

72. Feng, C.; Wang, H.; Lu, N.; Chen, T.; He, H.; Lu, Y.; Tu, X.M. Log-transformation and its implications for data analysis. *Shanghai Arch. Psychiatry* **2014**, *26*, 105–109. [CrossRef] [PubMed]

73. Lopes, A.; Machado, J.T.; Galhano, A. Empirical Laws and Foreseeing the Future of Technological Progress. *Entropy* **2016**, *18*, 217. [CrossRef]

74. Machado, J.A.T. Complex Dynamics of Financial Indices. *Nonlinear Dyn.* **2013**, *74*, 287–296. [CrossRef]

75. Machado, J.A.T. Relativistic Time Effects in Financial Dynamics. *Nonlinear Dyn.* **2014**, *75*, 735–744. [CrossRef]

76. Ubriaco, M.R. Entropies based on fractional calculus. *Phys. Lett. A* **2009**, *373*, 2516–2519. [CrossRef]

77. Machado, J.A.T. Fractional Order Generalized Information. *Entropy* **2014**, *16*, 2350–2361. [CrossRef]

78. Karci, A. Fractional order entropy: New perspectives. *Optik* **2016**, *127*, 9172–9177. [CrossRef]

79. Bagci, G.B. The third law of thermodynamics and the fractional entropies. *Phys. Lett. A* **2016**, *380*, 2615–2618. [CrossRef]

80. Xu, J.; Dang, C. A novel fractional moments-based maximum entropy method for high-dimensional reliability analysis. *Appl. Math. Model.* **2019**, *75*, 749–768. [CrossRef]

81. Xu, M.; Shang, P.; Qi, Y.; Zhang, S. Multiscale fractional order generalized information of financial time series based on similarity distribution entropy. *Chaos Interdiscip. J. Nonlinear Sci.* **2019**, *29*, 053108. [CrossRef] [PubMed]

82. Machado, J.A.T.; Lopes, A.M. Fractional Rényi entropy. *Eur. Phys. J. Plus* **2019**, *134*. [CrossRef]

83. Ferreira, R.A.C.; Machado, J.T. An Entropy Formulation Based on the Generalized Liouville Fractional Derivative. *Entropy* **2019**, *21*, 638. [CrossRef]

84. Matouk, A. Complex dynamics in susceptible-infected models for COVID-19 with multi-drug resistance. *Chaos Solitons Fractals* **2020**, *140*, 110257. [CrossRef]

Entropy **2020**, *22*, 1138

85. Matouk, A.; Khan, I. Complex dynamics and control of a novel physical model using nonlocal fractional differential operator with singular kernel. *J. Adv. Res.* **2020**, *24*, 463–474. [CrossRef]

86. Matouk, A.E. (Ed.) *Advanced Applications of Fractional Differential Operators to Science and Technology*; IGI Global: Hershey, PA, USA, 2020. [CrossRef]

87. Valério, D.; Trujillo, J.J.; Rivero, M.; Machado, J.T.; Baleanu, D. Fractional Calculus: A Survey of Useful Formulas. *Eur. Phys. J. Spec. Top.* **2013**, *222*, 1827–1846. [CrossRef]

Article

A Discretization Approach for the Nonlinear Fractional Logistic Equation

Mohammad Izadi [1,*] and Hari M. Srivastava [2,3,4]

[1]   Department of Applied Mathematics, Faculty of Mathematics and Computer,
       Shahid Bahonar University of Kerman, Kerman 76169-14111, Iran
[2]   Department of Mathematics and Statistics, University of Victoria, Victoria, BC V8W 3R4, Canada;
       harimsri@math.uvic.ca
[3]   Department of Medical Research, China Medical University Hospital, China Medical University,
       Taichung 40402, Taiwan
[4]   Department of Mathematics and Informatics, Azerbaijan University, 71 Jeyhun Hajibeyli Street,
       Baku AZ1007, Azerbaijan
*   Correspondence: izadi@uk.ac.ir

Received: 14 October 2020; Accepted: 17 November 2020; Published: 21 November 2020

Abstract: The present study aimed to develop and investigate the local discontinuous Galerkin method for the numerical solution of the fractional logistic differential equation, occurring in many biological and social science phenomena. The fractional derivative is described in the sense of Liouville-Caputo. Using the upwind numerical fluxes, the numerical stability of the method is proved in the L_∞ norm. With the aid of the shifted Legendre polynomials, the weak form is reduced into a system of the algebraic equations to be solved in each subinterval. Furthermore, to handle the nonlinear term, the technique of product approximation is utilized. The utility of the present discretization technique and some well-known standard schemes is checked through numerical calculations on a range of linear and nonlinear problems with analytical solutions.

Keywords: logistic differential equation; liouville-caputo fractional derivative; local discontinuous Galerkin methods; stability estimate

1. Introduction

In studies of elementary population dynamics the simplest model for the growth of a population is known as rate equation and structured by Malthus in (1798) [1]

$$\begin{cases} \dfrac{dM(t)}{dt} & = r\,M(t), \quad t > 0, \\ M(0) & = M_0, \end{cases} \tag{1}$$

where $M(t)$ denotes the population at time t, the non-zero parameter r equals to $r = \beta - \alpha$, where β and α are the per capita birth and death rates respectively. Here, M_0 is the population at time $t = 0$. The exact analytical solution of Malthus population model (1) is explained the constant population growth rate $M(t) = M_0 e^{rt}$. The Maithusian grow model is unrealistic over long times due to the fact that the solution of the rate equation is not included two main factors such as spread of diseases and the limitation on food supply. To model the effects of these factors in a population model, the logistic equation was considered by P. R. Verhulst in 1838 [2]

$$\frac{dN(t)}{dt} = rN(t)\left(1 - \frac{N(t)}{K}\right),$$

where the variable $N(t) = M(t)/M_{max}$ is the whole population and normalized to its maximum attainable value M_{max}, r denotes the intrinsic growth rate while the constant $K > 0$ known as the carrying capacity of the environment. By defining $X(t) := N(t)/K$ and $\sigma := rK$, the standard logistic equation can be rewritten as

$$\begin{cases} \dfrac{dX(t)}{dt} = \sigma X(t)\left(1 - X(t)\right), & t > 0, \\ X(0) = X_0. \end{cases} \tag{2}$$

where $X_0 = M(0)/M_{max}$. The exact solution of this equation can be easily obtained as

$$X(t) = \frac{X_0}{X_0 + (1 - X_0)e^{-\sigma t}}.$$

In the last decades, many efforts have been devoted to extend the integer-order models to the corresponding fractional-order models, which are more descriptive and can provide a powerful and valuable instrument for the explanation of hereditary and memory properties of several materials and process [3,4]. Replacing the classical derivative operator in (2) by a fractional one, the fractional logistic equation will be obtained. This model of population growth has been found applications in numerous disciplines of science and engineering. For instance, the growth of tumors in medicine [5] can be modelled as the fractional logistic equation (FLE). In addition, the milstone of various important mathematical models is based on the fractional logistic equation such as two models in Radar signals [6] and electroanalytical chemistry [7]. Several variations of the population growth model have been studied in the literature [8]. In the present study, we are going to investigate the following logistic population model of fractional order in the form

$$\begin{cases} {}^{LC}_{a}\mathcal{D}^{\nu}_{t} X(t) = \sigma X(t)\left(1 - X(t)\right) =: \sigma X(t)\, g(X(t)), & t > 0, \\ X(0) = X_0, \end{cases} \tag{3}$$

where the symbol ${}^{LC}_{a}\mathcal{D}^{\nu}_{t}$ denotes the fractional derivative operator of Liouville-Caputo type and $\nu \in (0,1]$. It should be emphasized that in (3) we have used the function $g(s) \equiv 1 - s$, which corresponds to the nonlinear logistic equation. However, to address the linear counterpart of this equation we also consider $g(s) \equiv 1$. The issue of existence and the uniqueness of the solution of (3) is discussed in detail in Reference [9].

It is known that for most fractional differential equations there is no possibility to find the exact solutions analytically. Consequently, exploring an approximate or numerical technique is of primary interest for such fractional equations. Many efforts have been made toward the exact analytical solution of the problem (3). The first one is proposed by West [10], which is based on the Carleman embedding technique. Later, it is shown that in Reference [11] the this analytical function is only very close to the numerical solutions of the FLE. The other analytical methods for the FLE include the fractional Taylor expansion method [12], a method based on Euler's numbers [13], and the varational iterative method [14]. Besides the analytical investigations, numerous computational approaches have been proposed for the nonlinear FLE. Let us mention the predictor-corrector approaches [9,15], the finite difference schemes [14,16], the spectral methods [17,18], the Bessel collocation method [19], the Chebyshev wavelet method [20], the Laguerre collocation method [21], and the fractional spline collocation method [22].

Many other numerical and approximation methods as well as computational approaches have been developed and applied for the FDEs which are based upon various closely-related models of real-world problems. For example, Baleanu et al. [23] made use of a Chebyshev spectral method based on operational matrices, a remarkable survey of numerical methods can be found in [24], a study of the fractional-order Bessel, Chelyshkov, and Legendre collocation schemes for the fractional Riccati equation was presented in [25], an operational matrix of fractional-order derivatives of Fibonacci polynomials was developed in [26], an introductory overview and recent developments involving FDEs was presented in [27], efficiency of the spectral collocation method in the dynamic simulation of the fractional-order epidemiological model of the Ebola virus was investigated in [28], the Jacobi collocation method and a spectral tau method based on shifted second-kind Chebyshev polynomilas for the approximate solution of some families of the fractional-order Riccati differential equations were discussed in [29,30], computational approaches to FDEs for the biological population model were discussed in [31], the generalized Chebyshev and Bessel colllocation approaches for fractional BVPs and multi-order FDEs were considered in [32,33], and a general wavelet quasi-linearization method for solving fractional-order population growth model was developed and applied in [34].

In this work, we take a further step towards proposing a numerical method for solving the FLE. We utilize a discontinuous finite element approach, i.e, the local discontinuous Galerkin (LDG) discretization approach for the FLE (3). To apply the LDG scheme, we must rewrite a given FDEs as a system of first-order ordinary differential equations (ODEs) with together a fractional integral. Hence, the discontinuous Galerkin (DG) method is employed to discretize the resulting system as well as the fractional integral. The first DG method was introduced by Reed and Hill [35] in 1973 for numerically solving neutron transport, that is, a time-independent linear hyperbolic equation. Since then the DG schemes have been well implemented for the classical ODEs was started by the work [36]. DG schemes as a subclass of finite element methods (FEMs) allow us to exploit discontinuous discrete basis functions. These local basis functions are usually selected as piecewise polynomials. Exploiting completely discontinuous basis functions offers great opportunities compared to traditional FEMs when used to discretize differential equations. In summary, the main gains of the DG methods are in terms of flexibility, accuracy as well as parallelizability, see cf. Reference [37].

To the best of our knowledge, the LDG approaches for the ODEs of fractional-order including one-term and multi-terms were first discussed in Reference [38] and then have been applied to many model problems [39–41]. It is worth mentioning that the success of LDG methods is based on the designing of appropriate numerical fluxes at the interface elements. In this work, we utilize the upwind numerical flux as natural choice for the FLE. By choosing the upwind fluxes we are able to prove the numerical stability of the LDG scheme.

The rest of this paper is organized as follows. In the next Section, we review some fractional calculus preliminaries and state some of their properties that will be used later on. The formulation of the LDG scheme for the logistic equation is established in Section 3. Hence, the algebraic form of the LDG scheme is obtained with the aid of shifted Legendre basis functions. The technique of product approximation is also applied to deal with the nonlinear term in the weak formulation. In Section 4 we establish the numerical stability of the scheme in the linear case and a discussion about the error estimation is made. In Section 5, the applicability and utility of the present numerical schemes are verified by performing several simulations on two linear and nonlinear population growth and logistic model problems. Finally, a conclusion is drawn in Section 6.

2. Fractional Calculus

Now, we present some fundamental and mathematical preliminaries of the fractional calculus theory to be utilized in our subsequent sections, see References [3,4,27].

Definition 1. *Let $v \geqslant 0$ is given. The (left) Riemann-Liouville fractional integral operator of order v is given by*

$$\mathcal{I}^v f(t) \equiv {}_a\mathcal{I}_t^v f(t) = \begin{cases} \dfrac{1}{\Gamma(v)} \displaystyle\int_a^t f(p)\,(t-p)^{v-1}\,dp, & v > 0,\ t > 0, \\ f(t), & v = 0. \end{cases}$$

The integral operator $\mathcal{I}^v$ has many properties. Among others, we make use of the following

(1) $\mathcal{I}^{\nu}\mathcal{I}^{\beta}f(t) = \mathcal{I}^{\nu+\beta}f(t)$,

(2) $\mathcal{I}^{\nu}\left(c_1 f(t) + c_2 g(t)\right) = c_1\mathcal{I}^{\nu}f(t) + c_2\mathcal{I}^{\nu}g(t)$, $\quad c_1, c_2 \in \mathbb{R}$,

(3) $\mathcal{I}^{\nu}t^{\gamma} = \frac{\Gamma(\gamma+1)}{\Gamma(\gamma+\nu+1)}t^{\nu+\gamma}$, $\quad \gamma > -1$.

The corresponding definition of the right Riemann-Liouville fractional integral on the interval $[t, b]$ instead of $[a, t]$ is given by

$$_t\mathcal{I}^{\nu}_b f(t) = \frac{1}{\Gamma(\nu)}\int_t^b f(p)\,(p-t)^{\nu-1}\,dp, \quad \nu > 0,\ t > 0.$$

Definition 2. *The fractional derivative $\mathcal{D}^{\nu}$ of $f(t)$ in the Liouville-Caputo's sense is defined as*

$$\mathcal{D}^{\nu}f(t) \equiv {}_a^{LC}\mathcal{D}^{\nu}_t f(t) = \begin{cases} \dfrac{1}{\Gamma(m-\nu)}\displaystyle\int_a^t \dfrac{f^{(m)}(p)}{(t-p)^{\nu-m+1}}\,dp, & m-1 < \nu < m, \quad t > 0, \\[4mm] f^{(m)}(t), & \nu = m, \quad m \in \mathbb{N}. \end{cases}$$

We make use of the following [4]:

$$\mathcal{D}^{\nu}(C) = 0 \quad (C \text{ is a constant}), \tag{4}$$

$$\mathcal{D}^{\nu}t^{\gamma} = \begin{cases} \dfrac{\Gamma(\gamma+1)}{\Gamma(\gamma+1-\nu)}t^{\gamma-\nu}, & \text{for } \gamma \in \mathbb{N}_0 \text{ and } \gamma \geqslant \lceil\nu\rceil, \text{ or } \gamma \notin \mathbb{N}_0 \text{ and } \gamma > \lfloor\nu\rfloor, \\[3mm] 0, & \text{for } \gamma \in \mathbb{N}_0 \text{ and } \gamma < \lceil\nu\rceil. \end{cases} \tag{5}$$

Here, we have used the ceiling and floor functions $\lceil\nu\rceil$, $\lfloor\nu\rfloor$ respectively. It should be noted that, two operators $\mathcal{I}^{\nu}$ and $\mathcal{D}^{\nu}$ are related through the following expression

$$\mathcal{D}^{\nu}f(t) = \mathcal{I}^{m-\nu}D^m f(t), \quad D = \frac{d}{dt}. \tag{6}$$

3. Discretized LDG Formulation

In order to formulate the LDG method for the logistic equation in (3), some basic notations will first be introduced.

Let us consider (3) on $L = (0, T)$ for some given $T > 0$. To rewrite (3) as a first-order system, we introduce two new variables $z_0(t) = X(t)$ and $z_1(t) = \frac{dX(t)}{dt}$ and use the relation (6) to get

$$\begin{cases} z_1(t) - \dfrac{dz_0(t)}{dt} = 0, \\[3mm] {}_0\mathcal{I}^{(1-\nu)}_t z_1(t) - \sigma z_0(t)\left(1 - z_0(t)\right) = 0, \\[3mm] z_0(0) - X_0 = 0, \end{cases} \tag{7}$$

with $\nu \in (0, 1]$ and $t \in L$. By Δ we denote a partitioning of the interval L into J subintervals $L_l = (t_{l-1}, t_l)$ for $l = 1, \ldots, J$. The grid points of Δ will be denoted as

$$0 =: t_0 < t_1 < \ldots < t_{J-1} < t_J := T.$$

By h_l we mean the length of each L_l, that is, $h_l = t_l - t_{l-1}$ for $l = 1, 2, \ldots, N$. The maximum length of these element is taken as $h := \max_{l=1}^{J} h_l$. We associate the mesh Δ with the broken Sobolev spaces

$$H_{\Delta}^1 = \{w : L \to \mathbb{R} \mid w|_{L_l} \in H^1(L_l),\ l = 1, 2, \ldots, J\}.$$

and

$$S_{\Delta} = \{w : L \to \mathbb{R} \mid w|_{L_l} \in L_2(L_l),\ l = 1, 2, \ldots, J\},$$

By using these function spaces, let assume that the solutions of system (7) belong to corresponding spaces

$$\left(z_0(t), z_1(t)\right) \in H_\Delta^1 \times S_\Delta.$$

It should be noted that the elements of space H_Δ^1 may be discontinuous in t at discrete time level t_1. In this respect, at the mesh grid points, defining the left-sided as well as the right-sided limits of a function w is necessary, where $w : L \to \mathbb{R}$ is a piecewise continuous function. By w_n^- and w_n^+, we let the left- and right-sided limits of w at t_l

$$w_l^+ = w^+(t_l) = w(t_l^+) := \lim_{s \to 0^+} w(t_n + s), \quad w_l^- = w^-(t_l) = w(t_l^-) := \lim_{t \to 0^-} w(t_n + s).$$

For any positive integer number r, we denote by $P_r(L_l)$ the space of polynomials of degree less or equal than r on the element $L_l \in \Delta$. We then let the approximate solutions $z_0(t), z_1(t)$ belong to a subspace $\mathcal{V}^{(r)} \subset H_\Delta^1$, which is a finite dimensional space. This subspace is defined as the space of discontinuous and piecewise polynomial functions

$$\mathcal{V}^{(r)} = \{w : L \to \mathbb{R} \mid w|_{L_l} \in P_r(L_l), \; l = 1, 2, \ldots, J\}.$$

We further define $\mathcal{Z}_0(t)$ and $\mathcal{Z}_1(t)$ as the DG approximations to the exact solutions $z_0(t)$ and $z_1(t)$ of the system (7) respectively on the element L_l. Below, we make use of the following notations

$$(w, v)_l := \int_{L_l} w\,v\,dt, \quad \langle w, v \rangle_l := \int_0^{t_l} w\,v\,dt, \quad \|w\|_l := \sqrt{\langle w, w \rangle_l}.$$

For obtaining the weak DG formulation, we first multiply the first equation in (7) by a test function $w_0 \in \mathcal{V}^{(r)}$ and integrate over L_l. By applying the integrating by parts we get

$$\left(\mathcal{Z}_1(t), w_0\right)_l + \left(\mathcal{Z}_0(t), \frac{dw_0}{dt}\right)_l - \mathcal{Z}_0(t_l^-)\,w_0(t_l^-) + \mathcal{Z}_0(t_{l-1}^+)\,w_0(t_{l-1}^+) = 0. \tag{8}$$

Hence, the second integral equation in (7) is multiplied by a test function $w_1 \in \mathcal{V}^{(r)}$ and integrate over L_l. To advance the solution in time, we replace $\mathcal{Z}_0(t_{l-1}^+)$ by the upwind flux $\mathcal{Z}_0(t_{l-1}^-)$ in (8). Thus, the discrete formulation for finding $\mathcal{Z}_0, \mathcal{Z}_1 \in \mathcal{V}^{(r)}$ takes the following form for all $w_0, w_1 \in \mathcal{V}^{(r)}$, and $l = 1, 2, \ldots, J$

$$\begin{cases} \left(\mathcal{Z}_1(t), w_0(t)\right)_l + \left(\mathcal{Z}_0(t), w_0'(t)\right)_l - \mathcal{Z}_0(t_l^-)\,w_0(t_l^-) + \mathcal{Z}_0(t_{l-1}^-)\,w_0(t_{l-1}^+) = 0, \\ \left(0\mathcal{I}_t^{(1-\nu)}\mathcal{Z}_1(t), w_1(t)\right)_l - \sigma\left(\mathcal{Z}_0(t), w_1(t)\right)_l + \sigma\left(\mathcal{Z}_0^2(t), w_1(t)\right)_l = 0, \\ \mathcal{Z}_0(t_0^- = 0) - X_0 = 0. \end{cases} \tag{9}$$

It should be noted that, to start computations on the first element $L_1 = (t_0, t_1)$ we use the given initial condition $\mathcal{Z}_0(t_0^-) = X_0$. Hence, by utilizing the upwind flux as the natural choice, we are able to solve the resultant equations element by element on each subinterval L_l for $l = 1, 2, \ldots, J$. In each element, we just need to invert a local matrix of size $(r + 1) \times (r + 1)$ in place of a global matrix of size $J(r + 1) \times J(r + 1)$.

Algebraic Formulation

Since the functions in $\mathcal{V}^{(r)}$ may be discontinuous across interfaces of the element, various local bases can be selected for finite element approximation in (9). Let us choose a basis in the space $P_r(L_l)$

formed by functions $\phi_0^l, \phi_1^l, \ldots, \phi_r^l$. Thus the numerical approximations $\mathcal{Z}_0$ of z_0 and $\mathcal{Z}_1$ of z_1 in every element L_l can be expressed as

$$\mathcal{Z}_0(t) = \sum_{i=0}^{q} \alpha_i^l \, \phi_i^l(t), \quad \mathcal{Z}_1(t) = \sum_{i=0}^{q} \beta_i^l \, \phi_i^l(t), \quad t \in L_l. \tag{10}$$

Here, the coefficients $\alpha_i^l, \beta_i^l, i = 0, \ldots, r$ denote the degrees of freedom to be sought in each $L_l, l = 1, \ldots, J$. To proceed, we take the test functions in each element L_l in the form $w_j = \phi_j^l(t)$ for $j = 0, 1, \ldots, r$ and $l = 0, 1, \ldots, J$. Now, by specifying the basis functions as we done below, the discrete LDG formulation (9) is reduced to a algebraic system of equations.

For practical implementation of the LDG scheme (9) for the FLE (3), we use the set of orthogonal Legendre polynomials for the space $\mathcal{V}^{(r)}$. Let us recall that, the i'th degree Legendre polynomials $P_i(s)$ can be generated by the well-known Rodriguez formula

$$P_i(s) = \frac{1}{2^i i!} \frac{d^i}{ds^i} (s^2 - 1)^i.$$

The Legendre polynomials satisfy the following relations [17]

$$\int_{-1}^{1} P_i(s) \, P_j(s) ds = \frac{2\delta_{ij}}{2i + 1}, \quad P_i(1) = 1, \quad P_i(s) = (-1)^i P_i(-s), \quad i, j \geqslant 0, \tag{11a}$$

$$(2i + 1) P_i(s) = \frac{dP_{i+1}(s)}{ds} - \frac{dP_{i-1}(s)}{ds}, \tag{11b}$$

where δ_{ij} denotes the Kronecker delta. The first property shows that these set of orthogonal polynomials are orthogonal with respect to weighting function $w(t) \equiv 1$ on $(-1, 1)$. The Legendre polynomial $P_i(s)$ of degree i can be explicitly expressed as follows

$$P_i(s) = \sum_{k=0}^{M_i} c_{ik} \, s^{i-2k}, \quad c_{ik} := \frac{1}{2^i} (-1)^k \binom{i}{k} \binom{2i - 2k}{i},$$

where $M_i = i/2$ or $(i - 1)/2$, whichever is an integer. Due to the fact that these polynomials are orthogonal on $[-1, 1]$, we map them onto the element L_l by using the following change of variable

$$s := \frac{2t - t_{l-1} - t_l}{h_l}, \quad t \in L_l.$$

Let the resultant shifted Legendre polynomials denoted by $\mathbb{L}_i(t)$. Thus, the explicit form of $\mathbb{L}_i(t)$ of degree i takes the form

$$\mathbb{L}_i(t) = \sum_{k=0}^{M_i} c_{ik} \left(\frac{2t - t_{l-1} - t_l}{h_l} \right)^{i-2k}.$$

By means of the binomial formula, one can further simplify the last expression as follows

$$\mathbb{L}_i(t) = \sum_{k=0}^{M_i} \sum_{m=0}^{i-2k} C_{ikm} \, t^m, \tag{12}$$

where the coefficients C_{ikm} are defined as

$$C_{ikm} := \frac{(-1)^{i+k+m} (2i - 2k)!}{2^i (i - k)! \, k! \, l! \, (i - 2k - m)!} \left(\frac{t_l + t_{l-1}}{t_l - t_{l-1}} \right)^{i-2k} \left(\frac{2}{t_l + t_{l-1}} \right)^m.$$

Now, we choose $\phi_i^l(t) = \mathbb{L}_i(t)$ in (10) for $l = 1, 2, \ldots, J$, where $\mathbb{L}_i$ is the shifted Legendre polynomial of degree i in t defined in (12). With this transformation, the unknown values α_i^l, β_i^l in (10) can be interpreted as the Legendre coefficients of the expansion of $\mathcal{Z}_0, \mathcal{Z}_1$. Hence, by the virtue of the Legendre properties (11) and inserting (10) into the discrete formulation (9) we have for $l = 1, \ldots, J$ as

$$
\sum_{i=0}^{r} \beta_i^l \left(\mathbb{L}_i(t), \mathbb{L}_j(t) \right)_l + \sum_{i=0}^{r} \alpha_i^l \left(\mathbb{L}_i(t), \mathbb{L}_j'(t) \right)_l - \sum_{i=0}^{r} \alpha_i^l + \sum_{i=0}^{r} \alpha_i^{l-1}(-1)^j = 0,
$$
$$
\sum_{i=0}^{r} \beta_i^l \left({}_0\mathcal{I}_t^{(1-\nu)} \mathbb{L}_i(t), \mathbb{L}_j(t) \right)_l - \sigma \sum_{i=0}^{r} \alpha_i^l \left(\mathbb{L}_i(t), \mathbb{L}_j(t) \right)_l + \sigma \left(\left[\sum_{i=0}^{r} \alpha_i^l \mathbb{L}_i(t) \right]^2, \mathbb{L}_j(t) \right)_l = 0,
$$

(13)

for $j = 0, \ldots, r$. To proceed, we need to deal with two main difficulties involving the integral and nonlinear terms that appear in (13). To tackle the integral term, the properties (1)–(3) of fractional integration in Section 2 is used to obtain

$$
{}_0\mathcal{I}_t^{(1-\nu)} \mathbb{L}_i(t) = \sum_{k=0}^{M_i} \sum_{m=0}^{i-2k} C_{ikm} \, {}_0\mathcal{I}_t^{(1-\nu)} t^m = \sum_{k=0}^{M_i} \sum_{m=0}^{i-2k} C_{ikm}' \, t^{m+1-\nu}, \quad C_{ikm}' := C_{ikm} \frac{\Gamma(m+1)}{\Gamma(m+2-\nu)}.
$$

Next, the explicit form (12) is utilized for $\mathbb{L}_j(t)$ and then ${}_0\mathcal{I}_t^{(1-\nu)} \mathbb{L}_i(t)$ will be inserted into the inner product. Now, by integration over L_l we obtain

$$
d_{i,j} := \left({}_0\mathcal{I}_t^{(1-\nu)} \mathbb{L}_i(t), \mathbb{L}_j(t) \right)_l = \sum_{k=0}^{M_i} \sum_{m=0}^{i-2k} \sum_{k'=0}^{M_j} \sum_{m'=0}^{j-2k'} C_{ikmjk'm'}'' \left(t_l^{m+m'+2-\nu} - t_{l-1}^{m+m'+2-\nu} \right), \quad (14)
$$

with the coefficients
$$
C_{ikmjk'm'}'' := C_{ikm}' C_{jk'm'} / (m + m' + 2 - \nu).
$$

The nonlinear term in (13) can be computed using the Legendre polynomials. For instance, if $r = 1$ we may write it as a product of two vectors

$$
n_j^{DC} := \left(\mathcal{Z}_0^2(t), \mathbb{L}_j(t) \right)_l = \left[[\alpha_0^l]^2, 2\alpha_0^l \alpha_1^l, [\alpha_1^l]^2 \right] \cdot \int_{L_l} \left[\mathbb{L}_0^2(t), \mathbb{L}_0(t) \mathbb{L}_1(t), \mathbb{L}_1^2(t) \right]^T \mathbb{L}_j(t) dt,
$$

for $j = 0, 1$. Therefore, it is not a difficult task to calculate n_j^{DC} by direct computation (D.C.) using the shifted Legendre polynomials on each L_l for different j. Of course one may exploit the symbolic toolbox in Matlab to facilitate the process of integration of these polynomials. Alternatively, to handle the nonlinear term in (13), the product approximation (P.A.) technique [42] is used in the following manner

$$
\mathcal{Z}_0^2(t) = \left[\sum_{i=0}^{r} \alpha_i^l \mathbb{L}_i(t) \right]^2 \approx \sum_{i=0}^{r} [\alpha_i^l]^2 \mathbb{L}_i(t).
$$

This technique enables us to write the nonlinear part as

$$
n_{i,j}^{PA} := \left(\mathcal{Z}_0^2(t), \mathbb{L}_j(t) \right)_l = \sum_{i=0}^{r} [\alpha_i^l]^2 \left(\mathbb{L}_i(t), \mathbb{L}_j(t) \right)_l. \quad (15)
$$

Now, it suffices to calculate the two first terms in (13). To this end, we compute the elements of the mass matrix as

$$
m_{i,j} := \left(\mathbb{L}_i(t), \mathbb{L}_j(t) \right)_l = \int_{L_l} \mathbb{L}_i(t) \mathbb{L}_j(t) dt = \begin{cases} \frac{h_l}{2i+1}, & i = j, \\ 0, & i \neq j. \end{cases} \quad (16)
$$

Finally, the entries of the stiffness matrix

$$s_{i,j} = \left(\mathbb{L}_i(t), \ \mathbb{L}_j'(t) \right)_l = \int_{L_l} \mathbb{L}_i(t)\mathbb{L}_j'(t)dt,$$

need to be calculated. In the new coordinate, we recursively employ the Legendre property (11b) to derive

$$\frac{h_l}{2}\mathbb{L}_{i+1}'(t) = (2i+1)\mathbb{L}_i(t) + (2(i-1)+1)L_{i-2}(t) + (2(i-4)+1)L_{i-4}(t) + \cdots .$$

By applying the orthogonality relation (11a) to the preceding equation and then simplifying the involved integral in $s_{i,j}$, we finally get

$$s_{i,j} = \begin{cases} 2, & \text{if } i > j \text{ and } (i+j) \text{ is even,} \\ 0, & \text{otherwise.} \end{cases} \tag{17}$$

Using (14)–(17), one may write (13) in the matrix-vector multiplication form for $l = 1, \ldots, J$ as follows

$$\begin{cases} M\boldsymbol{\beta}^l + (S - E)\boldsymbol{\alpha}^l = \boldsymbol{b}^l, \\ D\boldsymbol{\beta}^l - \sigma M(\boldsymbol{\alpha}^l - \boldsymbol{\alpha}^{2,l}) = 0, \end{cases} \tag{18}$$

where the unknown vectors $\boldsymbol{\alpha}^l, \boldsymbol{\beta}^l$, and $\boldsymbol{\alpha}^{2,l}$ are defined

$$\boldsymbol{\alpha}^l = \left(\alpha_0^l, \ldots, \alpha_r^l \right)^T, \qquad \boldsymbol{\beta}^l = \left(\beta_0^l, \ldots, \beta_r^l \right)^T, \qquad \boldsymbol{\alpha}^{2,l} = \left([\alpha_0^l]^2, \ldots, [\alpha_r^l]^2 \right)^T.$$

Note in (18) that the components of matrix E are $e_{i,j} := 1$ while that of M, S, N and D are $m_{i,j}, s_{i,j}, n_{i,j}$, and $d_{i,j}$ respectively for $i, j = 0, \ldots, r$ as defined above. Moreover, the components of the known vector $\boldsymbol{b}^l$ are

$$b_i := (-1)^{i+1} \mathcal{Z}_0(t_{l-1}^-), \quad i = 0, 1, \ldots, r.$$

Clearly, the value of $\mathcal{Z}_0(t_{l-1}^-)$ is already known from the preceding time interval L_{l-1}. Obviously this value at the first time interval is computed as X_0, the initial condition. Also, the obtained system (18) is a nonlinear algebraic system of equations have to be solved in each L_l for $l = 1, \ldots, J$. This system can be solved for example, via Newton type schemes. It is known that this method converges quadratically whenever the approximation is close to the actual solution of the given nonlinear system. Using the D.C. approach, we also arrive at a nonlinear system of equation in the general form $F(\boldsymbol{\alpha}^l, \boldsymbol{\beta}^l) = \boldsymbol{0}$ to be solved in each interval L_l. As we show in the numerical experiments, this approach is more accurate than the corresponding P.A. approach.

4. Numerical Stability and Error Estimates

Now, we are going to establish the stability of proposed LDG scheme when applied to the logistic equation in the linear case by considering $g(t) \equiv 1$ in (3). In this case we have

$$\begin{cases} {}_{a}^{LC}\mathcal{D}_t^\nu X(t) = \sigma X(t), \quad \nu \in (0,1). \\ X(0) = X_0. \end{cases} \tag{19}$$

Without loss of generality, let us assume that $\sigma < 0$. The numerical scheme of (19) is to find $\mathcal{Z}_0, \mathcal{Z}_1 \in \mathcal{V}^{(r)}$ such that

$$\begin{cases} \mathcal{Z}_0(t_l^-)\, w_0(t_l^-) - \mathcal{Z}_0(t_{l-1}^-)\, w_0(t_{l-1}^+) - \Big(\mathcal{Z}_1(t),\, w_0(t)\Big)_l - \Big(\mathcal{Z}_0(t),\, w_0'(t)\Big)_l = 0, \\[2mm] \Big({}_0\mathcal{I}_t^{(1-\nu)}\, \mathcal{Z}_1(t),\, w_1(t)\Big)_l = \sigma\Big(\mathcal{Z}_0(t),\, w_1(t)\Big)_l, \\[2mm] \mathcal{Z}_0(t_0^-) - X_0 = 0, \end{cases} \tag{20}$$

for all $w_0, w_1 \in \mathcal{V}^{(r)}$, and $l = 1, 2, \ldots, J$. Let us state the next lemma, which based on the semigroup properties of fractional integral operators and will be used below, a proof of which can be found in Reference [38].

Lemma 1. *Suppose that $\nu \in (0,1)$, then we have*

$$\Big\langle {}_0\mathcal{I}_t^{1-\nu} u,\, u \Big\rangle_l = \Big\langle {}_0\mathcal{I}_t^{\frac{1-\nu}{2}} u,\, {}_t\mathcal{I}_{t_l}^{\frac{1-\nu}{2}} u \Big\rangle_l = \cos\Big(\frac{(1-\nu)\pi}{2}\Big)\|u\|^2_{H^{\frac{1-\nu}{2}}([0,t_l])}.$$

Let us assume that $\widetilde{\mathcal{Z}}_0, \widetilde{\mathcal{Z}}_1 \in \mathcal{V}^{(r)}$ be the approximate solutions of $\mathcal{Z}_0, \mathcal{Z}_1$ respectively. Now, the numerical errors are defined as $E_{X_i} := \widetilde{\mathcal{Z}}_i - \mathcal{Z}_i$ for $i = 0,1$. It can be seen that $\widetilde{\mathcal{Z}}_0$ and $\widetilde{\mathcal{Z}}_1$ both satisfy (20). If we subtract Equation (20) from the same equations with $\widetilde{\mathcal{Z}}_0$ and $\widetilde{\mathcal{Z}}_1$, the following error equations will be obtained

$$\begin{cases} E_{X_0}(t_l^-)\, w_0(t_l^-) - E_{X_0}(t_{l-1}^-)\, w_0(t_{l-1}^+) - \Big(E_{X_1}(t),\, w_0(t)\Big)_l - \Big(E_{X_0}(t),\, w_0'(t)\Big)_l = 0, \\[2mm] -\dfrac{1}{\sigma}\Big({}_0\mathcal{I}_t^{(1-\nu)} E_{X_1}(t),\, w_1(t)\Big)_l = -\Big(E_{X_0}(t),\, w_1(t)\Big)_l, \end{cases} \tag{21}$$

which holds for all $w_0, w_1 \in \mathcal{V}^{(r)}$. Taking $w_0 = E_{X_0}$ and $w_1 = E_{X_1}$ in (21) followed by collecting these two equations, we conclude that

$$E_{X_0}^2(t_l^-) - E_{X_0}(t_{l-1}^-)\, E_{X_0}(t_{l-1}^+) - \Big(E_{X_0}(t),\, E_{X_0}'(t)\Big)_l - \frac{1}{\sigma}\Big({}_0\mathcal{I}_t^{(1-\nu)} E_{X_1}(t),\, E_{X_1}(t)\Big)_l = 0.$$

To deal with the third term, we utilize the identity $\Big(u, \frac{du}{dt}\Big)_l = (u^2(t_l^-) - u^2(t_{l-1}^+))/2$ with $u = E_{X_0}$. Hence, we multiply the preceding equation by two. Adding and subtracting $E_{X_0}^2(t_{l-1}^-)$ to the modified equation and rearranging the terms to obtain

$$\Big(E_{X_0}(t_{l-1}^+) - E_{X_0}(t_{l-1}^-)\Big)^2 + \Big(E_{X_0}^2(t_l^-) - E_{X_0}^2(t_{l-1}^-)\Big) - \frac{2}{\sigma}\Big({}_0\mathcal{I}_t^{(1-\nu)} E_{X_1}(t),\, E_{X_1}(t)\Big)_l = 0.$$

By summing over elements for $l = 1, \ldots, J$, we get

$$E_{X_0}^2(t_J^-) - E_{X_0}^2(t_0^-) + \sum_{l=1}^{J}\Big(E_{X_0}(t_{l-1}^+) - E_{X_0}(t_{l-1}^-)\Big)^2 - \frac{2}{\sigma}\Big\langle {}_0\mathcal{I}_t^{(1-\nu)} E_{X_1}(t),\, E_{X_1}(t)\Big\rangle_J = 0.$$

By using Lemma 1, we have established the following stability of the LDG in the L_∞ norm for (20) (see also References [38,40]):

Lemma 2. *We have the following L_∞ stability of the LDG scheme (20) and for the numerical errors hold*

$$E_{X_0}^2(t_J^-) = E_{X_0}^2(t_0^-) - \sum_{l=1}^{J}\Big(E_{X_0}(t_{l-1}^+) - E_{X_0}(t_{l-1}^-)\Big)^2 + \frac{2}{\sigma}\cos\Big(\frac{(1-\nu)\pi}{2}\Big)\|E_{X_1}\|^2_{H^{\frac{1-\nu}{2}}([0,t_J])} \tag{22}$$

We close this section by pointing out some facts about the order of convergence of the proposed LDG scheme. In Reference [38] it is shown that the solution can be calculated with optimal order of convergence $(r+1)$ in the L_2 norm. In this work the mechanism of superconvergence is also discussed. The authors observed the superconvergence of order $(r + 1) + \min\{r, \nu\}$ at downwind point of each element.

5. Numerical Results and Discussions

In this section, we present some results of computations using the proposed LDG scheme described in the preceding sections to test their accuracy and efficiency when applied to the logistic equation. To assess the accuracy of the present numerical algorithms, we calculate the difference between the true exact and numerical solutions whenever the exact solution is available. For this purpose, we also consider a linear fractional population model and then we solve the fractional logistic equation numerically.

In order to asses the numerical scheme more qualitatively, by EOC we denote the estimated order of convergence calculated through defining

$$EOC := \log_2 \left(\frac{E_a(h)}{E_a(h/2)} \right),$$

where $E_a(h)$ is the absolute error corresponding to the step-size h. Moreover, to test the validity and accuracy of proposed LDG method and to make a comparison between our numerical model results with the results of other existing methods, we employ the predictor-corrector PECE method of Adams-Bashforth-Moulton type considered in Reference [43] as well as the implicit product integration of trapezoidal type described in Reference [24]. All experimental computations have been done by using MATLAB R2017a.

5.1. Linear Model

In this section, we consider a linear test problem to show the effectiveness of the proposed LDG approach. For this purpose, we consider the fractional population growth

$$\begin{cases} {}^{LC}_a\mathcal{D}^\nu_t X(t) = \sigma^\nu X(t), \quad t > 0, \\ X(0) = X_0, \end{cases} \tag{23}$$

where $0 < \nu \leqslant 1$ and $\sigma > 0$. This model problem is previously studied in Reference [22] and can be considered as a generalization of the Malthusian model (1) to the fractional-order derivative. By the aid of the Laplace transform, the exact analytical solution of the initial-value problem can be obtained in terms of well-known Mittag-Leffler function [10]

$$X(t) = X_0 E_\nu(\sigma^\nu t^\nu), \quad E_\nu(z) = \sum_{k=0}^{\infty} \frac{z^k}{\Gamma(k\nu + 1)}.$$

Note that by taking $\nu = 1$ the exact solution becomes $X(t) = X_0 e^{\sigma t}$.

To start computation, we take $\sigma = 1$ for simplicity and set $X_0 = 3/4$. By considering $\nu = 1$ and $J = 1$, the approximate solutions for $r = 3, 6$, and $r = 9$ on the interval $0 \leqslant t \leqslant 2$ are obtained as follows

$\mathcal{Z}_{0,3}(t) = 0.4233870968\, t^3 - 0.1814516129\, t^2 + 1.0887096774\, t + 0.7016129032,$

$\mathcal{Z}_{0,6}(t) = 0.003185535427\, t^6 - 0.00147024712\, t^5 + 0.04410741361\, t^4 + 0.1140555342\, t^3 + 0.3795910747\, t^2$
$\qquad + 0.7492022902\, t + 0.7500339451,$

$\mathcal{Z}_{0,9}(t) = 0.00000608710804\, t^9 - 0.00000288336716\, t^8 + 0.0002076022472\, t^7 + 0.0009466489455\, t^6$
$\qquad + 0.006344760802\, t^5 + 0.0311919123\, t^4 + 0.1250209298\, t^3 + 0.3749959904\, t^2 + 0.7500003306\, t$
$\qquad + 0.7499999933.$

These approximations together with the corresponding absolute errors are depicted in Figure 1. Clearly, as r increased, more accurate results will be obtained. Note, in all cases, the step size is taken as $h = 2$. Moreover, we emphasize that numerical solutions for this model problem based on the fractional spline collocation scheme have been proposed in Reference [22] with achieved absolute errors larger than 1×10^{-4}, see Figure 2 in this paper. The parameters used in this approach related to $\nu = 1$ were $M_1 = 2^6, 2^7, 2^8, N_1 = 37, 69, 133$, which obviously are much more greater than our used parameters.

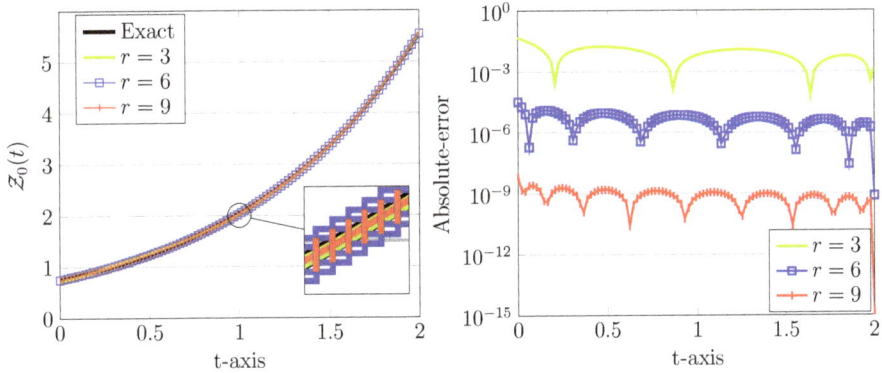

Figure 1. The approximated LDG with exact solutions (**left**) and the corresponding absolute errors (**right**) for $J = 1$, $\nu = 1$, $\sigma = 1$, $X_0 = 0.75$, and different $r = 3, 6, 9$.

Additionally, to justify our numerical model results, a comparison in Table 1 has been performed between the previous work on PECE [15,43] in terms of the number of (sub)intervals J is used in the computation. In this comparison, we compute the numerical solutions corresponding to $X(2)$ as well as absolute errors $|X(2) - \mathcal{Z}_0(2)|$ in these methods via different values of $J = 2^i$ for $i = 0, 1, \ldots 7$. For our LDG method we take $r = 2$ and $\nu = 1$. The last column in each method reports the corresponding EOC. The exact value of $X(2)$ up to 30 digits is

$$X(2) = 5.54179207419798736111715697916.$$

Table 1. Comparison of absolute errors in LDG with $r = 2$ and PECE for different number of interval J and $\nu = 1$. Numbers in bold show that the correct digits are obtained by the LDG.

	LDG			PECE		
J	$\mathcal{Z}_0(2)$	$\|X(2) - \mathcal{Z}_0(2)\|$	EOC	Numerical	Error	EOC
1	**5.**625000000000	8.3208_{-2}	—	3.750000000000	1.7918_{+0}	—
2	**5.5**43701171875	1.9091_{-3}	5.45	4.687500000000	0.8543_{+0}	1.07
4	**5.541**845071676	5.2998_{-5}	5.17	5.229675292969	0.3121_{+0}	1.45
8	**5.541793**647744	1.5735_{-6}	5.07	5.446685392454	9.5107_{-2}	1.71
16	**5.5417921**22228	4.8030_{-8}	5.03	5.515562177333	2.6230_{-2}	1.86
32	**5.54179207**5682	1.4842_{-9}	5.02	5.534910274764	6.8817_{-3}	1.93
64	**5.5417920742**44	4.6126_{-11}	5.01	5.540030137766	1.7619_{-3}	1.97
128	**5.5417920741**99	1.4380_{-12}	5.00	5.541346351966	4.4572_{-4}	1.98

The observed EOC seen for PECE in Table 1 is approximately 2 as was proved in Reference [43]. However, the superconvergence EOC about 5 ($\approx 2r + 1$) is clearly achieved for our results. This comparison indicates the thoroughness of the proposed method.

The numerical solutions for various values of $\nu = 0.65, 0.75, 0.85, 0.95$ using $r = 5$ and $J = 1$ are depicted in Figure 2, left plot. In all plots, the exact solutions are indicated by a solid line while the numerical counterpart are visualized by (coloured) dotted, dashed, and dash-dotted curves. Note that the computational domain is $[0, 1]$, which implies that the time step is $h = 1$. It can be seen from Figure 2 that the numerical solution obtained by the present LDG scheme has a good accuracy even using a relatively large time step and a low degree of the approximating polynomials. Furthermore, an appropriate choice of these computational parameters can improve the approximation accuracy.

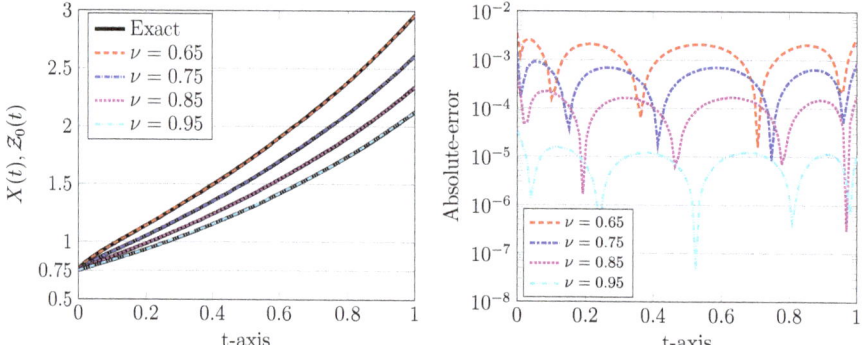

Figure 2. The approximated LDG with exact solutions (**left**) and the corresponding absolute errors (**right**) for $J = 1, r = 5, \sigma = 1, X_0 = 0.75$, and various values of $\nu = 0.65, 0.75, 0.85, 0.95$.

Finally, for the linear model problem (23), we investigate the standard L1 approximation method [44] and its variant known as the fast L1 method [45]. To implement these approaches, we use a uniform mesh with the step size $h = 1/1000$ on the interval $[0, 1]$. In the LDG scheme, we utilize $J = 1$ or $h = 1$ and $r = 5$ as the results shown in Figure 2. The numerical model results are presented in Table 2 for $\nu = 0.75$ and $\nu = 0.5$. For each ν, the corresponding exact solutions are also reported in the last column.

Table 2. Comparison of numerical solutions in LDG with $r = 5, h = 1$ and L1/fast L1 schemes with $h = 10^{-3}$ for some $t \in [0, 1]$ and $\nu = 0.75, 0.5$.

	$\nu = 0.75$				$\nu = 0.5$			
t	LDG	L1	Fast L1	Exact	LDG	L1	Fast L1	Exact
0.2	1.0536	1.0524	1.0524	1.053507	1.3420	1.3459	1.3345	1.349263
0.4	1.3512	1.3486	1.3486	1.350342	1.8370	1.8176	1.8176	1.822532
0.6	1.6963	1.6945	1.6945	1.697186	2.3489	2.3525	2.3525	2.359660
0.8	2.1128	2.1087	2.1087	2.112499	2.9957	2.9845	2.9845	2.994627
1.0	2.6134	2.6091	2.6091	2.614400	3.7385	3.7427	3.7427	3.756735

5.2. Nonlinear Model

We now consider the FLE (3) on $[0, 1]$ with the initial condition given by $X_0 = 1/2$ and the parameter $\sigma = 1/2$. Using $\nu = 1$, the analytical exact solution of the logistic equation is given by

$$X(t) = \frac{1}{1 + e^{-t/2}}.$$

The simulation results for this example can be found in Figures 3 and 4 for the number of elements equals to $J = 5$ and the polynomial degree $r = 2$. In Figure 3, we take $\nu = 1$ to compare the numerical results to the exact solution. Furthermore, we also use different approaches to treat the nonlinear term in the weak formulation, that is, the D.C. and P.A., which are utilized to compute n_j^{DC} and $n_{i,j}^{PA}$ in (15). As one can see that from Figure 3 that a slightly more accurate result is obtained by means of direct computation rather than product approximation, however, as mentioned it is more time-consuming. In order to observe the behaviour of numerical solutions more closely, a magnification of these solutions at $t = 0.4$ is done in Figure 3. The exact solution is depicted by a solid line.

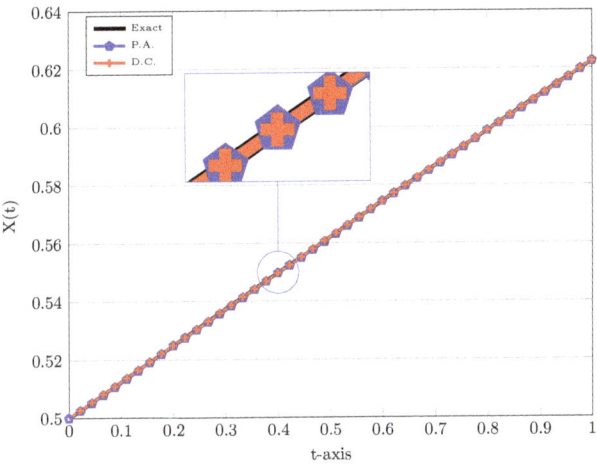

Figure 3. Numerical solutions of LDG scheme using P.A. and D.C. approaches with $h = 0.2$, $\sigma = 0.5$, $X_0 = 0.5$, and $\nu = 1.0$. The magnification of solutions at time $t = 0.4$ is plotted in the box. The exact solution is displayed by a solid line.

In the next experiment, we plot the absolute errors when utilizing two approaches D.C. and P.A., as one observes in Figure 4. The computational parameters are the same as those applied for Figure 3. In Figure 4, the left plot corresponds to the D.C. and the right plot is when we use P.A. technique. Note that in all plots we have divided further each interval L_l into ten subinterval uniformly to see the behaviour of the corresponding curves more precisely.

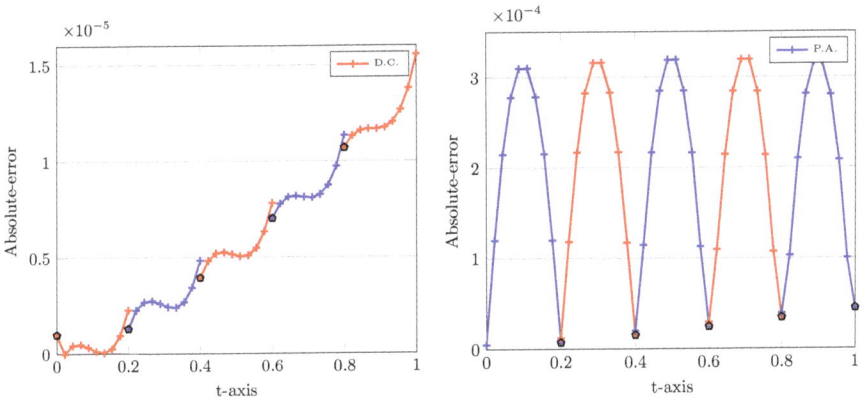

Figure 4. Absolute errors of LDG versus time using D.C. (**left**) and P.A. (**right**) approaches with $h = 0.2$, $\sigma = 0.5$, $X_0 = 0.5$, $\nu = 1.0$, and $r = 2$. In the left and right plots, the upwind and downwind points are highlighted by black pentagon.

Let us interpret the numerical errors depicted in Figure 4. On the left picture in which the P.A. technique is used, the smallest errors are obtained at upwind points. Almost the same magnitude of errors is achieved at downwind points. On the contrary, on the right picture without using the P.A. this process is reversed. This implies that the minimum values of absolute errors are achieved at downwind points and there exist considerable difference between them and the errors obtained at upwind points in each L_l. In the next experiments, we compare the numerical errors achieved at the final point $T = 1.0$, which is clearly a downwind point in the first approach.

In Tables 3 and 4, we summarize the numerical results related to $X(1)$ and its numerical approximation $\mathcal{Z}_0(1)$ are obtained by the LDG procedure (9). Here, we use $r = 1, 2$ and a different choice of the number of grid points $J = 1, 2, 4, 8$ and 16 are utilized. In these tables, we further compare the performance of two different D.C. and P.A. approaches. All calculations are shown with 10 decimal places of accuracy. In the last column of each table, the estimated order of convergence (EOC) is given. The exact value is $X(1) = 0.622459331201855$.

Table 3. Comparison of absolute errors in LDG with $r = 1$ using P.A. and D.C. for different number of interval J and $\nu = 1$. Numbers in bold show that the correct digits are obtained by the LDG.

	P.A.			D.C.		
J	$\mathcal{Z}_0(1)$	$|X(1) - \mathcal{Z}_0(1)|$	EOC	$\mathcal{Z}_0(1)$	$|X(1) - \mathcal{Z}_0(1)|$	EOC
1	0.6234038976	0.9445664060_{-3}	—	**0.6224**742460	0.1491482269_{-4}	—
2	0.6226973939	0.2380627190_{-3}	1.99	**0.6224**610781	0.1746857403_{-5}	3.09
4	0.6225290166	0.6968541429_{-4}	1.77	**0.6224595**421	0.2108842001_{-6}	3.05

Table 4. Comparison of absolute errors in LDG with $r = 2$ using P.A. and D.C. for different number of interval J and $\nu = 1$. Numbers in bold show that the correct digits are obtained by the LDG.

	P.A.			D.C.		
J	$\mathcal{Z}_0(1)$	$|X(1) - \mathcal{Z}_0(1)|$	EOC	$\mathcal{Z}_0(1)$	$|X(1) - \mathcal{Z}_0(1)|$	EOC
1	0.6233820141	0.9226828763_{-3}	—	**0.6224593**588	0.2759267670_{-7}	—
2	0.6226943815	0.2350503824_{-3}	1.97	**0.6224593**321	0.9149985214_{-9}	4.91
4	0.6225286311	0.6929984936_{-4}	1.76	**0.6224593**312	0.2863453918_{-10}	5.00

It can be seen from Tables 3 and 4 that using $r = 1$ and $r = 2$ in the D.C. approach, the results are accurate respectively to 6 and 10 decimal places for only $J = 4$ intervals. In other words, achieving an order of accuracy equal to 3 and 5 is possible if one uses the LDG scheme with $r = 1, 2$ degree of polynomials and for a small number of elements. These EOC are also confirmed the superconvergence order at downwind points previously reported in Reference [38]. Note that by utilizing the P.A. technique, the obtained EOC is equal to 2. We emphasize also that using the scheme PECE for the nonlinear logistic equation the EOC at most 2 will be achieved and of course a larger number of intervals J is required. In the next plot, we examine the behaviour of the absolute errors in the log scale for various polynomial degrees as well as with respect to the number of elements J, see Figure 5.

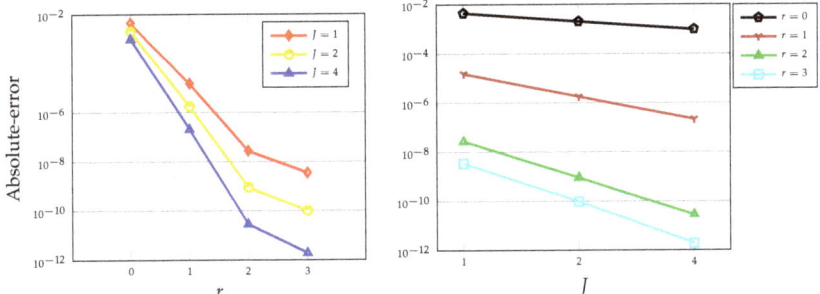

Figure 5. Absolute-errors versus polynomial degrees r for $J = 1, 2, 4$ (**left**) and against the number of elements J for $r = 0, 1, 2, 3$ (**right**) evaluated at $T = 1.0$ and for $\nu = 1$.

In the next experiment we show the impact of the fractional derivative on the approximated obtained solutions. In Figure 6 we present the approximated solutions at $J = 4, r = 3$ with different values of the fractional derivatives $\nu = 0.65, 0.75, 0.85, 0.95$ as well as $\nu = 1.0$. In these plots, we also compare the performance of two P.A. and D.C. approaches for these values of ν. In each case, for $\nu = 1.0$ the exact solution is also shown by a solid line. To justify our computed results, the implicit product-integration of trapezoidal (IPIT) rule with the step size $h = 1/256$ is used [24].

From both depictions in Figure 6, one can observe that the numerical solutions for $\nu \in (0, 1)$ are approaching to the solutions correspond to $\nu = 1$ for which the exact solution is known. Of course, more reliable results is obtained through the D.C. as previously tested for $\nu = 1$ in Tables 3 and 4.

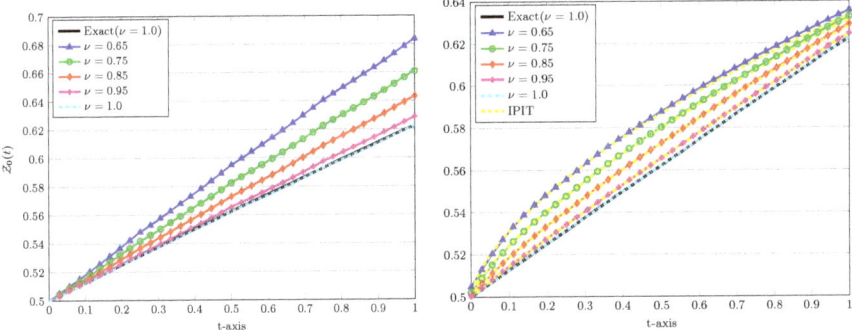

Figure 6. The approximated LDG solutions versus time using P.A. (**left**) and D.C. (**right**) approaches with $J = 4, r = 3, \sigma, X_0 = 0.5$, and various values of $\nu = 0.65, 0.75, 0.85, 0.95, 1.0$.

6. Conclusions

In this work, an approximation algorithm based on the LDG scheme is developed for the fractional-order logistic equation occurring in many biological and chemical phenomena. To be more precise, our numerical scheme based on discontinuous Galerkin finite element concept with Legendre basis functions yields to a set of nonlinear equations to be solved in each subinterval. The numerical stability in the linear case is proved and the order of convergence is also discussed. Beside the direct computation of the nonlinear term, the technique of product approximation is also utilized and then their performance are compared for various J, r and ν. We have tested the performance of the LDG scheme on the linear as well as nonlinear growth and logistic differential equations of fractional order. Comparing our numerical results with the PECE indicates that the present approaches produce an accurate approximation for the underlying model problems.

Entropy **2020**, *22*, 1328

Author Contributions: Conceptualization, M.I. and H.M.S.; Methodology, H.M.S.; Software, M.I.; Validation, H.M.S.; Writing—original draft, M.I.; Writing—review & editing, M.I. and H.M.S. Both authors have read and agreed to the published version of the manuscript.

Funding: This research received no external funding.

Conflicts of Interest: The authors declare no conflicts of interest.

References

1. Malthus, T.R. *Population: The First Essay (1798)*; University of Michigan Press: Ann Arbor, MI, USA, 1959.
2. Verhulst, P.F. Notice sur la loi que la population sint dons son accroissement. *Math. Phys.* **1838**, *10*, 113–121.
3. Kilbas, A.A.; Srivastava, H.M.; Trujillo, J.J. *Theory and Applications of Fractional Differential Equations, North-Holland Mathematical Studies*; Elsevier (North-Holland) Science Publishers: Amsterdam, The Netherlands; London, UK; New York, NY, USA, 2006; Volume 204.
4. Podlubny, I. *Fractional Differential Equations*; Academic Press: New York, NY, USA, 1999.
5. Foryś, U.; Marciniak-Czochra, A. Logistic equations in tumor growth modelling. *Int. J. Appl. Math. Comput. Sci.* **2003**, *13*, 317–325.
6. Krishna, B.T. Binary phase coded sequence generation using fractional order logistic equation. *Circuits Syst. Signal Process.* **2012**, *31*, 401–411.
7. Torresia, R.M.; de Torresib, S.I.C.; Gonzaleza, E.R. On the use of the quadratic logistic differential equation for the interpretation of electrointercalation processes. *J. Electroanal. Chem.* **1999**, *461*, 161–166.
8. Pastijn, H. Chaotic growth with the logistic model of P.-F. Verhulst in The Logistic Map and the Route to Chaos. In *Understanding Complex Systems*; Springer: Berlin, Germany, 2006; pp. 3–11.
9. El-Sayed, A.M.A.; El-Mesiry, A.E.M.; El-Saka, H.A.A. On the fractional-order logistic equation. *Appl. Math. Lett.* **2007**, *20*, 817–823.
10. West, B.J. Exact solution to fractional logistic equation. *Phys. A Stat. Mech. Its Appl.* **2015**, *429*, 103–108.
11. Area, I.; Losada, J.; Nieto, J.J. A note on the fractional logistic equation. *Phys. A Stat. Mech. Its Appl.* **2016**, *444*, 182–187.
12. Ortigueira, M.; Bengochea, G. A new look at the fractionalization of the logistic equation. *Phys. A Stat. Mech. Its Appl.* **2017**, *467*, 554–561.
13. D'Ovidio, M.; Loreti, P. Solutions of fractional logistic equations by Euler's numbers. *Phys. A Stat. Mech. Its Appl.* **2018**, *506*, 1081–1092.
14. Bhalekar, S.; Daftardar-Gejji, V. Solving fractional-order logistic equation using a new iterative method. *Int. J. Differ. Equ.* **2012**, *2012*, 975829.
15. Garrappa, R. On linear stability of predictor-corrector algorithms for fractional differential equations. *Int. J. Comput. Math.* **2010**, *87*, 2281–2290.
16. Khader, M.M. Numerical treatment for solving fractional logistic differential equation. *Differ. Equ. Dyn. Syst.* **2016**, *24*, 99–107.
17. Izadi, M. A comparative study of two Legendre-collocation schemes applied to fractional logistic equation. *Int. J. Appl. Comput. Math.* **2020**, *6*, 71.
18. Turalska, M.; West, B.J. A search for a spectral technique to solve nonlinear fractional differential equations. *Chaos Solitons Fractals* **2017**, *102*, 387–395.
19. Yuzbasi, S. A collocation method for numerical solutions of fractional-order logistic population model. *Int. J. Biomath.* **2016**, *9*, 1650031–1650045.
20. Khader, M.M.; Adel, M. Chebyshev wavelet procedure for solving FLDEs. *Acta Appl. Math.* **2018**, *158*, 1–10.
21. Khader, M.M.; Babatin, M.M. On approximate solutions for fractional logistic differential equation. *Math. Probl. Eng.* **2013**, *2013*, 391901.
22. Pitolli, F.; Pezza, L. A Fractional Spline Collocation Method for the Fractional Order Logistic Equation. In *Approximation Theory XV, San Antonio 2016*; Proceedings in Mathematics & Statistics; Fasshauer, G., Schumaker, L., Eds.; Springer: Cham, Switzerland, 2017; Volume 201, pp. 307–318.
23. Baleanu, D.; Shiri, B.; Srivastava, H.M.; Qurashi, M.A. A Chebyshev spectral method based on operational matrix for fractional differential equations involving non-singular Mittag-Leffler kernel. *Adv. Differ. Equ.* **2018**, *2018*, 353.

24. Garrappa, R. Numerical solution of fractional differential equations: A survey and a software tutorial. *Mathematics* **2018**, *6*, 16.

25. Izadi, M. Fractional polynomial approximations to the solution of fractional Riccati equation. *Punjab Univ. J. Math.* **2019**, *51*, 123–141.

26. Abd-Elhameed, W.M.; Youssri, Y.H. A novel operational matrix of Caputo fractional derivatives of Fibonacci polynomials: Spectral solutions of fractional differential equations. *Entropy* **2016**, *18*, 345.

27. Srivastava, H.M. Fractional-order derivatives and integrals: Introductory overview and recent developments. *Kyungpook Math. J.* **2020**, *60*, 73–116.

28. Srivastava, H.M.; Saad, K.M.; Khader, M.M. An efficient spectral collocation method for the dynamic simulation of the fractional epidemiological model of the Ebola virus. *Chaos Solitons Fractals* **2020**, *140*, 110174.

29. Singh, H.; Srivastava, H.M. Jacobi collocation method for the approximate solution of some fractional-order Riccati differential equations with variable coefficients. *Phys. A Statist. Mech. Appl.* **2019**, *523*, 1130–1149.

30. Abd-Elhameed, W.M.; Youssri, Y.H. Explicit shifted second-kind Chebyshev spectral treatment for fractional Riccati differential equation. *Comput. Model. Eng. Sci.* **2019**, *121*, 1029–1049.

31. Srivastava, H.M.; Dubey, V.P.; Kumar, R.; Singh, J.; Kumar, D.; Baleanu, D. An efficient computational approach for a fractional-order biological population model with carrying capacity. *Chaos Solitons Fractals* **2020**, *138*, 109880.

32. Izadi, M. An accurate approximation method for solving fractional order boundary value problems. *Acta Univ. M. Belii Ser. Math.* **2020**, *2020*, 52–67.

33. Izadi, M.; Cattani, C. Generalized Bessel polynomial for multi-order fractional differential equations. *Symmetry* **2020**, *12*, 1260.

34. Srivastava, H.M.; Shah, F.A.; Irfan, M. Generalized wavelet quasi-linearization method for solving population growth model of fractional order. *Math. Methods Appl. Sci.* **2020**, *43*, 8753–8762.

35. Reed, W.H.; Hill, T.R. *Triangular Mesh Methods for the Neutron Transport Equation*; Tech. Report LA-UR-73-479; Los Alamos Scientific Laboratory: Los Alamos, NM, USA, 1973.

36. Delfour, M.; Hager, W.; Trochu, F. Discontinuous Galerkin methods for ordinary differential equations. *Math. Comput.* **1981**, *36*, 455–473.

37. Cockburn, B.; Karniadakis, G.E.; Shu, C.W. (Eds.) *Discontinuous Galerkin Methods: Theory, Computation and Applications, Lecture Notes in Computational Science and Engineering*; Springer: Berlin, Germany, 2000; Volume 11.

38. Deng, W.; Hesthaven, J.S. Local discontinuous Galerkin method for fractional ordinary differential equations. *BIT Numer. Math.* **2015**, *55*, 967–985.

39. Izadi, M. Application of LDG scheme to solve semi-differential equations. *J. Appl. Math. Comput. Mech.* **2019**, *18*, 29–37.

40. Izadi, M.; Negar, M.R. Local discontinuous Galerkin approximations to fractional Bagley-Torvik equation. *Math. Meth. Appl. Sci.* **2020**, *43*, 4798–4813.

41. Izadi, M.; Afshar, M. Solving the Basset equation via Chebyshev collocation and LDG methods. *J. Math. Model.* **2020**. [CrossRef]

42. Christie, I.; Griffths, D.F.; Mitchell, A.R.; Sanz-Serna, J.M. Product approximations for nonlinear problems in finite element methods. *IMA J. Numer. Anal.* **1981**, *1*, 253–266.

43. Diethelm, K.; Freed, A.D. The Frac PECE Subroutine for the Numerical Solution of Differential Equations of Fractional Order. In *Forschung und Wissenschaftliches Rechnen 1998*; Heinzel, S., Plesser, T., Eds.; Gessellschaft fur Wissenschaftliche Datenverarbeitung: Göttingen, Germany, 1999; pp. 57–71.

44. Zhang, Y.; Sun, Z.; Liao, H. Finite difference methods for the time fractional diffusion equation on non-uniform meshes. *J. Comput. Phys.* **2014**, *265*, 195–210.

45. Jiang, S.; Zhang, J.; Zhang, Q.; Zhang, Z. Fast evaluation of the Caputo fractional derivative and its applications to fractional diffusion equations. *Commun. Comput. Phys.* **2017**, *21*, 650–678.

Publisher's Note: MDPI stays neutral with regard to jurisdictional claims in published maps and institutional affiliations.

MDPI

Review

Telegraphic Transport Processes and Their Fractional Generalization: A Review and Some Extensions

Jaume Masoliver

Department of Condensed Matter Physics and Complex Systems Institute (UBICS), University of Barcelona, 08007 Barcelona, Catalonia, Spain; jaume.masoliver@ub.edu

Abstract: We address the problem of telegraphic transport in several dimensions. We review the derivation of two and three dimensional telegrapher's equations—as well as their fractional generalizations—from microscopic random walk models for transport (normal and anomalous). We also present new results on solutions of the higher dimensional fractional equations.

Keywords: telegrapher's equations; fractional telegrapher's equation; continuous time random walk; transport problems

PACS: 02.50.Ey; 05.40.Fb; 05.40.Jc; 05.60.Cd

Citation: Masoliver, J. Telegraphic Transport Processes and Their Fractional Generalization: A Review and Some Extensions. *Entropy* **2021**, 23, 364. https://doi.org/10.3390/e23030364

Academic Editor: Bruce J. West

Received: 22 January 2021
Accepted: 12 March 2021
Published: 18 March 2021

1. Introduction

In many physical situations particle transport through continuous media is described by transport equations which are typically derived from general physical principles as, for instance, the conservation of energy and momentum [1]. Classical cases are provided by the transport of neutron in a reactor or the photon transport in a highly scattering medium [2]. In their most general form transport equations (one of the first and most paradigmatic example is the Boltzmann equation) are nonlinear integro-differential equations often with an incompletely known scattering kernel [1,2]. It is therefore very difficult, not to say impossible, to attain exact analytical solutions of the problem and even obtaining numerical solutions is not an easy task. Moreover, numerical solutions might not reproduce, or even detect, important qualitative characteristics of the transport process [2].

These difficulties have traditionally lead to the search of simpler and easier way to handle approximations. One of the most universal approximation is modeling the transport process by diffusion processes. Such approximation greatly simplifies the description of the transport process because in the absence of any field driving the particle and the usually complicated transport equation is reduced to the much simpler diffusion equation:

$$\frac{\partial p}{\partial t} = D\boldsymbol{\nabla}^2 p, \tag{1}$$

here $p(\mathbf{r}, t)$ is the probability density function (PDF) of the diffusing particle to be at $\mathbf{r}$ at time t and D is the diffusion coefficient.

Diffusion processes have two major characteristics: (i) the mean square deviation grows linearly with time,

$$\langle |\Delta\mathbf{r}(t)|^2 \rangle = Dt, \tag{2}$$

where $\Delta\mathbf{r}(t) = \mathbf{r}(t) - \langle\mathbf{r}(t)\rangle$; (ii) the PDF is Gaussian. Indeed, the solution to Equation (1) assuming the particle is initially at the origin, $p(\mathbf{r}, 0) = \delta(\mathbf{r})$, is

$$p(\mathbf{r}, t) = \frac{1}{(4\pi Dt)^{3/2}} e^{-r^2/4Dt}. \tag{3}$$

81

Despite its simplicity and the wide range of applications in countless areas of physical sciences, the diffusion approximation has, however, several limitations. We will here point out two of them. First, diffusion processes present an infinite velocity of propagation. This can be easily seen from Equation (3) where it is shown that the solution $p(\mathbf{r}, t) > 0$ never vanishes for any finite time and distances $r = |\mathbf{r}|$. There is, nonetheless, a nonzero probability (albeit small) of finding the diffusive particle, at any instant of time, arbitrarily far away from the initial position. In consequence diffusion models allow for arbitrary velocities, even larger than the speed of light in vacuum. This is contrary to the principles of relativity and certainly unsatisfactory from a conceptual point of view [3].

On the other hand, diffusion processes are also unable to account for ballistic motion and are rather useless in describing of early-time effects when ballistic motion may be important as well as near interfaces and in thin samples. This is certainly the case when modeling transport phenomena for which thermalization due to random collisions takes a finite time and the flux of ballistic particles might not be negligible, all of it resulting in anisotropic scattering along the forward direction. A particular but significant case is that of the photon migration through turbid media in which diffusion models are unable to account for ballistic photons and are inaccurate near boundaries [2,4–6]. A similar situation may arise in transport across membranes [7].

Telegraphic processes are a generalized form of diffusion processes in these two aspects. Thus (i) they allow for a finite velocity of propagation and (ii) they are nearly deterministic (i.e., ballistic) at short times while they are diffusive at long times when random collisions have been able to thermalize the motion. As a first approximation, the transport equation for telegraphic processes is the telegrapher's equation (TE):

$$\frac{\partial^2 p}{\partial t^2} + \frac{1}{\tau}\frac{\partial p}{\partial t} = v^2 \boldsymbol{\nabla}^2 p, \tag{4}$$

where $\tau > 0$ is a characteristic time, and $v > 0$ is a characteristic speed. From a mathematical point of view, this is a hyperbolic equation which, as $\tau \to \infty$ with v fixed, becomes the wave equation,

$$\frac{\partial^2 p}{\partial t^2} = v^2 \boldsymbol{\nabla}^2 p, \tag{5}$$

while as $\tau \to 0$ and $v \to \infty$ such that $v^2\tau \to D$ is finite it reduces to the diffusion Equation (1). The telegrapher's equation thus possesses wave and diffusion features describing "diffusion with finite propagation velocity" but also "wave motion with damping" [8]. Moreover, the limits to diffusion and wave equations are also achieved as time progresses. We can thus easily see by scaling time with τ that initially as $t \to 0$ (i.e., $t \ll \tau$), TE approaches to the wave equation while asymptotically as $t \to \infty$ ($t \gg \tau$) moves toward the diffusion equation. As a consequence [2,8]

$$\langle|\Delta\mathbf{r}(t)|^2\rangle \sim t^2, \ (t \to 0) \qquad \text{and} \qquad \langle|\Delta\mathbf{r}(t)|^2\rangle \sim t, \ (t \to \infty),$$

which heightens the duality of the TE and shows the transition from ballistic motion to diffusive motion as time progresses.

The TE appeared in the nineteenth century in the works of Kelvin and Heaviside related to the analysis of transmission of electromagnetic waves in telegraphic wires. In this context, the three dimensional telegrapher's equation can be derived by combining Maxwell's equations for homogeneous media [2,8]. TE can also be phenomenologically derived from thermodynamics by using Cattaneo's equation, a nonlocal generalization of Fick's law accounting for non instantaneous diffusions [9–11], and also from random walk theory where the one-dimensional TE is the master equation of the persistent random walk [12–15].

From a mesoscopic point of view (somewhere between the microscopic view of random walk models and the macroscopic approach of thermodynamics) telegraphic processes are closely related to Brownian motion. As was studied some years ago in Ref. [16], the telegrapher's equation, like the diffusion equation, can also be derived from the Chapman-

Kolmogorov equation, which is the master equation for Markovian processes [17]. It is worth noticing that such a derivation is obtained by retaining quadratic terms in the time expansion of the Chapman-Kolmogorov equation which sets a characteristic time scale and a characteristic velocity. The Markovian character of the process is assured for times greater than the characteristic time while a possible non-Markovian character for smaller times is still an unsettled question [16].

In the context of transport theory, the three-dimensional TE is the so-called P_1 approximation to the full transport equation for which the basic assumption is that the change in the direction of motion due to a single scattering event is small [1,2,18,19]. In a more recent approach [20] a three-dimensional TE model is obtained by a modification of the continuity equation for the probability current. The model is, however, limited to a discrete number of transport directions, which restricts possible applications. Other approaches suppose phenomenological generalizations, where a three dimensional TE is postulated for uniform isotropic media by assuming the same form as the one-dimensional TE, but with numerical corrections in the coefficients which guarantee correct ballistic ($t \to 0$) and diffusive ($t \to \infty$) behaviors in three dimensions [4–6]. The more fundamental and less phenomenological way of describing telegraphic processes is, however, based on random walk models since they try to reproduce the microscopic mechanism of transport.

Random walk models for describing telegraphic processes are modifications of the ordinary random walk because the latter, for long times and large distances (i.e., the so-called "fluid limit" [21]) leads to the diffusion equation but not to the telegrapher's equation [2,8,22]. However, and contrary to one dimension where the TE is readily obtained from the persistent random walk on the line [2,12,14], in higher dimensions obtaining the TE from microscopic models encounters serious difficulties. The main reason lies in the difficulty of generalizing persistence in dimensions greater than one [23–29].

We have recently solved this problem by obtaining the three-dimensional TE [30] and the two-dimensional TE [31] from random walk models (as we had done previously for the one dimensional case [32]). These models consist of a continuous version of two and three dimensional random walks with a continuum of states [33]. I will here review and enlarge these works.

For more than two decades, the so-called "anomalous transport" and "anomalous diffusion" have been the object of intense research with countless applications in many areas of physics, chemistry and natural and socio-economic sciences. There is an immense literature on the subject with many complete reports. As a necessarily short sample we may cite from early reviews in [34–39] to more recent reports [40–42] among many others. It is also worthwhile mentioning a less technical but excellent introduction in [43]. The concept first appeared from the theory of random processes, specifically within continuous time random walks, a powerful technique developed by Montroll and Weiss more than 50 years ago [22,44,45] (see a recent and updated review in Ref. [46]) and it was first applied to diffusion of charge carriers in organic semiconductors by Scher and Montroll in the 1970's [47,48].

Anomalous transport arises in motion through extremely disordered systems such as random media and fractal structures [49] and its most distinctive characteristic is that the mean square deviation follows the asymptotic law [35,36,50]

$$\langle |\Delta \mathbf{r}(t)|^2 \rangle \sim t^\alpha, \tag{6}$$

($t \to \infty$), where $\alpha > 0$ is any positive real number. When $0 < \alpha < 1$ the transport regime is subdiffusive, $\alpha = 1$ corresponds to diffusive transport while $\alpha > 1$ describes superdiffusion. Within the diffusive approximation and in the force-free case, the anomalous transport process is described by a fractional diffusion equation,

$$\frac{\partial^\alpha p}{\partial t^\alpha} = D \boldsymbol{\nabla}^{2\gamma} p \tag{7}$$

$(0 < \alpha \leq 1, 0 < \gamma \leq 1)$, $\partial^\alpha / \partial t^\alpha$ is the fractional Caputo derivative and $\nabla^{2\gamma}$ is the Riesz–Feller fractional Laplacian (see Section 5.2 for a definition of these operators). In the case of particles diffusing under the influence of an external field of force, Equation (7) is replaced by a fractional Fokker–Plank equation [36,38].

The mathematical properties of the solutions to the fractional diffusion Equation (7) have been thoroughly studied and very clearly exposed by Mainardi, Gorenflo and collaborators [51–53]. One of these properties is the scaling relation [21,35,53]

$$p(\mathbf{r}, t) = t^{-\alpha/2\gamma} f\left(\frac{\mathbf{r}}{t^{\alpha/2\gamma}}\right), \tag{8}$$

resulting in the mean square displacement [21]:

$$\langle r^2(t) \rangle = M t^{\alpha/\gamma}. \tag{9}$$

When $\gamma = 1$ but α is not an integer we have the 'time-fractional diffusion''; the case $0 < \alpha < 1$ corresponding to subdiffusion while $\alpha > 1$ to superdiffusion. When $\alpha = 1$ but γ is not integer, the fractional diffusion Equation (7) describes a Levy process, this case is always associated with superdiffusion and it is termed "space-fractional diffusion" [21,38].

As mentioned above, the original motivation for the fractional transport was devised from the continuous time random walk formalism [47,48]. As a result, the derivations of the fractional diffusion equation are mostly based on this formalism, although alternative approaches exist based on master equations or (fractional) Chapman–Kolmogorov expansions [36].

The fractional Equation (7) ignores changes in the dynamics of the diffusing particle as time increases. These changes account for ballistic motion and anisotropic scattering (among others) that are relevant in a number of experimental settings [54]. The TE explains some of these characteristics of transport which imply the transition form ballistic to diffusive motion asymptotically in time.

In a recent work [32] we have presented a derivation of the fractional telegrapher's equation (FTE) in one dimension based on a fractional generalization of the persistent random walk on the line. The continuous multistate model mentioned above allows for a fractional treatment which finally leads to fractional TEs in higher dimensions [30,31].

In this paper we review all these questions and present some new results. The paper is organized as follows. In Section 2 we present the continuous multistate random walk in three dimensions, which in homogeneous and isotropic cases, allows us to derive the three-dimensional telegrapher's equation (Section 3). In Section 4 we adapt the model to two dimensions and derive the corresponding telegrapher's equation. The rest of the paper is devoted to the fractional generalization of these matters. In Section 5 we set the general model for fractional telegraphic transport and obtain the space-time fractional telegrapher's equation in two and three dimensions along with the exact expression for the characteristic function. In Section 6 we study in detail the time-fractional telegrapher's equation, analyze its solution for any dimensionality, and obtain asymptotic results for the probability distribution and the moments of the distance travelled. Concluding remarks are presented in Section 7.

2. Continuous Multistate Random Walk in Three Dimensions

We review the microscopic model introduced in Ref. [30] for the transport of particles in continuous media. The model is based on a generalization of multistate random walks and assumes a continuum in the number of states [33]. In the traditional formulation of multistate random walks (see [15] for a recent review on multistate walks on the line) the walker can be in a discrete (but not necessarily finite) number of internal states. The transition between states is determined by a transition matrix with random Markovian elements. In order to model particle transport we will generalize the multistate random walk in two key features: (i) we assume that the walker (i.e., the particle) moves in three dimensions, and (ii) the model has internal states defined on a continuous set of values.

2.1. General Setting

Suppose a particle moving in the three dimensional space along a straight line determined by the bidimensional quantity $\Omega = (\theta, \varphi)$, where θ is the polar angle and φ is the azimuthal angle. The particular direction along which the particle is moving constitutes the "internal state" and since all possible direction form a continuous and denumerable set, the motion of the particle is thus described by a *continuous multistate random walk*.

At random instants of time the particle shifts direction and, hence, the duration of the motion along a given direction Ω (which is called a *sojourn*) is a random variable determined by a PDF denoted by $\psi(t|\Omega)$. The cumulative distribution

$$\Psi(t|\Omega) = \int_t^\infty \psi(t'|\Omega)dt', \tag{10}$$

gives the probability that the duration of a given sojourn is greater than t.

Let us denote by $h(\mathbf{r}, t|\Omega)$ the joint PDF for the displacement in a single sojourn along direction Ω to be equal to $\mathbf{r}$ and the sojourn duration to equal t. Let us also define $H(\mathbf{r}, t|\Omega)$ as the probability density for the displacement to be $\mathbf{r}$ when the duration is greater than t. Note that the duration PDF $\psi(t|\Omega)$ is the time marginal density of $h(\mathbf{r}, t|\Omega)$,

$$\int_{\mathbb{R}^3} h(\mathbf{r}, t|\Omega)d^3\mathbf{r} = \psi(t|\Omega), \tag{11}$$

while $\Psi(t|\Omega)$ is the marginal probability arising from $H(\mathbf{r}, t|\Omega)$,

$$\int_{\mathbb{R}^3} H(\mathbf{r}, t|\Omega)d^3\mathbf{r} = \Psi(t|\Omega). \tag{12}$$

At the end of a given sojourn, the particle moving along direction Ω' switches to direction Ω. We denote by $\beta(\Omega|\Omega')$ the PDF of this transition $\Omega' \to \Omega$ (note that $\beta(\Omega|\Omega')$ is the "scattering kernel" of the transport problem). In other words, the probability that a single scattering changes the direction of the particle from Ω' to a direction falling somewhere inside the angular region $(\Omega, \Omega + d\Omega)$ is given by

$$\text{Prob}\{\Omega' \to (\Omega, \Omega + d\Omega)\} = \beta(\Omega|\Omega')d^2\Omega, \tag{13}$$

where $d\Omega = (d\theta, d\varphi)$ and

$$d^2\Omega = \sin\theta d\theta d\varphi \tag{14}$$

is the surface element on the sphere of unit radius.

Note that in this model there is a nonvanishing probability of traveling along the same direction and in those cases where this probability is greater than $1/2$ the particle tends to persist in moving along the same direction. In this way the model can be seen as a higher dimensional generalization of the persistent random walk on the line [22].

Let us denote by $p(\mathbf{r}, \Omega, t)$ the joint PDF for the walker to be at $\mathbf{r}$ at time t while moving in direction Ω. Our final objective is, however, to know the density $p(\mathbf{r}, t)$ for the random walker to be at $\mathbf{r}$ at time t regardless the direction. The latter is the marginal density of the former,

$$p(\mathbf{r}, t) = \int p(\mathbf{r}, \Omega, t)d^2\Omega. \tag{15}$$

In order to evaluate $p(\mathbf{r}, \Omega, t)$ we define the auxiliary density $\rho(\mathbf{r}, \Omega, t)$ as

$$\rho(\mathbf{r}, \Omega, t)d^3\mathbf{r}dt = \text{Prob}\{\text{a sojourn in direction } \Omega \text{ ends}$$
$$\text{in the region } (\mathbf{r}, \mathbf{r} + d\mathbf{r}) \text{ at } (t, t + dt)\}.$$

This joint density describes the state of the process at the scattering points where the direction of the particle changes. Thus, if a scattering event happens at time t, it must either be the first one (assuming the initial one occurred at $t = 0$) or else an earlier change of direction $\Omega' \to \Omega$ [governed by $\beta(\Omega|\Omega')$] happened at any earlier time $t' < t$ with the random walker at some position $\mathbf{r}'$. It is not difficult to convince oneself that this renewal argument leads to the following integral equation for the auxiliary density:

$$\rho(\mathbf{r}, \Omega, t) = \beta(\Omega)h(\mathbf{r}, t|\Omega)$$
$$+ \int \beta(\Omega|\Omega')d^2\Omega' \int_0^t dt' \int_{\mathbb{R}^3} h(\mathbf{r} - \mathbf{r}', t - t'|\Omega)\rho(\mathbf{r}', \Omega', t')d^3\mathbf{r}', \quad (16)$$

where $\beta(\Omega)$ is the probability that the process starts moving in direction Ω.

In terms of the auxiliary density $\rho(\mathbf{r}, \Omega, t)$, the PDF $p(\mathbf{r}, \Omega, t)$ for the walker to be at $\mathbf{r}$ at time t while moving in direction Ω is

$$p(\mathbf{r}, \Omega, t) = \beta(\Omega)H(\mathbf{r}, \Omega, t)$$
$$+ \int \beta(\Omega|\Omega')d^2\Omega' \int_0^t dt' \int_{\mathbb{R}^3} H(\mathbf{r} - \mathbf{r}', t - t'|\Omega)\rho(\mathbf{r}', \Omega', t')d^3\mathbf{r}'. \quad (17)$$

The reasoning behind this equation is similar to the one given for obtaining Equation (16). Indeed, the displacement of the walker is either within the first sojourn, this given by βH, or else an earlier change of direction occurred at time $t' < t$ while the walker was at position $\mathbf{r}'$ and the time interval to the next scattering exceeded $t - t'$.

We thus see that in the most general case the solution to the problem, that is to say, obtaining the PDF $p(\mathbf{r}, t)$ (cf. Equation (15)) is given by first solving the integral Equation (16) for the auxiliary function ρ and afterwards substituting this solution into Equation (17) and the result into Equation (15). In the most general case, for arbitrary forms of $\beta(\Omega|\Omega')$, $h(\mathbf{r}, t, |\Omega)$ and $H(\mathbf{r}, t|\Omega)$, obtaining analytical expressions is out of reach, and one has to resort to numerical work.

2.2. Independent Scattering

In order to proceed further we assume that after each scattering the direction is randomized independently of the previous direction of the particle leading to the scattering kernel:

$$\beta(\Omega|\Omega') = \beta(\Omega). \quad (18)$$

The scattering process is thus an independent random process in the change of direction. In the context of fluctuations in laser fields this model corresponds to the so-called Burshtein model [55,56].

When the scattering kernel has the form given by Equation (18), Equations (16) and (17) reduce to

$$\rho(\mathbf{r}, \Omega, t) = \beta(\Omega)\left[h(\mathbf{r}, t|\Omega) + \int_0^t dt' \int_{\mathbb{R}^3} h(\mathbf{r} - \mathbf{r}', t - t'|\Omega)d^3\mathbf{r}' \int \rho(\mathbf{r}', t'|\Omega')d^2\Omega'\right], \quad (19)$$

and

$$p(\mathbf{r}, \Omega, t) = \beta(\Omega)\left[H(\mathbf{r}, t|\Omega) + \int_0^t dt' \int_{\mathbb{R}^3} H(\mathbf{r} - \mathbf{r}', t - t'|\Omega)d^3\mathbf{r}' \int \rho(\mathbf{r}', t'|\Omega')d^2\Omega'\right]. \quad (20)$$

Integrating Equations (19) and (20) with respect to all possible directions Ω, defining the direction-free densities (cf. Equation (15))

$$p(\mathbf{r}, t) = \int p(\mathbf{r}, \Omega, t)d^2\Omega, \qquad \rho(\mathbf{r}, t) = \int \rho(\mathbf{r}, \Omega, t)d^2\Omega, \quad (21)$$

and the averages

$$h(\mathbf{r}, t) = \int \beta(\Omega)h(\mathbf{r}, \Omega, t)d^2\Omega, \qquad H(\mathbf{r}, t) = \int \beta(\Omega)H(\mathbf{r}, \Omega, t)d^2\Omega, \quad (22)$$

we get a simpler integral equation for $\rho(\mathbf{r}, t)$:

$$\rho(\mathbf{r}, t) = h(\mathbf{r}, t) + \int_0^t dt' \int_{\mathbb{R}^3} h(\mathbf{r} - \mathbf{r}', t - t')\rho(\mathbf{r}', t')d^2\mathbf{r}', \tag{23}$$

and the PDF $p(\mathbf{r}, t)$ will be given by

$$p(\mathbf{r}, t) = H(\mathbf{r}, t) + \int_0^t dt' \int_{\mathbb{R}^3} H(\mathbf{r} - \mathbf{r}', t - t')\rho(\mathbf{r}', t')d^2\mathbf{r}'. \tag{24}$$

The problem can now be solved in Fourier–Laplace space. Thus, defining the joint Fourier and Laplace transform,

$$\hat{\bar{\rho}}(\boldsymbol{\omega}, s) = \int_0^\infty e^{-st}dt \int_{\mathbb{R}^3} e^{i\boldsymbol{\omega}\cdot\mathbf{r}}\rho(\mathbf{r}, t)d^3\mathbf{r},$$

the integral Equation (23) turns into a simple algebraic equation for $\hat{\bar{\rho}}$ whose solution can be readily obtained and reads

$$\hat{\bar{\rho}}(\boldsymbol{\omega}, s) = \frac{\hat{\bar{h}}(\boldsymbol{\omega}, s)}{1 - \hat{\bar{h}}(\boldsymbol{\omega}, s)}. \tag{25}$$

On the other hand, by transforming Equation (24) we get

$$\hat{\bar{p}}(\boldsymbol{\omega}, s) = \hat{\bar{H}}(\boldsymbol{\omega}, s)\left[1 + \hat{\bar{\rho}}(\boldsymbol{\omega}, s)\right],$$

which after substituting for (25) yields

$$\hat{\bar{p}}(\boldsymbol{\omega}, s) = \frac{\hat{\bar{H}}(\boldsymbol{\omega}, s)}{1 - \hat{\bar{h}}(\boldsymbol{\omega}, s)}, \tag{26}$$

The form of Equation (26) can be considered a generalization of the Montroll–Weiss equation [44,45] for higher dimensional continuous time random walks with independent directions.

2.3. The Isotropic and Uniform Random Walk

Equation (26) furnishes the formal solution to the transport problem for independent scattering in Fourier–Laplace space and it is valid for any form of the conditional densities $h(\mathbf{r}, t|\boldsymbol{\Omega})$ and $H(\mathbf{r}, t|\boldsymbol{\Omega})$ which describe the displacement inside a given sojourn in direction $\boldsymbol{\Omega}$. In other words, Equation (26) applies to any kind of motion inside a given sojourn and to any distribution of sojourn times. In order to proceed further and solve the problem in a specific way by obtaining the explicit expression for $p(\mathbf{r}, t)$ in real time and space, we first assume that the particle moves in an isotropic medium so that the pausing time density and its cumulative probability are independent of the direction,

$$\psi(t|\boldsymbol{\Omega}) = \psi(t), \qquad \Psi(t|\boldsymbol{\Omega}) = \Psi(t).$$

We next assume that inside any sojourn the motion is uniform with a constant speed c so that after each sojourn the velocity of the particle takes a different direction but with the same modulus and, hence, the kinetic energy is conserved. Despite its simplicity, the model describes the motion of non-interacting particles—such as, for instance, photons—undergoing elastic dispersion with fixed centers randomly distributed. The assumption of uniform motion leads to conclude that the conditional densities for the displacement inside a given sojourn have the form

$$h(\mathbf{r}, t|\boldsymbol{\Omega}) = \delta(\mathbf{r} - ct\mathbf{u})\psi(t), \qquad H(\mathbf{r}, t|\boldsymbol{\Omega}) = \delta(\mathbf{r} - ct\mathbf{u})\Psi(t), \tag{27}$$

where **u** is the unit vector pointing in direction $\mathbf{\Omega} = (\theta, \varphi)$, that is

$$\mathbf{u} = (\sin\theta\cos\varphi, \sin\theta\sin\varphi, \cos\theta). \tag{28}$$

The Fourier transforms of these densities read

$$\tilde{h}(\boldsymbol{\omega}, t|\mathbf{\Omega}) = \psi(t)e^{i(\boldsymbol{\omega}\cdot\mathbf{u})ct}, \qquad \tilde{H}(\boldsymbol{\omega}, t|\mathbf{\Omega}) = \Psi(t)e^{i(\boldsymbol{\omega}\cdot\mathbf{u})ct}. \tag{29}$$

In addition to the assumption that after each collision the new direction of the particle is randomized independently of the previous direction (cf. Equation (18)), we also suppose *complete isotropy* in the sense that all outgoing directions are equally likely. For the three dimensional motion this implies

$$\beta(\mathbf{\Omega}|\mathbf{\Omega}') = \beta(\mathbf{\Omega}) = \frac{1}{4\pi}. \tag{30}$$

The characteristic function of the displacement inside any sojourn independent of the direction is given by the average

$$\tilde{h}(\boldsymbol{\omega}, t) = \int \tilde{h}(\boldsymbol{\omega}, t|\mathbf{\Omega})\beta(\mathbf{\Omega})d^2\mathbf{\Omega}.$$

In the isotropic case and for uniform motion (cf. Equations (14), (29) and (30)) we have

$$\tilde{h}(\boldsymbol{\omega}, t) = \frac{1}{4\pi}\psi(t)\int e^{i(\boldsymbol{\omega}\cdot\mathbf{u})ct}d^2\mathbf{\Omega}.$$

That is,

$$\tilde{h}(\boldsymbol{\omega}, t) = \frac{1}{2}\psi(t)\int_0^\pi e^{i|\boldsymbol{\omega}|ct\cos\theta}\sin\theta d\theta,$$

which after integrating yields

$$\tilde{h}(\boldsymbol{\omega}, t) = \psi(t)\frac{\sin|\boldsymbol{\omega}|ct}{|\boldsymbol{\omega}|ct}. \tag{31}$$

Analogously

$$\tilde{H}(\boldsymbol{\omega}, t) = \Psi(t)\frac{\sin|\boldsymbol{\omega}|ct}{|\boldsymbol{\omega}|ct}. \tag{32}$$

In order to obtain the Fourier–Laplace transform of the PDF of the particle to be at position **r** at time t by means of Equation (26), we have to specify the form of the pausing time density $\psi(t)$. One of the most natural and universal assumptions consists in taking the random instants of time at which the scattering process occurs to be a Poissonian set of events which implies that time intervals inside any sojourn are exponentially distributed [57]. Thus

$$\psi(t) = \lambda e^{-\lambda t} \quad \Rightarrow \quad \Psi(t) = e^{-\lambda t},$$

where λ^{-1} is the average time interval between two consecutive scattering events (i.e., the mean sojourn duration). We have

$$\tilde{h}(\boldsymbol{\omega}, t) = \lambda e^{-\lambda t}\frac{\sin|\boldsymbol{\omega}|ct}{|\boldsymbol{\omega}|ct}, \qquad \tilde{H}(\boldsymbol{\omega}, t) = \frac{1}{\lambda}\tilde{h}(\boldsymbol{\omega}, t).$$

We next take the Laplace transform of these expressions. Recalling that [58]

$$\mathcal{L}\left\{\frac{\sin|\boldsymbol{\omega}|ct}{t}\right\} = \arctan\left(\frac{|\boldsymbol{\omega}|c}{s}\right),$$

and the property $\mathcal{L}\{e^{-\lambda t}f(t)\} = \hat{f}(\lambda + s)$, we get

$$\hat{\hat{h}}(\omega, s) = \frac{\lambda}{|\omega|c} \arctan\left(\frac{|\omega|c}{\lambda + s}\right) \tag{33}$$

and

$$\hat{\hat{H}}(\omega, s) = \frac{1}{|\omega|c} \arctan\left(\frac{|\omega|c}{\lambda + s}\right). \tag{34}$$

Substituting Equations (33) and (34) into Equation (26) we finally get

$$\hat{\hat{p}}(\omega, s) = \frac{\arctan[|\omega|c/(\lambda + s)]}{|\omega|c - \lambda \arctan[|\omega|c/(\lambda + s)]}, \tag{35}$$

which constitutes the exact solution of the homogeneous and isotropic model and the starting point for deriving the three dimensional telegrapher's equation as we will see next. It is worth mentioning that a similar expression was obtained some years ago by Claes and Van den Broeck [59] in the context of modeling the end-to-end distance of polymer chains, although they used a different approach.

3. Telegrapher's Equation

The homogeneous and isotropic random walk reviewed is a microscopic model of particle transport. We can construct the TE from this model by coarse graining the dynamics to the fluid limit approximation.

3.1. Fluid Limit Approximation

The fluid limit approximation consists in rewriting the model for large times and distances [21,51]. Because of Tauberian theorems [60,61], large times and distances, $t \to \infty$ and $|r| \to \infty$, correspond to small Laplace and Fourier variables, $s \to 0$ and $|\omega| \to 0$. Note that to achieve such a limit, i.e., to get an approximate expression for the transformed PDF $\hat{\hat{p}}(\omega, s)$ for small values of s and $|\omega|$, we have two different and equivalent ways of proceeding. We can thus proceed either through the direct expansion of $\hat{\hat{p}}$ given by Equation (35) or else through the expansions of $\hat{\hat{h}}$ and $\hat{\hat{H}}$ (cf. Equations (33) and (34)) as $s \to 0$ and $|\omega| \to 0$ and their subsequent substitution in Equation (26). Obviously both procedures yield the same result but, albeit longer, we follow the second approach since it turns to be instrumental for the fractional generalization of the random walk.

We thus start off with Equation (33) and first perform the long-distance limit ($|\omega| \to 0$) and postpone for a moment the long-time limit ($s \to 0$). As $|\omega| \to 0$ we have the following expansion

$$\begin{aligned} \arctan\left(\frac{|\omega|c}{\lambda + s}\right) &= \frac{|\omega|c}{\lambda + s} - \frac{1}{3}\left(\frac{|\omega|c}{\lambda + s}\right)^3 + O(|\omega|^5) \\ &= \frac{|\omega|c}{(\lambda + s)^3}\left[(\lambda + s)^2 - \frac{1}{3}(|\omega|c)^2 + O(|\omega|^4)\right]. \end{aligned} \tag{36}$$

From Equations (33), (34) and (36) we write

$$\hat{\hat{h}}(\omega, s) = \frac{\lambda}{(\lambda + s)^3}\left[(\lambda + s)^2 - \frac{1}{3}|\omega|^2 c^2 + O(|\omega|^4)\right], \tag{37}$$

and

$$\hat{\hat{H}}(\omega, s) = \frac{1}{(\lambda + s)^3}\left[(\lambda + s)^2 - \frac{1}{3}|\omega|^2 c^2 + O(|\omega|^4)\right]. \tag{38}$$

Hence

$$
\begin{aligned}
1 - \hat{\tilde{h}}(\boldsymbol{\omega}, s) &= 1 - \frac{\lambda}{(\lambda + s)^3}\left[(\lambda + s)^2 - \frac{1}{3}|\boldsymbol{\omega}|^2 c^2 + O(|\boldsymbol{\omega}|^4)\right] \\
&= \frac{1}{(\lambda + s)^3}\left[(\lambda + s)^3 - \lambda(\lambda + s)^2 + \frac{\lambda}{3}|\boldsymbol{\omega}|^2 c^2 + O(|\boldsymbol{\omega}|^4)\right] \\
&= \frac{1}{(\lambda + s)^3}\left[s(\lambda + s)^2 + \frac{\lambda}{3}|\boldsymbol{\omega}|^2 c^2 + O(|\boldsymbol{\omega}|^4)\right],
\end{aligned}
$$

and as $s \to 0$, we may write

$$
1 - \hat{\tilde{h}}(\boldsymbol{\omega}, s) = \frac{1}{(\lambda + s)^3}\left[s(\lambda^2 + 2\lambda s) + \frac{\lambda}{3}|\boldsymbol{\omega}|^2 c^2 + O(s^3, |\boldsymbol{\omega}|^4)\right]. \tag{39}
$$

Substituting Equations (38) and (39) into Equation (26) yields

$$
\hat{\tilde{p}}(\boldsymbol{\omega}, s) = \frac{(\lambda + s)^2 - (c|\boldsymbol{\omega}|)^2/3 + O(|\boldsymbol{\omega}|^4)}{s(\lambda^2 + 2\lambda s) + \lambda(c|\boldsymbol{\omega}|)^2/3 + O(s^3, |\boldsymbol{\omega}|^4)}. \tag{40}
$$

In order to ensure the stability of Equation (40) under Fourier-Laplace inversion [and, hence, for the existence of a valid approximation for $p(\mathbf{r}, t)$], it is necessary that the powers of s and $|\boldsymbol{\omega}|$ which appear in the numerator of Equation (40) be less than the corresponding powers of the denominator [60]. We, therefore, write

$$
\hat{\tilde{p}}(\boldsymbol{\omega}, s) = \frac{\lambda^2 + 2\lambda s + O(s^2, |\boldsymbol{\omega}|^2)}{s(\lambda^2 + 2\lambda s) + \lambda(c|\boldsymbol{\omega}|)^2/3 + O(s^3, |\boldsymbol{\omega}|^4)},
$$

that is,

$$
\hat{\tilde{p}}(\boldsymbol{\omega}, s) = \frac{\lambda/2 + s + O(s^2, |\boldsymbol{\omega}|^2)}{s(\lambda/2 + s) + c^2|\boldsymbol{\omega}|^2/6 + O(s^3, |\boldsymbol{\omega}|^4)},
$$

and take as a fluid limit approximation of the PDF the expression

$$
\hat{\tilde{p}}(\boldsymbol{\omega}, s) = \frac{s + \lambda/2}{s(s + \lambda/2) + c^2|\boldsymbol{\omega}|^2/6}. \tag{41}
$$

3.2. The Three-Dimensional Telegrapher's Equation

Equation (41) is the starting point for deriving the two-dimensional TE. We next obtain the associated partial differential equation for $p(\mathbf{r}, t)$ whose solution, in Fourier–Laplace space and with appropriate initial conditions, is precisely given by Equation (41). To this end we multiply both sides of Equation (41) by the denominator and rewrite the result as

$$
s^2 \hat{\tilde{p}}(\boldsymbol{\omega}, s) - s + \frac{\lambda}{2}[s\hat{\tilde{p}}(\boldsymbol{\omega}, s) - 1] = -\frac{c^2}{6}|\boldsymbol{\omega}|^2 \hat{\tilde{p}}(\boldsymbol{\omega}, s).
$$

We now proceed to Fourier inversion. Taking into account

$$
\mathcal{F}^{-1}\{|\boldsymbol{\omega}|^2 \hat{\tilde{p}}(\boldsymbol{\omega}, s)\} = -\boldsymbol{\nabla}^2 \hat{p}(\mathbf{r}, s), \qquad \mathcal{F}^{-1}\{1\} = \delta(\mathbf{r}),
$$

the Fourier inversion yields

$$
s^2 \hat{p}(\mathbf{r}, s) - s\delta(\mathbf{r}) + \frac{\lambda}{2}[s\hat{p}(\mathbf{r}, s) - \delta(\mathbf{r})] = \frac{c^2}{6}\boldsymbol{\nabla}^2 \hat{p}(\mathbf{r}, s).
$$

Let us next address Laplace inversion. With the standard initial conditions [62]

$$
p(\mathbf{r}, 0) = \delta(\mathbf{r}), \qquad \left.\frac{\partial p(\mathbf{r}, t)}{\partial t}\right|_{t=0} = 0, \tag{42}
$$

and the Laplace inversion formulas [58]

$$\mathcal{L}^{-1}\left\{s^2\hat{p}(\mathbf{r},s) - s\delta(\mathbf{r})\right\} = \frac{\partial^2 p(\mathbf{r},t)}{\partial t^2},$$

$$\mathcal{L}^{-1}\left\{s\hat{p}(\mathbf{r},s) - \delta(\mathbf{r})\right\} = \frac{\partial p(\mathbf{r},t)}{\partial t},$$

we find that $p(\mathbf{r},t)$ satisfies the three-dimensional TE

$$\frac{\partial^2 p}{\partial t^2} + \frac{1}{\tau}\frac{\partial p}{\partial t} = v^2\boldsymbol{\nabla}^2 p, \tag{43}$$

with

$$\tau = 1/(2\lambda) \quad \text{and} \quad v = c/\sqrt{6}, \tag{44}$$

as characteristic time and velocity respectively.

TE (43) enjoys both wave and diffusion characteristics. This duality becomes even more apparent as time progresses. Thus, as $t \to 0$ Equation (43) reduces to the wave equation while as $t \to \infty$ it goes to the diffusion equation. Indeed, scaling time with τ one can easily see that [15,30]

$$\frac{\partial^2 p}{\partial t^2} \simeq v^2\boldsymbol{\nabla}^2 p \quad (t \to 0), \qquad \frac{\partial p}{\partial t} \simeq D\boldsymbol{\nabla}^2 p \quad (t \to \infty)$$

$(D = v^2\tau)$ which leads to

$$\langle|\mathbf{r}(t)|^2\rangle \sim t^2 \quad (t \to 0), \qquad \langle|\mathbf{r}(t)|^2\rangle \sim t \quad (t \to \infty),$$

showing the transition from ballistic motion to diffusive motion as time increases.

4. The Two Dimensional Case

Up to this point we have developed the telegraphic approximation to transport in three dimensions. We will now briefly report on how to treat the problem in lower dimensions.

In one dimension the standard derivation of the TE is based on the persistent random walk on the line [14]. In this model there is only one possible direction and the walker has two possible states since it can move either to the left or to the right with equal probability which is the isotropic case for the one dimensional motion. We do not present the details for the one-dimensional case here, but instead refer the interested reader to Ref. [15] for a recent and rather complete report. Note that the TE obtained is

$$\frac{\partial^2 p}{\partial t^2} + \frac{1}{\tau}\frac{\partial p}{\partial t} = v^2\frac{\partial^2 p}{\partial x^2}, \tag{45}$$

where in this case $v = c$ coincides with the velocity of the moving particle and, as in three dimensions, $\tau = (2\lambda)^{-1}$ (recall that λ^{-1} is the mean sojourn time when switching times are Poissonian).

4.1. General Model

The two-dimensional case has been recently developed in Ref. [31]. This microscopic model for transport in planar media has many similarities (but some particular differences) with the three dimensional model presented above. Let us note that now the direction of the particle is not given by the solid angle $\boldsymbol{\Omega} = (\theta, \varphi)$ but by the planar angle φ. Therefore, the equations for the continuous multistate random walk in two dimensions will be the same as those in three dimensions (cf. Section 2) with the replacements

$$\boldsymbol{\Omega} \longrightarrow \varphi, \qquad \int d^2\boldsymbol{\Omega} \longrightarrow \int_0^{2\pi} d\varphi, \qquad \int_{\mathbb{R}^3} d^3\mathbf{r} \longrightarrow \int_{\mathbb{R}^2} d^2\mathbf{r}. \tag{46}$$

Thus, in the most general case the bidimensional model will be described by Equations (16) and (17) with these replacements in which the change of direction is governed by the transition density

$$\beta(\varphi|\varphi')d\varphi = \text{Prob}\{\varphi' \to \varphi + d\varphi\},$$

with similar definitions as those of the three dimensional walk for the densities $\psi(t|\varphi)$, $\Psi(t|\varphi)$, $h(\mathbf{r},t|\varphi)$ and $H(\mathbf{r},t|\varphi)$ (cf. Section 2.1).

For the case of independent scattering (Section 2.2)

$$\beta(\varphi|\varphi') = \beta(\varphi).$$

As in three dimensions, we can now also define direction-free densities by means of Equations (21) and (22) through replacements (46). Finally, the Fourier–Laplace transform of the PDF,

$$\hat{p}(\omega,s) = \int_0^\infty e^{-st} dt \int_{\mathbb{R}^2} e^{i\omega\cdot\mathbf{r}} p(\mathbf{r},t) d^2\mathbf{r},$$

is explicitly given by the generalization of Montroll–Weiss Equation (26),

$$\hat{p}(\omega,s) = \frac{\hat{\tilde{H}}(\omega,s)}{1 - \hat{\tilde{h}}(\omega,s)}, \tag{47}$$

where $\hat{\tilde{h}}(\omega,s)$ is given by the average over all possible directions φ (cf. Equation (22))

$$\hat{\tilde{h}}(\omega,s) = \int_0^{2\pi} \beta(\varphi)\hat{\tilde{h}}(\omega,s|\varphi)d\varphi,$$

and a similar expression for $\hat{\tilde{H}}(\omega,s)$.

4.2. The Isotropic and Uniform Case

In an isotropic medium (cf. Section 2.3) the pausing time densities are independent of the direction taken by the particle, $\psi(t|\varphi) = \psi(t)$ and $\Psi(t|\varphi) = \Psi(t)$, and for uniform motion we have [cf. Equations (27)–(29)]

$$h(\mathbf{r},t|\varphi) = \delta(\mathbf{r} - ct\mathbf{u})\psi(t), \qquad H(\mathbf{r},t|\varphi) = \delta(\mathbf{r} - ct\mathbf{u})\Psi(t), \tag{48}$$

and the Fourier transforms are

$$\tilde{h}(\omega,t|\varphi) = \psi(t)e^{i(\omega\cdot\mathbf{u})ct}, \qquad \tilde{H}(\omega,t|\varphi) = \Psi(t)e^{i(\omega\cdot\mathbf{u})ct}, \tag{49}$$

where $\mathbf{u}$ is the unit vector pointing in direction φ,

$$\mathbf{u} = (\cos\varphi, \sin\varphi).$$

Assuming that all directions are equally likely (i.e., complete isotropy), we have

$$\beta(\varphi) = \frac{1}{2\pi}$$

and

$$
\begin{aligned}
\tilde{h}(\omega,t) &= \int_0^{2\pi} b(\varphi)\tilde{h}(\omega,t|\varphi)d\varphi = \frac{\psi(t)}{2\pi}\int_0^{2\pi} e^{ict|\omega|\cdot\mathbf{u}}d\varphi \\
&= \frac{\psi(t)}{2\pi}\int_0^{2\pi} e^{ict|\omega|\cos\varphi}d\varphi = \frac{\psi(t)}{\pi}\int_0^{\pi} \cos(ct|\omega|\cos\varphi)d\varphi.
\end{aligned}
$$

From the integral representation of the Bessel function $J_0(z)$ [63],

$$J_0(ct|\boldsymbol{\omega}|) = \frac{1}{\pi} \int_0^\pi \cos(ct|\boldsymbol{\omega}|\cos\varphi)d\varphi, \tag{50}$$

we get

$$\tilde{h}(\boldsymbol{\omega},t) = \psi(t)J_0(ct|\boldsymbol{\omega}|), \tag{51}$$

and analogously

$$\tilde{h}(\boldsymbol{\omega},t) = \Psi(t)J_0(ct|\boldsymbol{\omega}|). \tag{52}$$

For exponentially distributed sojourn intervals $\psi(t) = \lambda e^{-\lambda t}$ and $\Psi(t) = e^{-\lambda t}$, we write

$$\tilde{h}(\boldsymbol{\omega},t) = \lambda e^{-\lambda t}J_0(ct|\boldsymbol{\omega}|), \qquad \tilde{H}(\boldsymbol{\omega},t) = \frac{1}{\lambda}\tilde{h}(\boldsymbol{\omega},t).$$

Using the Laplace transformation formula [58]

$$\mathcal{L}\{J_0(ct|\boldsymbol{\omega}|)\} = \frac{1}{\sqrt{s^2 + c^2t^2|\boldsymbol{\omega}|^2}},$$

and the standard property

$$\mathcal{L}\{e^{-\lambda t}f(t)\} = \hat{f}(\lambda + s),$$

we get

$$\hat{\tilde{h}}(\boldsymbol{\omega},s) = \frac{\lambda}{\sqrt{(\lambda+s)^2 + c^2|\boldsymbol{\omega}|^2}}$$

and $\hat{\tilde{h}}(\boldsymbol{\omega},s) = \hat{\tilde{h}}(\boldsymbol{\omega},s)/\lambda$. Finally, from the Montroll–Weiss Equation (47) we obtain the exact solution to the homogeneous and isotropic random walk on the plane,

$$\hat{p}(\boldsymbol{\omega},s) = \frac{1}{\sqrt{(\lambda+s)^2 + c^2|\boldsymbol{\omega}|^2} - \lambda}. \tag{53}$$

Notice the completely different form for the exact PDF of the planar model compared to that of the three dimensional case given by Equation (35).

4.3. Fluid Limit Approximation and Telegrapher's Equation

As we have done in the three dimensional transport, in order to get the two-dimensional equation we first make the fluid limit approximation of the planar model, that is, the long-distance and long-time limits of the exact PDF (53). By mimicking the steps done in Section 3.1 to obtain the fluid limit approximation in the three dimensional case, we can easily see that in two dimensions we obtain the same result but with the replacement

$$c^2/6 \longrightarrow c^2/4. \tag{54}$$

Thus, the approximation for the PDF reads (cf. Equation (41))

$$\hat{p}(\boldsymbol{\omega},s) = \frac{s + \lambda/2}{s(s+\lambda/2) + c^2|\boldsymbol{\omega}|^2/4}, \tag{55}$$

and similar expressions for the quantities $\hat{h}$ and $\hat{H}$. Assuming the initial conditions given in Equation (42), inverting Equation (55) and following the same procedure as in the three dimensional case we finally obtain the two-dimensional TE,

$$\frac{\partial^2 p}{\partial t^2} + \frac{1}{\tau}\frac{\partial p}{\partial t} = v^2\boldsymbol{\nabla}^2 p, \tag{56}$$

where as in three dimensions $\tau = (2\lambda)^{-1}$ but now

$$v = c/2.$$

Before proceeding further and explaining the fractional generalizations of telegraphic transport, let us point the significant issue of boundary conditions which are instrumental in first-passage, escape and survival problems. The question is far from being simple specially for telegraphic processes and in one and higher dimensions it has, to my knowledge, not being settled yet. In the transport of particles the problem of survival is closely related to the question of when the particle is absorbed (and, hence, disappears) if it reaches a certain critical region of boundary S_c. For diffusion processes, absorption at S_c corresponds to $p(\mathbf{r}, t|\mathbf{r}_0) = 0$ when $\mathbf{r} \in S_c$ (or $p(\mathbf{r}, t|\mathbf{r}_0) = 0$ when $\mathbf{r}_0 \in S_c$). That is, if the particle reaches S_c (or starts at S_c) disappears. For telegraphic processes (and in the context of particle transport, at least for one-dimensional processes) the situation is more complex because of the property of persistence inherent in the telegrapher's equation [14]. In this context persistence, which is analogous to the physical property of momentum, makes necessary, in deriving boundary conditions for absorption, to take into account the direction in which the particle is traveling. For if the particle starts at S_c, or at time t reaches S_c, will disappear (that is, it will be absorbed) only if the direction of the velocity is the appropriate one, otherwise the particle will escape.

For one dimensional processes we studied this situation some years ago [64,65] and refer the interested reader to these works for more information. In higher dimensions the situation may be even more involved. There are, however, problems which are not related to the escape out of some region (which implies absorption at the boundary of the region) but only on the first arrival to some region S_c. It can be shown that in these cases the boundary condition is $p(\mathbf{r}, t|\mathbf{r}_0) = 0$ (if $\mathbf{r} \in S_c$ or $\mathbf{r}_0 \in S_c$), regardless the direction of the velocity at this particular location (see [66] for a problem of this sort in one dimension).

5. Fractional Transport

Likewise the one dimensional case, the two and three dimensional telegraphic transport processes described above are ordinary (i.e., non-fractional) processes in the sense that for small time ($t \ll \tau$) they behave like an ordinary wave front while for long times ($t \gg \tau$) they act like an ordinary diffusion processes,

$$\langle |\mathbf{r}(t)|^2 \rangle \sim t^2 \quad (t \to 0), \qquad \langle |\mathbf{r}(t)|^2 \rangle \sim t \quad (t \to \infty).$$

However, in transport through highly disordered systems as, for instance, random media or fractal structures, ordinary diffusion becomes anomalous, that is

$$\langle |\mathbf{r}(t)|^2 \rangle \sim t^\alpha,$$

where α is any positive real number. Two questions arise: (i) How does this circumstance affect telegraphic transport? and (ii) what is the fractional TE ruling such kind of processes? In one dimension we addressed these questions by setting a fractional version of the persistent random walk on the line [32]. In higher dimensions we followed this path and generalized the continuous and isotropic walks described in previous sections [30,31]. Let us review these findings.

5.1. The Fractional Isotropic Walk

We will first work on the three dimensional case and obtain a fractional version of the homogenous and isotropic random walk in the fluid limit approximation. To this end we generalize the expressions of $\hat{\tilde{h}}(\omega, s)$ and $\hat{\tilde{H}}(\omega, s)$—given in Section 3.1 in the fluid limit approximation—to include fractional behavior.

Let us start with Equation (39) which when $s \to 0$ yields

$$1 - \hat{\tilde{h}}(\omega, s) = \frac{1}{(\lambda + s)^3}\left[\lambda^2 s + 2\lambda s + \frac{\lambda}{3}|\omega|^2 c^2 + O(s^3, |\omega|^4)\right],$$

we further approximate the denominator by $(\lambda + s)^3 = \lambda^3 + O(s)$, so that

$$1 - \hat{\tilde{h}}(\omega, s) \simeq \frac{s}{\lambda} + 2\left(\frac{\lambda}{2}\right)^2 + \frac{1}{3\lambda^2}|\omega|^2 c^2 \cdots.$$

We thus take as a fluid limit approximation for the sojourn density $\hat{\tilde{h}}$ the expression

$$\hat{\tilde{h}}(\omega, s) \simeq 1 - \frac{s}{\lambda} - 2\left(\frac{s}{\lambda}\right)^2 - \frac{1}{3\lambda^2}|\omega|^2 c^2 \cdots \tag{57}$$

We next obtain an appropriate fluid limit approximation for the sojourn probability $\hat{\tilde{H}}$. Thus starting from Equation (34) and following the same approximation scheme we get

$$\begin{aligned}
\hat{\tilde{H}}(\omega, s) &= \frac{1}{(\lambda + s)^3}\left[(\lambda + s)^2 - \frac{1}{3}|\omega|^2 c^2 + O(|\omega|^4)\right] \\
&= \frac{1}{(\lambda + s)^3}\left[\lambda^2 + 2\lambda s - \frac{1}{3}|\omega|^2 c^2 + O(s^2, |\omega|^4)\right]. \\
&\simeq \frac{1}{\lambda^3}\left(\lambda^2 + 2\lambda s\right) \cdots,
\end{aligned}$$

That is,

$$\hat{\tilde{H}}(\omega, s) = \frac{1}{\lambda}\left(1 + \frac{2s}{\lambda}\right) \cdots \tag{58}$$

Let us incidentally note that substituting Equations (57) and (58) into Montroll–Weiss Equation (26) yields the fluid limit solution (41) for the PDF $\hat{\tilde{p}}(\omega, s)$ which has been the starting point of the derivation of the TE.

We are now ready to construct a fractional generalization of the three-dimensional isotropic random walk. Thus, and looking at Equation (57), we propose the following expansion for the sojourn density in the fluid limit:

$$\hat{\tilde{h}}(\omega, s) = 1 - (Ts)^\alpha - 2(Ts)^{2\alpha} - \frac{1}{3}(L|\omega|)^{2\gamma} \cdots \tag{59}$$

$(s, |\omega| \to 0)$, where $0 < \alpha \le 1, 0 < \gamma \le 1$ and $T > 0$ and $L > 0$ are arbitrary parameters, T defines a characteristic time and L a characteristic length.

In addition to the fractional approximation for $\hat{\tilde{h}}(\omega, s)$ we also assume a fractional expansion for the function $\hat{\tilde{H}}(\omega, s)$ consistent with Equation (59). To this end, we return to Section 2 and average Equations (11) and (12) over all directions Ω, with the result (in Laplace space)

$$\int_{\mathbb{R}^3} \hat{h}(\mathbf{r}, s)d^3\mathbf{r} = \hat{\psi}(s), \qquad \int_{\mathbb{R}^3} \hat{H}(\mathbf{r}, s)d^3\mathbf{r} = \hat{\Psi}(s), \tag{60}$$

where the sojourn PDF's independent of direction are

$$\hat{h}(\mathbf{r}, s) = \int \hat{h}(\mathbf{r}, s|\Omega)\beta(\Omega)d^2\Omega, \quad \text{and} \quad \hat{\psi}(s) = \int \hat{\psi}(s|\Omega)\beta(\Omega)d^2\Omega,$$

and similar expressions for $\hat{H}(\mathbf{r}, s)$ and $\hat{\Psi}(s)$. Note that in terms of the Fourier transform we may write

$$\hat{\tilde{h}}(\omega = 0, s) = \hat{\psi}(s), \quad \text{and} \quad \hat{\tilde{H}}(\omega = 0, s) = \hat{\Psi}(s).$$

However, Laplace transforming Equation (10) we see that $\hat{\Psi}(s) = [1 - \hat{\psi}(s)]/s$, consequently,

$$\hat{H}(\omega = 0, s) = \frac{1}{s}[1 - \hat{h}(\omega = 0, s)].$$

Inserting Equation (59) into this expression yields

$$\hat{H}(\omega = 0, s) = T^{\alpha}s^{\alpha-1} + 2T^{2\alpha}s^{2\alpha-1},$$

which leads us to conjecture the following fluid limit approximation:

$$\hat{H}(\omega, s) \simeq T^{\alpha}s^{\alpha-1} + 2T^{2\alpha}s^{2\alpha-1}\cdots \tag{61}$$

$(s \to 0, |\omega| \to 0)$. Let us stress that this is simply a conjecture because the approximation given by Equation (61) might have depended on $|\omega|$ as well [32].

Substituting Equations (59) and (61) into Montroll–Weiss Equation (26) and reorganizing terms yields

$$\hat{p}(\omega, s) = \frac{2T^{2\alpha}s^{\alpha-1}[s^{\alpha} + 1/2T^{\alpha}]}{2T^{2\alpha}[s^{2\alpha} + s^{\alpha}/2T^{\alpha} + |\omega|^{2\gamma}(L^{2\gamma}/6T^{2\alpha})]},$$

that is,

$$\hat{p}(\omega, s) = \frac{s^{\alpha-1}(s^{\alpha} + 1/\tau)}{s^{2\alpha} + s^{\alpha}/\tau + v^2|\omega|^{2\gamma}}, \tag{62}$$

where

$$\tau = 2T^{\alpha}, \qquad v = \frac{1}{\sqrt{6}}(L^{\gamma}/T^{\alpha}), \tag{63}$$

$(0 < \alpha \le 1, 0 < \gamma \le 1)$. The parameters τ and v can be considered as a fractional time and a fractional characteristic velocity, respectively.

5.2. Fractional Telegrapher's Equation in Three Dimensions

To derive the fractional telegrapher's equation (FTE) in three dimensions for the fractional isotropic model we first need to introduce some mathematical formalism concerning fractional derivatives.

The Caputo fractional derivative of order $\beta > 0$ of a function $\phi(t)$ is defined by the functional [21,52,53,67,68]

$$\frac{\partial^{\beta}\phi(t)}{\partial t^{\beta}} = \begin{cases} \dfrac{1}{\Gamma(n - \beta)}\displaystyle\int_0^t \frac{\phi^{(n)}(t')dt'}{(t - t')^{1+\beta-n}}, & n - 1 < \beta < n, \\ \phi^{(n)}(t), & \beta = n, \end{cases} \tag{64}$$

$(n = 1, 2, 3, \dots)$. Using this definition we can readily obtain the Laplace transform of the Caputo derivative. Indeed, Laplace transforming Equation (64) and using the convolution theorem we obtain

$$\mathcal{L}\left\{\frac{\partial^{\beta}\phi(t)}{\partial t^{\beta}}\right\} = \frac{1}{\Gamma(n - \beta)}\mathcal{L}\left\{\phi^{(n)}(t)\right\}\mathcal{L}\left\{t^{n-\beta-1}\right\},$$

where $\mathcal{L}\{\cdot\}$ stands for the Laplace transform. With the explicit forms [58]

$$\mathcal{L}\left\{\phi^{(n)}(t)\right\} = s^n\hat{\phi}(s) - \sum_{k=0}^{n-1} s^{n-1-k}\phi^{(k)}(0),$$

and

$$\mathcal{L}\left\{t^{n-\beta-1}\right\} = \Gamma(n - \beta)s^{\beta-n},$$

the Laplace transform of the Caputo derivative is found to be

$$\mathcal{L}\left\{\frac{\partial^{\beta}\phi(t)}{\partial t^{\beta}}\right\} = s^{\beta}\hat{\phi}(s) - s^{\beta-1}\phi(0) - \sum_{k=1}^{n-1} s^{\beta-1-k}\phi^{(k)}(0). \tag{65}$$

The second kind of fractional operator we need is the Riesz–Feller fractional Laplacian of order β ($0 < \beta \le 2$) of a function $g(x)$ vanishing at $x \to \pm\infty$. There are several equivalent ways to define it [68], although one of the simplest and most operative definitions is obtained using Fourier analysis. We thus define [21]:

$$\nabla^{\beta}g(\mathbf{r}) = \mathcal{F}^{-1}\left\{-|\omega|^{\beta}\tilde{g}(\omega)\right\}, \tag{66}$$

($0 < \beta \le 2$), where $\mathcal{F}^{-1}\{\cdot\}$ stands for the inverse Fourier transform, and

$$\tilde{g}(\omega) = \int_{\mathbb{R}^3} e^{i\omega\cdot\mathbf{r}}g(\mathbf{r})d^3\mathbf{r},$$

is the direct transform.

We are now ready to derive the three-dimensional FTE. We begin with Equation (62) which we rewrite as

$$\left(s^{2\alpha} + \frac{1}{\tau}s^{\alpha} + v^2|\omega|^{2\gamma}\right)\hat{p}(\omega,s) = s^{2\alpha-1} + \frac{1}{\tau}s^{\alpha-1}.$$

Taking into account the definition of the Riesz–Feller Laplacian, Equation (66), and recalling that $\mathcal{F}^{-1}\{1\} = \delta(\mathbf{r})$, the Fourier inversion yields

$$\left(s^{2\alpha} + \frac{1}{\tau}s^{\alpha} - v^2\nabla^{2\gamma}\right)\hat{p}(\mathbf{r},s) = \left(s^{2\alpha-1} + \frac{1}{\tau}s^{\alpha-1}\right)\delta(\mathbf{r}),$$

and, after reorganizing terms, we have

$$s^{2\alpha}\hat{p}(\mathbf{r},s) - s^{2\alpha-1}\delta(\mathbf{r}) + \frac{1}{\tau}\left[s^{\alpha}\hat{p}(\mathbf{r},s) - s^{\alpha-1}\delta(\mathbf{r})\right] = v^2\nabla^{2\gamma}\hat{p}. \tag{67}$$

In order to Laplace invert this equation, and thus obtaining an equation for $p(\mathbf{r},t)$, we first evaluate the Laplace transforms of the fractional derivatives $\partial^{\alpha}p/\partial^{\alpha}t$ and $\partial^{2\alpha}p/\partial^{2\alpha}t$ using Equation (65). We must distinguish the cases $\beta = \alpha$ and $\beta = 2\alpha$.

(i) Set $\beta = \alpha$ in Equation (65). Since $0 < \alpha \le 1$, we see that $n = 1$. Hence

$$\mathcal{L}\left\{\frac{\partial^{\alpha}p(\mathbf{r},t)}{\partial t^{\alpha}}\right\} = s^{\alpha}\hat{p}(\mathbf{r},s) - s^{\alpha-1}p(\mathbf{r},0).$$

However, $p(\mathbf{r},0) = \delta(\mathbf{r})$ (cf. Equation (42)). Therefore

$$\frac{\partial^{\alpha}p(\mathbf{r},t)}{\partial t^{\alpha}} = \mathcal{L}^{-1}\left\{s^{\alpha}\hat{p}(\mathbf{r},s) - s^{\alpha-1}\delta(\mathbf{r})\right\}. \tag{68}$$

(ii) When $\beta = 2\alpha$ ($0 < \alpha \le 1$) we need to distinguish the cases (a) $0 < \alpha \le 1/2$ and (b) $1/2 < \alpha \le 1$. For case (a) we have $0 < 2\alpha \le 1$, which reproduces the conditions leading to Equation (68), That is,

$$\mathcal{L}\left\{\frac{\partial^{2\alpha}p(\mathbf{r},t)}{\partial t^{2\alpha}}\right\} = s^{2\alpha}\hat{p}(\mathbf{r},s) - s^{2\alpha-1}\delta(\mathbf{r}).$$

In case (b) we have $1 < 2\alpha \le 2$ and from Equation (65) with $n = 2$ we write

$$\mathcal{L}\left\{\frac{\partial^{2\alpha}p(\mathbf{r},t)}{\partial t^{2\alpha}}\right\} = s^{2\alpha}\hat{p}(\mathbf{r},s) - s^{2\alpha-1}\delta(\mathbf{r}) - s^{2(\alpha-1)}\left.\frac{\partial p(\mathbf{r},t)}{\partial t}\right|_{t=0}.$$

Since $\partial p/\partial t|_{t=0} = 0$ (cf. Equation (42)) we see that this case coincides with case (a) above. Therefore,

$$\frac{\partial^{2\alpha} p}{\partial t^{2\alpha}} = \mathcal{L}^{-1}\left\{s^{2\alpha}\hat{p}(\mathbf{r},s) - s^{2\alpha-1}\delta(\mathbf{r})\right\}, \tag{69}$$

$(0 < \alpha \leq 1)$. Returning to Equation (67) and taking the inverse transform we find

$$\mathcal{L}^{-1}\left\{s^{2\alpha}\hat{p}(\mathbf{r},s) - s^{2\alpha-1}\delta(\mathbf{r})\right\} + 2\lambda\mathcal{L}^{-1}\left\{s^{\alpha}\hat{p}(\mathbf{r},s) - s^{\alpha-1}\delta(\mathbf{r})\right\} = v^2\boldsymbol{\nabla}^{2\gamma} p.$$

Using Equations (68) and (69), we readily obtain

$$\frac{\partial^{2\alpha} p}{\partial t^{2\alpha}} + \frac{1}{\tau}\frac{\partial^{\alpha} p}{\partial t^{\alpha}} = v^2\boldsymbol{\nabla}^{2\gamma} p, \tag{70}$$

which is the fractional telegrapher's equation in three dimensions where τ is the fractional time and v the fractional velocity (cf. Equation (63)).

As is well known, and as we have remarked in previous sections, the ordinary TE enjoys both wave and diffusion characteristics. We now extend this duality to the fractional TE. To this end we take the limit $\tau \to 0$ in Equation (70) and also letting $v \to \infty$ such that $\tau v^2 \to D$ finite. This results in the fractional diffusion equation (cf. Equation (7))

$$\frac{\partial^{\alpha} p}{\partial t^{\alpha}} = D\boldsymbol{\nabla}^{2\gamma} p. \tag{71}$$

Let us see that for any values of τ and v the fractional diffusion equation is also the asymptotic (in time) limit of the fractional TE (recall that a similar situation occurs with the ordinary TE). Indeed, by passing to the limit $s \to 0$ in the fluid-limit expression of the PDF (cf. Equation (62)) the small s approximation for $\hat{p}(\boldsymbol{\omega},s)$ is readily found to be

$$\hat{p}(\boldsymbol{\omega},s) \simeq \frac{s^{\alpha-1}}{s^{\alpha} + (\tau v^2)|\boldsymbol{\omega}|^{2\gamma}},$$

which after Fourier-Laplace inversion yields Equation (71) with $D = \tau v^2$. Therefore, by virtue of Tauberian theorems the fractional diffusion Equation (71) is the long-time approximation of the fractional TE.

The fractional TE also contains the fractional wave equation as a special case. Thus letting $\tau \to \infty$ with v finite in Equation (70) we get

$$\frac{\partial^{2\alpha} p}{\partial t^{2\alpha}} = v^2\boldsymbol{\nabla}^{2\gamma} p. \tag{72}$$

Note that when $\alpha = 1/2$ and $\gamma = 1$ this equation reduces to the ordinary diffusion equation. In this regard Mainardi's terminology "fractional diffusion-wave equation" [52] is more precise than "fractional wave equation". Let us finally observe that the fractional diffusion-wave equation is the small-time limit of the fractional TE. Indeed, the limit $s \to \infty$ in Equation (62) yields

$$\hat{p}(\boldsymbol{\omega},s) \simeq \frac{s^{2\alpha-1}}{s^{2\alpha} + v^2|\boldsymbol{\omega}|^{2\gamma}},$$

and the Fourier–Laplace inversion results in Equation (72). Again, due to Tauberian theorems we see that the fractional diffusion-wave equation is the short-time limit of the fractional TE.

We thus see from the preceding discussion that the fractional TE embraces two different dynamics. one of them, at small times, representing fractional wavelike behavior, and another one which at long times enhances fractional diffusion-like behavior. This constitutes the fractional generalization of the dual character between waves and diffusions showed by the ordinary TE.

5.3. Lower Dimensional Cases

We next address lower dimensional problems and will see that in one and two dimensions the fractional TE has formally the same form than in three dimensions.

5.3.1. One Dimension

As we know the one-dimensional fractional case is based the fractional generalization of the continuous-time persistent random walk on the line which is a discrete two-state model and whose derivation from the continuous multistate model described above, although similar in many aspects, it is not straightforward. We will here state just the main result and refer the interested reader to [32] or the review [15] for details. Thus, the one-dimensional fractional TE has formally the same appearance that the three-dimensional Equation (70)

$$\frac{\partial^{2\alpha} p}{\partial t^{2\alpha}} + \frac{1}{\tau}\frac{\partial^\alpha p}{\partial t^\alpha} = v^2 \frac{\partial^{2\gamma} p}{\partial x^{2\gamma}}, \tag{73}$$

where the Caputo derivatives with respect to time are equal to those of the three dimensional Equation (70) and

$$\frac{\partial^{2\gamma} p}{\partial x^{2\gamma}} = \mathcal{F}^{-1}\left\{-|\omega|^{2\gamma}\tilde{p}(\omega,t)\right\},$$

is the Riesz–Feller fractional derivative (cf. Equation (66)), where

$$\tilde{p}(\omega,t) = \int_{-\infty}^{\infty} e^{i\omega x} p(x,t)dx,$$

is the Fourier transform of the one-dimensional PDF $p(x,t)$ and $\mathcal{F}^{-1}\{\cdot\}$ is the inverse transform.

As the reader can easily check, the solution of Equation (73) with the standard initial conditions:

$$p(x,0) = \delta(x), \qquad \left.\frac{\partial p(x,t)}{\partial t}\right|_{t=0} = 0,$$

in Fourier–Laplace space reads

$$\hat{\tilde{p}}(\omega,s) = \frac{s^{\alpha-1}(s^\alpha + 1/\tau)}{s^{2\alpha} + s^\alpha/\tau + v^2|\omega|^{2\gamma}}. \tag{74}$$

which is the one-dimensional version of Equation (62).

5.3.2. Two Dimensions

As we have discussed in Section 4.3 most expressions of the two-dimensional case are the same than in three dimensions after the replacement $c^2/6 \to c^2/4$ (cf. Equation (54)). Thus, for instance, the sojourn densities $\hat{h}$ and $\hat{H}$—which in three dimensions and in the fluid limit approximations are given by Equations (57) and (58)—now read

$$\hat{h}(\omega,s) \simeq 1 - \frac{s}{\lambda} - 2\left(\frac{s}{\lambda}\right)^2 - \frac{1}{2\lambda^2}|\omega|^2 c^2 \cdots, \qquad \hat{H}(\omega,s) = \frac{1}{\lambda}\left(1 + \frac{2s}{\lambda}\right)\cdots \tag{75}$$

$(s \to 0, |\omega| \to 0)$ which mimicking the discussion of Section 5.1 leads to the following fractional generalization of these functions [see Equations (59) and (61)]

$$\hat{h}(\omega,s) = 1 - (Ts)^\alpha - 2(Ts)^{2\alpha} - \frac{1}{2}(L|\omega|)^{2\gamma}\cdots, \qquad \hat{H}(\omega,s) \simeq T^\alpha s^{\alpha-1} + 2T^{2\alpha}s^{2\alpha-1}\cdots \tag{76}$$

Substituting Equation (76) into Montroll–Weiss Equation (26) and reorganizing terms yields

$$\hat{\tilde{p}}(\omega,s) = \frac{s^{\alpha-1}(s^\alpha + 1/\tau)}{s^{2\alpha} + s^\alpha/\tau + v^2|\omega|^{2\gamma}}, \tag{77}$$

where

$$\tau = 2T^\alpha, \qquad v = \frac{1}{\sqrt{2}}(L^\gamma/T^\alpha). \tag{78}$$

Equation (77) gives the transformed PDF of the fractional two-dimensional isotropic random walk in the fluid limit approximation. Let us note that this expression has the same form as that of the three-dimensional case (cf. Equation (62)) with the same time parameter τ but a different velocity parameter v (cf. Equation (63)). Therefore, following exactly the same procedure detailed in the previous section we find the two-dimensional TE

$$\frac{\partial^{2\alpha} p}{\partial t^{2\alpha}} + \frac{1}{\tau}\frac{\partial^\alpha p}{\partial t^\alpha} = v^2 \nabla^{2\gamma} p, \tag{79}$$

which has the same form as the fractional TE (70) of the three dimensional case and with the same limiting behavior regarding fractional diffusion and wave-like performances.

5.4. Characteristic Function

Solving the fractional telegrapher's equation and thus obtaining the exact analytical form of the PDF $p(\mathbf{r}, t)$ seems to be out of reach for any dimension. It is, however, possible to obtain regardless dimensionality, a close and exact expression of the characteristic function $\tilde{p}(\omega, t)$ (i.e., the Fourier transform of the PDF $p(\mathbf{r}, t)$) of the space-time fractional telegrapher's Equation (70). To this end we will perform the Laplace inversion of the joint Fourier–Laplace $\hat{\tilde{p}}(\omega, s)$. Since this function has formally the same form in one, two and tree dimensions (cf. Equations (74), (77) and (62) respectively) the differences between them only arise when Fourier inverting and the characteristic function will be formally the same in all cases. This similarity also shows that the time structure of any average will be the same regardless the number of spatial dimension (we will see this fact explicitly in our discussion on the moments of time-fractional processes to be discussed in the next section).

We start off with Equation (77). Let us first note that taking into account the factorization

$$s^{2\alpha} + s^\alpha/\tau + v^2|\omega|^{2\gamma} = \left[s^\alpha + \frac{1}{2\tau} - \eta(\omega)\right]\left[s^\alpha + \frac{1}{2\tau} + \eta(\omega)\right],$$

where

$$\eta(\omega) = \sqrt{1/4\tau^2 - v^2|\omega|^{2\gamma}}, \tag{80}$$

Equation (77) can be written as

$$\hat{\tilde{p}}(\omega, s) = \frac{s^{\alpha-1}}{2\eta(\omega)}\left[\frac{1/2\tau + \eta(\omega)}{s^\alpha + 1/2\tau - \eta(\omega)} - \frac{1/2\tau - \eta(\omega)}{s^\alpha + 1/2\tau + \eta(\omega)}\right]. \tag{81}$$

Further manipulations yield

$$\frac{s^{\alpha-1}}{s^\alpha + 1/2\tau \pm \eta(\omega)} = \frac{s^{-1}}{1 + [1/2\tau \pm \eta(\omega)]s^{-\alpha}} = \sum_{n=0}^{\infty}(-1)^n\left[1/2\tau \pm \eta(\omega)\right]^n s^{-1-n\alpha}.$$

We next proceed to Laplace inversion. Since [58]

$$\mathcal{L}^{-1}\left\{s^{-1-n\alpha}\right\} = \frac{t^{n\alpha}}{\Gamma(1 + n\alpha)},$$

we have

$$\mathcal{L}^{-1}\left\{\frac{s^{\alpha-1}}{s^\alpha + 1/2\tau \pm \eta(\omega)}\right\} = \sum_{n=0}^{\infty}(-1)^n\frac{([1/2\tau \pm \eta(\omega)]t^\alpha)^n}{\Gamma(1 + n\alpha)} = \mathrm{E}_\alpha\left(-[1/2\tau \pm \eta(\omega)]t^\alpha\right),$$

where $E_\alpha(\cdot)$ is the Mittag–Leffler function [69]

$$E_\alpha(z) = \sum_{n=0}^{\infty} \frac{z^n}{\Gamma(1+n\alpha)}. \tag{82}$$

For α not an integer, the Mittag–Leffler function $E_\alpha(z)$ can be regarded as a kind of "fractional generalization" of the exponential function. Indeed when $\alpha = 1$ and since $\Gamma(1+n) = n!$ we immediately see from Equation (82) that $E_1(z) = e^z$.

Taking the inverse Laplace transform of Equation (81) and using the above intermediate results we finally obtain the characteristic function of the space-time fractional transport process

$$\tilde{p}(\omega, t) = \frac{1}{2\eta(\omega)} \Big\{ [1/2\tau \ + \ \eta(\omega)] E_\alpha\big(-[1/2\tau - \eta(\omega)]t^\alpha\big)$$
$$- \ [1/2\tau - \eta(\omega)] E_\alpha\big(-[1/2\tau + \eta(\omega)]t^\alpha\big) \Big\}, \tag{83}$$

which we recall is valid for any dimension of the underlying space.

In the wave-like limit when v is finite but $\tau \to \infty$ the fractional TE (70) reduces to the fractional wave-diffusion Equation (72). In this case (cf. Equation (80))

$$\eta(\omega) = iv|\omega|^\gamma,$$

and the characteristic function reads

$$\tilde{p}(\omega, t) = \frac{1}{2} \big[E_\alpha\big(-iv|\omega|^\gamma t^\alpha\big) + E_\alpha\big(iv|\omega|^\gamma t^\alpha\big) \big], \tag{84}$$

a solution already obtained by Mainardi for the wave-diffusion equation [52].

In the diffusion-like limit $\tau \to 0$ and $v \to \infty$ such that $2\tau v^2 = D$ (finite) and from Equation (80) we see that

$$2\tau\eta(\omega) = \sqrt{1 - 4\tau^2 v^2 |\omega|^{2\gamma}} \longrightarrow 1,$$

and

$$\frac{1}{2\tau} \mp \eta(\omega) = \frac{1}{2\tau} \Big(1 \mp \sqrt{1 - 2\tau D|\omega|^{2\gamma}} \Big) = \frac{D|\omega|^{2\gamma}}{1 \pm \sqrt{1 - 2\tau D|\omega|^{2\gamma}}} \longrightarrow D|\omega|^{2\gamma}.$$

From Equation (83) we get

$$\tilde{p}(\omega, t) = E_\alpha\big(-Dt^\alpha |\omega|^{2\gamma}\big), \tag{85}$$

a well known result which corresponds to a Levy density with fractional time [36,38]. When $\alpha = 1$ this result reduces to the ordinary Levy distribution with zero mean,

$$\tilde{p}(\omega, t) = e^{-Dt|\omega|^{2\gamma}}. \tag{86}$$

6. Time-Fractional Telegraphic Transport

In the last section we have developed the fractional telegraphic transport in its most general form assuming that both time and space are fractional. This leads, as the master equation of the process, to the space-time fractional telegrapher's equation which is formally the same in one, two and three dimensions. We have also seen that in both cases the general space-time fractional TE reduces to the space-time fractional wave equation at short times and to the space-time fractional diffusion equation at long times. This dual character is even more manifest for the time-fractional equation when the spatial exponent $\gamma = 1$ and only time is fractional. For fractional diffusion this particular case has been extensively studied in the literature and has many applications [35–37,39,41–43].

In any dimension the time-fractional TE is

$$\frac{\partial^{2\alpha} p}{\partial t^{2\alpha}} + \frac{1}{\tau}\frac{\partial^{\alpha} p}{\partial t^{\alpha}} = v^2 \nabla^2 p \tag{87}$$

$(0 < \alpha \leq 1)$, where ∇^2 is either the two or the three dimensional Laplacian and in one dimension is the second spacial derivative.

For the time-fractional TE we can obtain more analytical results than for the space-time TE. Results that turn out to be very useful because they clearly mark the similarities and dissimilarities between telegraphic transport processes in different dimensions. For one dimension we had already obtained in [32] some of the results presented here but not in higher dimensions. We now fill this gap in which the analogies and differences among different dimensions are clearly manifested.

Our starting point is again the Fourier–Laplace solution of the fractional TE (cf. Equations (62), (74) or (77) with $\gamma = 1$)

$$\hat{\tilde{p}}(\boldsymbol{\omega}, s) = \frac{s^{\alpha-1}(s^{\alpha} + 1/\tau)}{s^{2\alpha} + s^{\alpha}/\tau + v^2|\boldsymbol{\omega}|^2}. \tag{88}$$

The basic idea is the following: since time is now the only fractional variable but not space, it is possible to Fourier invert $\hat{\tilde{p}}(\boldsymbol{\omega}, s)$ and obtain a closed expression for the Laplace transform $\tilde{p}(\mathbf{r}, s)$ which compels us to treat different dimensions separately. Once we get the expression for $\tilde{p}(\mathbf{r}, s)$, the use of Tauberian theorems will allow us to obtain asymptotic expressions of the PDF $p(\mathbf{r}, t)$ at long and short times. Even though the one dimensional problem was fully addressed in [32], we present here all three dimensions and compare each result.

6.1. Laplace Transform of the PDF

Let us proceed to Fourier invert the expression (88) for $\hat{\tilde{p}}(\boldsymbol{\omega}, s)$. To this end we need to treat each dimension separately.

6.1.1. One Dimension

Recall that in one dimension the expression (88) of the transformed density $\hat{\tilde{p}}$ remains valid although in this case $|\omega|$ is not the modulus of a vector but the absolute value of a single variable. By virtue of the symmetry of $\hat{\tilde{p}}$ with respect to ω, the Fourier inversion will be given by

$$\tilde{p}(x, s) = \frac{1}{2\pi}\int_{-\infty}^{\infty} e^{-i\omega x}\hat{\tilde{p}}(\omega, s)d\omega = \frac{1}{\pi}\int_{0}^{\infty}\hat{\tilde{p}}(\omega, s)\cos\omega x d\omega.$$

Substituting for Equation (88) yields

$$\tilde{p}(x, s) = \frac{1}{\pi s}\left(s^{2\alpha} + s^{\alpha}/\tau\right)\int_{0}^{\infty}\frac{\cos\omega x}{s^{2\alpha} + s^{\alpha}/\tau + v^2\omega^2}d\omega,$$

and recalling the integral [70]

$$\int_{0}^{\infty}\frac{\cos\omega x}{a^2 + b^2\omega^2}d\omega = \frac{1}{2ab}e^{-a|x|/b},$$

we get

$$\tilde{p}(x, s) = \frac{1}{2vs}\sqrt{s^{2\alpha} + s^{\alpha}/\tau}\exp\left\{-\frac{|x|}{v}\sqrt{s^{2\alpha} + s^{\alpha}/\tau}\right\}. \tag{89}$$

For $\alpha = 1$ the fractional TE (87) reduces to the the ordinary TE and in this one-dimensional case the Laplace transform (89) can be inverted yielding the exact PDF $p(x, t)$ in terms of modified Bessel functions. We refer the interested reader to [32] for more details. For the fractional case when $\alpha \neq 1$, the exact analytical inversion of Equation (89) seems to be out of reach. However, as we will see below, we can obtain approximate solutions for large values of time.

6.1.2. Two Dimensions

In this case the Fourier inversion formula yields for the PDF in two dimensions

$$
\begin{aligned}
\hat{p}(\mathbf{r},s) &= \frac{1}{(2\pi)^2}\int_{\mathbb{R}^2} e^{-i\boldsymbol{\omega}\cdot\mathbf{r}}\hat{p}(\boldsymbol{\omega},s)d^2\omega = \frac{1}{(2\pi)^2}\int_0^\infty\int_0^{2\pi} e^{-i\omega r\cos\varphi}\hat{p}(\omega,s)\omega d\omega d\varphi \\
&= \frac{1}{(2\pi)^2}\int_0^\infty \omega\hat{p}(\omega,s)d\omega\int_0^{2\pi} e^{-i\omega r\cos\varphi}d\varphi,
\end{aligned}
$$

where $\omega = |\boldsymbol{\omega}|$ and we have taken into account that $\hat{p}(\boldsymbol{\omega},s)$ depends only on the modulus $|\boldsymbol{\omega}| = \omega$ [see Equation (88)].

From the integral representation of the Bessel function of zero order [63],

$$
J_0(\omega r) = \frac{1}{2\pi}\int_0^{2\pi} e^{-i\omega r\cos\varphi}d\varphi,
$$

we write

$$
\hat{p}(r,s) = \frac{1}{2\pi}\int_0^\infty \omega J_0(\omega r)\hat{p}(\omega,s)d\omega.
$$

Substituting for (88) we have

$$
\hat{p}(x,s) = \frac{1}{2\pi s}\left(s^{2\alpha}+s^\alpha/\tau\right)\int_0^\infty \frac{\omega J_0(\omega r)}{s^{2\alpha}+s^\alpha/\tau+v^2\omega^2}d\omega,
$$

and taking into account the integral [70]

$$
\int_0^\infty \frac{\omega J_0(a\omega)}{b^2+\omega^2}d\omega = K_0(ab),
$$

$(a > 0, \operatorname{Re} b > 0)$, where $K_0(\cdot)$ is a modified Bessel function, we finally obtain

$$
\hat{p}(r,s) = \frac{1}{2\pi v^2 s}\left(s^{2\alpha}+s^\alpha/\tau\right)K_0\!\left(\frac{r}{v}\sqrt{s^{2\alpha}+s^\alpha/\tau}\right). \tag{90}
$$

6.1.3. Three Dimensions

Bearing in mind that $\hat{p}(\boldsymbol{\omega},s)$ depends only on the modulus $|\boldsymbol{\omega}| = \omega$ (cf. Equation (88)), the Fourier inversion of $\hat{p}$ is

$$
\begin{aligned}
\hat{p}(\mathbf{r},s) &= \frac{1}{(2\pi)^3}\int_{\mathbb{R}^3} e^{-i\boldsymbol{\omega}\cdot\mathbf{r}}\hat{p}(\boldsymbol{\omega},s)d^3\omega \\
&= \frac{1}{(2\pi)^3}\int_0^\infty\int_0^\pi\int_0^{2\pi} e^{-i\omega r\cos\theta}\hat{p}(\boldsymbol{\omega},s)\omega^2\sin\theta d\omega d\theta d\varphi \\
&= \frac{1}{(2\pi)^2}\int_0^\infty \omega^2\hat{p}(\boldsymbol{\omega},s)d\omega\int_0^\pi e^{-i\omega r\cos\theta}\sin\theta d\theta.
\end{aligned}
$$

Since

$$
\int_0^\pi e^{-i\omega r\cos\theta}\sin\theta d\theta = \frac{2}{\omega r}\sin\omega r,
$$

we have

$$
\hat{p}(r,s) = \frac{1}{2\pi^2 r}\int_0^\infty \omega\sin\omega r\hat{p}(\boldsymbol{\omega},s)d\omega.
$$

Substituting for Equation (88) yields

$$
\hat{p}(x,s) = \frac{1}{2\pi^2 rs}\left(s^{2\alpha}+s^\alpha/\tau\right)\int_0^\infty \frac{\omega\sin\omega r}{s^{2\alpha}+s^\alpha/\tau+v^2\omega^2}d\omega,
$$

and taking into account the integral [70]

$$\int_0^\infty \frac{\omega \sin a\omega}{b^2 + \omega^2} d\omega = \frac{\pi}{2} e^{-ab},$$

$(a \geq 0, \mathrm{Re}\, b > 0)$, we obtain

$$\hat{p}(r,s) = \frac{1}{4\pi r v^2 s} \left(s^{2\alpha} + s^\alpha/\tau \right) \exp\left\{ -\frac{r}{v} \sqrt{s^{2\alpha} + s^\alpha/\tau} \right\}, \tag{91}$$

which is the exact PDF in three dimensions. Notice the different form taken by $\hat{p}(r,s)$ in one, two and three dimensions (cf. Equations (89)–(91), respectively).

Let us also observe that, like in the general space-time fractional cases, for the time-fractional case ($\alpha \neq 1$, $\gamma = 1$) the expressions for $\hat{p}$ given by Equations (89)–(91) are very difficult, not to say impossible, to invert analytically. Thus, obtaining the exact analytical form of the PDF $p(r,t)$ in real time seems to be beyond reach. We can, however, obtain approximate solutions, valid for large values of time, using Tauberian theorems which relate the small s behavior of $\hat{p}(r,s)$ with the large t behavior of $p(r,t)$ [61,71].

6.2. Long-Time Asymptotic Expressions

For the asymptotic analysis we rely on Tauberian theorems which allow us to infer the behavior of $p(r,t)$ for long times out of the expression for $\hat{p}(r,s)$ for small values of the Laplace variable s [61,71]. We work again each dimension separately.

6.2.1. One Dimension

We briefly summarize only the main results in one dimension and refer the reader to [32] for details. For long times such that $t \gg \tau^{1/\alpha}$ we have shown that [32]

$$p(x,t) \simeq \frac{t^{-\alpha/2}}{2v\sqrt{\tau}} M_{\alpha/2}\left(\frac{|x|t^{-\alpha/2}}{2v\sqrt{\tau}} \right), \qquad (t \gg \tau^{1/\alpha}), \tag{92}$$

where $M_{\alpha/2}(\cdot)$ is the Mainardi function defined by the power series [52,72]

$$M_\beta(z) = \sum_{n=0}^\infty \frac{(-1)^n z^n}{n! \Gamma(-\beta n + 1 - \beta)}. \tag{93}$$

The function $M_\beta(z)$ is an entire function for $0 < \beta < 1$ [52] being a special case of the Wright function [69,72] (see below) which is, in turn, closely related to Fox function frequently used in the anomalous diffusion literature [36]. We incidentally note that after the replacement $v\sqrt{\tau} \to D$, the asymptotic expression (92) becomes the exact solution to the time fractional diffusion equation (cf. Equation (71) with $\gamma = 1$), solution obtained by Mainardi some years ago [52].

From Equation (93) we see that $M_{\alpha/2}(z) \to 1/\Gamma(1 - \alpha/2)$ as $z \to 0$ and Equation (92) yields the asymptotic power law

$$p(x,t) \sim t^{-\alpha/2}, \qquad (t \to \infty). \tag{94}$$

6.2.2. Two Dimensions

Noticing that as $s \to 0$ (specifically, if $s \ll \tau^{-1/\alpha}$)

$$s^{2\alpha} + s^\alpha/\tau = (s^\alpha/\tau)(\tau s^\alpha + 1) \simeq s^\alpha/\tau, \tag{95}$$

we write for the two dimensional density (90)

$$\hat{p}(r,s) \simeq \frac{s^{\alpha-1}}{2\pi v^2 \tau} K_0\left(\frac{r s^{\alpha/2}}{v\sqrt{\tau}} \right), \qquad (s \ll \tau^{-1/\alpha}). \tag{96}$$

On the other hand [63]

$$K_0(z) = -\left[\gamma + \ln(z/2)\right] I_0(z) + 2 \sum_{n=1}^{\infty} \frac{1}{n} I_{2n}(z),$$

($\gamma = 0.5772 \cdots$ is the Euler constant and $I_\nu(z)$ are modified Bessel functions), but [63]

$$I_\nu(z) = \sum_{n=0}^{\infty} \frac{(z/2)^{\nu+2n}}{\Gamma(\nu+n+1)} = O(z^\nu) \qquad (z \to 0),$$

thus

$$K_0(z) = -\left[\gamma + \ln(z/2)\right] + O(z^2 \ln z). \tag{97}$$

Hence

$$K_0\left(\frac{rs^{\alpha/2}}{v\sqrt{\tau}}\right) = -\left[\gamma + \ln\left(\frac{r}{2v\sqrt{\tau}}\right)\right] - \frac{\alpha}{2}\ln s + O(s^\alpha \ln s),$$

which substituting into Equation (96) yields the approximate expressions valid for small values of s (i.e., when $s \ll \tau^{-1/\alpha}$)

$$\hat{p}(r,s) \simeq \frac{-1}{2\pi v^2 \tau}\left\{\left[\gamma + \ln\left(\frac{r}{2v\sqrt{\tau}}\right)\right]\frac{1}{s^{1-\alpha}} + \frac{\alpha/2}{s^{1-\alpha}}\ln s\right\}. \tag{98}$$

We next proceed to Laplace inverting this small s expression for $\hat{p}(r,s)$ which by virtue of Tauberian theorems will provide an approximate expression of $p(r,t)$ suitable for long times. Taking into account the Laplace inversion formulae [32,58]

$$\mathcal{L}\left\{\frac{1}{s^\beta}\right\} = \frac{t^{\beta-1}}{\Gamma(\beta)} \qquad \text{and} \qquad \mathcal{L}\left\{\frac{\ln s}{s^\beta}\right\} = \frac{t^{\beta-1}}{\Gamma(\beta)}\left[\psi(\beta) - \ln t\right]$$

[$\beta > 0$ and $\psi(z) = \Gamma'(z)/\Gamma(z)$] we have

$$\begin{aligned}
p(r,t) &\simeq \frac{-1}{2\pi v^2 \tau}\left\{\left[\gamma + \ln\left(\frac{r}{2v\sqrt{\tau}}\right)\right]\frac{t^{-\alpha}}{\Gamma(1-\alpha)} + \frac{\alpha}{2}\frac{t^{-\alpha}}{\Gamma(1-\alpha)}\left[\psi(1-\alpha) - \ln t\right]\right\} \\
&= \frac{1}{2\pi v^2 \tau}\frac{t^{-\alpha}}{\Gamma(1-\alpha)}\left[\frac{\alpha}{2}\ln t - \ln\left(\frac{r}{2v\sqrt{\tau}}\right) - \gamma - \psi(1-\alpha)\right] \\
&= \frac{1}{2\pi v^2 \tau}\frac{t^{-\alpha}}{\Gamma(1-\alpha)}\left[\ln\left(\frac{2v t^{\alpha/2}\sqrt{\tau}}{r}\right) - \gamma - \psi(1-\alpha)\right],
\end{aligned}$$

and neglecting constant terms we finally get

$$p(r,t) \simeq \frac{1}{2\pi v^2 \tau}\frac{t^{-\alpha}}{\Gamma(1-\alpha)}\ln\left(\frac{2v t^{\alpha/2}\sqrt{\tau}}{r}\right), \tag{99}$$

($t \gg \tau^{1/\alpha}$). Therefore, in two dimensions the PDF of the time fractional TE obeys the asymptotic logarithmic power law

$$p(x,t) \sim t^{-\alpha}\ln t \qquad (t \to \infty). \tag{100}$$

6.2.3. Three Dimensions

In three dimensions the starting point of our asymptotic analysis is Equation (91) which using the small s approximation given in Equation (95) yields

$$\hat{p}(r,s) \simeq \frac{s^{\alpha-1}}{4\pi r v^2 \tau}e^{-rs^{\alpha/2}/v\sqrt{\tau}} \qquad (s \ll \tau^{-1/\alpha}), \tag{101}$$

and expanding the exponential we write

$$\hat{p}(r,s) \simeq \frac{s^{\alpha-1}}{4\pi r v^2 \tau} \sum_{n=0}^{\infty} \frac{1}{n!} \left(\frac{-r}{v\sqrt{\tau}} \right)^n s^{-1+(1+n/2)\alpha} \qquad (s \ll \tau^{-1/\alpha}).$$

Recall again that because of Tauberian theorems the inversion of this expression for $\hat{p}(r,s)$, valid for small values of s, will provide an asymptotic expression for $p(r,t)$ suitable for large values of t. Thus, taking into account the Laplace inversion formula [32,52]

$$\mathcal{L}^{-1}\left\{ s^{\delta} \right\} = \frac{t^{-1-\delta}}{\Gamma(-\delta)} \tag{102}$$

(where $\delta \neq 0$ and not a positive integer) we obtain for $t \gg \tau^{1/\alpha}$

$$p(r,t) \simeq \frac{1}{4\pi r v^2 \tau} \sum_{n=0}^{\infty} \frac{1}{n!} \left(\frac{-r}{v\sqrt{\tau}} \right)^n \frac{t^{-(1+n/2)\alpha}}{\Gamma[1-(1+n/2)\alpha]},$$

that is

$$p(r,t) \simeq \frac{t^{-\alpha}}{4\pi r v^2 \tau} \sum_{n=0}^{\infty} \frac{1}{n!} \frac{1}{\Gamma[1-(1+n/2)\alpha]} \left(\frac{-rt^{-\alpha/2}}{v\sqrt{\tau}} \right)^n, \qquad (t \gg \tau^{1/\alpha}). \tag{103}$$

This asymptotic expression for $p(r,t)$ can be written in a more compact form by using the Wright function defined by [69,72]

$$W_{\lambda,\mu}(z) = \sum_{n=0}^{\infty} \frac{z^n}{n!\Gamma(\mu+\lambda n)}, \tag{104}$$

($\lambda > -1$ and μ and z arbitrary complex numbers). It is an entire function originally proposed by Wright in the 1930's for the asymptotic theory of partitions [69]. When $\lambda = 1$ the Wright function can be written in terms of the Bessel function of order $\mu - 1$ [69,72]. Moreover, Mainardi function $M_{\beta}(z)$ defined in Equation (93) is a particular case of the Wright function. Indeed,

$$M_{\beta}(z) = W_{-\beta,1-\beta}(-z).$$

From Equations (103) and (104) we see that the asymptotic PDF for the three dimensional case can be written as

$$p(r,t) \simeq \frac{t^{-\alpha}}{4\pi r v^2 \tau} W_{\alpha/2,1-\alpha} \left(\frac{-rt^{-\alpha/2}}{v\sqrt{\tau}} \right), \qquad (t \gg \tau^{1/\alpha}). \tag{105}$$

Finally, from Equation (104) we see that $W_{\lambda,\mu}(z) \to 1/\Gamma(\mu)$ as $z \to 0$ and Equation (105) yields the asymptotic power law

$$p(r,t) \sim t^{-\alpha}, \qquad (t \to \infty). \tag{106}$$

6.3. Moments of the Effective Distance Travelled

One of the magnitudes of greatest interest in transport problems is the distance covered by the particle from the starting point. The evaluation of the actual distance is very involved due to the random turnarounds of the trajectory. We will take as an estimate of it the effective distance travelled (taking into account that the transport processes starts at the origin of the coordinate system) which is the quantity $|x(t)|$ in one dimension, and $|\mathbf{r}(t)| = r(t)$ in two and three dimensions. We will thus work each dimension separately, although, as stated in Section 5.4, moments are essentially the same for each dimension.

Let us note that for space-time fractional processes, the moments of the distance travelled may be infinite (as in the case of the Levy processes). However, for time-fractional processes these moments are finite and we can get analytical expressions for them using the forms of the PDF obtained above. Moments also make explicit the dual character of the fractional telegraphic transport between fractional wave transport and fractional diffusion transport which generalizes the same duality presented by the ordinary TE.

6.3.1. One Dimension

Moments are defined by

$$\langle |x(t)|^n \rangle = \int_{-\infty}^{\infty} |x|^n p(x,t), \qquad (n = 1, 2, \dots),$$

and recalling that $p(x,t)$ is an even function of x, the Laplace transform can be written as

$$\mathcal{L}\{\langle |x(t)|^n \rangle\} = 2 \int_0^{\infty} x^n \hat{p}(x,s) dx.$$

Substituting for Equation (89) yields

$$\mathcal{L}\{\langle |x(t)|^n \rangle\} = \frac{\sqrt{\beta(s)}}{vs} \int_0^{\infty} x^n e^{-x\sqrt{\beta(s)}/v} dx = \frac{v^n}{s[\beta(s)]^{n/2}} \int_0^{\infty} z^n e^{-z} dz = \frac{v^n n!}{s[\beta(s)]^{n/2}},$$

where $\beta(s) = s^{2\alpha} + s^{\alpha}/\tau$. Hence

$$\mathcal{L}\{\langle |x(t)|^n \rangle\} = \frac{n! v^n}{s(s^{2\alpha} + s^{\alpha}/\tau)^{n/2}}, \qquad (n = 1, 2, \dots). \tag{107}$$

Recall that when $\tau \to \infty$ we recover the fractional wave equation. In this case from Equation (107) we have the "wave limit"

$$\mathcal{L}\{\langle |x(t)|^n \rangle\} = \frac{n! v^n}{s^{1+n\alpha}} \qquad \Rightarrow \qquad \langle |x(t)|^n \rangle = \frac{n! v^n}{\Gamma(1+n\alpha)} t^{n\alpha}. \tag{108}$$

On the other hand when $\tau \to 0$ but $v \to \infty$ such that $v^2\tau = D$ (finite) we recover the fractional diffusion equation and have the "diffusion limit"

$$\mathcal{L}\{\langle |x(t)|^n \rangle\} = \frac{n! D^{n/2}}{s^{1+n\alpha/2}} \qquad \Rightarrow \qquad \langle |x(t)|^n \rangle = \frac{n! D^{n/2}}{\Gamma(1+n\alpha/2)} t^{n\alpha/2}. \tag{109}$$

Let us also recall that the wave limit is the one we recover from TE as $t \to 0$, whereas the diffusion limit corresponds to the long time limit. Indeed, taking into account that

$$(s^{2\alpha} + s^{\alpha}/\tau)^{n/2} \simeq s^{n\alpha} \ (s \to \infty) \quad \text{and} \quad (s^{2\alpha} + s^{\alpha}/\tau)^{n/2} \simeq (s^{\alpha}/\tau)^n \ (s \to 0)$$

and bearing in mind Tauberian theorems, we see from Equation (107)

$$\mathcal{L}\{\langle |x(t)|^n \rangle\} \simeq \frac{n! v^n}{s^{1+n\alpha}} \ (s \to \infty) \ \Rightarrow \ \langle |x(t)|^n \rangle \simeq \frac{n! v^n}{\Gamma(1+n\alpha)} t^{n\alpha} \ (t \to 0), \tag{110}$$

and

$$\mathcal{L}\{\langle |x(t)|^n \rangle\} \simeq \frac{n! (v\sqrt{\tau})^n}{s^{1+n\alpha/2}} \ (s \to 0) \ \Rightarrow \ \langle |x(t)|^n \rangle = \frac{n! (v\sqrt{\tau})^n}{\Gamma(1+n\alpha/2)} t^{n\alpha/2} \ (t \to \infty). \tag{111}$$

6.3.2. Two Dimensions

We now have

$$\mathcal{L}\{\langle |\mathbf{r}(t)|^n \rangle\} = \int_{\mathbb{R}^2} |\mathbf{r}|^n \hat{p}(\mathbf{r},s) d^2\mathbf{r} = \int_0^{\infty} \int_0^{2\pi} |\mathbf{r}|^n \hat{p}(\mathbf{r},s) r dr d\varphi,$$

that is,

$$\mathcal{L}\{\langle r^n(t)\rangle\} = 2\pi \int_0^\infty r^{n+1}\hat{p}(r,s)dr.$$

Substituting for Equation (90) and a simple change of variables yields

$$\mathcal{L}\{\langle r^n(t)\rangle\} = \frac{v^n}{s(s^{2\alpha} + s^\alpha/\tau)^{n/2}} \int_0^\infty z^{n+1}K_0(z)dz,$$

but [70]

$$\int_0^\infty z^{n+1}K_0(z)dz = 2^n\Gamma^2(1+n/2),$$

so that

$$\mathcal{L}\{\langle r^n(t)\rangle\} = \frac{2^n\Gamma^2(1+n/2)v^n}{s(s^{2\alpha} + s^\alpha/\tau)^{n/2}}, \qquad (n=1,2,\dots). \tag{112}$$

Observe that this two-dimensional expression is equal to the one-dimensional moment (107) except for a mere numerical factor and, therefore, all two-dimensional expressions can be recovered from the one dimensional ones under the replacement $n! \to 2^n\Gamma^2(1+n/2)$. Thus, in particular, we see from Equations (110) and (111) that

$$\langle r^n(t)\rangle \sim t^{n\alpha} \quad (t\to 0) \qquad \text{and} \qquad \langle r^n(t)\rangle \sim t^{n\alpha/2} \quad (t\to\infty). \tag{113}$$

6.3.3. Three Dimensions

In the three dimensional case we have

$$\mathcal{L}\{\langle |\mathbf{r}(t)|^n\rangle\} = \int_{\mathbb{R}^3} |\mathbf{r}|^n\hat{p}(\mathbf{r},s)d^3\mathbf{r} = \int_0^\infty \int_0^\pi \int_0^{2\pi} |\mathbf{r}|^n\hat{p}(\mathbf{r},s)r^2\sin\theta dr d\theta d\varphi,$$

that is,

$$\mathcal{L}\{\langle r^n(t)\rangle\} = 4\pi \int_0^\infty r^{n+2}\hat{p}(r,s)dr.$$

Substituting for Equation (91) and elementary integration yields

$$\mathcal{L}\{\langle r^n(t)\rangle\} = \frac{(n+1)!v^n}{s(s^{2\alpha} + s^\alpha/\tau)^{n/2}}, \qquad (n=1,2,\dots), \tag{114}$$

which has the same structure as the one and two dimensional cases (cf. Equations (107) and (112)) and, as a consequence, the asymptotic expressions for moments will also be given by Equation (113).

7. Concluding Remarks

We have reviewed the main aspects of telegraphic transport processes which account for "diffusion with finite velocity" [8] and whose master equation is the telegrapher's equation instead of the diffusion equation. The main part of this report is a comprehensive account of our previous works [30–32], on the derivation, out of random walks models, of the telegrapher's equation (ordinary as well as fractional) in one, two and three dimensions.

We have mostly focussed on two and three dimensions because, for one hand, early attempts to derive higher dimensional TE's from random walk models had been fruitless and, on the other hand, higher dimensional models are usually more relevant for transport problems than any one-dimensional model. We thus present models that are two and three dimensional generalizations of the persistent random walk on the line. The models are based on multistate random walks with a continuous number of states representing the different directions the particle can take. We set the general integral equations for the probability density function of the particle evolution on the plane or in the space. When at every point all possible directions are independent and do not depend on the orientation and position (isotropy and homogeneity), the general equations can be exactly solved in Fourier-Laplace space. The isotropic and homogeneous models are suitable for addressing the transport of particles experiencing elastic collisions with fixed centers randomly distributed such as photons moving in turbid media [2].

These continuous models constitute a microscopic description for transport in which we statistically count the (elastic) collisions of the particles. If we zoom out this microscopic description by implementing the fluid-limit approximation of large times and distances—and, thus, going to a mesoscopic description of the process—we end up with the telegrapher's equation as the master equation of the transport processes.

We have also generalized the telegrapher's equation to account for anomalous transport in several dimensions. To this end the isotropic and homogeneous random walk has been extended to allow for fractional behavior both in time and space variables. The dual character of the ordinary TE between wave and diffusion behaviors is also manifest in the space-time fractional TE where at small times this equation reduces to the fractional diffusion-wave equation while at long times it does to the anomalous diffusion equation.

The two different dynamics governing the fractional transport—one of them, ruling transport at small times, is given by fractional wave behavior, while at large time the dynamics is dictated by fractional diffusion behavior—are even more apparent for the time-fractional transport when only the time variable is fractional. In this case all moments of the distance to the initial position (the effective distance travelled by the particle) exist and have an analytical expression in terms of their Laplace transforms. For small and large times these moments are approximated by

$$\langle r^n(t) \rangle \sim t^{n\alpha} \quad (t \to 0), \qquad \langle r^n(t) \rangle \sim t^{n\alpha/2} \quad (t \to \infty),$$

($n = 1, 2, \dots$). When $0 < \alpha < 1/2$ there is a transition from two different subdiffusive regimes, while if $1/2 < \alpha < 1$ the transition is from superdiffusion at small times to subdiffusion at large times. This fact generalizes the passage from ballistic motion to normal diffusion shown by the ordinary telegrapher's equation.

The exact solution for the characteristic function $\tilde{p}(\omega, t)$ of the fractional transport has also been obtained regardless the dimensionality of the process, and approximate expressions for wave and diffusion regimes are attained as well. These variety of expressions have been explored by Mainardi and collaborators [52,53,67,72] on solutions for fractional diffusion and fractional wave-diffusion equations (see also Mainardi's recent and useful survey appeared in this special issue [73]). Additionally, Orsingher and collaborators [74–78] have proposed several kinds of solutions to the fractional TE and explored their properties.

For the time-fractional transport we have been able to go one step further and obtain the exact form of the Laplace transform $\hat{p}(r, s)$ in one, two and three dimensions. From these expressions it is possible to get analytical forms of the PDF $p(r, t)$ valid for sufficiently long times which, in one and three dimensions are written in terms of Wright functions (cf. Equations (92) and (103)), while in two dimensions by a logarithmic function (cf. Equation (99)). From these expressions we have obtained, as $t \to \infty$, the asymptotic power laws

$$p(x,t) \sim t^{-\alpha/2}, \qquad \text{(one dimension);} \qquad\qquad p(x,t) \sim t^{-\alpha}, \qquad \text{(three dimensions);}$$

and the logarithmic power law

$$p(x,t) \sim t^{-\alpha} \ln t, \qquad \text{(two dimensions).}$$

Let us finish by recalling that a substantial part of this paper is a review of previous works but a significant part is, to my knowledge, new. This is the case of the higher dimensional extension of the characteristic function for the space-time fractional TE (cf. Section 5.4), as well as the whole Section 6 on higher dimensional time-fractional telegraphic processes.

Funding: This research received no external funding.

Institutional Review Board Statement: Not applicable.

Informed Consent Statement: Not applicable.

Acknowledgments: J.M. acknowledges partial financial supports from MINECO under Contract No. FIS2016-78904-C3-2-P and from AGAUR under Contract No. 2017SGR1064.

Conflicts of Interest: The author declares no conflict of interest.

References

1. Duderstadt, J.J.; Martin, W.R. *Transport Theory*; J. Wiley: New York, NY, USA, 1979.
2. Weiss, G.H. Some applications of the persistent random walks and the telegrapher's equation. *Phys. A* **2020**, *311*, 381–410. [CrossRef]
3. Shlesinger, M.; Klafter, J.; Zumofen, G. Lévy flights: Chaotic, turbulent and relatisvistic. *Fractals* **1995**, *3*, 491. [CrossRef]
4. Durian, D.J.; Rudnick, J. Photon migration at short times and distances and in cases of strong absorption. *J. Opt. Soc. Am. A* **1997**, *14*, 235. [CrossRef]
5. Lemieux, P.A.; Vera, M.U.; Durian, D.J. Diffusing-light spectroscopy beyond the diffusion limit: The role of ballistic transport and anisotropic scattering. *Phys. Rev. E* **1998**, *57*, 4498. [CrossRef]
6. Durian, D.J.; Rudnick, J. Spatially resolved backscattering: Implementation of extrapolation boundary condition and exponential source. *J. Opt. Soc. Am. A* **1999**, *16*, 837. [CrossRef]
7. Wang, J.; Dlamini, D.S.; Mishra, S.J.; Pendergast, M.T.M.; Wong, M.C.Y.; Mamba, B.B.; Freger, V.; Verliefde, A.R.D.; Hoek, E.M.V. A critical review of transport through osmotic membranes. *J. Membr. Sci.* **2014**, *454*, 516–537. [CrossRef]
8. Masoliver, J.; Weiss, G.H. Finite-velocity diffiusion. *Eur. J. Phys.* **1996**, *17*, 190. [CrossRef]
9. Joseph, D.D.; Preziosi, L. Heat waves. *Rev. Mod. Phys.* **1989**, *61*, 41. [CrossRef]
10. Jou, D.; Casas-Vázquez, J.; Lebon, G. *Extended Irreversible Thermodynamics*, 4th ed.; Springer: Berlin, Germany, 2010.
11. Méndez, V.; Campos, D.; Horsthemke, W. Growth and dispersal with inertia: Hyperbolic reaction-transport systems. *Phys. Rev. E* **2014**, *90*, 042114. [CrossRef] [PubMed]
12. Goldstein, S. On diffusion by discontinuous movements and on the telegraph equation. *Q. J. Mech. Appl. Math.* **1951**, *4*, 129. [CrossRef]
13. Kac, M. A stochastic model related to the telegrapher's equation. *Rocky Mt. J. Math.* **1974**, *4*, 497. [CrossRef]
14. Masoliver, J.; Lindenberg, K.; Weiss, G.H. A continuous time generalization of the persistent random walk. *Phys. A* **1989**, *182*, 891. [CrossRef]
15. Masoliver, J.; Lindenberg, K. Continuous-time persistent random walk: A review and some generalizations. *Eur. Phys. J. B* **2017**, *90*, 107. [CrossRef]
16. Olivares-Robles, M.A.; García-Colín, L.S. Mesoscopic derivation of hyperbolic transport equations. *Phys. Rev. E* **1994**, *50*, 2451. [CrossRef] [PubMed]
17. Masoliver, J. *Random Processes: First-Passage and Escape*; World Scientific: Singapore, 2018.
18. Ishimaru, A.J. Diffusion of light in turbid material. *Appl. Opt.* **1989**, *28*, 2210. [CrossRef]
19. Heizler, S.I. Asymptotic telegrapher's equation (P_1) approximation for the transport equation. *Nucl. Sci. Eng.* **2010**, *166*, 17. [CrossRef]
20. Plyukhin, A.V. Stochastic processes leading to wave equations in dimensions higher than one. *Phys. Rev. E* **2010**, *81*, 021113. [CrossRef]
21. Balescu, R. V-Langevin equations, continuous-time persistent random walks and fractional diffusion. *Chaos Solitons Fractals* **2007**, *34*, 62. [CrossRef]
22. Weiss, G.H. *Aspects and Applications of the Random Walk*; North-Holland: Amsterdam, The Netherlands, 1994.
23. Masoliver, J.; Porrà, J.M.; Weiss, G.H. The continuum limit of a two-dimensional persistent random walk. *Phys. A* **1992**, *182*, 593. [CrossRef]
24. Porrà, J.M.; Masoliver, J.; Weiss, G.H. A diffusion model incorporating anisotropic properties. *Phys. A* **1995**, *218*, 229. [CrossRef]
25. Boguñá, M.; Porrà, J.M.; Masoliver, J. Generalization of the persistent random walk to dimensions greater than one. *Phys. Rev. E* **1998**, *58*, 6992. [CrossRef]
26. Godoy, S.; García-Colín, L.S. Nonvalidity of the telegrapher's diffusion equation in two and three dimensions for crystalline solids. *Phys. Rev. E* **1997**, *55*, 2127. [CrossRef]
27. Kolesnik, A.D.; Orsingher, E. A planar random motion with an infinite number of directions controlled by the damped wave equation. *J. Appl. Prob.* **2005**, *42*, 1168. [CrossRef]
28. Orsingher, E.; De Gregorio, A. Random flights in higher spaces. *J. Theor. Prob.* **2007**, *20*, 769. [CrossRef]
29. Kolesnik, A.; Pinsky, M.A. Isotropic random motion at finite speed with K-Erlang distributed direction alternatives. *J. Stat. Phys.* **2011**, *142*, 828.
30. Masoliver, J. Three dimensional telegrapher's equation and its fractional generalization. *Phys. Rev. E* **2017**, *96*, 022101. [CrossRef]
31. Masoliver, J.; Lindenberg, K. Two-dimensional telegraphic processes and their fractional generalization. *Phys. Rev. E* **2020**, *101*, 012137. [CrossRef]
32. Masoliver, J. Fractional telegrapher's equation from fractional persistent random walks. *Phys. Rev. E* **2016**, *93*, 052107. [CrossRef]
33. Masoliver, J. Mean first-passage time for non-Markovian continuous noise. *Phys. Rev A* **1992**, *45*, 2256. [CrossRef]
34. Havlin, S.; Ben-Avraham, D. Diffusion in disorderd media. *Adv. Phys.* **1987**, *36*, 695. [CrossRef]

35. Bouchaud, J.P.; Georges, A. Anomalous diffusion behavior on disordered media: Statistical mechanics, models and physical applications. *Phys. Rep.* **1990**, *195*, 127. [CrossRef]
36. Metzler, R.; Klafter, J. The random walk guide to anomalous diffusion: A fractional dynamics approach. *Phys. Rep.* **2000**, *339*, 1–77. [CrossRef]
37. West, B.J.; Bologna, M.; Grigolini, P. *Physics of Fractal Operators*; Springer: Berlin, Germany, 2003.
38. Metzler, R.; Klafter, J. The restaurant at the end of the random walk: Recent developments in the description of anomalous transport by fractional dynamics. *J. Phys. A* **2004**, *37*, R161–R208. [CrossRef]
39. Balescu, R. *Aspects of Anomalous Transport in Plasmas*; Taylor & Francis: London, UK, 2005.
40. Eliazar, I.I.; Shlesinger, M.F. Fractional motions. *Phys. Rep.* **2013**, *527*, 101–129. [CrossRef]
41. West, B.J. Fractional view of complexity: A tutorial. *Rev. Mod. Phys.* **2014**, *86*, 1169–1184. [CrossRef]
42. West, B.J. *Fractional Calculus View of Complexity: Tomorrow's Science*; CRC Press: Boca Raton, FL, USA, 2016.
43. Klafter, J.; Sokolov, I. Anomalous diffusion spreads its wings. *Phys. World* **2005**, *18*, 29. [CrossRef]
44. Montroll, E.W.; Weiss, G. H. Random walks on lattices II. *J. Math. Phys.* **1965**, *6*, 167–181. [CrossRef]
45. Montroll, E.W.; Shlesinger, M.F. The wonderful world of random walks. In *Studies in Statistical Mechanics*; Lebowitz, J.L., Montroll, E.W., Eds.; North-Holland: Amsterdam, The Netherlands, 1984; Volume 11.
46. Kutner, R.; Masoliver, J. The continuous-time random walk still trendy: Fifty-year history, state of the art and outlook. *Eur. Phys. J. B* **2017**, *90*, 50. [CrossRef]
47. Scher, H.; Montroll, E.W. Random walks on lattices IV. *J. Stat. Phys.* **1973**, *9*, 101.
48. Scher, H.; Montroll, E.W. Anomalous transit-time dispersion in amorphous solids. *Phys. Rev. B* **1975**, *12*, 2455. [CrossRef]
49. ben-Avraham, D.; Havlin, S. *Diffusion and Reactions in Fractals and Disordered Systems*; Cambridge University Press: Cambridge, UK, 2000.
50. Castiglione, P.; Mazzino, A.; Muratore-Ginanneschi, P.; Vulpiani, A. On strong anomalous diffusion. *Phys. D* **1999**, *134*, 75. [CrossRef]
51. Gorenflo, R.; Mainardi, F.; Vivoli, A. Continuous-time random walk and parametric subordination in fractional diffusion. *Chaos Solitons Fractals* **2007**, *34*, 87. [CrossRef]
52. Mainardi, F. The fundamental solution for the fractional diffusion-wave equation. *Appl. Math. Lett.* **1996**, *9*, 23. [CrossRef]
53. Mainardi, F.; Luchko, Y.; Pagnini, G. The fundamental solution of the space-time fractional diffusion equation. *Fract. Calc. Appl. Anal.* **2001**, *4*, 153.
54. Rebenshtok, A.; Denisov, S.; Hänggi, P.; Barkai, E. Infinite densities for Lévy walks. *Phys. Rev. E* **2014**, *90*, 062135. [CrossRef]
55. Burshtein, I.; Zharikov, A.A.; Temkin, S.I. Response of a two-level system to a random modulation of the resonance with an arbitrary strong external field. *J. Phys. B* **1988**, *21*, 1907. [CrossRef]
56. Kofman, A.G.; Zabel, R.; Levine, A.M.; Prior, Y. Non-Markovian stochastic jump processes I. Input field analysis. *Phys. Rev. A* **1990**, *41*, 6434. [CrossRef] [PubMed]
57. Kingman, J.F.C. *Poisson Processes*; Oxford University Press: Oxford, UK, 2002.
58. Roberts, G.E.; Kaufman, H. *Table of Laplace Transforms*; W. B. Saunders: Philadelphia, PA, USA, 1966.
59. Claes, I.; Van den Broeck, C. Random walks with persistence. *J. Stat. Phys.* **1987**, *49*, 383. [CrossRef]
60. Feller, W. *An Introduction to Probability Theory and Its Applications*; J. Wiley: New York, NY, USA, 1971; Volume II.
61. Pitt, H.R. *Tauberian Theorems*; Oxford University Press: Oxford, UK, 1958.
62. Masoliver, J.; Porrà, J.M.; Weiss, G.H. Solution to the telegrapher's equation in the presence of reflecting and partly reflecting boundaries. *Phys. Rev. E* **1993**, *48*, 939. [CrossRef]
63. Magnus, W.; Oberhettinger, F.; Soni, R.P. *Formulas and Theorems for the Special Functions of Mathematical Physics*; Springer: Berlin, Germany, 1966.
64. Weiss, G.H. First passage times for correlated random walks and some generalizations. *J. Stat. Phys.* **1984**, *37*, 325. [CrossRef]
65. Masoliver, J.; Weiss, G.H. First passage times for generalized telegrapher's equation. *Phys. A* **1992**, *183*, 537. [CrossRef]
66. Masoliver, J. Telegraphic processes with stochastic resetting. *Phys. Rev. E* **2019**, *99*, 012121. [CrossRef]
67. Gorenflo, R.; Mainardi, F. Fractional calculus. In *Fractals and Fractional Calculus in Continuum Mechanics*; Carpinteri, A., Mainardi, F., Eds.; Springer: Berlin, Germany, 1997.
68. Podlubny, I. *Fractional Differential Equations*; Academic Press: San Diego, CA, USA, 1999.
69. Erdelyi, A.; Magnus, W.; Oberhettinger, F.; Tricomi, F.G. *Higher Transcendental Functions*; McGraw-Hill: New York, NY, USA, 1953; Volume 3.
70. Gradshteyn, I.S.; Ryzhik, I.M. *Table of Integrals, Series and Products*, 7th ed.; Elsevier: Amsterdam, The Netherlands, 2007.
71. Handelsman, R.A.; Lew, J.S. Asymptotic expansions of the Laplace convolutions for large argument and fat tail densities for certain sums of random variables. *SIAM J. Math. Anal.* **1974**, *5*, 425–451. [CrossRef]
72. Mainardi, F.; Pagnini, G. The role of Fox-Wright functions in fractional sub-diffusion of distributed order. *J. Comp. Appl. Math.* **2007**, *207*, 24. [CrossRef]
73. Mainardi, F. Why the Mittag-Leffler function can be considered the queen function of the fractional calculus? *Entropy* **2020**, *22*, 1359. [CrossRef]
74. Orsingher, E.; Zhao, X. The space-fractional telegraph equation and the related fractional telegraph process. *Chin. Ann. Math.* **2003**, *B24*, 1. [CrossRef]

75. Orsingher, E.; Beghin, L. Time-fractional telegraph equation and telegraph processes with Brownian time. *Probab. Theory Relat. Fields* **2004**, *128*, 141.
76. D'Ovidio, M.; Orsingher, E.; Toaldo, B. Time changed processes governed by space-time fractional telegraph equations. *Stoch. Anal. Appl.* **2014**, *32*, 1009. [CrossRef]
77. Orsingher, E.; Toaldo, B. Space-time fractional equations and the related stable processes at random time. *J. Theor. Probab.* **2017**, *30*, 1–26. [CrossRef]
78. Lafracte, F.; Orsingher, E. On the fractional wave equation. *Mathematics* **2020**, *8*, 874. [CrossRef]

Review

Why the Mittag-Leffler Function Can Be Considered the Queen Function of the Fractional Calculus?

Francesco Mainardi

Dipartimento di Fisica e Astronomia, Università di Bologna, Via Irnerio 46, I-40126 Bologna, Italy;
francesco.mainardi@bo.infn.it

Received: 14 November 2020; Accepted: 26 November 2020; Published: 30 November 2020

Abstract: In this survey we stress the importance of the higher transcendental Mittag-Leffler function in the framework of the Fractional Calculus. We first start with the analytical properties of the classical Mittag-Leffler function as derived from being the solution of the simplest fractional differential equation governing relaxation processes. Through the sections of the text we plan to address the reader in this pathway towards the main applications of the Mittag-Leffler function that has induced us in the past to define it as the *Queen Function of the Fractional Calculus*. These applications concern some noteworthy stochastic processes and the time fractional diffusion-wave equation We expect that in the future this function will gain more credit in the science of complex systems. Finally, in an appendix we sketch some historical aspects related to the author's acquaintance with this function.

Keywords: fractional calculus; Mittag-Leffler functions; Wright functions; fractional relaxation; diffusion-wave equation; Laplace and Fourier transform; fractional Poisson process complex systems

1. Introduction

For few decades, the special transcendental function known as the Mittag-Leffler function has attracted increasing attention of researchers because of its key role in treating problems related to integral and differential equations of fractional order.

His function was introduced in 1903–1905 by the Swedish mathematician Mittag-Leffler and at the beginning of the last century up to the 1990s, this function was seldom considered by mathematicians and applied scientists.

Before the 1990s, from a mathematical point of view, we recall the 1930 paper by Hille and Tamarkin [1] on the solutions of the Abel integral equation of the second kind, and the books by Davis [2], Sansone & Gerretsen [3], Dzherbashyan [4] (unfortunately in Russian), and finally Samko et al. [5]. Particular mention would be for the 1955 Handbook of High Transcendental Functions of the Bateman project [6], where this function was treated in Volume 3, in the chapter devoted to miscellaneous functions. For former applications we recall an interesting note by Davis [2] reporting previous research by Dr. Kenneth S. Cole in connection with nerve conduction, and the papers by Cole & Cole [7], Gross [8] and Caputo & Mainardi [9,10], where the Mittag-Leffler function was adopted to represent the responses in dielectric and viscoelastic media. More information are found in the Appendix of this survey.

In the 1960's the Mittag-Leffler function started to emerge from the realm of miscellaneous functions because it was considered as a special case of the general class of Fox *H* functions, that can exhibit an arbitrary number of parameters in their integral Mellin-Barnes representation. However, in our opinion, this classification in a too general framework has, to some extent, obscured the relevance and the applicability of this function in applied sciences. In fact, most mathematical models are based on a small number of parameters, say 1 or 2 or 3, so that a general theory may be confusing whereas the adoption of a generalized Mittag-Leffler function with 2 or 3 indices may be sufficient.

Nowadays it is well recognized that the Mittag-Leffler function plays a fundamental role in Fractional Calculus even if with a single parameter (as originally introduced by Mittag-Leffler) just to be worthy of being referred to as the *Queen Function of Fractional Calculus*, see Mainardi & Gorenflo [11]. We find some information on the Mittag-Leffler functions in any treatise on Fractional Calculus but for more details we refer the reader to the surveys of Haubold, Mathai and Saxena [12] and by Van Mieghem [13] and to the treatise by Gorenflo et al. [14], devoted just to Mittag-Leffler functions, related topics and applications.

The plan of this survey is the following. We start to give in Section 2 the main definitions and properties of the Mittag-Leffler function in one parameter with related Laplace transforms. Then in Section 3 we describe its use in the simplest fractional relaxation equation pointing out its compete monotonicity. The asymptotic properties are briefly discussed in Section 4. In Section 5 we briefly discuss the so called generalized Mittag-Leffler function, that is the 2-parameter Mittag-Leffler function. Of course further generalization to 3 and more parameter will be referred to specialized papers and books. Then in the following sections we discuss the application of the Mittag-Leffler function in some noteworthy stochastic processes. We start in Section 6 with the fractional Poisson process, and then in Section 7 with its application of the thinning of renewal processes. The main application are dealt in Section 8 where we discuss the continuous time random walks (CTRW) and then in Section 9 we point out the asymptotic universality. In Section 10 we discuss the time fractional diffusion-wave processes pointing out the role of the Mittag-Leffler functions in two parameters and their connection with the basic Wright functions. In Appendix A we find it worthwhile to report the acquaintance of the author with the Mittag-Leffler functions started in the late 1960s and continued up to nowadays.

We recall that Sections 3–10 are taken from several papers by the author, published alone and with colleagues and former students. Furthermore we have not considered other applications of the Mittag-Leffler functions including, for example, anomalous diffusion theory in terms of fractional and generalized Langevin equations. On this respect we refer the readers to the articles of the author, see References [15,16], and to the recent book by Sandev and Tomovski [17] and references therein. For many items related to the Mittag-Leffler functions we refer again to the treatise by Gorenflo et al. [14].

2. The Mittag-Leffler Functions: Definitions and Laplace Transforms

The Mittag-Leffler function is defined by the following power series, convergent in the whole complex plane,

$$E_\alpha(z) := \sum_{n=0}^{\infty} \frac{z^n}{\Gamma(\alpha n + 1)}, \quad \alpha > 0, \quad z \in \mathbb{C}. \tag{1}$$

We recognize that it is an entire function of order $1/\alpha$ providing a simple generalization of the exponential function $\exp(z)$ to which it reduces for $\alpha = 1$. For detailed information on the Mittag-Leffler-type functions and their Laplace transforms the reader may consult e.g., [6,18,19] and the recent treatise by Gorenflo et al. [14].

We also note that for the convergence of the power series in (1) the parameter α may be complex provided that $\Re(\alpha) > 0$. The most interesting properties of the Mittag-Leffler function are associated with its asymptotic expansions as $z \to \infty$ in various sectors of the complex plane.

In this paper we mainly consider the Mittag-Leffler function of order $\alpha \in (0, 1)$ on the negative real semi-axis where is known to be completely monotone (CM) due a classical result by Pollard [20], see also Feller [21].

Let us recall that a function $\phi(t)$ with $t \in \mathbb{R}^+$ is called a completely monotone (CM) function if it is non-negative, of class C^∞, and $(-1)^n \phi^{(n)}(t) \geq 0$ for all $n \in \mathbb{N}$. Then a function $\psi(t)$ with $t \in \mathbb{R}^+$ is called a Bernstein function if it is non-negative, of class C^∞, with a CM first derivative. These functions play fundamental roles in linear hereditary mechanics to represent relaxation and creep processes, see, for example, Mainardi [22]. For mathematical details we refer the interested reader to the survey paper by Miller and Samko [23] and to the treatise by Schilling et al. [24].

In particular we are interested in the function

$$e_\alpha(t) := E_\alpha(-t^\alpha) = \sum_{n=0}^{\infty} (-1)^n \frac{t^{\alpha n}}{\Gamma(\alpha n + 1)}, \quad t > 0, \quad 0 < \alpha \le 1, \tag{2}$$

whose Laplace transform pair reads

$$e_\alpha(t) \div \frac{s^{\alpha-1}}{s^\alpha + 1}, \quad \alpha > 0. \tag{3}$$

Here we have used the notation $\div$ to denote the juxtaposition of a function of time $f(t)$ with its Laplace transform

$$\tilde{f}(s) = \int_0^\infty e^{-st} f(t) \, dt.$$

The pair (3) can be proved by transforming term by term the power series representation of $e_\alpha(t)$ in the R.H.S of (2). Similarly we can prove the following Laplace transform pair for its time derivative

$$e'_\alpha(t) = \frac{d}{dt} E_\alpha(-t^\alpha) \div -\frac{1}{s^\alpha + 1}, \quad \alpha > 0. \tag{4}$$

For this Laplace transform pair we can simply apply the usual rule for the Laplace transform for the first derivative of a function, that reads

$$\frac{d}{dt} f(t) \div s \tilde{f}(s) - f(0^+). $$

3. The Mittag-Leffler Function in Fractional Relaxation Processes

For readers' convenience let us briefly outline the topic concerning the generalization via fractional calculus of the first-order differential equation governing the phenomenon of (exponential) relaxation. Recalling (in non-dimensional units) the initial value problem

$$\frac{du}{dt} = -u(t), \quad t \ge 0, \quad \text{with} \quad u(0^+) = 1 \tag{5}$$

whose solution is

$$u(t) = \exp(-t), \tag{6}$$

the following two alternatives with $\alpha \in (0,1)$ are offered in the literature:

$$(a) \quad \frac{du}{dt} = -D_t^{1-\alpha} u(t), \quad t \ge 0, \quad \text{with} \quad u(0^+) = 1, \tag{7}$$

$$(b) \quad {}_* D_t^\alpha u(t) = -u(t), \quad t \ge 0, \quad \text{with} \quad u(0^+) = 1, \tag{8}$$

where $D_t^{1-\alpha}$ and ${}_* D_t^\alpha$ denote the fractional derivative of order $1 - \alpha$ in the Riemann-Liouville sense and the fractional derivative of order α in the Caputo sense, respectively.

For a generic order $\mu \in (0,1)$ and for a sufficiently well-behaved function $f(t)$ with $t \in \mathbb{R}^+$ the above derivatives are defined as follows, see for example, Gorenflo and Mainardi [18], Podlubny [19],

$$(a) \quad D_t^\mu f(t) = \frac{1}{\Gamma(1-\mu)} \frac{d}{dt} \left[\int_0^t \frac{f(\tau)}{(t-\tau)^\mu} \, d\tau \right], \tag{9}$$

$$(b) \quad {}_* D_t^\mu f(t) = \frac{1}{\Gamma(1-\mu)} \int_0^t \frac{f'(\tau)}{(t-\tau)^\mu} \, d\tau. \tag{10}$$

Between the two derivatives we have the relationship

$$_*D_t^\mu f(t) = D_t^\mu f(t) - f(0^+) \frac{t^{-\mu}}{\Gamma(1-\mu)} = D_t^\mu \left[f(t) - f(0^+) \right] . \tag{11}$$

Both derivatives in the limit $\mu \to 1^-$ reduce to the standard first derivative but for $\mu \to 0^+$ we have

$$D_t^\mu f(t) \to f(t), \quad _*D_t^\mu f(t) = f(t) - f(0^+), \quad \mu \to 0^+ . \tag{12}$$

In analogy to the standard problem (5), we solve the problems (7) and (8) with the Laplace transform technique, using the rules pertinent to the corresponding fractional derivatives, that we recall hereafter for a generic order $\mu \in (0,1)$,

$$(a) \quad D_t^\mu f(t) \div s^\mu \widetilde{f}(s) - g(0^+), \quad g(0^+) = \frac{1}{\Gamma(1-\mu)} \lim_{t \to 0^+} \int_0^t (t-\tau)^{-\mu} f(\tau) \, d\tau . \tag{13}$$

$$(b) \quad _*D_t^\mu f(t) \div s^\mu \widetilde{f}(s) - f(0^+) . \tag{14}$$

We note that it is generally more cumbersome to use the Laplace transform pair for the Riemann Liouville derivative (13) than for the Caputo derivative (14). Indeed the rule (13) requires the initial value of the fractional integral of $f(t)$ whereas the rule (14) simply requires the initial value of $f(t)$. For this property the Caputo derivative is mostly used in physical problems where finite initial values are given.

Then we recognize that the problems (a) and (b) are equivalent since the Laplace transform of the solution in both cases comes out as

$$\widetilde{u}(s) = \frac{s^{\alpha-1}}{s^\alpha + 1} , \tag{15}$$

that yields, in virtue of the Laplace transform pair (3),

$$u(t) = e_\alpha(t) := E_\alpha(-t^\alpha) . \tag{16}$$

We thus recognize that the Mittag-Leffler function provides the solution to the fractional relaxation equation, as outlined, for example, by Gorenflo and Mainardi [18], Mainardi and Gorenflo [11], and Mainardi [22].

Furthermore, by anti-transforming the R.H.S of (3) by using the complex Bromwich formula, and taking into account for $0 < \alpha < 1$ the contribution from branch cut on the negative real semi-axis (the denominator $s^\alpha + 1$ does not vanish in the cut plane $-\pi \leq \arg s \leq \pi$), we get, see the survey by Gorenflo and Mainardi [18],

$$e_\alpha(t) = \int_0^\infty e^{-rt} K_\alpha(r) \, dr , \tag{17}$$

where

$$K_\alpha(r) = \mp \frac{1}{\pi} \mathrm{Im} \left\{ \frac{s^{\alpha-1}}{s^\alpha + 1} \bigg|_{s \, = \, r \, e^{\pm i\pi}} \right\} = \frac{1}{\pi} \frac{r^{\alpha-1} \sin(\alpha\pi)}{r^{2\alpha} + 2 r^\alpha \cos(\alpha\pi) + 1} \geq 0. \tag{18}$$

We note that this formula was obtained as a simple exercise of complex analysis without being aware of the Titchmarsh formula for inversion of Laplace transforms [25], revised by Gross and Levi [26] and by Gross [27]. This formula is rarely outlined in books on Laplace transforms so we refer the reader for example to Apelblat's book [28] for its presence. Since $K_\alpha(r)$ is non-negative for all r in the integral, the above formula proves that $e_\alpha(t)$ is a CM function in view of the Bernstein theorem. This theorem provides a necessary and sufficient condition for a CM function as a real Laplace transform of non-negative measure.

However, the CM property of $e_\alpha(t)$ can also be seen as a consequence of the result by Pollard [20] because the transformation $x = t^\alpha$ is a Bernstein function for $\alpha \in (0,1)$. In fact it is known that a

CM function can be obtained by composing a CM with a Bernstein function based on the following theorem: *Let $\phi(t)$ be a CM function and let $\psi(t)$ be a Bernstein function, then $\phi[\psi(t)]$ is a CM function.*

As a matter of fact, $K_\alpha(r)$ provides an interesting spectral representation of $e_\alpha(t)$ in frequencies. With the change of variable $\tau = 1/r$ we get the corresponding spectral representation in relaxation times, namely

$$e_\alpha(t) = \int_0^\infty e^{-t/\tau} H_\alpha(\tau)\, d\tau\,, \quad H_\alpha(\tau) = \tau^{-2} K_\alpha(1/\tau)\,, \tag{19}$$

that can be interpreted as a continuous distributions of elementary (i.e., exponential) relaxation processes. As a consequence we get the identity between the two spectral distributions, that is

$$K_\alpha(r) = H_\alpha(\tau) = \frac{1}{\pi} \frac{\tau^{\alpha-1}\, \sin(\alpha\pi)}{\tau^{2\alpha} + 2\,\tau^\alpha\, \cos(\alpha\pi) + 1}\,, \tag{20}$$

a surprising fact pointed out in Linear Viscoelasticity by the author in his book [22]. This kind of universal/scaling property seems a peculiar one for our Mittag-Leffler function $e_\alpha(t)$.

In Figure 1, we show $K_\alpha(r)$ for some values of the parameter α. Of course for $\alpha = 1$ the Mittag-Leffler function reduces to the exponential function $\exp(-t)$ and the corresponding spectral distribution is the Dirac delta generalized function centred at $r = 1$, namely $\delta(r-1)$.

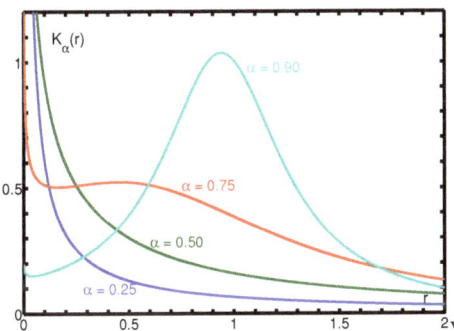

Figure 1. The spectral function $K_\alpha(r)$ for $\alpha = 0.25, 0.50, 0.75, 0.90$ in the frequency range $0 \le r \le 2$.

In Figure 2, we show some plots of $e_\alpha(t)$ for some values of the parameter α. It is worthwhile to note the different rates of decay of $e_\alpha(t)$ for small and large times. In fact the decay is very fast as $t \to 0^+$ and very slow as $t \to +\infty$.

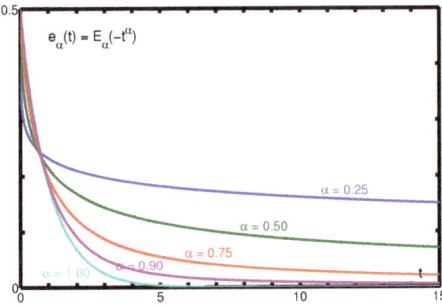

Figure 2. The Mittag-Leffler function $e_\alpha(t)$ for $\alpha = 0.25, 0.50, 0.75, 0.90, 1.$ in the time range $0 \le t \le 15$.

The Mittag-Leffler function turns out the basic function in relaxation processes of physical interest occurring in viscoelastic and dielectric materials. We refer the readers for viscoelasticity, that is, to the contribution of the author including References [22,29,30] whereas for dielectric materials to the survey by Garrappa et al. [31]. For the pioneers who have pointed out the role of the Mittagf-Leffler function in mechanical and dielectric relaxation processes we refer to the recent survey by Mainardi and Consiglio [32].

4. Asymptotic Approximations to the Mittag-Lefler Function

We now report the two common asymptotic approximations of our Mittag-Leffler function. Indeed, it is common to point out that the function $e_\alpha(t)$ matches for $t \to 0^+$ with a stretched exponential with an infinite negative derivative, whereas as $t \to \infty$ with a negative power law. The short time approximation is derived from the convergent power series representation (2). In fact,

$$e_\alpha(t) = 1 - \frac{t^\alpha}{\Gamma(1+\alpha)} + \cdots \sim \exp\left[-\frac{t^\alpha}{\Gamma(1+\alpha)}\right], \quad t \to 0. \tag{21}$$

The long time approximation is derived from the asymptotic power series representation of $e_\alpha(t)$ that turns out to be, see [6]

$$e_\alpha(t) \sim \sum_{n=1}^{\infty} (-1)^{n-1} \frac{t^{-\alpha n}}{\Gamma(1-\alpha n)}, \quad t \to \infty, \tag{22}$$

so that, at the first order,

$$e_\alpha(t) \sim \frac{t^{-\alpha}}{\Gamma(1-\alpha)}, \quad t \to \infty. \tag{23}$$

As a consequence the function $e_\alpha(t)$ interpolates for intermediate time t between the stretched exponential and the negative power law. The stretched exponential models the very fast decay for small time t, whereas the asymptotic power law is due to the very slow decay for large time t. In fact, we have the two commonly stated asymptotic representations:

$$e_\alpha(t) \sim \begin{cases} e_\alpha^0(t) := \exp\left[-\frac{t^\alpha}{\Gamma(1+\alpha)}\right], & t \to 0; \\ \\ e_\alpha^\infty(t) := \frac{t^{-\alpha}}{\Gamma(1-\alpha)} = \frac{\sin(\alpha\pi)}{\pi} \frac{\Gamma(\alpha)}{t^\alpha}, & t \to \infty. \end{cases} \tag{24}$$

The stretched exponential replaces the rapidly decreasing expression $1 - t^\alpha/\Gamma(1+\alpha)$ from (21). Of course, *for sufficiently small and for sufficiently large values of t we have the inequality*

$$e_\alpha^0(t) \le e_\alpha^\infty(t), \quad 0 < \alpha < 1. \tag{25}$$

In Figures 3 and 4, we compare for $\alpha = 0.25, 0.5, 0.75, 0.90$ in logarithmic scales the function $e_\alpha(t)$ (continuous line) and its asymptotic representations, the stretched exponential $e_\alpha^0(t)$ valid for $t \to 0$ (dashed line) and the power law $e_\alpha^\infty(t)$ valid for $t \to \infty$ (dotted line). We have chosen the time range $10^{-5} \le t \le 10^{+5}$.

We note from Figures 3 and 4 that, whereas the plots of $e_\alpha^0(t)$ remain always under the corresponding ones of $e_\alpha(t)$, the plots of $e_\alpha^\infty(t)$ start above those of $e_\alpha(t)$ but, at a certain point, an intersection may occur so changing the sign of the relative errors. The interested reader may consul the plots of the relative errors in the 2014 paper by the author [33] from which, in particular, Figures 1–4 have been extracted.

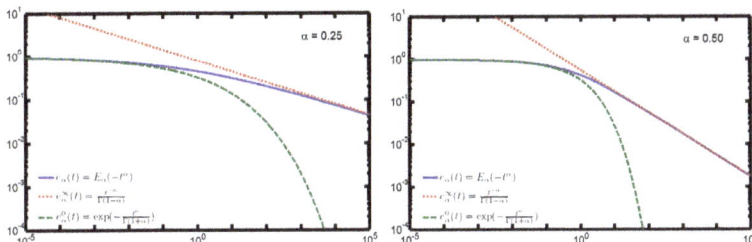

Figure 3. Approximations $e_\alpha^0(t)$ (dashed line) and $e_\alpha^\infty(t)$ (dotted line) to $e_\alpha(t)$ in $10^{-5} \leq t \leq 10^{+5}$ for $\alpha = 0.25$ (LEFT) and for $\alpha = 0/50$ (RIGHT).

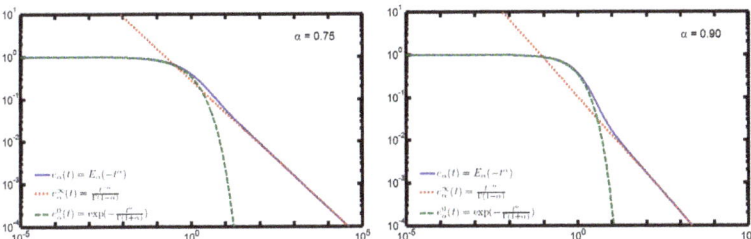

Figure 4. Approximations $e_\alpha^0(t)$ (dashed line) and $e_\alpha^\infty(t)$ (dotted line) to $e_\alpha(t)$ (LEFT) and the corresponding relative errors (RIGHT) in $10^{-5} \leq t \leq 10^{+5}$ for $\alpha = 0/75$ (LEFT) and for $\alpha = 0.90$ (RIGHT).

5. The Generalized Mittag-Leffler Function

In this survey we will devote our attention mainly to the classical Mittag-Leffler function in one parameter α as introduced by Mittag-Leffler in 1903 and defined by the power series in (1). We have just learned from the instructive E-print by Van Mieghem [13] that the series (1) was discussed by Hadamard in 1893, that is 10 years earlier than Mittag-Leffler himself.

As a matter of fact a straightforward generalization of the classical Mittag-Leffler function is obtained by replacing the additive constant 1 in the argument of the Gamma function in (1) by an arbitrary complex parameter β. It was formerly considered in 1905 by Reference [34] and soon later by Mittag-leffler himself, almost incidentally in one of his notes. Later, in the 1950's, such generalization was investigated by Humbert and Agarwal, with respect to the Laplace transformation, see References [35–37]. Usually, when dealing with Laplace transform pairs, the parameter β is required to be real and positive like α.

For this function we agree to use the notation

$$E_{\alpha,\beta}(z) := \sum_{n=0}^{\infty} \frac{z^n}{\Gamma(\alpha n + \beta)}, \quad \Re(\alpha) > 0, \ \beta \in \mathbb{C}, \ z \in \mathbb{C}. \tag{26}$$

Of course $E_{\alpha,1}(z) \equiv E_\alpha(z)$. The series is still convergent for all the complex plane $\mathbb{C}$ so the function (26) is still entire for $\Re(\alpha) > 0$ for any $\beta \in \mathbb{C}$ with order $1/\Re(\alpha)$ so the additional parameter play any role on this respect. However the Laplace transform pairs concerning the Mittag-Leffler function (26) and its derivative are known to be with $\alpha, \beta > 0$ and $\Re(s) > |\lambda|^{1\alpha}$, see, for example, Refs. [14,19,22],

$$t^{\beta-1} E_{\alpha,\beta}(-\lambda t^\alpha) \div \frac{s^{\alpha-\beta}}{s^\alpha + \lambda} = \frac{s^{-\beta}}{1 + \lambda s^{-\alpha}}. \tag{27}$$

and

$$t^{\alpha k + \beta - 1} E^{(k)}_{\alpha,\beta}(\lambda t^\alpha) \; \div \; \frac{k! \, s^{\alpha - \beta}}{(s^\alpha - \lambda)^{k+1}}, \quad k = 0, 1, 2, \ldots . \tag{28}$$

We also note the following relation concerning the first derivative of the classical Mittag-Leffler function with the two-parameter Mittag-Leffler function usually overlooked by several authors but easily proved:

$$\phi_\alpha(t) := t^{-(1-\alpha)} E_{\alpha,\alpha}\left(-t^\alpha\right) = -\frac{d}{dt} E_\alpha\left(-t^\alpha\right), \quad t \geq 0, \quad 0 < \alpha < 1. \tag{29}$$

We report the plot of the function $\phi_\alpha(t)$ herewith in Figure 5.

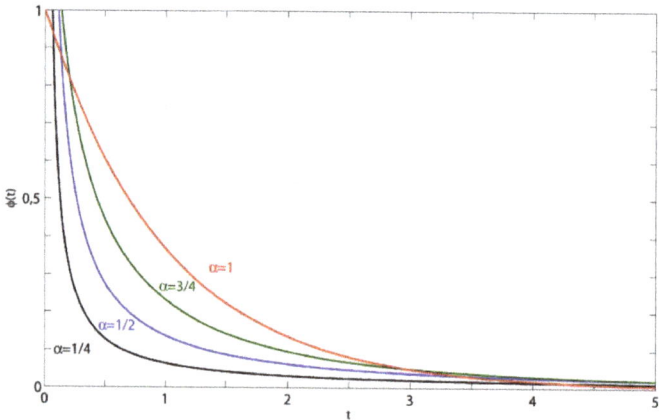

Figure 5. Plots of $\phi_\alpha(t)$ with $\alpha = 1/4, 1/2, 3/4, 1$ versus t; for $0 \leq t \leq 5$.

We note that Mittag-Leffler functions with more than two parameters were also dealt with by several authors as pointed out in [14]. In particular, for the 3-parameter Mittag-Leffler function (known as Prabhakar function) and related operators we refer the reader to the recent survey by Giusti et al. [38] and references therein. Kiryakova has dealt in a number of papers the multi-index Mittag-Leffler functions, see for example [39].

6. The Fractional Poisson Process and the Mittag-Leffler Function

Hereafter we describe how the Mittag-Leffler function enters into the so-called fractional Poisson process. We are following the original approach by Mainardi et al. in [40] where the fractional Poisson process is referred to as the renewal process of the Mittag-Leffler type. However, an independent approach to the fractional Poisson process was given for example, by Laskin in [41].

6.1. Essentials of Renewal Theory

The concept of *renewal process* has been developed as a stochastic model for describing the class of counting processes for which the times between successive events are independent identically distributed (*iid*) non-negative random variables, obeying a given probability law. These times are referred to as waiting times or inter-arrival times. For more details see, for example, the classical treatises by Cox [42], Feller [21].

For a renewal process having waiting times $T_1, T_2, \ldots$, let

$$t_0 = 0, \quad t_k = \sum_{j=1}^{k} T_j, \quad k \geq 1. \tag{30}$$

That is $t_1 = T_1$ is the time of the first renewal, $t_2 = T_1 + T_2$ is the time of the second renewal and so on. In general t_k denotes the kth renewal.

The process is specified if we know the probability law for the waiting times. In this respect we introduce the *probability density function (pdf)* $\phi(t)$ and the (cumulative) distribution function $\Phi(t)$ so defined:

$$\phi(t) := \frac{d}{dt} \Phi(t), \quad \Phi(t) := P\,(T \leq t) = \int_0^t \phi(t')\,dt'. \tag{31}$$

When the non-negative random variable represents the lifetime of technical systems, it is common to refer to $\Phi(t)$ as to the *failure probability* and to

$$\Psi(t) := P\,(T > t) = \int_t^\infty \phi(t')\,dt' = 1 - \Phi(t), \tag{32}$$

as to the *survival probability*, because $\Phi(t)$ and $\Psi(t)$ are the respective probabilities that the system does or does not fail in $(0, T]$. A relevant quantity is the *counting function* $N(t)$ defined as

$$N(t) := \max\{k | t_k \leq t, \ k = 0, 1, 2, \ldots\}, \tag{33}$$

that represents the effective number of events before or at instant t. In particular we have $\Psi(t) = P\,(N(t) = 0)$. Continuing in the general theory we set $F_1(t) = \Phi(t)$, $f_1(t) = \phi(t)$, and in general

$$F_k(t) := P\,(t_k = T_1 + \cdots + T_k \leq t), \ f_k(t) = \frac{d}{dt} F_k(t), \ k \geq 1, \tag{34}$$

thus $F_k(t)$ represents the probability that the sum of the first k waiting times is less or equal t and $f_k(t)$ its density. Then, for any fixed $k \geq 1$ the normalization condition for $F_k(t)$ is fulfilled because

$$\lim_{t \to \infty} F_k(t) = P\,(t_k = T_1 + \cdots + T_k < \infty) = 1. \tag{35}$$

In fact, the sum of k random variables each of which is finite with probability 1 is finite with probability 1 itself. By setting for consistency $F_0(t) \equiv 1$ and $f_0(t) = \delta(t)$, where for the Dirac delta generalized function in $\mathbb{R}^+$ we assume the *formal representation*

$$\delta(t) := \frac{t^{-1}}{\Gamma(0)}, \quad t \geq 0,$$

we also note that for $k \geq 0$ we have

$$P\,(N(t) = k) := P\,(t_k \leq t, \ t_{k+1} > t) = \int_0^t f_k(t')\,\Psi(t - t')\,dt'. \tag{36}$$

We now find it convenient to introduce the simplified $*$ notation for the Laplace convolution between two causal well-behaved (generalized) functions $f(t)$ and $g(t)$

$$\int_0^t f(t')\,g(t - t')\,dt' = (f * g)\,(t) = (g * f)\,(t) = \int_0^t f(t - t')\,g(t')\,dt'.$$

Being $f_k(t)$ the *pdf* of the sum of the k *iid* random variables $T_1, \ldots, T_k$ with *pdf* $\phi(t)$, we easily recognize that $f_k(t)$ turns out to be the k-fold convolution of $\phi(t)$ with itself,

$$f_k(t) = \left(\phi^{*k}\right)(t),\tag{37}$$

so Equation (36) simply reads:

$$P\left(N(t) = k\right) = \left(\phi^{*k} * \Psi\right)(t).\tag{38}$$

Because of the presence of Laplace convolutions a renewal process is suited for the Laplace transform method. Throughout this paper we will denote by $\widetilde{f}(s)$ the Laplace transform of a sufficiently well-behaved (generalized) function $f(t)$ according to

$$\mathcal{L}\left\{f(t); s\right\} = \widetilde{f}(s) = \int_0^{+\infty} e^{-st} f(t)\, dt, \quad s > s_0,$$

and for $\delta(t)$ consistently we will have $\widetilde{\delta}(s) \equiv 1$. Note that for our purposes we agree to take s real. We recognize that (38) reads in the Laplace domain

$$\mathcal{L}\{P\left(N(t) = k\right); s\} = \left[\widetilde{\phi}(s)\right]^k \widetilde{\Psi}(s),\tag{39}$$

where, using (32),

$$\widetilde{\Psi}(s) = \frac{1 - \widetilde{\phi}(s)}{s}.\tag{40}$$

6.2. The Classical Poisson Process as a Renewal Process

The most celebrated renewal process is the Poisson process characterized by a waiting time *pdf* of exponential type,

$$\phi(t) = \lambda\, e^{-\lambda t}, \quad \lambda > 0, \quad t \geq 0.\tag{41}$$

The process has *no memory*. Its moments turn out to be

$$\langle T \rangle = \frac{1}{\lambda}, \quad \langle T^2 \rangle = \frac{1}{\lambda^2}, \quad \ldots, \quad \langle T^n \rangle = \frac{1}{\lambda^n}, \quad \ldots,\tag{42}$$

and the *survival probability* is

$$\Psi(t) := P\left(T > t\right) = e^{-\lambda t}, \quad t \geq 0.\tag{43}$$

We know that the probability that k events occur in the interval of length t is

$$P\left(N(t) = k\right) = \frac{(\lambda t)^k}{k!}\, e^{-\lambda t}, \quad t \geq 0, \quad k = 0, 1, 2, \ldots.\tag{44}$$

The probability distribution related to the sum of k *iid* exponential random variables is known to be the so-called *Erlang distribution* (of order k). The corresponding density (the *Erlang pdf*) is thus

$$f_k(t) = \lambda\, \frac{(\lambda t)^{k-1}}{(k-1)!}\, e^{-\lambda t}, \quad t \geq 0, \quad k = 1, 2, \ldots,\tag{45}$$

so that the Erlang distribution function of order k turns out to be

$$F_k(t) = \int_0^t f_k(t')\, dt' = 1 - \sum_{n=0}^{k-1} \frac{(\lambda t)^n}{n!}\, e^{-\lambda t} = \sum_{n=k}^{\infty} \frac{(\lambda t)^n}{n!}\, e^{-\lambda t}, \quad t \geq 0.\tag{46}$$

In the limiting case $k = 0$ we recover $f_0(t) = \delta(t)$, $F_0(t) \equiv 1$, $t \geq 0$.

The results (44)–(46) can easily obtained by using the technique of the Laplace transform sketched in the previous section noting that for the Poisson process we have:

$$\widetilde{\phi}(s) = \frac{\lambda}{\lambda + s}, \quad \widetilde{\Psi}(s) = \frac{1}{\lambda + s}, \tag{47}$$

and for the Erlang distribution:

$$\widetilde{f}_k(s) = [\widetilde{\phi}(s)]^k = \frac{\lambda^k}{(\lambda + s)^k}, \quad \widetilde{F}_k(s) = \frac{[\widetilde{\phi}(s)]^k}{s} = \frac{\lambda^k}{s(\lambda + s)^k}. \tag{48}$$

We also recall that the survival probability for the Poisson renewal process obeys the ordinary differential equation (of relaxation type)

$$\frac{d}{dt}\Psi(t) = -\lambda\Psi(t), \quad t \geq 0; \quad \Psi(0^+) = 1. \tag{49}$$

6.3. The Renewal Process of Mittag-Leffler Type

A "fractional" generalization of the Poisson renewal process is simply obtained by generalizing the differential Equation (49) replacing there the first derivative with the integro-differential operator $_*D_t^\beta$ that is interpreted as the fractional derivative of order β in Caputo's sense, see Section 2. We write, taking for simplicity $\lambda = 1$,

$$_*D_t^\beta \Psi(t) = -\Psi(t), \quad t > 0, \quad 0 < \beta \leq 1; \quad \Psi(0^+) = 1. \tag{50}$$

We also allow the limiting case $\beta = 1$ where all the results of the previous section (with $\lambda = 1$) are expected to be recovered.

For our purpose we need to recall the Mittag-Leffler function as the natural "fractional" generalization of the exponential function, that characterizes the Poisson process. We again recall that the Mittag-Leffler function of parameter β is defined in the complex plane by the power series

$$E_\beta(z) := \sum_{n=0}^{\infty} \frac{z^n}{\Gamma(\beta n + 1)}, \quad \beta > 0, \quad z \in \mathbb{C}, \tag{51}$$

as stated in Section 2 where the parameter was denoted by α.

The solution of Equation (50) is known to be, see Section 3

$$\Psi(t) = E_\beta(-t^\beta), \quad t \geq 0, \quad 0 < \beta \leq 1, \tag{52}$$

so

$$\phi(t) := -\frac{d}{dt}\Psi(t) = -\frac{d}{dt}E_\beta(-t^\beta), \quad t \geq 0, \quad 0 < \beta \leq 1. \tag{53}$$

Then, the corresponding Laplace transforms read

$$\widetilde{\Psi}(s) = \frac{s^{\beta-1}}{1 + s^\beta}, \quad \widetilde{\phi}(s) = \frac{1}{1 + s^\beta}, \quad 0 < \beta \leq 1. \tag{54}$$

Hereafter, we find it convenient to summarize the most relevant features of the functions $\Psi(t)$ and $\phi(t)$ when $0 < \beta < 1$. We begin to quote their series expansions convergent in all of $\mathbb{R}$ suitable for $t \to 0^+$ and their asymptotic representations for $t \to \infty$,

$$\Psi(t) = \sum_{n=0}^{\infty} (-1)^n \frac{t^{\beta n}}{\Gamma(\beta n + 1)} \sim \frac{\sin(\beta\pi)}{\pi} \frac{\Gamma(\beta)}{t^\beta}, \tag{55}$$

and

$$\phi(t) = \frac{1}{t^{1-\beta}} \sum_{n=0}^{\infty} (-1)^n \frac{t^{\beta n}}{\Gamma(\beta n + \beta)} \sim \frac{\sin(\beta\pi)}{\pi} \frac{\Gamma(\beta+1)}{t^{\beta+1}} . \tag{56}$$

In contrast to the Poissonian case $\beta = 1$, in the case $0 < \beta < 1$ for large t the functions $\Psi(t)$ and $\phi(t)$ no longer decay exponentially but algebraically. As a consequence of the power-law asymptotics the process turns out to be no longer Markovian but of long-memory type. However, we recognize that for $0 < \beta < 1$ both functions $\Psi(t)$, $\phi(t)$ keep the "completely monotonic" character of the Poissonian case. as can be simply derived from Section 2. We recall that *complete monotonicity* of our functions $\Psi(t)$ and $\phi(t)$ means

$$(-1)^n \frac{d^n}{dt^n} \Psi(t) \geq 0, \quad (-1)^n \frac{d^n}{dt^n} \phi(t) \geq 0, \quad n = 0, 1, 2, \ldots, \quad t \geq 0, \tag{57}$$

or equivalently, their representability as real Laplace transforms of non-negative generalized functions (or measures).

For the generalizations of Equations (44)–(46), characteristic of the Poisson and Erlang distributions respectively, we must point out the Laplace transform pair

$$t^{\beta k} E_\beta^{(k)}(-t^\beta) \div \frac{k! \, s^{\beta-1}}{(1+s^\beta)^{k+1}}, \quad \beta > 0, \quad k = 0, 1, 2, \ldots, \tag{58}$$

with $E_\beta^{(k)}(z) := \dfrac{d^k}{dz^k} E_\beta(z)$, that can be deduced from the book by Podlubny, see Equation (1.80) in Reference [19]. Then, by using the Laplace transform pairs (25) and Equations (52), (53), (58) in Equations (37) and (38), we have the *generalized Poisson distribution*,

$$P(N(t) = k) = \frac{t^{k\beta}}{k!} E_\beta^{(k)}(-t^\beta), \quad k = 0, 1, 2, \ldots \tag{59}$$

and the *generalized Erlang pdf's* (of order $k \geq 1$),

$$f_k(t) = \beta \frac{t^{k\beta-1}}{(k-1)!} E_\beta^{(k)}(-t^\beta). \tag{60}$$

The *generalized Erlang distribution functions* turn out to be

$$F_k(t) = \int_0^t f_k(t')\, dt' = 1 - \sum_{n=0}^{k-1} \frac{t^{n\beta}}{n!} E_\beta^{(n)}(-t^\beta) = \sum_{n=k}^{\infty} \frac{t^{n\beta}}{n!} E_\beta^{(n)}(-t^\beta). \tag{61}$$

7. The Gnedenko-Kovalenko Theory of Thinning and the Mittag-Leffler Function

The *thinning* theory for a renewal process has been considered in detail by Gnedenko and Kovalenko [43] in the first edition of their book on Queue theory of 1968. However, the connection with the Laplace transform of the Mittag-Leffler function outlined at the end of this section in Equations (71) and (72), see also [44] and [45], is surprisingly not present in the second edition of the book by Gnedenko & Kovalenko in 1989.

We must note that other authors, like Szántai [46,47] speak of *rarefaction* in place of thinning.

Let us sketch here the essentials of this theory: in the interest of transparency and readability we avoid the possible decoration of the relevant power law by multiplying it with a *slowly varying function*.

Denoting by t_n, $n = 1, 2, 3, \ldots$ the time instants of events of a renewal process, assuming $0 = t_0 < t_1 < t_2 < t_3 < \ldots$, with *i.i.d.* waiting times $T_1 = t_1$, $T_k = t_k - t_{k-1}$ for $k \geq 2$, (generically denoted by T), *thinning (or rarefaction)* means that for each positive index k a decision is made: the event happening in the instant t_k is deleted with probability p or it is maintained with probability

$q = 1 - p, 0 < q < 1$. This procedure produces a *thinned* or *rarefied* renewal process with fewer events (very few events if q is near zero, the case of particular interest) in a moderate span of time.

To compensate for this loss we change the unit of time so that we still have not very few but still a moderate number of events in a moderate span of time. Such change of the unit of time is equivalent to rescaling the waiting time, multiplying it with a positive factor τ so that we have waiting times $\tau T_1, \tau T_2, \tau T_3, \ldots$, and instants $\tau t_1, \tau t_2, \tau t_3, \ldots$, in the rescaled process. Our intention is, vaguely speaking, to dispose on τ in relation to the rarefaction parameter q in such a way that for q near zero in some sense the "average" number of events per unit of time remains unchanged. In an asymptotic sense we will make these considerations precise.

Denoting by $F(t) = P(T \leq t)$ the probability distribution function of the (original) waiting time T, by $f(t)$ its density ($f(t)$ is a generalized function generating a probability measure) so that $F(t) = \int_0^t f(t')\, dt'$, and analogously by $F_k(t)$ and $f_k(t)$ the distribution and density, respectively, of the sum of k waiting times, we have recursively

$$f_k(t) = \int_0^t f_{k-1}(t - t')\, dF(t')\,, \text{ for } k \geq 2\,. \tag{62}$$

Observing that after a maintained event the next one of the original process is kept with probability q but dropped in favour of the second-next with probability $p\,q$ and, generally, $n - 1$ events are dropped in favour of the n-th-next with probability $p^{n-1}\,q$, we get for the waiting time density of the thinned process the formula

$$g_q(t) = \sum_{n=1}^{\infty} q\, p^{n-1} f_n(t)\,. \tag{63}$$

With the modified waiting time $\tau\,T$ we have

$$P(\tau T \leq t) = P(T \leq t/\tau) = F(t/\tau)\,,$$

hence the density $f(t/\tau)/\tau$, and analogously for the density of the sum of n waiting times $f_n(t/\tau)/\tau$. The density of the waiting time of the rescaled (and thinned) process now turns out as

$$g_{q,\tau}(t) = \sum_{n=1}^{\infty} q\, p^{n-1} f_n(t/\tau)/\tau\,. \tag{64}$$

In the Laplace domain we have $\widetilde{f}_n(s) = \left(\widetilde{f}(s) \right)^n$, hence (using $p = 1 - q$)

$$\widetilde{g}_q(s) = \sum_{n=1}^{\infty} q\, p^{n-1} \left(\widetilde{f}(s) \right)^n = \frac{q\, \widetilde{f}(s)}{1 - (1 - q)\, \widetilde{f}(s)}\,, \tag{65}$$

from which by Laplace inversion we can, in principle, construct the waiting time density of the thinned process. By rescaling we get

$$\widetilde{g}_{q,\tau}(s) = \sum_{n=1}^{\infty} q\, p^{n-1} \left(\widetilde{f}(\tau s) \right)^n = \frac{q\, \widetilde{f}(\tau s)}{1 - (1 - q)\, \widetilde{f}(\tau s)}\,. \tag{66}$$

Being interested in stronger and stronger thinning (*infinite thinning*) let us now consider a scale of processes with the parameters τ (of *rescaling*) and q (of *thinning*), with q tending to zero *under a scaling relation* $q = q(\tau)$ *yet to be specified*.

We have essentially two cases for the waiting time distribution: its expectation value is finite or infinite. In the first case we put

$$\lambda = \int_0^{\infty} t'\, f(t')\, dt' < \infty\,. \tag{67}$$

In the second case we assume a queue of power law type (dispensing with a possible decoration by a function slowly varying at infinity)

$$\Psi(t) := \int_t^\infty f(t')\, dt' \sim \frac{c}{\beta} t^{-\beta},\ t \to \infty \quad \text{if}\quad 0 < \beta < 1. \tag{68}$$

Then, by the Karamata theory (see References [21,48]) the above conditions mean in the Laplace domain

$$\widetilde{f}(s) = 1 - \lambda\, s^\beta + o\left(s^\beta\right), \quad \text{for}\quad s \to 0^+, \tag{69}$$

with a positive coefficient λ and $0 < \beta \le 1$. The case $\beta = 1$ obviously corresponds to the situation with finite first moment (2.6a), whereas the case $0 < \beta < 1$ is related to a power law queue with $c = \lambda\, \Gamma(\beta+1)\, \sin(\beta\pi)/\pi$.

Now, passing to the limit of $q \to 0$ of infinite thinning under the scaling relation

$$q = \lambda\, \tau^\beta,\quad 0 < \beta \le 1, \tag{70}$$

between the positive parameters q and τ, the Laplace transform of the rescaled density $\widetilde{g_{q,\tau}}(s)$ in (66) of the thinned process tends for fixed s to

$$\widetilde{g}(s) = \frac{1}{1 + s^\beta}, \tag{71}$$

which corresponds to the Mittag-Leffler density

$$g(t) = -\frac{d}{dt} E_\beta(-t^\beta) = \phi^{ML}(t). \tag{72}$$

Let us remark that Gnedenko and Kovalenko obtained (71) as the Laplace transform of the limiting density but did not identify it as the Laplace transform of a Mittag-Leffler type function. Observe that in the special case $\lambda < \infty$ we have $\beta = 1$, hence as the limiting process the Poisson process, as formerly shown in 1956 by Rényi [49].

8. The Continuous Time Random Walk (CTRW) and the Mittag-Leffler Function

The name *continuous time random walk* (CTRW) became popular in physics after Montroll and Weiss (just to cite the pioneers) published a celebrated series of papers on random walks for modelling diffusion processes on lattices, see, for example, Reference [50], and the book by Weiss [51] with references therein. CTRWs are rather good and general phenomenological models for diffusion, including processes of anomalous transport, that can be understood in the framework of the classical renewal theory. In fact a CTRW can be considered as a compound renewal process (a simple renewal process with reward) or a random walk *subordinated* to a simple renewal process. Hereafter we will mainly follow the approach by Gorenflo & Mainardi, see, for example, Reference [52].

A spatially one-dimensional CTRW is generated by a sequence of independent identically distributed (*iid*) positive random waiting times $T_1, T_2, T_3, \ldots$, each having the same probability density function $\phi(t)$, $t > 0$, and a sequence of *iid* random jumps $X_1, X_2, X_3, \ldots$, in $\mathbb{R}$, each having the same probability density $w(x)$, $x \in \mathbb{R}$.

Let us remark that, for ease of language, we use the word density also for generalized functions in the sense of Gel'fand & Shilov [53], that can be interpreted as probability measures. Usually the *probability density functions* are abbreviated by *pdf*. We recall that $\phi(t) \ge 0$ with $\int_0^\infty \phi(t)\, dt = 1$ and $w(x) \ge 0$ with $\int_{-\infty}^{+\infty} w(x)\, dx = 1$.

Setting $t_0 = 0$, $t_n = T_1 + T_2 + \ldots T_n$ for $n \in \mathbb{N}$, the wandering particle makes a jump of length X_n in instant t_n, so that its position is $x_0 = 0$ for $0 \le t < T_1 = t_1$, and $x_n = X_1 + X_2 + \ldots X_n$, for $t_n \le t < t_{n+1}$. We require the distribution of the waiting times and that of the jumps to be independent of each other. So, we have a compound renewal process (a renewal process with reward), compare Reference [42].

By natural probabilistic arguments we arrive at the *integral equation* for the probability density $p(x,t)$ (a density with respect to the variable x) of the particle being in point x at instant t,

$$p(x,t) = \delta(x)\,\Psi(t) + \int_0^t \phi(t-t') \left[\int_{-\infty}^{+\infty} w(x-x')\,p(x',t')\,dx' \right] dt', \tag{73}$$

in which $\delta(x)$ denotes the Dirac generalized function, and the *survival function*

$$\Psi(t) = \int_t^\infty \phi(t')\,dt' \tag{74}$$

denotes the probability that at instant t the particle is still sitting in its starting position $x = 0$. Clearly, Equation (73) satisfies the initial condition $p(x,0^+) = \delta(x)$.

Note that the *special choice*

$$w(x) = \delta(x-1) \tag{75}$$

gives the *pure renewal process*, with position $x(t) = N(t)$, denoting the *counting function*, and with jumps all of length 1 in positive direction happening at the renewal instants.

For many purposes the integral Equation (73) of CTRW can be easily treated by using the Laplace and Fourier transforms. Writing these as

$$\mathcal{L}\{f(t);s\} = \widetilde{f}(s) := \int_0^\infty e^{-st}\,f(t)\,dt, \quad \mathcal{F}\{g(x);\kappa\} = \widehat{g}(\kappa) := \int_{-\infty}^{+\infty} e^{+i\kappa x}\,g(x)\,dx,$$

then in the Laplace-Fourier domain Equation (73) reads

$$\widehat{\widetilde{p}}(\kappa,s) = \frac{1-\widetilde{\phi}(s)}{s} + \widetilde{\phi}(s)\,\widehat{w}(\kappa)\,\widehat{\widetilde{p}}(\kappa,s)\,. \tag{76}$$

Introducing formally in the Laplace domain the auxiliary function

$$\widetilde{H}(s) = \frac{1-\widetilde{\phi}(s)}{s\,\widetilde{\phi}(s)} = \frac{\widetilde{\Psi}(s)}{\widetilde{\phi}(s)}, \quad \text{hence} \quad \widetilde{\phi}(s) = \frac{1}{1+s\widetilde{H}(s)}, \tag{77}$$

and assuming that its Laplace inverse $H(t)$ exists, we get, following Mainardi et al. [54], in the Laplace-Fourier domain the equation

$$\widetilde{H}(s)\left[s\widehat{\widetilde{p}}(\kappa,s) - 1 \right] = [\widehat{w}(\kappa) - 1]\,\widehat{\widetilde{p}}(\kappa,s)\,, \tag{78}$$

and in the space-time domain the generalized Kolmogorov-Feller equation

$$\int_0^t H(t-t')\,\frac{\partial}{\partial t'} p(x,t')\,dt' = -p(x,t) + \int_{-\infty}^{+\infty} w(x-x')\,p(x',t)\,dx', \tag{79}$$

with $p(x,0) = \delta(x)$, where $H(t)$ acts as a *memory function*.

If the Laplace inverse $H(t)$ of the formally introduced function $\widetilde{H}(s)$ does not exist, we can formally set $\widetilde{K}(s) = 1/\widetilde{H}(s)$ and multiply (78) with $\widetilde{K}(s)$. Then, if $K(t)$ exists, we get in place of (79) the alternative form of the generalized Kolmogorov-Feller equation

$$\frac{\partial}{\partial t} p(x,t) = \int_0^t K(t-t') \left[-p(x,t') + \int_{-\infty}^{+\infty} w(x-x')\,p(x',t')\,dx' \right] dt', \tag{80}$$

with $p(x,0) = \delta(x)$ and $K(t)$ acts as a *memory function*

Special choices of the memory function $H(t)$ are (**i**) and (**ii**), see Equations (81) and (85):

$$\text{(i)} \quad H(t) = \delta(t) \quad \text{corresponding to} \quad \widetilde{H}(s) = 1, \tag{81}$$

giving the *exponential waiting time* with

$$\widetilde{\phi}(s) = \frac{1}{1+s}, \quad \phi(t) = \Psi(t) = e^{-t}. \tag{82}$$

In this case we obtain in the Fourier-Laplace domain

$$s\widehat{\widetilde{p}}(\kappa, s) - 1 = [\widehat{w}(\kappa) - 1]\,\widehat{\widetilde{p}}(\kappa, s), \tag{83}$$

and in the space-time domain the *classical Kolmogorov-Feller equation*

$$\frac{\partial}{\partial t} p(x,t) = -p(x,t) + \int_{-\infty}^{+\infty} w(x-x')\,p(x',t)\,dx', \quad p(x,0) = \delta(x). \tag{84}$$

$$\text{(ii)} \quad H(t) = \frac{t^{-\beta}}{\Gamma(1-\beta)}, \ 0 < \beta < 1, \text{ corresponding to } \widetilde{H}(s) = s^{\beta-1}, \tag{85}$$

giving the *Mittag-Leffler waiting time* with

$$\widetilde{\phi}(s) = \frac{1}{1+s^\beta}, \quad \phi(t) = -\frac{d}{dt} E_\beta(-t^\beta) = \phi^{ML}(t), \quad \Psi(t) = E_\beta(-t^\beta). \tag{86}$$

In this case we obtain in the Fourier-Laplace domain

$$s^{\beta-1}\left[s\widehat{\widetilde{p}}(\kappa,s) - 1\right] = [\widehat{w}(\kappa) - 1]\,\widehat{\widetilde{p}}(\kappa,s), \tag{87}$$

and in the space-time domain the *time fractional Kolmogorov-Feller equation*

$$_*D_t^\beta\, p(x,t) = -p(x,t) + \int_{-\infty}^{+\infty} w(x-x')\,p(x',t)\,dx', \quad p(x,0^+) = \delta(x), \tag{88}$$

where $_*D_t^\beta$ denotes the fractional derivative of of order β in the Caputo sense, see Section 3.

The time fractional Kolmogorov-Feller equation can be also expressed via the Riemann-Liouville fractional derivative $D_t^{1-\beta}$, see again Section 3, that is

$$\frac{\partial}{\partial t} p(x,t) = D_t^{1-\beta}\left[-p(x,t) + \int_{-\infty}^{+\infty} w(x-x')\,p(x',t)\,dx'\right], \tag{89}$$

with $p(x,0^+) = \delta(x)$. The equivalence of the two forms (88) and (89) is easily proved in the Fourier-Laplace domain by multiplying both sides of Equation (87) with the factor $s^{1-\beta}$.

We note that the choice (**i**) may be considered as a limit of the choice (**ii**) as $\beta = 1$. In fact, in this limit we find $\widetilde{H}(s) \equiv 1$ so $H(t) = t^{-1}/\Gamma(0) \equiv \delta(t)$ so that Equations (78)–(79) reduce to Equations (83)–(84), respectively. In this case the order of the Caputo derivative reduces to 1 and that of the R-L derivative to 0, whereas the Mittag-Leffler waiting time law reduces to the exponential.

In the sequel we will formally unite the choices (**i**) and (**ii**) by defining what we call the Mittag-Leffler memory function

$$H^{ML}(t) = \begin{cases} \dfrac{t^{-\beta}}{\Gamma(1-\beta)}, & \text{if } 0 < \beta < 1, \\ \delta(t), & \text{if } \beta = 1, \end{cases} \tag{90}$$

whose Laplace transform is

$$\widetilde{H}^{ML}(s) = s^{\beta-1}, \quad 0 < \beta \le 1. \tag{91}$$

Thus we will consider the whole range $0 < \beta \le 1$ by extending the Mittag-Leffler waiting time law in (86) to include the exponential law (82).

Remark 1. *Equation (79) clearly may be supplemented by an arbitrary initial probability density $p(x,0) = f(x)$. The corresponding replacement of $\delta(x)$ by $f(x)$ in (73) then requires in (76) multiplication of the term $(1 - \widetilde{\phi}(s))/s$ by $\widehat{f}(\kappa)$ and in (78) replacement of the LHS by $\widetilde{H}(s) \left[s\widehat{\widetilde{p}}(\kappa,s) - \widehat{f}(\kappa) \right]$. With $p(x,0) = \delta(x)$ we obtain $p(x,t)$ the fundamental solution of Equation (79).*

Note: The probability density function for the waiting time distribution in terms of the Mittag-Leffler function was formerly given since 1995 by Hilfer [55–57]. In these papers the waiting time density was given with the Mittag-Leffler function in two parameters without noting the relation with the first derivative of the classical Mittag-Leffler function as stated in Equation (29). We also note that 10 years earlier Balakrishnan [58] had given a similar expression without recognizing the Mittag-Leffler function. Like in the case of the thinning process dealt by Gnedenko-Kowalenko (see Section 7) once again the Mitag-Leffler function was unknown to the authors.

Manipulations: Rescaling and Respeeding

We now consider two types of manipulations on the CTRW by acting on its governing Equation (79) in its Laplace-Fourier representation (78).
(**A**): rescaling the waiting time, hence the whole time axis;
(**B**): respeeding the process.

(**A**) means change of the unit of time (measurement). We replace the random waiting time T by a waiting time τT, with the positive *rescaling factor* τ. Our idea is to take $0 < \tau \ll 1$ in order to bring into near sight the distant future so that in a moderate span of time we will have a large number of jump events. For $\tau > 0$ we get the rescaled waiting time density

$$\widetilde{\phi}_\tau(s) = \widetilde{\phi}(\tau s). \tag{92}$$

By decorating also the density p with an index τ we obtain the rescaled integral equation of the CTRW in the Laplace-Fourier domain as

$$\widetilde{H}_\tau(s) \left[s\widehat{\widetilde{p}}_\tau(\kappa,s) - 1 \right] = [\widehat{w}(\kappa) - 1] \, \widehat{\widetilde{p}}_\tau(\kappa,s), \tag{93}$$

where, in analogy to (77),

$$\widetilde{H}_\tau(s) = \frac{1 - \widetilde{\phi}(\tau s)}{s\,\widetilde{\phi}(\tau s)}. \tag{94}$$

(**B**) means multiplying the quantity representing $\dfrac{\partial}{\partial t}p(x,t)$ by a factor $1/a$, where $a > 0$ is the *respeeding factor*: $a > 1$ means *acceleration*, $0 < a < 1$ means *deceleration*. In the Laplace-Fourier representation this means multiplying the RHS of Equation (78) by the factor a since the expression $\left[s\widehat{\widetilde{p}}(\kappa,s) - 1 \right]$ corresponds to $\dfrac{\partial}{\partial t}p(x,t)$.

We now chose to consider the procedures of rescaling and respeeding in their combination so that the equation in the transformed domain of the rescaled and respeeded process has the form

$$\widetilde{H}_\tau(s) \left[s\widehat{\widetilde{p}}_{\tau,a}(\kappa,s) - 1 \right] = a\,[\widehat{w}(\kappa) - 1]\,\widehat{\widetilde{p}}_{\tau,a}(\kappa,s), \tag{95}$$

Clearly, the two manipulations can be discussed separately: the choice $\{\tau > 0, a = 1\}$ means *pure rescaling*, the choice $\{\tau = 1, a > 0\}$ means *pure respeeding* of the original process. In the special case $\tau = 1$ we only respeed the original system; if $0 < \tau \ll 1$ we can counteract the compression effected by rescaling to again obtain a moderate number of events in a moderate span of time by respeeding (decelerating) with $0 < a \ll 1$. These vague notions will become clear as soon as we consider power law waiting times.

Defining

$$\tilde{H}_{\tau,a}(s) := \frac{\tilde{H}_\tau(s)}{a} = \frac{1 - \tilde{\phi}(\tau s)}{a s \, \tilde{\phi}(\tau s)}. \tag{96}$$

we finally get, in analogy to (78), the equation

$$\tilde{H}_{\tau,a}(s) \left[s \widehat{\tilde{p}}_{\tau,a}(\kappa, s) - 1 \right] = [\widehat{w}(\kappa) - 1] \, \widehat{\tilde{p}}_{\tau,a}(\kappa, s). \tag{97}$$

What is the combined effect of rescaling and respeeding on the waiting time density? In analogy to (77) and taking account of (96) we find

$$\tilde{\phi}_{\tau,a}(s) = \frac{1}{1 + s \tilde{H}_{\tau,a}(s)} = \frac{1}{1 + s \dfrac{1 - \tilde{\phi}(\tau s)}{a s \, \tilde{\phi}(\tau s)}}, \tag{98}$$

and so, for the deformation of the waiting time density, the *essential formula*

$$\frac{a \, \tilde{\phi}(\tau s)}{1 - (1 - a) \tilde{\phi}(\tau s)}. \tag{99}$$

Remark 2. *The formula (99) has the same structure as the thinning formula (66) in Section 5 (just devoted to the thinning theory) by identification of a with q. In both problems we have a rescaled process defined by a time scale τ, and we send the relevant factors τ, a and q to zero under a proper relationship. However in the thinning theory the relevant independent parameter going to 0 is that of thinning (actually respeeding) whereas in the present problem it is the rescaling parameter τ.*

9. Power Laws and Asymptotic Universality of the Mittag-Leffler Waiting Time Density

We have essentially two different situations for the waiting time distribution according to its first moment (the expectation value) being finite or infinite. In other words we assume for the waiting time *pdf* $\phi(t)$ either

$$\rho := \int_0^\infty t' \, \phi(t') \, dt' < \infty, \quad \text{labelled as } \beta = 1, \tag{100}$$

or

$$\phi(t) \sim c \, t^{-(\beta+1)} \text{ for } t \to \infty \quad \text{hence } \Psi(t) \sim \frac{c}{\beta} t^{-\beta}, \, 0 < \beta < 1, \, c > 0. \tag{101}$$

For convenience we have dispensed in (101) with decorating by a slowly varying function at infinity with an asymptotic power law. Then, by the standard Tauberian theory (see References [21,48]) the above conditions (100)–(101) mean in the Laplace domain the (comprehensive) asymptotic form

$$\tilde{\phi}(s) = 1 - \lambda s^\beta + o(s^\beta) \quad \text{for} \quad s \to 0^+, \quad 0 < \beta \le 1, \tag{102}$$

where we have

$$\lambda = \rho, \quad \text{if} \quad \beta = 1; \, \lambda = c\Gamma(-\beta) = \frac{c}{\Gamma(\beta+1)} \frac{\pi}{\sin(\beta\pi)}, \, \text{if } 0 < \beta < 1. \tag{103}$$

Then, *fixing s* as required by the continuity theorem of probability theory for Laplace transforms, taking

$$a = \lambda \tau^\beta, \tag{104}$$

and *sending τ to zero*, we obtain in the limit the Mittag-Leffler waiting time law. In fact, Equations (99) and (102) imply as $\tau \to 0$ with $0 < \beta \leq 1$,

$$\widetilde{\phi}_{\tau,\lambda\tau^\beta}(s) = \frac{\lambda\tau^\beta \left[1 - \lambda\tau^\beta s^\beta + o(\tau^\beta s^\beta)\right]}{1 - (1 - \lambda\tau^\beta)\left[1 - \lambda\tau^\beta s^\beta + o(\tau^\beta s^\beta)\right]} \rightarrow \frac{1}{1 + s^\beta}, \tag{105}$$

the Laplace transform of $\phi^{ML}(t)$. This formula expresses **the asymptotic universality of the Mittag-Leffler waiting time law** that includes the exponential law for $\beta = 1$. It can easily be generalized to the case of power laws decorated with slowly varying functions, thereby using the Tauberian theory by Karamata (see again References [21,48]).

Comment: The formula (105) says that our general power law waiting time density is gradually deformed into the Mittag-Leffler waiting time density as τ tends to zero.

Remark 3. *Let us stress here the distinguished character of the Mittag-Leffler waiting time density* $\phi^{ML}(t) = -\frac{d}{dt}E_\beta(-t^\beta)$. *Considering its Laplace transform*

$$\widetilde{\phi}^{ML}(s) = \frac{1}{1 + s^\beta}, \quad \phi^{ML}(t) = -\frac{d}{dt}E_\beta(-t^\beta), \; 0 < \beta \leq 1, \tag{106}$$

we can easily prove the identity

$$\widetilde{\phi}^{ML}_{\tau,a}(s) = \widetilde{\phi}^{ML}(\tau s/a^{1/\beta}) \quad \text{for all} \quad \tau > 0, \quad a > 0. \tag{107}$$

Note that Equation (107) states the *self-similarity* of the combined operation *rescaling-respeeding* for the Mittag-Leffler waiting time density. In fact, (107) implies $\phi^{ML}_{\tau,a}(t) = \phi^{ML}(t/c)/c$ with $c = \tau/a^{1/\beta}$, which means replacing the random waiting time T^{ML} by $c\,T^{ML}$. As a consequences, choosing $a = \tau^\beta$ we have

$$\widetilde{\phi}^{ML}_{\tau,\tau^\beta}(s) = \widetilde{\phi}^{ML}(s) \quad \text{for all} \quad \tau > 0. \tag{108}$$

Hence *the Mittag-Leffler waiting time density is invariant against combined rescaling with τ and respeeding with $a = \tau^\beta$.*

Observing (105) we can say that $\phi^{ML}(t)$ is a $\tau \to 0$ attractor for any power law waiting time (101) under simultaneous rescaling with τ and respeeding with $a = \lambda\tau^\beta$. In other words, this attraction property of the Mittag-Leffler probability distribution with respect to power law waiting times (with $0 < \beta \leq 1$) is a kind of analogy to the attraction of sums of power law jump distributions by stable distributions.

10. The Mittag-Leffler Functions W.R.T the Time Fractional Diffusion-Wave Equations and the Wright Functions

In this section we show the relations of the Mittag-Leffler function with the Wright function via Laplace and Fourier transformations, in order to provide other arguments to outline the role of the Mittag-Leffler in the Fractional Calculus. For this purpose, because of the necessity to work with two independent parameters we first recall the proper definitions of the Mittag-Leffler and the Wright function. Then we will consider the time fractional diffusion-wave equation with its fundamental solutions to the basic boundary value problem that turn out to be expressed in terms of special cases of the Wright functions, the so called F and M functions. Finally we pay attention to some noteworthy formulas for the M-Wright function, including its connections with the Mittag-Leffler function.

10.1. Definitions and Main Properties of the Wright Functions

The classical *Wright function*, that we denote by $W_{\lambda,\mu}(z)$, is defined by the series representation convergent in the whole complex plane,

$$W_{\lambda,\mu}(z) := \sum_{n=0}^{\infty} \frac{z^n}{n!\,\Gamma(\lambda n + \mu)}, \quad \lambda > -1, \quad \mu \in \mathbb{C}, \tag{109}$$

As a consequence $W_{\lambda,\mu}(z)$ is an *entire function* for all $\lambda \in (-1, +\infty)$. Originally Wright assumed $\lambda \geq 0$ in connection with his investigations on the asymptotic theory of partition [59,60] and only in 1940 he considered $-1 < \lambda < 0$, [61]. We note that in the Vol 3, Chapter 18 of the handbook of the Bateman Project [6], presumably for a misprint, the parameter λ is restricted to be non-negative, whereas the Wright functions remained practically ignored in other handbooks. In 1993 the present author, being aware only of the Bateman handbook, proved that the Wright function is entire also for $-1 < \lambda < 0$ in his approaches to the time fractional diffusion equation, as outlined in his papers published from 1994 to 1997, [62–66]. For other earlier treatments of this function we refer to the 1999 paper by Gorenflo, Luchko and Mainardi [67]).

In view of the asymptotic representation in the complex domain and of the Laplace transform the Wright functions were distinguished by the author in *first kind* ($\lambda \geq 0$) and *second kind* ($-1 < \lambda < 0$) as outlined e.g., in the Appendix F of his book [22].

We note that the Wright functions are entire of order $1/(1 + \lambda)$ hence only the first kind functions ($\lambda \geq 0$) are of exponential order whereas the second kind functions ($-1 < \lambda < 0$) are not of exponential order. The case $\lambda = 0$ is trivial since $W_{0,\mu}(z) = e^z/\Gamma(\mu)$.

Following the proofs in Appendix F in Reference [22] we get the following Laplace transform pairs of the Wright functions in terms of the Mittag-Leffler functions in two parameters, where r can be the time variable $t > 0$ or the space variable $x > 0$)
for the first kind ($\lambda \geq 0$)

$$W_{\lambda,\mu}(\pm r) \;\div\; \frac{1}{s}\,E_{\lambda,\mu}\left(\pm\frac{1}{s}\right), \quad \lambda > 0, \tag{110}$$

for the second kind ($\lambda = -\nu$, $0 < \nu < 1$)

$$W_{-\nu,\mu}(-r) \;\div\; E_{\nu,\mu+\nu}(-s), \quad 0 < \nu < 1. \tag{111}$$

The Wright functions of the first kind are useful to find the solutions of some (linear and non-linear) differential equations of fractional order as recently shown by Garra and Mainardi, [68].

Since the pioneering works in 1990's by the author, noteworthy cases of Wright functions of the second kind, known as *auxiliary functions F* and *M* play fundamental roles in solving the Signalling problem and the Cauchy value problem, respectively for the time fractional diffusion-wave equation.

We first recall hereafter these auxiliary functions in terms of the Wright functions of the second kind, following their power series representations. They read

$$F_\nu(z) := W_{-\nu,0}(-z), \quad 0 < \nu < 1, \tag{112}$$

and

$$M_\nu(z) := W_{-\nu,1-\nu}(-z), \quad 0 < \nu < 1, \tag{113}$$

interrelated through

$$F_\nu(z) = \nu\, z\, M_\nu(z). \tag{114}$$

The *series representations* of our auxiliary functions are derived from those of $W_{\lambda,\mu}(z)$ in (109). We have:

$$F_\nu(z) = \sum_{n=1}^{\infty} \frac{(-z)^n}{n!\,\Gamma(-\nu n)} = -\frac{1}{\pi} \sum_{n=1}^{\infty} \frac{(-z)^n}{n!}\,\Gamma(\nu n + 1)\,\sin(\pi\nu n),\tag{115}$$

and

$$M_\nu(z) = \sum_{n=0}^{\infty} \frac{(-z)^n}{n!\,\Gamma[-\nu n + (1-\nu)]} = \frac{1}{\pi} \sum_{n=1}^{\infty} \frac{(-z)^{n-1}}{(n-1)!}\,\Gamma(\nu n)\,\sin(\pi\nu n),\tag{116}$$

where we have used the well-known reflection formula for the Gamma function,

$$\Gamma(\zeta)\,\Gamma(1-\zeta) = \pi/\sin\,\pi\zeta.$$

10.2. The Time-Fractional Diffusion-Wave Equation and the Related Green Functions

For the reader's convenience let us recall the main formulas for the time fractional diffusion equations and their fundamental solutions (also referred to as the Green functions) for the Cauchy and Signalling problems. For more details we refer to References [69,70].

Denoting as usual x, t the space and time variables, and $r = r(x,t)$ the response variable, the family of these evolution equations reads

$$\frac{\partial^\beta r}{\partial t^\beta} = a\,\frac{\partial^2 r}{\partial x^2}, \quad 0 < \beta \le 2,\tag{117}$$

where *the time derivative of order β is intended in the Caputo sense,* namely is the operator $_*D_t^\beta$, introduced in Section 3, but for order less than 1, see Equation (10), and a is a positive constant of dimension $L^2\,T^{-\beta}$. Thus we must distinguish the cases $0 < \beta \le 1$ and $1 < \beta \le 2$. We have

$$\frac{\partial^\beta r}{\partial t^\beta} := \begin{cases} \dfrac{1}{\Gamma(1-\beta)} \displaystyle\int_0^t \left[\dfrac{\partial}{\partial\tau}\,r(x,\tau)\right] \dfrac{d\tau}{(t-\tau)^\beta}, & 0 < \beta < 1, \\[4mm] \dfrac{\partial r}{\partial t}, & \beta = 1; \end{cases}\tag{118}$$

$$\frac{\partial^\beta r}{\partial t^\beta} := \begin{cases} \dfrac{1}{\Gamma(2-\beta)} \displaystyle\int_0^t \left[\dfrac{\partial^2}{\partial\tau^2}\,r(x,\tau)\right] \dfrac{d\tau}{(t-\tau)^{\beta-1}}, & 1 < \beta < 2, \\[4mm] \dfrac{\partial^2 r}{\partial t^2}, & \beta = 2. \end{cases}\tag{119}$$

It should be noted that in both cases $0 < \beta \le 1$, $1 < \beta \le 2$, the time fractional derivative in the L.H.S. of Equation (117) can be removed by a suitable fractional integration, leading to alternative forms where the necessary initial conditions at $t = 0^+$ explicitly appear.

For this purpose we apply to Equation (117) the fractional integral operator of order β, namely

$$J_t^\beta f(t) := \frac{1}{\Gamma(\beta)} \int_0^t (t-\tau)^{\beta-1}\, f(\tau)\, d\tau.$$

For $\beta \in (0,1]$ we have:

$$J_t^\beta \circ {_*D_t^\beta}\, r(x,t) = J_t^\beta \circ J_t^{1-\beta}\, D_t^1\, r(x,t) = J_t^1\, D_t^1\, r(x,t) = r(x,t) - r(x,0^+).$$

For $\beta \in (1,2]$ we have:

$$J_t^\beta \circ {_*D_t^\beta}\, r(x,t) = J_t^\beta \circ J_t^{2-\beta}\, D_t^2\, r(x,t) = J_t^2\, D_t^2\, r(x,t) = r(x,t) - r(x,0^+) - t\,r_t(x,0^+).$$

Then, as a matter fact, we get the integro-differential equations:

if $0 < \beta \le 1$:

$$r(x,t) = r(x,0^+) + \frac{a}{\Gamma(\beta)} \int_0^t \left(\frac{\partial^2 r}{\partial x^2}\right)(t-\tau)^{\beta-1} d\tau ; \tag{120}$$

if $1 < \beta \le 2$:

$$r(x,0^+) + t\frac{\partial}{\partial t} r(x,t)|_{t=0^+} + \frac{a}{\Gamma(\beta)} \int_0^t \left(\frac{\partial^2 r}{\partial x^2}\right)(t-\tau)^{\beta-1} d\tau. \tag{121}$$

Denoting by $f(x)$, $x \in \mathbb{R}$ and $h(t)$, $t \in \mathbb{R}^+$ sufficiently well-behaved functions, the basic boundary-value problems are thus formulated as following, assuming $0 < \beta \le 1$,

(a) Cauchy problem

$$r(x,0^+) = f(x), \quad -\infty < x < +\infty ; \; r(\mp\infty, t) = 0, \; t > 0; \tag{122}$$

(b) Signalling problem

$$r(x,0^+) = 0, \; x > 0; \; r(0^+,t) = h(t), \; r(+\infty, t) = 0, \; t > 0. \tag{123}$$

If $1 < \beta < 2$, we must add into (122) and (123) the initial values of the first time derivative of the field variable, $r_t(x,0^+)$, since in this case the corresponding fractional derivative is expressed in terms of the second order time derivative. To ensure the continuous dependence of our solution with respect to the parameter β also in the transition from $\beta = 1^-$ to $\beta = 1^+$, we agree to assume

$$\frac{\partial}{\partial t} r(x,t)|_{t=0^+} = 0, \; \text{for } 1 < \beta \le 2, \tag{124}$$

as it turns out from the integral forms (120)–(121).

In view of our subsequent analysis we find it convenient to set

$$\nu := \beta/2, \quad \text{so} \quad \begin{cases} 0 < \nu \le 1/2, \iff 0 < \beta \le 1, \\ 1/2 < \nu \le 1, \iff 1 < \beta \le 2, \end{cases} \tag{125}$$

and from now on to add the parameter ν to the independent space-time variables x, t in the solutions, writing $r = r(x,t;\nu)$.

For the Cauchy and Signalling problems we introduce the so-called *Green functions* $\mathcal{G}_c(x,t;\nu)$ and $\mathcal{G}_s(x,t;\nu)$, which represent the respective fundamental solutions, obtained when $f(x) = \delta(x)$ and $h(t) = \delta(t)$. As a consequence, the solutions of the two basic problems are obtained by a space or time convolution according to

$$r(x,t;\nu) = \int_{-\infty}^{+\infty} \mathcal{G}_c(x-\xi,t;\nu) f(\xi) d\xi , \tag{126}$$

$$r(x,t;\nu) = \int_{0^-}^{t^+} \mathcal{G}_s(x,t-\tau;\nu) h(\tau) d\tau . \tag{127}$$

It should be noted that in (126) $\mathcal{G}_c(x,t;\nu) = \mathcal{G}_c(|x|,t;\nu)$ because the Green function of the Cauchy problem turns out to be an even function of x. According to a usual convention, in (127) the limits of integration are extended to take into account the possibility of impulse functions centred at the extremes.

Now we recall the results obtained in 1990's by the author that allow us to express the two Green functions in terms of the auxiliary functions $F_\nu(\xi)$ and $M_\nu(\xi)$ where, for $x > 0, t > 0$

$$\xi := x/(\sqrt{a}\, t^\nu) > 0 \tag{128}$$

acts as *similarity variable*. Then we obtain the Green functions in the space-time domain in the form

$$\mathcal{G}_c(x, t; \nu) = \frac{1}{2\nu x} F_\nu(\xi) = \frac{1}{2\sqrt{a}\, t^\nu} M_\nu(\xi), \tag{129}$$

$$\mathcal{G}_s(x, t; \nu) = \frac{1}{t} F_\nu(\xi) = \frac{\nu x}{\sqrt{a}\, t^{1+\nu}} M_\nu(\xi). \tag{130}$$

We also recognize the following *reciprocity relation* for the original Green functions,

$$2\nu x\, \mathcal{G}_c(x, t; \nu) = t\, \mathcal{G}_s(x, t; \nu) = F_\nu(\xi) = \nu\xi\, M_\nu(\xi). \tag{131}$$

Now $F_\nu(\xi)$, $M_\nu(\xi)$ are the *auxiliary functions* for the general case $0 < \nu \le 1$, which generalize those well known for the standard (Fourier) diffusion equation and for the standard (D'alembert) wave equation derived for $\nu = 1/2$ and for $\nu = 1$, respectively.

10.3. Some Noteworthy Results for the M_ν Wright Function

In this survey we find worthwhile to concentrate our attention on a single auxiliary function, the M-Wright function, sometimes referred to as the *Mainardi function*. Indeed this function is indeed referred with this name in the 1999 book by Podlubny [19], that is one of the most cited treatises on fractional calculus. Then this name is found in several successive papers and books related to fractional diffusion and wave processes, see for example, the relevant 2015 paper by Sandev et al. [71].

Let us now recall some interesting analytic results related to the so-called Mainardi function. One reason for the major attention is due to its straightforward generalization of the Gaussian probability density obtained for $\nu = 1/2$, that is the fundamental solution of the Cauchy problem for the standard diffusion equation. Furthermore it allows an impressive visualization of the evolution with the order $\nu \in (0, 1)$ of the Green function of the Cauchy problem of the fractional diffusion wave Equation (129) as shown in the next figures with $a = 1$ and taking $t = 1$.

The readers are invited to look at the YouTube video by my former student Armando Consiglio whose title is "Simulation of the $M-$Wright function", in which the author has shown the evolution of this function as the parameter ν changes between 0 and 0.85 in the interval $(-5 < x < +5)$ of $\mathbb{R}$ centered in $x = 0$ represented herewith in Figures 6 and 7 at fixed time $t = 1$.

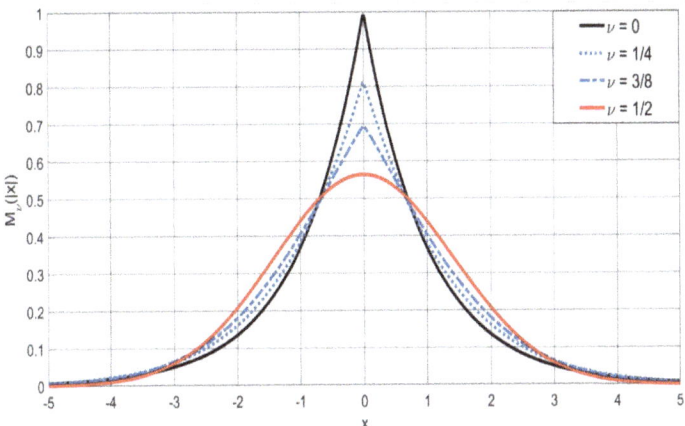

Figure 6. Plot of the symmetric $M-$Wright function $M_\nu(|x|)$ for $0 \le \nu \le 1/2$. Note that the M-Wright function becomes a Gaussian density for $\nu = 1/2$.

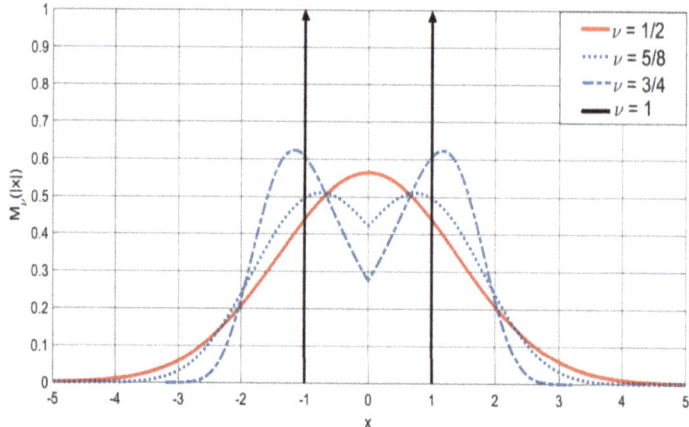

Figure 7. Plot of the symmetric $M-$Wright type function $M_\nu(|x|)$ for $1/2 \leq \nu \leq 1$. Note that the MWright function becomes a a sum of two delta functions centered in $x = \pm 1$ for $\nu = 1$.

The readers interested to have more details on the classical Wright functions should consult the recent survey by Luchko [72] and references therein.

In view of time-fractional diffusion processes related to time-fractional diffusion equations it is worthwhile to introduce the function in two variables

$$\mathbb{M}_\nu(x,t) := t^{-\nu} M_\nu(x t^{-\nu}), \quad 0 < \nu < 1, \quad x, t \in \mathbb{R}^+, \tag{132}$$

which defines a spatial probability density in x evolving in time t with self-similarity exponent $H = \nu$. Of course for $x \in \mathbb{R}$ we have to consider the symmetric version of the M-Wright function. obtained from (132) multiplying by $1/2$ and replacing x by $|x|$.

Hereafter we provide a list of the main properties of this function, which can be derived from the Laplace and Fourier transforms for the corresponding Wright M-function in one variable presented in papers by Mainardi and recalled in the Appendix F of Reference [22].

For the Laplace transform of $\mathbb{M}_\nu(x,t)$ with respect to $t > 0$ and $x > 0$ we get respectively:

$$\mathcal{L}\{\mathbb{M}_\nu(x,t); t \to s\} := \int_0^\infty e^{-st} t^{-\nu} M_\nu(x t^{-\nu}) \, dt = s^{\nu-1} e^{-x s^\nu}; \tag{133}$$

$$\mathcal{L}\{\mathbb{M}_\nu(x,t); x \to s\} := \int_0^\infty e^{-sx} t^{-\nu} M_\nu(x t^{-\nu}) \, dx = E_{\nu,1}(-st^\nu). \tag{134}$$

For the Fourier transforms with respect to the spatial variable x we have for $\mathbb{M}_\nu(x,t)$ with $x \in \mathbb{R}^+$,

$$\mathcal{F}_C\{\mathbb{M}_\nu(x,t); x \to \kappa\} := \int_0^\infty \cos(\kappa x) t^{-\nu} M_\nu(x t^{-\nu}) \, dx = E_{2\nu,1}(-\kappa^2 t^{2\nu}),$$

$$\mathcal{F}_S\{\mathbb{M}_\nu(x,t); x \to \kappa\} := \int_0^\infty \sin(\kappa x) t^{-\nu} M_\nu(x t^{-\nu}) \, dx = \kappa^\nu E_{2\nu,\nu+1}(-\kappa^2 t^{2\nu}), \tag{135}$$

so that for the symmetric function $\mathbb{M}_\nu(|x|,t)$ we get

$$\mathcal{F}\{\mathbb{M}_\nu(|x|,t); x \to \kappa\} = 2\int_0^\infty \cos(\kappa x) t^{-\nu} M_\nu(x t^{-\nu}) \, dx = 2E_{2\nu,1}\left(-\kappa^2 t^{2\nu}\right). \tag{136}$$

Restricting our attention at the known analytic expressions of the M_ν functions versus x at fixed time $t = 1$ we recall the following results for some special rational values of the parameter ν: $\nu = 1/3$ (see Reference [22])

$$M_{1/3}(x) = 3^{2/3} \text{Ai}(x/3^{1/3}), \tag{137}$$

Entropy **2020**, *22*, 1359

$\nu = 1/2$ (see Reference [22])

$$M_{1/2}(x) = \frac{1}{\sqrt{\pi}}e^{-x^2/4},$$ (138)

$\nu = 2/3$ (see Reference [73])

$$M_{2/3}(x) = 3^{-2/3}\left[3^{1/3}\,x\,\mathrm{Ai}\left(x^2/3^{4/3}\right) - 3\mathrm{Ai}'\left(x^2/3^{4/3}\right)\right]e^{-2x^3/27}.$$ (139)

In the above equations Ai and Ai$'$ denote the *Airy function* and its first derivative.

Funding: This research received no external funding.

Acknowledgments: The work of the author has been carried out in the framework of the activities of the National Group of Mathematical Physics (GNFM, INdAM). The author would like to thank the anonymous reviewers for their helpful and constructive comments and Bruce West for the invitation.

Conflicts of Interest: The author declares no conflict of interest.

Appendix A. My Acquaintance with the Mittag-Leffler Function Since the Late 1960's

I was formerly acquainted with the Mittag-Leffler function from the pioneering 1947 paper by Gross on creep and relaxation in linear viscoelasticity. It was during my PhD studies at the University of Bologna under the supervision of Prof Caputo in the year 1969. Indeed I was asked to apply in the framework of anelastic materials the derivative of non-integer order introduced by Prof Caputo in [74,75]. More recently this fractional derivative was named after him thanks the suggestions of Gotrenflo and Mainardi [18] and Podlubny [19]. I understood that the Mittag-Leffler function proposed by Gross both in creep and relaxation processes could be used in the corresponding processes in the fractional Zener model. Because Gross had computed and plotted only the spectra, see Figure 1 in this article, I was interested to plot the Mittag-Leffler function on which I was addressed in the Third volume of the Handbook of the Bateman Project published in 1955 [6]. Carrying out the plot of the Mittag-Leffler function $E_\alpha(-t^\alpha)$ using a Fortran program was not easy for me using its power series representation, so I limited the time interval to [0.5] with ordinate in logarithmic scale. As far as I know this was the former plot of this function, see References [9,10] where the results of my PhD thesis were published in 1971 jointly with my supervisor. Later I was acquainted with the viscoelastic model by Rabotnov in 1948 [76] and with the Russian school of Meskov and Rossikhin who used the so-called Rabotnov function, indeed related to the Mittag-Leffler function, and consequently with results similar to some extent to those in References [9,10]. However, our work was totally independent from the Russian school (incidentally published in Russian), as outlined in the Notes to the chapter 3 of my 2010 book, see pp. 74–76 in [22]. More later, in the 1980 I was acquainted with the results by Bagley-Torvik and by Koeller that confirmed the relevant role of the Mittag-Leffler functions in linear viscoelastic models governed by constitutive laws of fractional order. Once again their results crossed with those in References [9,10]. However, I have to confess that, when in conferences of those years I dealt with fractional derivatives in rheology, the audience remained indifferent if not hostile and laughable so I left this topic preferring to transfer my research interests to wave phenomena, in particular on the effects of dissipation on linear dispersive waves.

Incidentally, in 1980's, I was also aware of the nice treatise by Harold T. Davis on the Theory of Linear operators published in 1936 [2], where the author gave information about the fractional calculus and the Mittag-Leffler function. It was my honor to publish a recent survey on the contributions by Davis and Gross (already recalled in Introduction), whom I consider the pioneers of fractional relaxation processes in viscoelastic and dielectric materials [32]. In the firsts years of 1990s under the push of fractals, the relevance of fractional derivatives (used not always in a correct way) was outlined in several papers. For this I was induced to come back to fractional calculus. It was just this occasion for me to devote my research interests to the application of fractional calculus in relaxation, oscillation phenomena governed by fractional ODEs and diffusion, wave phenomena governed by fractional PDEs. Once again I understood the relevance of the Mittag-Leffler functions but also that of the Wright functions, both of them classified as miscellaneous functions in the handbook of Bateman project. I must note that, as far as I known, the Bateman handbook was the only one published in English to deal with these special function, and therefore accessible to me.

The year 1994 was the golden year for me as far as my acquaintance with fractional calculus and related special functions is concerned. Indeed I took profit by the acquaintance in three different conferences with the late Prof Gorenflo and Prof. Nigmatullin (in Bordeaux, France), with Prof. Podlubny and Prof. Caputo (in Atlanta, USA), and with Prof Virginia Kiryakova and the late Prof. Stankovic (in Sofia, Bulgaria), among other authorities of the fractional calculus. But it was with Prof Gorenflo that I started a collaboration for more than 20 years (1995–2015) motivated by our common interest towards the potential of the Mittag-Leffler functions in the applications of the fractional calculus.

Then, since 1997, I was interested in the emerging science of Econophysics thanks mainly to my younger colleague Enrico Scalas. With Gorenflo, Scalas and his student Raberto we published some papers on the advent of fractional Calculus in Econophysics, see e.g., [54] and my historical survey in Mathematics [77]. In 2007, on the occasion of the 80-birthday of Prof. Caputo, I published with Gorenflo a survey in Fractional Calculus and Applied Analysis [11] where I took the liberty to propose for the Mittag-Leffler function the (successful) title of *the Queen Function of the Fractional Calculus*. Some years earlier, Gorenflo had contacted the American Mathematical Society to give a specific number to the Mittag-Leffler function, that is 33E12, in the MSC classification.

Gorenflo and I promoted the Mittag-Lffler functions in several Conferences and Workshops in all the world. In particular, I would like to recall my lectures in India (under invitation of Prof Mathai, director of the Center of Mathematical Sciences, in Brazil (under invitation of Prof Edmundo Capelas de Oliveira, Campinas University) and in US (under invitation of Prof. Karniadakis, Brown University, see Reference [78]).

I like to outline my gratitude to Professor Michele Caputo (1927) and Rudolf Gorenflo (1930–2017) for having provided me with useful advice in earlier and later times, respectively. It is my pleasure to enclose a photo showing the author between them, taken in Bologna, April 2002.

Figure A1. F. Mainardi between R. Gorenflo (left) and M. Caputo (right).

Unfortunately, I lost Gorenflo's guidance and collaboration in 2015 when he suffered strong health troubles that led him to his death on 20 October 2017 at 87 years. He was Emeritus Professor of Mathematics at the Free University of Berlin since his retirement in 1998.

Entropy **2020**, *22*, 1359

Nowadays I am quite interested to promote the special functions of the Mittag-Leffler and Wright type with the second edition of the treatise by Gorenflo et al. [14] and my surveys [79,80], including the present review.

References

1. Hille, E.; Tamarkin, J.D. On the theory of linear integral equations. *Ann. Math.* **1930**, *31*, 479–528. [CrossRef]
2. Davis, H.T. *The Theory of Linear Operators*; The Principia Press: Bloomington, Indiana, 1936.
3. Sansone, G.; Gerretsen, J. *Lectures on the Theory of Functions of a Complex Variable*; Holomorphic Functions: Nordhoff, Groningen, 1960; Volume I.
4. Dzherbashyan, M.M. *Integral Transforms and Representations of Functions in the Complex Plane*; Nauka: Moscow, Russia, 1966. (In Russian)
5. Samko, S.G.; Kilbas, A.A.; Marichev, O.I. *Fractional Integrals and Derivatives, Theory and Applications*; Gordon and Breach: Amsterdam, The Netherlands, 1993.
6. Erdélyi, A.; Magnus, W.; Oberhettinger, F.; Tricomi, F. *Higher Transcendental Functions*; McGraw-Hill: New York, NY, USA, 1955; Volume 3.
7. Cole, K.S.; Cole, R.H. Dispersion and absorption in dielectrics, II. Direct current characteristics. *J. Chem. Phys.* **1942**, *10*, 98–105. [CrossRef]
8. Gross, B. On creep and relaxation. *J. Appl. Phys.* **1947**, *18*, 212–221. [CrossRef]
9. Caputo, M.; Mainardi, F. A new dissipation model based on memory mechanism. *Pure Appl. Geophys. (PAGEOPH)* **1971**, *91*, 134–147; Reprinted in *Fract. Calc. Appl. Anal.* **2007**, *10*, 309–324. [CrossRef]
10. Caputo, M.; Mainardi, F. Linear models of dissipation in anelastic solids. *Riv. Nuovo Cimento* **1971**, *1*, 161–198. [CrossRef]
11. Mainardi, F.; Gorenflo, R. Time-Fractional Derivatives in Relaxation Processes: A Tutorial Survey. *Fract. Calc. Appl. Anal.* **2007**, *10*, 269–308.
12. Haubold, H.J.; Mathai, A.M.; Saxena, R.K. Mittag-Leffler functions and their applications. *J. Appl. Math.* **2011**, *2011*, 298628. [CrossRef]
13. Van Mieghem, P. The Mittag-Leffler funcytion. *arXiv* **2005**, arXiv:2005.13330.
14. Gorenflo, R.; Kilbas, A.A.; Mainardi, F.; Rogosin, S. *Mittag-Leffler Functions. Related Topics and Applications*, 2nd ed.; Springer: Berlin, Germany, 2020.
15. Mainardi, F.; Pironi, P. The fractional Langevin equation: Brownian motion revisited. *Extr. Math.* **1996**, *11*, 140–154.
16. Mainardi, F.; Mura, A.; Tampieri, F. Brownian motion and anomalous diffusion revisited via a fractional Langevin equation. *Mod. Probl. Stat. Phys.* **2009**, *8*, 3–23.
17. Sandev, T.; Tomovoski, Ž. *Fractional Equations and Models. Theory and Applications*; Springer: Cham, Switzerland, 2019.
18. Gorenflo, R.; Mainardi, F. Fractional Calculus: Integral and Differential Equations of Fractional Order. In *Fractals and Fractional Calculus in Continuum Mechanics*; Carpinteri, A., Mainardi, F., Eds.; Springer: New York, NY, USA, 1997; pp. 223–276.
19. Podlubny, I. *Fractional Differential Equations*; Academic Press: San Diego, CA, USA, 1999.
20. Pollard, H. The completely monotonic character of the Mittag-Leffler function $E_a(-x)$. *Bull. Am. Math. Soc.* **1948**, *54*, 1115–1116. [CrossRef]
21. Feller, W. *An Introduction to Probability Theory and Its Applications*, 2nd ed.; Wiley: New York, NY, USA, 1971; Volume II; First edition (1966)
22. Mainardi, F. *Fractional Calculus and Waves in Linear Viscoelasticity*; Imperial College Press: London, UK, 2010; Second edition in preparation.
23. Miller, K.S.; Samko, S.G. Completely monotonic functions. *Integr. Transf. Spec. Funct.* **2001**, *12*, 389–402. [CrossRef]
24. Schilling, R.L.; Song, R.; Vondracek, Z. *Bernstein Functions. Theory and Applications*, 2nd ed.; De Gruyter: Berlin, Germny, 2012.
25. Titchmarsh, E.C. *Introduction to the Theory of Fourier Integrals*; Oxford University Press: Oxford, UK, 1937.
26. Gross, B.; Levi, B. Sobra el calculo de la transformacio inverse de Laplace. *Math. Notae* **1946**, *6*, 213–224.
27. Gross, B. Note on the inversion of the Laplace transform. *Philos. Mag.* **1950**, *41*, 543–544. [CrossRef]

28. Apelblat, A. *Integral Transforms and Volterra Functions*; Nova Publisher: New York, NY, USA, 2011.

29. Mainardi, F. Fractional viscoelasticity. In *Handbook of Fractional Calculus with Applications*; Tarasov, V., Ed.; De Gruyter GmbH: Berlin, Germany, 2019; Volume 5, pp. 153–182.

30. Mainardi, F.; Spada, G. Creep, relaxation and viscosity properties for basic fractional models in rheology. *Eur. Phys. J.* **2011**, *193*, 133–160. [CrossRef]

31. Garrappa, R.; Mainardi, F.; Maione, G. Models of dielectric relaxation based on completely monotone functions. *Fract. Calc. Appl. Anal.* **2016**, *19*, 1105–1160. [CrossRef]

32. Mainardi, F.; Consiglio, A. The pioneers of the Mittag-Leffler functions in dielectrical and mechanical relaxation processes. *WSEAS Trans. Math.* **2020**, *19*, 289–300. [CrossRef]

33. Mainardi, F. On some properties of the Mittag-Leffler function $E_\alpha(-t^\alpha)$, completely monotone for $t > 0$ with $0 < \alpha < 1$. *Discret. Contin. Dyn. Syst. Ser. B* **2014**, *19*, 2267–2278. [CrossRef]

34. Wiman, A. Über den Fundamentalsatz der Theorie der Funkntionen $E_\alpha(x)$. *Acta Math.* **1905**, *29*, 191–201. [CrossRef]

35. Humbert, P. Quelques résultats relatifs à la fonction de Mittag-Leffler. *C. R. Acad. Sci. Paris* **1953**, *236*, 1467–1468.

36. Agarwal, R.P. A propos d'une note de M. Pierre Humbert. *C. R. Acad. Sci. Paris* **1953**, *236*, 2031–2032.

37. Humbert, P.; Agarwal, R.P. Sur la fonction de Mittag-Leffler et quelques-unes de ses généralisations. *Bull. Sci. Math.* **1953**, *77*, 180–185.

38. Giusti, A.; Colombaro, I.; Garra, R.; Garrappa, R.; Polito, F.; Popolizio, M.; Mainardi, F. A Guide to Prabhakar functions and operators. *Fract. Calc. Appl. Anal.* **2020**, *23*, 9–54. [CrossRef]

39. Kiryakova, V. The multi-index Mittag-Leffler functions as an important class of special functions of fractional calculus. *Comp. Math. Appl.* **2010**, *59*, 1885–1895. [CrossRef]

40. Mainardi, F.; Gorenflo, R.; Scalas, E. A fractional generalization of the Poisson processes. *Vietnam J. Math.* **2004**, *32*, 53–64.

41. Laskin, N. Fractional Poisson processes. *Comm. Nonlinear Sci. Num. Sim.* **2003**, *8*, 201–213. [CrossRef]

42. Cox, D.R. *Renewal Theory*, 2nd ed.; Methuen: London, UK, 1967.

43. Gnrdenko, B.V.; Kowalenko, I.N. *Introduction to Queueing Theory*; Israel Program for Scientific Translations: Jerusalem, Israel, 1968.

44. Gorenflo, R.; Mainardi, F. Continuous time random walk, Mittag-Leffler waiting time and fractional diffusion: Mathematical aspects, Chapter 4. In *Anomalous Transport: Foundations and Applications*; Klages, R., Radons, G., Sokolov, I.M., Eds.; Wiley-VCH: Weinheim, Germany, 2008; pp. 93–127.

45. Gorenflo, R.; Mainardi, F. The Mittag-Leffler function in the thinning theory for renewal processes. *Theory Probab. Math. Stat.* **2018**, *98*, 100–108. [CrossRef]

46. Szàntai, T. Limiting distribution for the sums of random number of random variables concerning the rarefaction of recurrent events. *Stud. Sci. Math. Hung.* **1971**, *6*, 443–452.

47. Szàntai, T. On an invariance problem related to different rarefactions of recurrent events. *Stud. Sci. Math. Hung.* **1971**, *6*, 453–456.

48. Widder, D.V. *The Laplace Transform*; Princeton University Press: Princeton, NJ, USA, 1946.

49. Renyi, A. A characteristic of the Poisson stream. *Proc. Math. Inst. Hung. Acad. Sci.* **1956**, *1*, 563–570. (In Hungarian)

50. Montroll, E.W.; Weiss, G.H. Random walks on lattices, II. *J. Math. Phys.* **1965**, *6*, 167–181. [CrossRef]

51. Weiss, G.H. *Aspects and Applications of Random Walks*; North-Holland: Amsterdam, The Netherlands, 1994.

52. Gorenflo, R.; Mainardi, F. Parametric Subordination in Fractional Diffusion Processes. In *Fractional Dynamics, Recent Advances*; Klafter, J., Lim, S.C., Metzler, R., Eds.; World Scientific: Singapore, 2012; Chapter 10, pp. 227–261.

53. Gelf, I.M.; Shilov, G.E. *Generalized Functions*; Academic Press: New York, NY, USA, 1964; Volume 1.

54. Mainardi, F.; Raberto, M.; Gorenflo, R.; Scalas, E. Fractional calculus and continuous-time finance II: The waiting time distribution. *Phys. A* **2000**, *287*, 468–481. [CrossRef]

55. Hilfer, R. Exact solutions for a class of fractal time random walks. *Fractals* **1995**, *3*, 211–216. [CrossRef]

56. Hilfer, R. On fractional diffusion and continuous time random walks. *Phys. A* **2003**, *329*, 35–39. [CrossRef]

57. Hilfer, R.; Anton, L. Fractional master equations and fractal time random walks. *Phys. Rev. E* **1995**, *51*, R848–R851. [CrossRef]

58. Balakrishnan, V. Anomalous diffusion in one dimension. *Phys. A* **1985**, *132*, 569–580. [CrossRef]

59. Wright, E.M. On the coefficients of power series having exponential singularities. *J. Lond. Math. Soc.* **1933**, *8*, 71–79. [CrossRef]
60. Wright, E.M. The asymptotic expansion of the generalized Bessel function. *Proc. Lond. Math. Soc. (Ser. II)* **1935**, *38*, 257–270. [CrossRef]
61. Wright, E.M. The generalized Bessel function of order greater than one. *Quart. J. Math. Oxf. Ser.* **1940**, *11*, 36–48. [CrossRef]
62. Mainardi, F. On the initial value problem for the fractional diffusion-wave equation. In *Waves and Stability in Continuous Media*; Rionero, S., Ruggeri, T., Eds.; World Scientific: Singapore, 1994; pp. 246–251.
63. Mainardi, F. The Time Fractional Diffusion-Wave-Equation. *Radiophys. Quantum Electron.* **1995**, *38*, 20–36. [CrossRef]
64. Mainardi, F. The Fundamental Solutions for the Fractional Diffusion-Wave Equation. *Appl. Math. Lett.* **1996**, *9*, 23–28. [CrossRef]
65. Mainardi, F. Fractional Relaxation-Oscillation and Fractional Diffusion-Wave Phenomena. *Chaos Solitons Fractals* **1996**, *7*, 1461–1477. [CrossRef]
66. Mainardi, F. Fractional Calculus: Some Basic Problems in Continuum and Statistical Mechanics. In *Fractals and Fractional Calculus in Continuum Mechanics*; Carpinteri, A., Mainardi, F., Eds.; Springer: New York, NY, USA, 1997; pp. 291–348.
67. Gorenflo, R.; Luchko, Y.; Mainardi, F. Analytical Properties and Applications of the Wright Function. *Fract. Calc. Appl. Anal.* **1999**, *2*, 383–414.
68. Garra, R.; Mainardi, F. Some aspects of Wright functions in fractional differential equations. *arXiv* **2020**, arXiv:2007.13340.
69. Mainardi, F.; Luchko, Y.; Pagnini, G. The Fundamental Solution of the Space-Time Fractional Diffusion Equation. *Fract. Calc. Appl. Anal.* **2001**, *4*, 153–192.
70. Luchko, Y.; Mainardi, F. Fractional diffusion-wave hhenomena. In *Handbook of Fractional Calculus with Applications*; Tarasov, V., Ed.; De Gruyter GmbH: Berlin, Germany, 2019; Volume 5, pp. 71–98.
71. Sandev, T.; Chechkin, A.V.; Korabel, N.; Kantz, H.; Sokolov, I.M.; Metzler, R. Distributed-order diffusion equations and multifractality: Models and solutions. *Phys. Rev. E* **2015**, *92*, 042117. [CrossRef]
72. Luchko, Y. The Wright function and its applications. In *Handbook of Fractional Calculus with Applications*; Kochubei, A., Luchko, Y., Eds.; De Gruyter GmbH: Berlin, Germany, 2019; Volume 1, pp. 241–268.
73. Hanyga, A. Multidimensional solutions of time-fractional diffusion-wave equations. *Proc. R. Soc. Lond. Ser. A* **2002**, *458*, 933–957. [CrossRef]
74. Caputo, M. Linear models of dissipation whose *Q* is almost frequency independent, Part II. *Geophys. J. R. Astr. Soc.* **1967**, *13*, 529–539; Reprinted in *Fract. Calc. Appl. Anal.* **2008**, *11*, 4–14. [CrossRef]
75. Caputo, M. *Elasticità e Dissipazione*; Zanichelli: Bologna, Italy, 1969. (In Italian)
76. Rabotnov, Y.N. Equilibrium of an elastic medium with after effect. *Prikl. Matem. i Mekh. (PMM)* **1948**, *12*, 81–91. (In Russian) [CrossRef]
77. Mainardi, F. On the Advent of Fractional Calculus in Econophysics via Continuous-Time Random Walk. *Mathematics* **2020**, *8*, 641. [CrossRef]
78. Mainardi, F. A Course on Fractional Calculus. Available online: www.brown.edu/academics/applied-mathematics/teaching-schedule/fractional-calculus-lecture-notes (accessed on 29 November 2020).
79. Mainardi, F. A tutorial on the basic special functions of Fractional Calculus. *WSEAS Trans. Math.* **2020**, *19*, 74–98. [CrossRef]
80. Mainardi, F.; Consiglio, A. The Wright functions of the second kind in Mathematical Physics. *Mathematics* **2020**, *8*, 884. [CrossRef]

Publisher's Note: MDPI stays neutral with regard to jurisdictional claims in published maps and institutional affiliations.

Article

Why Do Big Data and Machine Learning Entail the Fractional Dynamics?

Haoyu Niu [1,†], YangQuan Chen [2,*,†] and Bruce J. West [3]

1 Electrical Engineering and Computer Science Department, University of California, Merced, CA 95340, USA; hniu2@ucmerced.edu
2 Mechanical Engineering Department, University of California, Merced, CA 95340, USA
3 Office of the Director, Army Research Office, Research Triangle Park, NC 27709, USA; brucejwest213@gmail.com
* Correspondence: ychen53@ucmerced.edu; Tel.: +1-209-2284672
† MESA Lab address: Room 22, 4225 Hospital Road, Atwater, CA 95301, USA.

Abstract: Fractional-order calculus is about the differentiation and integration of non-integer orders. Fractional calculus (FC) is based on fractional-order thinking (FOT) and has been shown to help us to understand complex systems better, improve the processing of complex signals, enhance the control of complex systems, increase the performance of optimization, and even extend the enabling of the potential for creativity. In this article, the authors discuss the fractional dynamics, FOT and rich fractional stochastic models. First, the use of fractional dynamics in big data analytics for quantifying big data variability stemming from the generation of complex systems is justified. Second, we show why fractional dynamics is needed in machine learning and optimal randomness when asking: "is there a more optimal way to optimize?". Third, an optimal randomness case study for a stochastic configuration network (SCN) machine-learning method with heavy-tailed distributions is discussed. Finally, views on big data and (physics-informed) machine learning with fractional dynamics for future research are presented with concluding remarks.

Keywords: fractional calculus; fractional dynamics; fractional-order thinking; heavytailedness; big data; machine learning; variability; diversity

Citation: Niu, H.; Chen, Y.; West, B.J. Why Do Big Data and Machine Learning Entail the Fractional Dynamics? *Entropy* **2021**, *23*, 297. https://doi.org/10.3390/e23030297

Academic Editor: Jose A. Tenreiro Machado

Received: 2 February 2021
Accepted: 24 February 2021
Published: 28 February 2021

1. Fractional Calculus (FC) and Fractional-Order Thinking (FOT)

Fractional calculus (FC) is the quantitative analysis of functions using non-integer-order integration and differentiation, where the order can be a real number, a complex number or even the function of a variable. The first recorded query regarding the meaning of a non-integer order differentiation appeared in a letter written in 1695 by Guillaume de l'Hôpital to Gottfried Wilhelm Leibniz, who at the same time as Isaac Newton, but independently of him, co-invented the infinitesimal calculus [1]. Numerous contributors have provided definitions for fractional derivatives and integrals [2] since then, and the theory along with the applications of FC have been expanded greatly over the centuries [3–5]. In more recent decades, the concept of **fractional dynamics** has merged and gained followers in the statistical and chemical physics communities [6–8]. For example, optimal image processing has improved through the use of fractional-order differentiation and fractional-order partial differential equations as summarized in Chen et al. [9–11]. Anomalous diffusion was described using fractional-diffusion equations in [12,13], and Metzler et al. used fractional Langevin equations to model viscoelastic materials [14].

Today, big data and machine learning (ML) are two of the hottest topics of applied scientific research, and they are closely related to one another. To better understand them, we also need fractional dynamics, as well as fractional-order thinking (FOT). Section 2 is devoted to the discussion of the relationships between big data, variability, and fractional dynamics, as well as to fractional-order data analytics (FODA) [15]. The topics touched on in this section include the Hurst parameter [16,17], fractional Gaussian noise (fGn), fractional Brownian motion (fBm), the fractional autoregressive integrated moving average (FARIMA) [18], the formalism of continuous time random walk (CTRW) [19], unmanned aerial vehicles (UAVs) and precision agriculture (PA) [20]. In Section 3, how to learn efficiently (optimally) for ML algorithms is investigated. The key to developing an efficient learning process is the method of optimization. Thus, it is important to design an efficient or perhaps optimal optimization method. The derivative-free methods, and the gradient-based methods, such as the Nesterov accelerated gradient descent (NAGD) [21], are both discussed. Furthermore, the authors propose designing and analyzing the ML algorithms in an S or Z transform domain in Section 3.3. FC is used in optimal randomness in the methods of stochastic gradient descent (SGD) [22] and random search, and in implementing the internal model principle (IMP) [23].

FOT is a way of thinking using FC. For example, there are non-integers between the integers; between logic 0 and logic 1, there is the fuzzy logic [24]; compared with integer-order splines, there are fractional-order splines [25]; between the high-order integer moments, there are non-integer-order moments, etc. FOT has been entailed by many research areas, for example, self-similar [26,27], scale-free or scale-invariant, power-law, long-range-dependence (LRD) [28,29], and $1/f^\alpha$ noise [30,31]. The terms porous media, particulate, granular, lossy, anomaly, disorder, soil, tissue, electrodes, biology [32], nano, network, transport, diffusion, and soft matters are also intimately related to FOT. However, in the present section, we mainly discuss **complexity and inverse power laws (IPL)**.

1.1. Complexity and Inverse Power Laws

When studying complexity, it is fair to ask, what does it mean to be complex? When do investigators begin identifying a system, network or phenomenon as complex [33,34]? There is an agreement among a significant fraction of the scientific community that when the distribution of the data associated with the process of interest obeys an IPL, the phenomenon is complex; see Figure 1. On the left side of the figure, the complexity "bow tie" [35–38] is the phenomenon of interest, thought to be a complex system. On the right side of the figure is the spectrum of system properties associated with IPL probability density functions (PDFs): the system has one or more of the properties of being scale-free, having a heavy tail, having a long-range dependence, and/or having a long memory [39,40]. In the book by West and Grigolini [41], there is a table listing a sample of the empirical power laws and IPLs uncovered in the past two centuries. For example, in scale-free networks, the degree distributions follow an IPL in connectivity [42,43]; in the processing of signals containing pink noise, the power spectrum follows an IPL [29]. For other examples, such as the probability density function (PDF), the autocorrelation function (ACF) [44], allometry ($Y = aX^b$) [45], anomalous relaxation (evolving over time) [46], anomalous diffusion (mean squared dissipation versus time) [13], and self-similarity can all be described by the IPL "bow tie" depicted in Figure 1.

The power law is usually described as:

$$f(x) = ax^k,\tag{1}$$

when k is negative, $f(x)$ is an IPL. One important characteristic of this power law is scale invariance [47] determined by:

$$f(cx) = a(cx)^k = c^k f(x) \propto f(x).\tag{2}$$

Note that when x is the time, the scaling depicts a property of the system dynamics. However, when the system is stochastic, the scaling is a property of the PDF (or correlation structure) and is a constraint on the collective properties of the system.

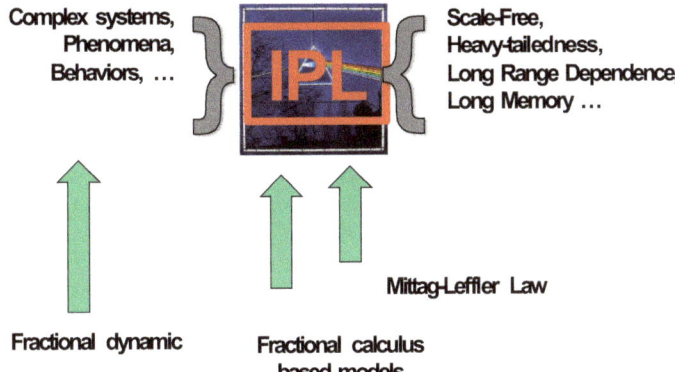

Figure 1. Inverse power law (complexity "bow tie"): On the left are the systems of interest that are thought to be complex. In the center panel, an aspect of the empirical data is characterized by an inverse power law (IPL). The right panel lists the potential properties associated with systems with data that have been processed and yield an IPL property. See text for more details.

FC is entailed by complexity, since an observable phenomenon represented by a fractal function has integer-order derivatives that diverge. Consequently, for the complexity characterization and regulation, we ought to use the fractional dynamics point of view because the fractional derivative of a fractal function is finite. Thus, complex phenomena, no matter whether they are natural or carefully engineered, ought to be described by fractional dynamics. Phenomena in complex systems in many cases should be analyzed using FC-based models, where mathematically, the IPL is actually the "Mittag–Leffler law" (MLL), which asymptotically becomes an IPL (Figure 2), known to have heavy-tail behavior.

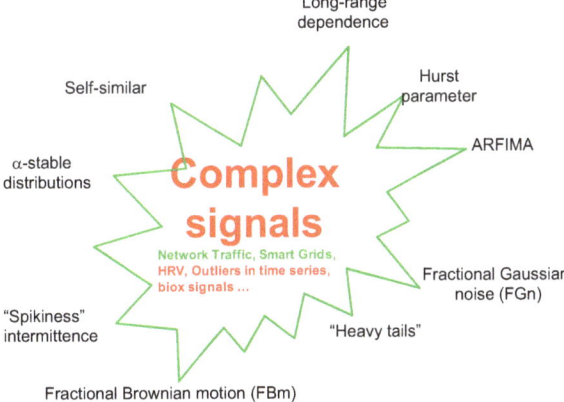

Figure 2. Complex signals (IPL): Here, the signal generated by a complex system is depicted. Exemplars of the systems are given as are the potential properties arising from the systems' complexity.

When an IPL results from processing data, one should think about how the phenomena can be connected to the FC. In [48], Gorenflo et al. explained the role of the FC in generating stable PDFs by generalizing the diffusion equation to one of fractional order. For the Cauchy problem, they considered the space-fractional diffusion equation:

$$\frac{\partial u}{\partial t} = D(\alpha)\frac{\partial^{\alpha} u}{\partial |x|^{\alpha}},$$

(3)

where $-\infty < x < \infty$, $t \geq 0$ with $u(x,0) = \delta(x)$, $0 < \alpha \leq 2$, and $D(\alpha)$ is a suitable diffusion coefficient. The fractional derivative in the diffusion variable is of the Riesz–Feller form, defined by its Fourier transform to be $|k|^{a}$. For the signalling problem, they considered the so-called time-fractional diffusion equation [49]:

$$\frac{\partial^{2\beta} u}{\partial t^{2\beta}} = D(\beta)\frac{\partial^{2} u}{\partial x^{2}},$$

(4)

where $x \geq 0$, $t \geq 0$ with $u(0,t) = \delta(t)$, $0 < \beta < 1$, and $D(\beta)$ is a suitable diffusion coefficient. Equation (4) has also been investigated in [50–52]. Here, the Caputo fractional derivative in time is used.

There are rich forms in stochasticity [22], for example, heavytailedness, which corresponds to fractional-order master equations [53]. In Section 1.2, heavy-tailed distributions are discussed.

1.2. Heavy-Tailed Distributions

In probability theory, heavy-tailed distributions are PDFs whose tails do not decay exponentially [54]. Consequently, they have more weight in their tails than does an exponential distribution. In many applications, it is the right tail of the distribution that is of interest, but a distribution may have a heavy left tail, or both tails may be heavy. Heavy-tailed distributions are widely used for modeling in different disciplines, such as finance [55], insurance [56], and medicine [57]. The distribution of a real-valued random variable X is said to have a heavy right tail if the tail probabilities $P(X > x)$ decay more slowly than those of any exponential distribution:

$$\lim_{x \to \infty}\left(\frac{P(X > x)}{e^{-\lambda x}}\right) = \infty,$$

(5)

for every $\lambda > 0$ [58]. For the heavy left tail, an analogous definition can be constructed [59]. Typically, there are three important subclasses of heavy-tailed distributions: fat-tailed, long-tailed and subexponential distributions.

1.2.1. Lévy Distribution

A Lévy distribution, named after the French mathematician Paul Lévy, can be generated by a random walk whose steps have a probability of having a length determined by a heavy-tailed distribution [60]. As a fractional-order stochastic process with heavy-tailed distributions, a Lévy distribution has better computational characteristics [61]. A Lévy distribution is stable and has a PDF that can be expressed analytically, although not always in closed form. The PDF of Lévy flight [62] is:

$$p(x,\mu,\gamma) = \begin{cases} \frac{\sqrt{\frac{\gamma}{2\pi}}}{e^{\frac{\gamma}{2(x-\mu)}}(x-\mu)^{3/2}}, & x > \mu, \\ 0, & x \leq \mu, \end{cases}$$

(6)

where μ is the location parameter and γ is the scale parameter. In practice, the Lévy distribution is updated by

$$L\acute{e}vy(\beta) = \frac{u}{|v|^{1/\beta}},$$

(7)

where u and v are random numbers generated from a normal distribution with a mean of 0 and standard deviation of 1 [63]. The stability index β ranges from 0 to 2. Moreover, it is interesting to point out that the well-known Gaussian and Cauchy distributions are special cases of the Lévy PDF when the stability index is set to 2 and 1, respectively.

1.2.2. Mittag–Leffler PDF

The Mittag–Leffler PDF [64] for the time interval between events can be written as a mixture of exponentials with a known PDF for the exponential rates:

$$E_\theta(-t^\theta) = \int_0^\infty exp(-\mu t)g(\mu)d\mu, \tag{8}$$

with a weight for the rates given by:

$$g(\mu) = \frac{1}{\pi} \frac{\sin(\theta\pi)}{\mu^{1+\theta} + 2\cos(\theta\pi)\mu + \mu^{1-\theta}}. \tag{9}$$

The most convenient expression for the random time interval was proposed by [65]:

$$\tau_\theta = -\gamma_t (\ln u \frac{\sin(\theta\pi)}{\tan(\theta\pi v)} - \cos(\theta\pi))^{1/\theta}, \tag{10}$$

where $u, v \in (0,1)$ are independent uniform random numbers, γ_t is the scale parameter, and τ_θ is the Mittag–Leffler random number. In [66], Wei et al. used the Mittag–Leffer distribution for improving the Cuckoo Search algorithm, which did show an improved performance.

1.2.3. Weibull Distribution

A random variable is described by a Weibull distribution function F:

$$F(x) = e^{-(x/k)^{\lambda_w}}, \tag{11}$$

where $k > 0$ is the scale parameter, and $\lambda_w > 0$ is the shape parameter [67]. If the shape parameter is $\lambda_w < 1$, the Weibull distribution is determined to be heavy tailed.

1.2.4. Cauchy Distribution

A random variable is described by a Cauchy PDF if its cumulative distribution is [68,69]:

$$F(x) = \frac{1}{\pi}\arctan(\frac{2(x - \mu_c)}{\sigma}) + \frac{1}{2}, \tag{12}$$

where μ_c is the location parameter and σ is the scale parameter. Cauchy distributions are examples of fat-tailed distributions, which have been empirically encountered in a variety of areas including physics, earth sciences, economics and political science [70]. Fat-tailed distributions include those whose tails decay like an IPL, which is a common point of reference in their use in the scientific literature [71].

1.2.5. Pareto Distribution

A random variable is said to be described by a Pareto PDF if its cumulative distribution function is

$$F(x) = \begin{cases} 1 - (\frac{b}{x})^a, x \geq b, \\ 0, x < b, \end{cases} \tag{13}$$

where $b > 0$ is the scale parameter and $a > 0$ is the shape parameter (Pareto's index of inequality) [72] (Figure 3).

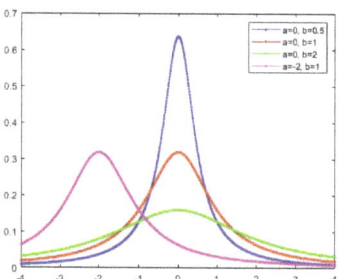

Figure 3. Cauchy distributions are examples of fat-tailed distributions. The parameter a is the location parameter; the parameter b is the scale parameter.

1.2.6. The α-Stable Distribution

A PDF is said to be stable if a linear combination of two independent random variables, each with the same distribution, has the same distribution for the conjoined variable. This PDF is also called the Lévy α-stable distribution [73,74]. Since the normal distribution, Cauchy distribution and Lévy distribution all have the above property, one can consider them to be special cases of stable distributions. Stable distributions have $0 < \alpha \leq 2$, with the upper bound corresponding to the normal distribution, and $\alpha = 1$, to the Cauchy distribution (Figure 4). The PDFs have undefined variances for $\alpha < 2$, and undefined means for $\alpha \leq 1$. Although their PDFs do not admit a closed-form formula in general, except in special cases, they decay with an IPL tail and the IPL index determines the behavior of the PDF. As the IPL index gets smaller, the PDF acquires a heavier tail. An example of an IPL index analysis is given in Section 1.4.

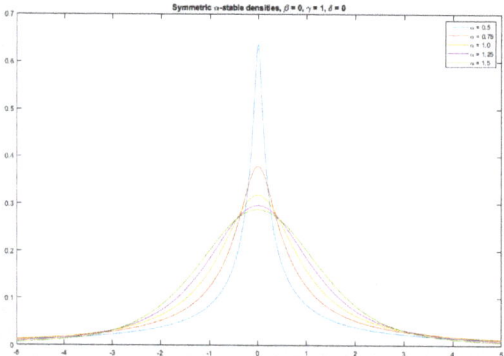

Figure 4. Symmetric α-stable distributions with unit scale factor. The most narrow PDF shown has the smallest IPL index and, consequently, the most weight in the tail regions.

1.3. Mixture Distributions

A mixture distribution is derived from a collection of other random variables. First, a random variable is selected by chance from the collection according to given probabilities of selection. Then, the value of the selected random variable is realized. The mixture PDFs are complicated in terms of simpler PDFs, which provide a good model for certain datasets. The different subsets of the data can exhibit different characteristics. Therefore, the mixed PDFs can effectively characterize the complex PDFs of certain real-world datasets. In [75], a robust stochastic configuration network (SCN) based on a mixture of Gaussian and Laplace PDFs was proposed. Thus, Gaussian and Laplace distributions are mentioned in this section for comparison purposes.

1.3.1. Gaussian Distribution

A random variable X has a Gaussian distribution with the mean μ_G and variance σ_G^2 ($-\infty < \mu_G < \infty$ and $\sigma_G > 0$) if X has a continuous distribution for which the PDF is as follows [76]:

$$f(x|\mu_G, \sigma_G^2) = \frac{1}{(2\pi)^{1/2}\sigma_G}e^{-\frac{1}{2}(\frac{x-\mu_G}{\sigma_G})^2}, for \; -\infty < x < \infty. \tag{14}$$

1.3.2. Laplace Distribution

The PDF of the Laplace distribution can be written as follows [75]:

$$F(x|\mu_l, \eta) = \frac{1}{(2\eta^2)^{1/2}}e^{(-\frac{\sqrt{2}|x-\mu_l|}{\eta})}, \tag{15}$$

where μ_l and η represent the location and scale parameters, respectively.

1.4. IPL Tail-Index Analysis

There are two approaches to the problem of the IPL tail-index estimation: the parametric [77] and the nonparametric [78]. To estimate the tail index using the parametric approach, some researchers employ a generalized extreme value (GEV) distribution [79] or Pareto distribution, and they may apply the maximum-likelihood estimator (MLE).

The stochastic gradient descent (SGD) has been widely used in deep learning with great success because of the computational efficiency [80,81]. The gradient noise (GN) in the SGD algorithm is often considered to be Gaussian in the large data regime by assuming that the classical central limit theorem (CLT) kicks in. The machine-learning tasks are usually considered as solving the following optimization problem:

$$w^* = argmin\{f(w) \triangleq \frac{1}{n}\sum_{i=1}^{n} f^{(i)}(w)\}, \tag{16}$$

where w denotes the weights of the neural network, f denotes the loss function, and n denotes the total number of instances. Then, the SGD is calculated based on the following iterative scheme:

$$w_{k+1} = w_k - \eta\nabla f_k(w_k), \tag{17}$$

where k means the iteration number, and $\nabla f_k(w_k)$ denotes the stochastic gradient at iteration k.

Since the gradient noise might not be Gaussian, the use of Brownian motion would not be appropriate to represent its behavior. Therefore, Şimşekli et al. replaced the gradient noise with the α-stable Lévy motion [82], whose increments have an α-stable distribution [83]. Because of the heavy-tailed nature of the α-stable distribution, the Lévy motion might incur large, discontinuous jumps [84], and therefore, it would exhibit a fundamentally different behavior than would Brownian motion (Figure 5):

Figure 6 shows that there are two distinct phases of SGD (in this configuration, before and after iteration 1000). At first, the loss decreases very slowly, the accuracy slightly increases, and more interestingly, α rapidly decreases. When α reaches its lowest level, which means a longer tail distribution, there is a significant jump, which causes a sudden decrease in accuracy. Beyond this point, the process recovers again, and we see stationary behavior in α and an increasing behavior in the accuracy.

(a)   (b)

Figure 5. (**a**) Brownian motion; (**b**) Lévy motion. Note that both figures are at the same size scale.

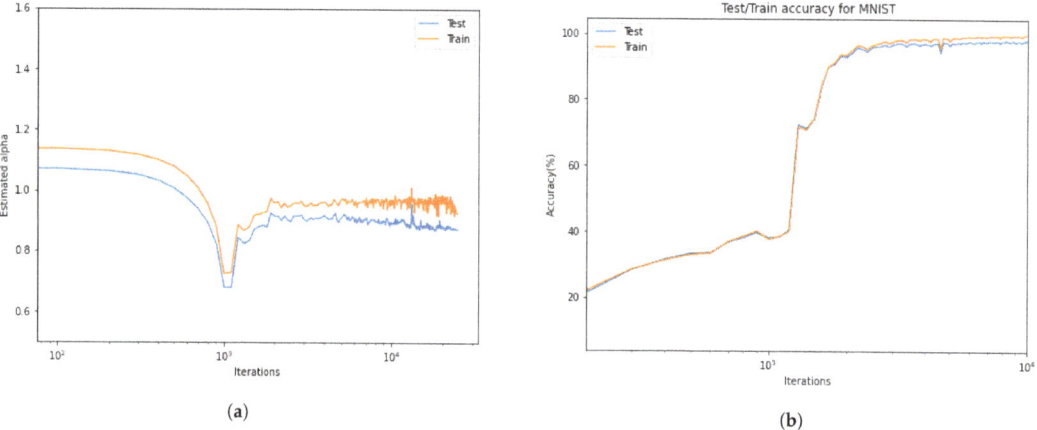

(a)   (b)

Figure 6. (**a**) The behavior of tail-index α during the iterations; (**b**) The training and testing accuracy. At first, the α decreases very slowly; when α reaches its lowest level, which means longer tail distribution, there is a significant jump, which causes a sudden decrease in accuracy. Beyond this point, the process recovers again, and we see stationary behavior in α and an increasing behavior in the accuracy.

2. Big Data, Variability and FC

The term "big data" started showing up in the early 1990s. The world's technological per capita capacity to store information has roughly doubled every 40 months since the 1980s [85]. Since 2012, there have been 2.5 exabytes (2.5×2^{60} bytes) of data generated every day [86]. According to data report predictions, there will be 163 zettabytes of data by 2025 [87]. Firican proposed, in [88], ten characteristics (properties) of big data to prepare for both the challenges and advantages of big data initiatives (Table 1). In this article, **variability** is the most important characteristic being discussed. Variability refers to several properties of big data. First, it refers to the number of inconsistencies in the data, which need to be understood by using anomaly- and outlier-detection methods for any meaningful analytics to be performed. Second, variability can also refer to diversity [89,90], resulting from disparate data types and sources, for example, healthy or unhealthy [91,92]. Finally, variability can refer to multiple research topics (Table 2).

Considering variability, Xunzi (312 BC–230 BC), who was a Confucian philosopher, made a useful observation: "Throughout a thousand acts and ten thousand changes, his way remains one and the same" [93]. Therefore, we ask: what is the "one and the same" for big data? This is the **variability**, which refers to the behavior of the dynamic system. The ancient Greek philosopher Heraclitus (535 BC–475 BC) also realized the importance of variability, prompting him to say: "The only thing that is constant is change"; "It is in changing that we find purpose"; "Nothing endures but change"; "No man ever steps in the same river twice, for it is not the same river and he is not the same man".

Heraclitus actually recognized the (fractional-order) dynamics of the river without modern scientific knowledge (in nature). After two thousand years, the integer-order calculus was invented by Sir Issac Newton and Gottfried Wilhelm Leibniz, whose main purpose was to quantify that change [94,95]. From then, scientists started using integer-order calculus to depict dynamic systems, differential equations, modelling, etc. In the 1950s, Scott Blair, who first introduced the FC into rheology, pointed out that the integer-order dynamic view of change is only for our own "convenience" (a little bit selfish). In other words, denying fractional calculus is equivalent to denying the existence of non-integers between the integers!

Table 1. The 10 Vs of big data.

Characteristics	Description
1. Volume	Best known characteristic of big data; more than 90 percent of the whole data were created in the past couple of years.
2. Velocity	The speed at which data are being generated.
3. Variety	Processing structured, unstructured and semistructured data.
4. Variability	Inconsistent speed of data loading, multitude of data dimensions, and number of inconsistencies.
5. Veracity	Confidence or trust in the data.
6. Validity	Refers to how accurate and correct the data are.
7. Vulnerability	Security concerns, data breaches.
8. Volatility	Design policy for data currency, availability, and rapid retrieval of information when required.
9. Visualization	Develop new tools considering the complex relationships between the above properties.
10. Value	The most important of the 10 Vs; substantial value must be found.

Table 2. Variability in multiple research topics.

Topics	Description
1. Climate variability	Changes in the components of the climate system and their interactions.
2. Genetic variability	Measurements of the tendencies of individual genotypes between regions.
3. Heart rate variability	Physiological phenomenon where the time interval between heart beats varies.
4. Human variability	Measurements of the characteristics, physical or mental, of human beings.
5. Spatial variability	Measurements at different spatial points exhibit different values.
6. Statistical variability	A measure of dispersion in statistics.

Blair said [96]: "We may express our concepts in Newtonian terms if we find this convenient but, if we do so, we must realize that we have made a translation into a language which is foreign to the system which we are studying (1950)".

Therefore, variability exists in big data. However, how do we realize the modeling, analysis and design (MAD) for the variability in big data within complex systems? We need fractional calculus! In other words, big data are at the nexus of complexity and FC. Thus, we first proposed fractional-order data analytics (FODA) in 2015. Metrics based on using the fractional-order signal processing techniques should be used for quantifying the generating dynamics of observed or perceived variability [15].

2.1. Hurst Parameter, fGn, and fBm

The Hurst parameter or Hurst exponent (H) was proposed for the analysis of the long-term memory of time series. It was originally developed to quantify the long-term storage capacity of reservoirs for the Nile river's volatile rain and drought conditions more than a half century ago [16,17]. To date, the Hurst parameter has also been used to measure the intensity of long range dependence (LRD) in time series [97], which requires accurate modeling and forecasting. The self-similarity and the estimation of the statistical parameters of LRD have commonly been investigated recently [98]. The Hurst parameter has also been used for characterizing the LRD process [97,99]. A LRD time series is defined as a stationary process that has long-range correlations if its covariance function $C(n)$ decays slowly as:

$$\lim_{n \to \infty} \frac{C(n)}{n^{-\alpha}} = c, \tag{18}$$

where $0 < \alpha < 1$, which relates to the Hurst parameter according to $\alpha = 2 - 2H$ [100,101]. The parameter c is a finite, positive constant. When the value of n is large, $C(n)$ behaves as the IPL c/n^α [102]. Another definition for an LRD process is that the weakly stationary time-series $X(t)$ is said to be LRD if its power spectral density (PSD) follows:

$$f(\lambda) \sim C_f |\lambda|^{-\beta}, \tag{19}$$

as $\lambda \to 0$, for a given $C_f > 0$ and a given real parameter $\beta \in (0,1)$, which corresponds to $H = (1 + \beta)/2$ [103]. When $0 < H < 0.5$, it indicates that the time intervals constitute a negatively correlated process. When $0.5 < H < 1$, it indicates that time intervals constitute a positively correlated process. When $H = 0.5$, it indicates that the process is uncorrelated.

Two of the most common LRD processes are fBm [104] and fGn [105]. The fBm process with $H(0 < H < 1)$ is defined as:

$$B_H(t) = \frac{1}{\Gamma(H+1/2)} \left\{ \int_{-\infty}^{0} [(t-s)^{H-1/2} - (-s)^{H-1/2}]dW(s) + \int_{0}^{t}(t-s)^{H-1/2}dW(s) \right\}, \tag{20}$$

where W denotes a Wiener process defined on $(-\infty, \infty)$ [106]. The fGn process is the increment sequences of the fBm process, defined as:

$$X_k = Y(k+1) - Y(k), \tag{21}$$

where $Y(k)$ is a fBm process [107].

2.2. Fractional Lower-Order Moments (FLOMs)

The FLOM is based on α-stable PDFs. The PDFs of an α-stable distribution decay in the tails more slowly than a Gaussian PDF does. Therefore, for sharp spikes or occasional bursts in signals, an α-stable PDF can be used for characterizing signals more frequently than Gauss-distributed signals [108]. Thus, the FLOM plays an important role in impulsive processes [109], equivalent to the role played by the mean and variance in a Gaussian processes. When $0 < \alpha \leq 1$, the α-stable processes have no finite first- or higher-order moments; when $1 < \alpha < 2$, the α-stable processes have a finite first-order moment and all the FLOMs with moments of fractional order that is less than 1. The correlation between the FC and FLOM was investigated in [110,111]. For the Fourier-transform pair $p(x)$ and $\phi(\mu)$, the latter is the characteristic function and is the Fourier transform of the PDF; a complex FLOM can have complex fractional lower orders [110,111]. A FLOM-based fractional power spectrum includes a covariation spectrum and a fractional low-order covariance spectrum [112]. FLOM-based fractional power spectrum techniques have been successfully used in time-delay estimation [112].

2.3. Fractional Autoregressive Integrated Moving Average (FARIMA) and Gegenbauer Autoregressive Moving Average (GARMA)

A continuous-time linear time-invariant (LTI) system can be characterized using a linear difference equation, which is known as an autoregression and moving average (ARMA) model [113,114]. The process X_t of ARMA(p,q) is defined as:

$$\Phi(B)X_t = \Theta(B)\epsilon_t, \tag{22}$$

where ϵ_t is white Gaussian noise (wGn), and B is the backshift operator. However, the ARMA model can only describe a short-range dependence (SRD) property. Therefore, based on the Hurst parameter analysis, more suitable models, such as FARIMA [115,116] and fractional integral generalized autoregressive conditional heteroscedasticity (FIGARCH) [117], were designed to more accurately analyze the LRD processes. The most important feature of these models is the long memory characteristic. The FARIMA and FIGARCH can capture both the short- and the long-memory nature of time series. For example, the FARIMA process X_t is usually defined as [118]:

$$\Phi(B)(1-B)^d X_t = \Theta(B)\epsilon_t, \tag{23}$$

where $d \in (-0.5, 0.5)$, and $(1-B)^d$ is a fractional-order difference operator. The locally stationary long-memory FARIMA model has the same equation as that of Equation (23), except that d becomes d_t, which is a time-varying parameter [119]. The locally stationary long-memory FARIMA model captures the local self-similarity of the system.

The generalized locally stationary long-memory process FARIMA model was investigated in [119]. For example, a generalized FARIMA model, which is called the Gegenbauer autoregressive moving average (GARMA), was introduced in [120]. The GARMA model is defined as:

$$\Phi(B)(1-2uB+B^2)^d X_t = \Theta(B)\epsilon_t, \tag{24}$$

where $u \in [-1, 1]$, which is a parameter that can control the frequency at which the long memory occurs. The parameter d controls the rate of decay of the autocovariance function. The GARMA model can also be extended to the so-called "k-factor GARMA model", which allows for long-memory behaviors to be associated with each of k frequencies (Gegenbauer frequencies) in the interval [0, 0.5] [121].

2.4. Continuous Time Random Walk (CTRW)

The CTRW model was proposed by Montroll and Weiss as a generalization of diffusion processes to describe the phenomenon of anomalous diffusion [19]. The basic idea is to calculate the PDF for the diffusion process by replacing the discrete steps with continuous time, along with a PDF for step lengths and a waiting-time PDF for the time intervals between steps. Montroll and Weiss applied random intervals between the successive steps in the walking process to account for local structure in the environment, such as traps [122]. The CTRW has been used for modeling multiple complex phenomena, such as chaotic dynamic networks [123]. The correlation between CTRW and diffusion equations with fractional time derivatives has also been established [124]. Meanwhile, time-space fractional diffusion equations can be treated as CTRWs with continuously distributed jumps or continuum approximations of CTRWs on lattices [125].

2.5. Unmanned Aerial Vehicles (UAVs) and Precision Agriculture

As a new remote-sensing platform, researchers are more and more interested in the potential of small UAVs for precision agriculture [126–136], especially for heterogeneous crops, such as vineyards and orchards [137,138]. Mounted on UAVs, lightweight sensors, such as RGB cameras, multispectral cameras and thermal infrared cameras, can be used to collect high-resolution images. The higher temporal and spatial resolutions of the images, relatively low operational costs, and nearly real-time image acquisition make the UAVs an ideal platform for mapping and monitoring the variability of crops and trees. UAVs can create big data and demand the FODA due to the "complexity" and, thus, variability inherent in the life process. For example, Figure 7 shows the normalized difference vegetation index (NDVI) mapping of a pomegranate orchard at a USDA ARS experimental field. Under different irrigation levels, the individual trees can show strong variability during the analysis of water stress. Life is complex! Thus, it entails variability, which as discussed above, in turn, entails fractional calculus. UAVs can then become "Tractor 2.0" for farmers in precision agriculture.

Figure 7. Normalized difference vegetation index (NDVI) mapping of pomegranate trees.

3. Optimal Machine Learning and Optimal Randomness

Machine learning (ML) is the science (and art) of programming computers so they can learn from data [139]. A more engineering-oriented definition was given by Tom Mitchell in 1997 [140], "A computer program is said to learn from experience E with respect to some task T and some performance measure P, if its performance on T, as measured by P, improves with experience E".

Most ML algorithms perform training by solving optimization problems that rely on first-order derivatives (Jacobians), which decide whether to increase or decrease weights. For huge speed boosts, faster optimizers are being used instead of the regular gradient descent optimizer. For example, the most popular boosters are momentum optimization [141], Nesterov acelerated gradient [21], AdaGrad [142], RMSProp [143] and Adam optimization [144]. The second-order (Hessian) optimization methods usually find the solutions with faster rates of convergence but with higher computational costs. Therefore, the answer to the following question is important: what is a more optimal ML algorithm? What if the derivative is fractional order instead of integer order? In this section, we discuss some applications of fractional-order gradients to optimization methods in machine-learning algorithms and investigate the accuracy and convergence rates.

As mentioned in the big data section, there is a huge amount of data in human society and nature. During the learning process of ML, we care not only about the speed, but also the accuracy of the data the machine is learning (Figure 8). The learning algorithm is important; otherwise, the data labeling and other labor costs will exhaust people beyond their abilities. When applying the accoladed artificial intelligence (AI) to an algorithm, a strong emphasis is on artificial, only followed weakly by intelligence. Therefore, the key to ML is what optimization methods are being applied. The convergence rate and global searching are two important parts of the optimization method.

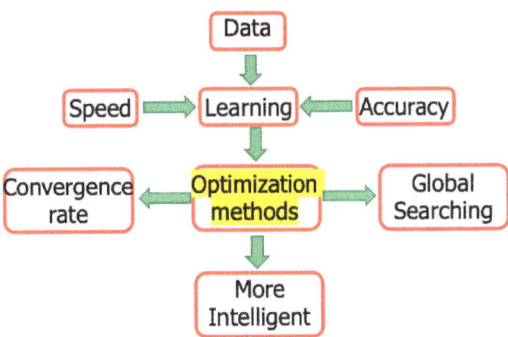

Figure 8. Data analysis in nature.

Reflection: ML is, today, a hot research topic and will probably remain so into the near future. How a machine can learn efficiently (optimally) is always important. The key for the learning process is the optimization method. Thus, in designing an efficient optimization method, it is necessary to answer the following three questions:

- What is the optimal way to optimize?
- What is the **more optimal** way to optimize?
- Can we demand **"more optimal machine learning"**, for example, deep learning with the minimum/smallest labeled data)?

Optimal randomness: In the section on the Lévy PDF, the Lévy flight is the search strategy for food the albatross has developed over millions of years of evolution. Admittedly, this is a slow optimization procedure [84]. From this perspective, we should call "Lévy distribution" an optimized or learned randomness used by albatrosses for searching for food. Therefore, we pose the question: "can the search strategy be more optimal than Lévy flight?" The answer is yes if one adopts the FC [145]! Optimization is a very complex area of study. However, a few studies have investigated using FC to obtain a better optimization strategy.

Theoretically, there are two broad optimization categories; these are derivative-free and gradient-based. For the derivative-free methods, there are the direct-search methods, consisting of particle swarm optimization (PSO) [146,147], etc. For the gradient-based methods, there are gradient descent and its variants. Both of the two categories have shown better performance when using the FC as demonstrated below.

3.1. Derivative-Free Methods

For derivative-free methods, there are single agent search and swarm-based search methods (Figure 9). Exploration is often achieved by randomness or random numbers in terms of some predefined PDFs. Exploitation uses local information such as gradients to search local regions more intensively, and such intensification can enhance the rate of convergence. Thus, a question was posed: what is the optimal randomness? Wei et al. [148] investigated the optimal randomness in a swarm-based search. Four heavy-tailed PDFs have been used for sample path analysis (Figure 10). Based on the experimental results, the randomness-enhanced cuckoo search (CS) algorithms [66,149,150] can identify the unknown specific parameters of a fractional-order system with better effectiveness and robustness. The randomness-enhanced CS algorithms can be considered as a promising tool for solving real-world complex optimization problems. The reason is that optimal randomness is applied with fractional-order noise during the exploration, which is more optimal than the "optimized PSO", CS. The fractional-order noise refers to the stable PDFs [48]. In other words, when we are discussing optimal randomness, we are discussing fractional calculus!

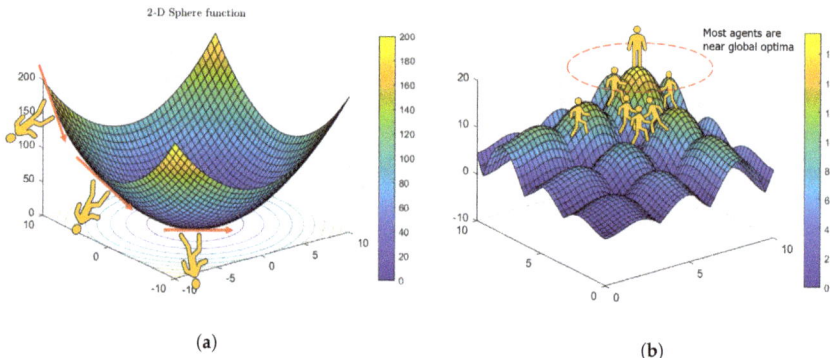

Figure 9. The 2-D Alpine function for derivative-free methods; there are (**a**) single agent search and (**b**) swarm-based search methods.

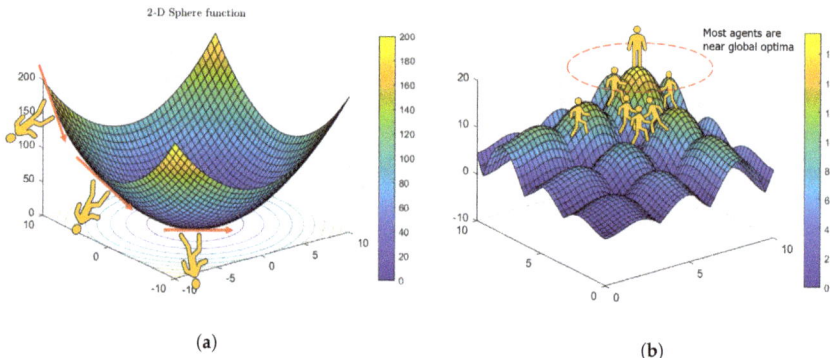

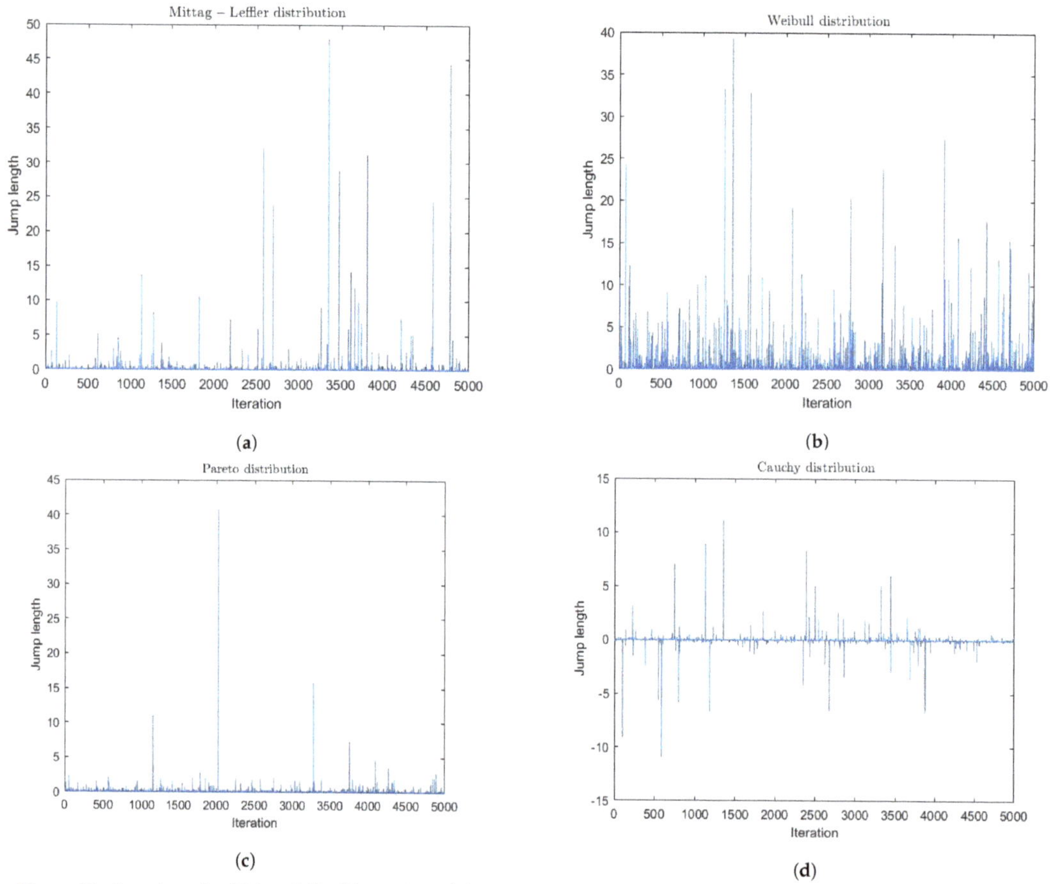

Figure 10. Sample paths. Wei et al. [148] investigated the optimal randomness in a swarm-based search. Four heavy-tailed PDFs were used for sample path analysis; there are (**a**) Mittag-Leffler distribution, (**b**) Weibull distribution, (**c**) Pareto distribution, and (**d**) Cauchy distribution. The Long steps, referring to the jump length, frequently happened for all distributions, which showed strong heavy-tailed performance. For more details, please refer to [148].

3.2. The Gradient-Based Methods

The gradient descent (GD) is a very common optimization algorithm, which can find the optimal solutions by iteratively tweaking parameters to minimize the cost function. The stochastic gradient descent (SGD) randomly selects times during the training process. Therefore, the cost function bounces up and down, decreasing on average, which is good for escape from local optima. Sometimes, noise is added into the GD method, and usually, such noise follows a Gaussian PDF in the literature. We ask, "why not heavy-tailed PDFs"? The answer to this question could lead to interesting future research.

Nesterov Accelerated Gradient Descent (NAGD)

There are many variants of GD analysis as suggested in Figure 11. One of the most popular methods is the NAGD [21]:

$$\begin{cases} y_{k+1} = ay_k - \mu\nabla f(x_k), \\ x_{k+1} = x_k + y_{k+1} + by_k, \end{cases} \tag{25}$$

where by setting $b = -a/(1+a)$, one can derive the NAGD. When $b = 0$, one can derive the momentum GD. The NAGD can also be formulated as:

$$\begin{cases} x_k = y_{k-1} - \mu\nabla f(y_{k-1}), \\ y_k = x_k + \frac{k-1}{k+2}(x_k - x_{k-1}). \end{cases} \tag{26}$$

Set $t = k\sqrt{\mu}$, and one can, in the continuous limit, derive the corresponding differential equation:

$$\ddot{X} + \frac{3}{t}\dot{X} + \nabla f(X) = 0. \tag{27}$$

The main idea of Jordan's work [151] is to analyze the iteration algorithm in the continuous-time domain. For differential equations, one can use the Laypunov or variational method to analyze the properties; for example, the convergence rate is $O(\frac{1}{t^2})$. One can also use the variational method to derive the master differential equation for an optimization method, such as the least action principle [152], Hamilton's variational principle [153] and the quantum-mechanical path integral approach [154]. Wilson et al. [151] built a Euler–Lagrange function to derive the following equation:

$$\ddot{X}_t + 2\gamma\dot{X}_t + \frac{\gamma^2}{\mu}\nabla f(X_t) = 0. \tag{28}$$

which is in the same form as the master differential equation of NAGD.

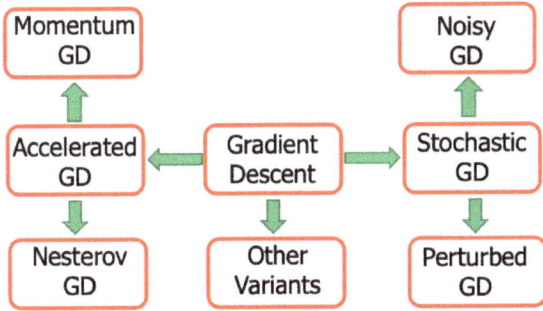

Figure 11. Gradient descent and its variants.

Jordan's work revealed that one can transform an iterative (optimization) algorithm to its continuous-time limit case, which can simplify the analysis (Laypunov methods). One can directly design a differential equation of motion (EOM) and then discretize it to derive an iterative algorithm (variational method). The key is to find a suitable Laypunov functional to analyze the stability and convergent rate. The new exciting fact established by Jordan is that optimization algorithms can be systematically synthesized using Lagrangian mechanics (Euler–Lagrange) through EOMs.

Thus, is there an optimal way to optimize using optimization algorithms stemming from Equation (28)? Obviously, why not an equation such as Equation (28) of fractional order? Considering the $\dot{X}_t$ as $X_t^{(\alpha)}$, it will provide us with more research possibilities, such as the fractional-order calculus of variation (FOCV) and fractional-order Euler–Lagrange (FOEL) equation. For the SGD, optimal randomness using the fractional-order noises can also offer better than the best performance, similarly shown by Wei et al. [148].

3.3. What Can the Control Community Offer to ML?

In the IFAC 2020 World Congress Pre-conference Workshop, Eric Kerrigan proposed "The Three Musketeers" that the control community can contribute to ML [155]. These three are the IMP [23], the Nu-Gap metric [156] and model discrimination [157]. Herein, we focused on the IMP. Kashima et al. [158] transferred the convergence problem of numerical algorithms into a stability problem of a discrete-time system. An et al. [159] explained that the commonly used SGD-momentum algorithm in ML is a PI controller and designed a PID algorithm. Motivated by [159] but differently from M. Jordan's work, we proposed designing and analyzing the algorithms in the S or Z domain. Remember that GD is a first-order algorithm:

$$x_{k+1} = x_k - \mu \nabla f(x_k), \tag{29}$$

where $\mu > 0$ is the step size (or learning rate). Using the Z transform, one can achieve:

$$X(z) = \frac{\mu}{z-1}[-\nabla f(x_k)]_z. \tag{30}$$

Approximate the gradient around the extreme point x^*, and one can obtain:

$$\nabla f(x_k) \approx A(x_k - x^*), \text{ with } A = \nabla^2 f(x^*). \tag{31}$$

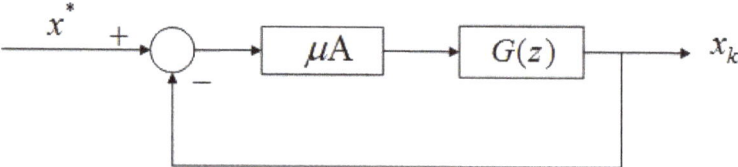

Figure 12. The integrator model (embedded in $G(z)$). The integrator in the forward loop eliminates the tracking steady-state error for a constant reference signal (internal model principle (IMP)).

For the plain GD in Figure 12, we have $G(z) = 1/(z-1)$, which is an integrator. For fractional-order GD (FOGD), the updating term of x_k in Equation (29) can be treated as a filtered gradient signal. In [160], Fan et al. shared similar thoughts: "Accelerating the convergence of the moment method for the Boltzmann equation using filters". The integrator in the forward loop eliminates the tracking error for a constant reference signal according to the internal model principle (IMP). Similarly, the GD momentum (GDM) designed to accelerate the conventional GD, which is popularly used in ML, can be analyzed using Figure 12 by:

$$\begin{cases} y_{k+1} = \alpha y_k - \mu \nabla f(x_k), \\ x_{k+1} = x_k + y_{k+1}, \end{cases} \tag{32}$$

where y_k is the accumulation of the history gradient and $\alpha \in (0, 1)$ is the rate of the moving average decay. Using the Z transform for the update rule, one can derive:

$$\begin{cases} zY(z) = \alpha Y(z) - \mu[\nabla f(x_k)]_z, \\ zX(z) = X(z) + zY(z). \end{cases} \tag{33}$$

Then, after some algebra, one obtains the following equation:

$$X(z) = \frac{\mu z}{(z-1)(z-\alpha)}[-\nabla f(x_k)]_z. \tag{34}$$

For the GD momentum, we have $G(z) = \frac{z}{(z-1)(z-\alpha)}$ in Figure 12, with an integrator in the forward loop. The GD momentum is a second-order ($G(z)$) algorithm with an additional pole at $z = \alpha$ and one zero at $z = 0$. The "second-order" refers to the order of $G(z)$, which makes it different from the algorithm using the *Hessian* matrix information. Moreover, NAGD can be simplified as:

$$\begin{cases} y_{k+1} = x_k - \mu \nabla f(x_k), \\ x_{k+1} = (1-\lambda)y_{k+1} + \lambda y_k, \end{cases} \tag{35}$$

where μ is the step size and λ is a weighting coefficient. Using the Z transform for the update rule, one can derive:

$$\begin{cases} zY(z) = X(z) - \mu[\nabla f(x_k)]_z, \\ zX(z) = (1-\lambda)zY(z) + \lambda Y(z). \end{cases} \tag{36}$$

Different from the GD momentum, and after some algebra, one obtains:

$$X(z) = \frac{-(1-\lambda)z - \lambda}{(z-1)(z+\lambda)}\mu[\nabla f(x_k)]_z = \frac{z + \frac{\lambda}{1-\lambda}}{(z-1)(z+\lambda)}\mu(1-\lambda)[-\nabla f(x_k)]_z. \tag{37}$$

For NAGD, we have $G(z) = \frac{z + \frac{\lambda}{1-\lambda}}{(z-1)(z+\lambda)}$, again, with an integrator in the forward loop (Figure 12). NAGD is a second-order algorithm with an additional pole at $z = -\lambda$ and a zero at $z = \frac{-\lambda}{1-\lambda}$.

"Can $G(z)$ be of higher order or fractional order"? Of course it can! As shown in Figure 12, a necessary condition for the stability of an algorithm is that all the poles of the closed-loop system are within the unit disc. If the Lipschitz continuous gradient constant L is given, one can replace A with L, and then, the condition is sufficient. For each $G(z)$, there is a corresponding iterative optimization algorithm. $G(z)$ can be a third- or higher-order system. Apparently, $G(z)$ can also be a fractional-order system. Considering a general second-order discrete system:

$$G(z) = \frac{z+b}{(z-1)(z-a)}, \tag{38}$$

the corresponding iterative algorithm is Equation (25). As mentioned earlier, when setting $b = -a/(1+a)$, one can derive the NAGD. When $b = 0$, one can derive the momentum GD. The iterative algorithm can be viewed as a state-space realization of the corresponding system. Thus, it may have many different realizations (all are equivalent). Since two parameters a and b are introduced for a general second-order algorithm design, we used the integral squared error (ISE) as the criterion to optimize the parameters. This is because for different target functions $f(x)$, the Lipschitz continuous gradient constant is different. Thus, the loop forward gain is defined as $\rho := \mu A$.

Table 3. General second-order algorithm design. The parameter ρ is the loop forward gain; see text for more details.

ρ	0.4	0.8	1.2	1.6	2.0	2.4
a	−0.6	−0.2	0.2	0.6	1	1.4
b	1.5	0.25	−0.1667	−0.3750	−0.5	−0.5833

According to the experimental results (Table 3), interestingly, it is found that the optimal a and b satisfy $b = -a/(1+a)$, which is the same design as NAGD. Other criteria such as the IAE and ITAE were used to find other optimal parameters, but the results are the same as for the ISE. Differently from for NAGD, the parameters were determined by search optimization rather than by mathematical design, which can be extended to more general cases. The algorithms were then tested using the MNIST dataset (Figure 13). It is obvious that for different zeros and poles, the performance of the algorithms is different. One finds that both the $b = -0.25$ and $b = -0.5$ cases perform better than does the SGD momentum. Additionally, both $b = 0.25$ and $b = 0.5$ perform worse. It is also shown that an additional zero can improve the performance, if adjusted properly. It is interesting to observe that both the method and the Nesterov method give an optimal choice of the zero, which is closely related to the pole ($b = -a/(1+a)$).

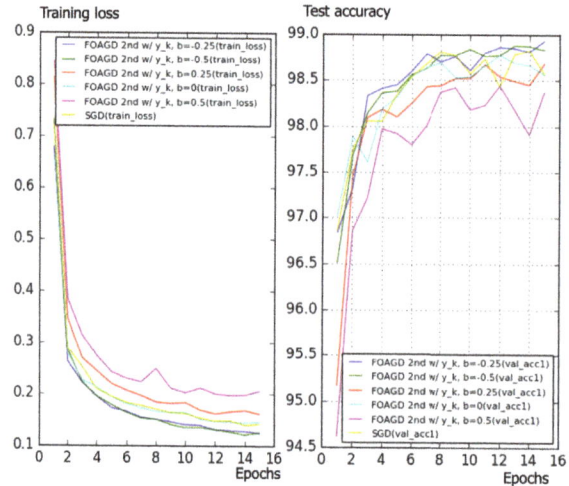

Figure 13. Training loss (**left**); test accuracy (**right**). It is obvious that for different zeros and poles, the performance of the algorithms is different. One finds that both the $b = -0.25$ and $b = -0.5$ cases perform better than does the stochastic gradient descent (SGD) momentum. Additionally, both $b = 0.25$ and $b = 0.5$ perform worse. It is also shown that an additional zero can improve the performance, if adjusted carefully.

Now, let us consider a general third-order discrete system:

$$G(z) = \frac{z^2 + cz + d}{(z-1)(z^2 + az + b)}. \tag{39}$$

Set $b = d = 0$; it will reduce to the second-order algorithm discussed above. Compared with the second-order case, the poles can now be complex numbers. More generally, a higher-order system can contain more internal models. If all the poles are real, then:

$$G(z) = \frac{1}{(z-1)} \frac{(z-c)}{(z-a)} \frac{(z-d)}{(z-b)}, \tag{40}$$

whose corresponding iterative optimization algorithm is

$$
\begin{cases}
y_{k+1} = y_k - \mu \nabla f(x_k), \\
z_{k+1} = az_k + y_{k+1} - cy_k, \\
x_{k+1} = bx_k + z_{k+1} - dz_k.
\end{cases}
\tag{41}
$$

Table 4. General third-order algorithm design, with parameters defined by Equation (41).

ρ	0.4	0.8	1.2	1.6	2.0	2.4
a	0.6439	0.5247	−0.4097	−0.5955	−1.0364	−1.4629
b	0.0263	0.0649	0.0419	−0.0398	0.0364	0.0880
c	1.5439	0.5747	−0.3763	−0.3705	−0.5364	−0.6462
d	0.0658	0.0812	0.0350	−0.0408	0.0182	0.0367

After some experiments (Table 4), it was found that since the ISE was used for tracking a step signal (it is quite simple), the optimal poles and zeros are the same as for the second-order case with a pole-zero cancellation. This is an interesting discovery. In this optimization result, all the poles and zeros are real, and the resulting performance is not very good, as expected. Compare this with the second-order case; the only difference is that in the latter, complex poles can possibly appear. Thus, the question arises: "how do complex poles play a role in the design?" The answer is obvious: by fractional calculus!

Inspired by M. Jordan's idea in the frequency domain, a continuous time fractional-order system was designed:

$$
G(s) = \frac{1}{s(s^\alpha + \beta)},
\tag{42}
$$

where $\alpha \in (0, 2)$, $\beta \in (0, 20]$ at first. It was then found that the optimal parameters were obtained by searching using the ISE criterion (Table 5).

Table 5. The continuous time fractional-order system.

ρ	0.3	0.5	0.7	0.9
α	1.8494	1.6899	1.5319	1.2284
β	20	20	20	20

Equation (42) encapsulates the continuous-time design, and one can use the numerical inverse Laplace transform (NILP) [161] and Matlab command **stmcb( )** [162] to derive its discrete form. After the complex poles are included, one can have:

$$
G(z) = \frac{(z+c)}{(z-1)} \left(\frac{1}{z-a+jb} + \frac{1}{z-a-jb} \right)
\tag{43}
$$

whose corresponding iterative algorithm is:

$$
\begin{cases}
y_{k+1} = ay_k - bz_k - \mu \nabla f(x_k), \\
z_{k+1} = az_k + by_k, \\
x_{k+1} = x_k + y_{k+1} + cy_k.
\end{cases}
\tag{44}
$$

Then, the algorithms were tested again using the MNIST dataset, and the results were compared with the SGD's. For the fractional order, $\rho = 0.9$ was used, $a = 0.6786$, $b = 0.1354$, and different values for zero c were used. When $c = 0$, the result was similar to that for the second-order SGD. When c was not equal to 0, the result was similar to that for the second-order NAGD. For the SGD, α was set to be 0.9, and the learning rate was 0.1

(Figure 14). Both $c = 0$ and $c = 0.283$ perform better than the SGD momentum; generally, with appropriate values of c, better performance can be achieved than in the second-order cases. The simulation results demonstrate that fractional calculus (complex poles) can potentially improve the performance, which is closely related to the learning rate.

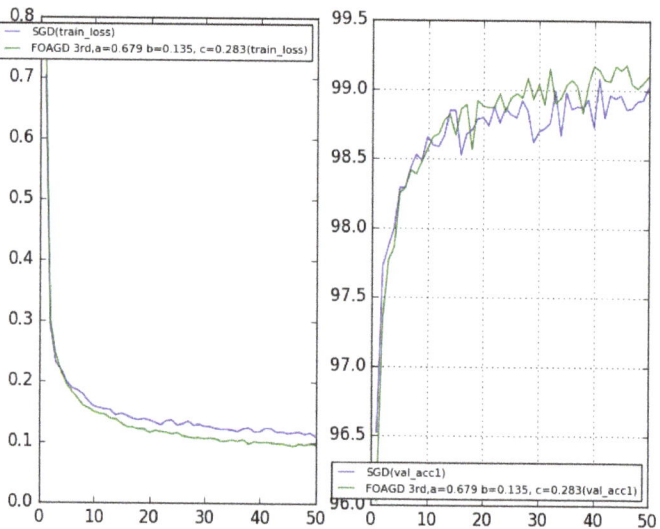

Figure 14. Training loss (**left**); test accuracy (**right**).

In general, M. Jordan asked the question: "is there an optimal way to optimize?". Our answer is a resounding yes, by limiting dynamics analysis and discretization and SGD with other randomness, such as Langevin motion. Herein, the question posed was: "is there a more optimal way to optimize?". Again, the answer is yes, but it requires the fractional calculus to be used to optimize the randomness in SGD, random search and the IMP. There is more potential for further investigations along this line of ideas.

4. A Case Study of Machine Learning with Fractional Calculus: A Stochastic Configuration Network with Heavytailedness

4.1. Stochastic Configuration Network (SCN)

The SCN model is generated incrementally by using stochastic configuration (SC) algorithms [163]. Compared with the existing randomized learning algorithms for single-layer feed-forward neural networks (SLFNNs) [164], the SCN can randomly assign the input weights (w) and biases (b) of the hidden nodes in a supervisory mechanism, which is selecting random parameters with an inequality constraint and assigning the scope of the random parameters adaptively. It can ensure that the built randomized learner models have a universal approximation property. Then, the output weights are analytically evaluated in either a constructive or selective manner [163]. In contrast with the known randomized learning algorithms, such as the randomized radial basis function (RRBF) networks [165] and the random vector functional link (RVFL) [166], the SCN can provide good generalization performance at a faster speed. Concretely, there are three types of SCN algorithms, which are labeled for convenience as SC-I, SC-II and SC-III.

The SC-I algorithm uses a constructive scheme to evaluate the output weights only for the newly added hidden node [167]. All of the previously obtained output weights are kept the same. The SC-II algorithm recalculates part of the current output weights by analyzing a local-least-squares problem with a user-defined shifting window size. The SC-III algorithm finds all the output weights together by solving a global-least-squares problem. The SCN has better performance than other randomized neural networks in terms of fast learning, the scope of the random parameters, and the required human intervention. Therefore, it has already been used in many data-processing projects, such as [134,168,169].

4.2. SCN with Heavy-Tailed PDFs

For the original SCN algorithms, weights and biases are randomly generated using a uniform PDF. Randomness plays a significant role in both exploration and exploitation. A good neural network architecture with randomly assigned weights can easily outperform a more deficient architecture with finely tuned weights [170]. Therefore, it is critical to discuss the optimal randomness for the weights and biases in SCN algorithms. Heavy-tailed PDFs have shown optimal randomness for finding targets [171,172], which plays a significant role in exploration and exploitation [148]. Therefore, herein, heavy-tailed PDFs were used to randomly update the weights and biases in the hidden layers to determine if the SCN models display improved performance. Some of the key parameters of the SCN models are listed in Table 6. For example, the maximum times of random configuration T_{max} are set as 200. The scale factor lambda in the activation function, which directly determines the range for the random parameters, was examined by using different settings (0.5–200). The tolerance was set as 0.05. Most of the parameters for the SCN with heavy-tailed PDFs were kept the same with the original SCN algorithms for comparison purposes. For more details, please refer to [163] and Appendix A.

Table 6. Stochastic configuration networks (SCNs) with key parameters.

Properties	Values
Name:	"Stochastic Configuration Networks"
Version:	"1.0 beta"
L:	hidden node number
W:	input weight matrix
b:	hidden layer bias vector
Beta:	output weight vector
r:	regularization parameter
tol:	tolerance
Lambda:	random weight range
L_{max}:	maximum number of hidden neurons
T_{max}:	maximum times of random configurations
nB:	number of node being added in one loop

4.3. A Regression Model and Parameter Tuning

The dataset of the regression model was generated by a real-valued function [173]:

$$f(x) = 0.2e^{-(10x-4)^2} + 0.5e^{-(80x-40)^2} + 0.3e^{-(80x-20)^2}, \tag{45}$$

where $x \in [0, 1]$. There were 1000 points randomly generated from the uniform distribution on the unit interval [0, 1] in the training dataset. The test set had 300 points generated from a regularly spaced grid on [0, 1]. The input and output attributes were normalized into [0, 1], and all the results reported in this research represent averages over 1000 independent trials. The settings of the parameters were similar to for the SCN in [163].

Heavy-tailed PDF algorithms have user-defined parameters, for example, the power-law index for SCN-Lévy, and location and scale parameters for SCN-Cauchy and SCN-Weibull, respectively. Thus, to illustrate the effect of parameters on the optimization results and to offer reference values for the proposed SCN algorithms, parameter analysis was conducted, and corresponding experiments were performed. Based on the experimental results, for the SCN-Lévy algorithm, the most optimal power-law index is 1.1 for achieving the minimum number of hidden nodes. For the SCN-Weibull algorithm, the optimal location parameter α and scale parameter β for the minimum number of hidden nodes are 1.9

and 0.2, respectively. For the SCN-Cauchy algorithm, the optimal location parameter α and scale parameter β for the minimum number of hidden nodes are 0.9 and 0.1, respectively.

Performance Comparison among SCNs with Heavy-Tailed PDFs

In Table 7, the performance of SCN, SCN-Lévy, SCN-Cauchy, SCN-Weibull and SCN-Mixture are shown, in which mean values are reported based on 1000 independent trials. Wang et al. [163] used time cost to evaluate the SCN algorithms' performance. In the present study, the authors used the mean hidden node numbers to evaluate the performance. The number of hidden nodes is associated with modeling accuracy. Therefore, herein, the analysis determined if an SCN with heavy-tailed PDFs used fewer hidden nodes to generate high performance, which would make the NNs less complex. According to the numerical results, the SCN-Cauchy used the lowest number of mean hidden nodes, 59, with an root mean squared error (RMSE) of 0.0057. The SCN-Weibull had a mean number of 63 hidden nodes, with an RMSE of 0.0037. The SCN-Mixture had a mean number of 70 hidden nodes, with an RMSE of 0.0020. The mean number of hidden nodes for SCN-Lévy was also 70. The original SCN model had a mean number of 75 hidden nodes. A more detailed training process is shown in Figure 15. With fewer hidden node numbers, the SCN models with heavy-tailed PDFs can be faster than the original SCN model. The neural network structure is also less complicated than the SCN. Our numerical results for the regression task demonstrate remarkable improvements in modeling performance compared with the current SCN model results.

Table 7. Performance comparison of SCN models for regression problem.

Models	Mean Hidden Node Number	RMSE
SCN	75 ± 5	0.0025
SCN-Lévy	70 ± 6	0.0010
SCN-Cauchy	59 ± 3	0.0057
SCN-Weibull	63 ± 4	0.0037
SCN-Mixture	70 ± 5	0.0020

4.4. MNIST Handwritten Digit Classification

The handwritten digit dataset contains 4000 training examples and 1000 testing examples, a subset of the MNIST handwritten digit dataset. Each image is a 20-by-20-pixel grayscale image of the digit (Figure 16). Each pixel is represented by a number indicating the grayscale intensity at that location. The 20-by-20 grid of pixels is "unrolled" into a 400-dimensional vector.

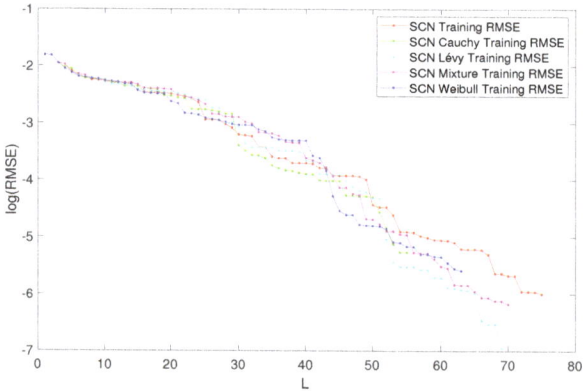

Figure 15. Performance of SCN, SCN-Lévy, SCN-Weibull, SCN-Cauchy and SCN-Mixture. The parameter L is the hidden node number.

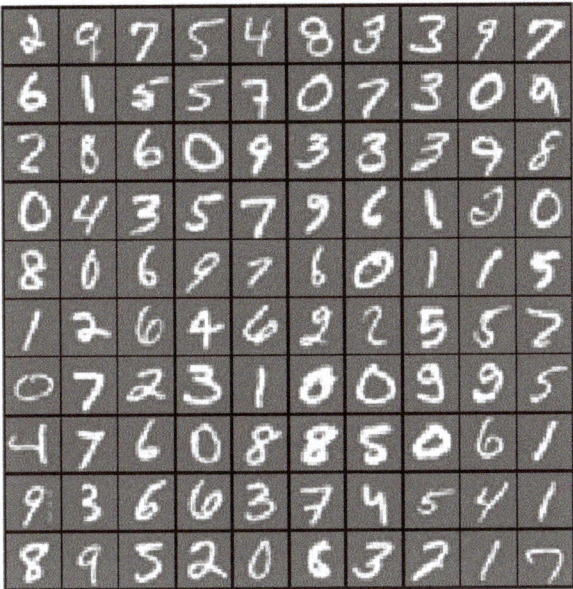

Figure 16. The handwritten digit dataset example.

Similar to the parameter tuning for the regression model, parameter analysis was conducted to illustrate the impact of parameters on the optimization results and to offer reference values for the MNIST handwritten digit classification SCN algorithms. Corresponding experiments were performed. According to the experimental results, for the SCN-Lévy algorithm, the most optimal power law index is 1.6 for achieving the best RMSE performance. For the SCN-Cauchy algorithm, the optimal location parameter α and scale parameter β for the lowest RMSE are 0.2 and 0.3, respectively.

Performance Comparison among SCNs on MNIST

The performance of the SCN, SCN-Lévy, SCN-Cauchy and SCN-Mixture are shown in Table 8. Based on the experimental results, the SCN-Cauchy, SCN-Lévy and SCN-Mixture have better performance in training and test accuracy, compared with the original SCN model. A detailed training process is shown in Figure 17. Within around 100 hidden nodes, the SCN models with heavy-tailed PDFs perform similarly to the original SCN model. When the number of the hidden nodes is greater than 100, the SCN models with heavy-tailed PDFs have lower RMSEs. Since more parameters for weights and biases are initialized in heavy-tailed PDFs, this may cause an SCN with heavy-tailed PDFs to converge to the optimal values at a faster speed. The experimental results for the MNIST handwritten classification problem demonstrate improvements in modeling performance. They also show that SCN models with heavy-tailed PDFs have a better search ability for achieving lower RMSEs.

Table 8. Performance comparison of SCNs.

Models	Training Accuracy	Test Accuracy
SCN	$94.0 \pm 1.9\%$	$91.2 \pm 6.2\%$
SCN-Lévy	$94.9 \pm 0.8\%$	$91.7 \pm 4.5\%$
SCN-Cauchy	$95.4 \pm 1.3\%$	$92.4 \pm 5.5\%$
SCN-Mixture	$94.7 \pm 1.1\%$	$91.5 \pm 5.3\%$

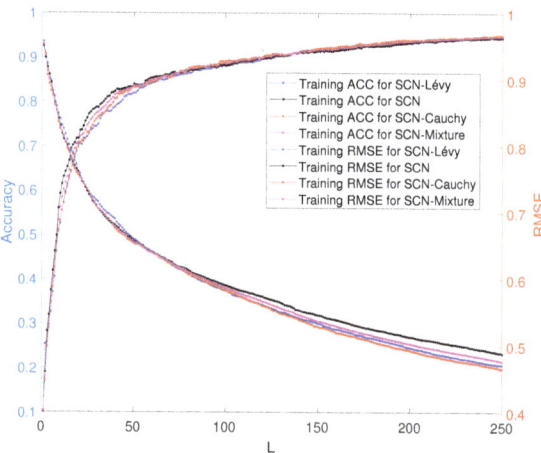

Figure 17. Classification performance of SCNs.

5. Take-Home Messages and Looking into the Future: Fractional Calculus Is Physics Informed

Big data and machine learning (ML) are two of the hottest topics of applied scientific research, and they are closely related to one another. To better understand them, in this article, we advocate fractional calculus (FC), as well as fractional-order thinking (FOT), for big data and ML analysis and applications. In Section 2, we discussed the relationships between big data, variability and FC, as well as why fractional-order data analytics (FODA) should be used and what it is. The topics included the Hurst parameter, fractional Gaussian noise (fGn), fractional Brownian motion (fBm), the fractional autoregressive integrated moving average (FARIMA), the formalism of continuous time random walk (CTRW), unmanned aerial vehicles (UAVs) and precision agriculture (PA).

In Section 3, how to learn efficiently (optimally) for ML algorithms is discussed. The key to developing an efficient learning process is the method of optimization. Thus, it is important to design an efficient optimization method. The derivative-free methods, as well as the gradient-based methods, such as the Nesterov accelerated gradient descent (NAGD), are discussed. Furthermore, it is shown to be possible, following the internal model principle (IMP), to design and analyze the ML algorithms in the S or Z transform domain in Section 3.3. FC is used in optimal randomness in the methods of stochastic gradient descent (SGD) and random search. Nonlocal models have commonly been used to describe physical systems and/or processes that cannot be accurately described by classical approaches [174]. For example, fractional nonlocal Maxwell's equations and the corresponding fractional wave equations were applied in [175] for fractional vector calculus [176]. The nonlocal differential operators [177], including nonlocal analogs of the gradient/Hessian, are the key of these nonlocal models, which could lead to very interesting research with FC in the near future.

Fractional dynamics is a response to the need for a more advanced characterization of our complex world to capture structure at very small or very large scales that had previously been smoothed over. If one wishes to obtain results that are better than the best possible using integer-order calculus-based methods, or are "more optimal", we advocate applying FOT and going fractional! In this era of big data, decision and control need FC, such as fractional-order signals, systems and controls. The future of ML should be physics-informed, scientific (cause–effect embedded or cause–effect discovery) and involving the use of FC, where the modeling is closer to nature. Laozi (unknown, around the 6th century to 4th century BC), the ancient Chinese philosopher, is said to have written

Entropy **2021**, *23*, 297

a short book *Dao De Jing(Tao Te Ching)*, in which he observed: "The Tao that can be told is not the eternal Tao" [178]. People over thousands of years have shared different understandings of the meaning of the Tao. Our best understanding of the Tao is nature, whose rules of complexity can be explained in a non-normal way. Fractional dynamics, FC and heavytailedness may well be that non-normal way (Figure 18), at least for the not-too-distant future.

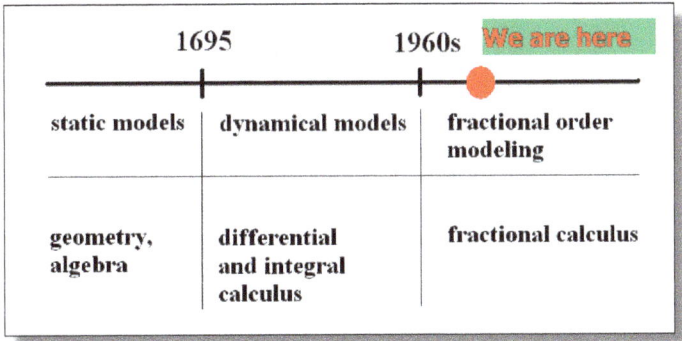

Figure 18. Timeline of FC (courtesy of Professor Igor Podlubny).

Author Contributions: H.N. drafted the original manuscript based on numerous talks/discussions with Y.C. in the past several years plus a seminar (http://mechatronics.ucmerced.edu/news/2020/why-big-data-and-machine-learning-must-meet-fractional-calculus) (accessed on 2 February 2021). Y.C. and B.J.W. contributed to the result interpretation, discussions and editing of the manuscript. All authors have read and agreed to the published version of the manuscript.

Funding: This research received no external funding.

Institutional Review Board Statement: Not applicable.

Informed Consent Statement: Not applicable.

Data Availability Statement: Not applicable.

Acknowledgments: Thanks go to Jiamin Wei, Yuquan Chen, Guoxiang Zhang, Tiebiao Zhao, Lihong Guo, Zhenlong Wu, Yanan Wang, Panpan Gu, Jairo Viola, Jie Yuan, etc., for walks, chats and tea/coffee breaks at Castle, Atwater, CA, before the COVID-19 era. In particular, Yuquan Chen performed computation in various IMP-based GD schemes, and Jiamin Wei performed the computation in cuckoo searches using four different heavy-tailed randomnesses. YangQuan Chen would like to thank Justin Dianhui Wang for many fruitful discussions in the past years on SCN, in particular, and machine learning in general. We gratefully acknowledge the support of NVIDIA Corporation with the donation of the Titan X Pascal GPU used for this research. Last but not least, we thank the helpful reviewers for constructive comments.

Conflicts of Interest: The authors declare no conflict of interest.

Abbreviations

The following abbreviations are used in this manuscript:

ACF	Auto-Correlation Function
AI	Artificial Intelligence
ARMA	Autoregression and Moving Average
CLT	Classical Central Limit Theorem
CS	Cuckoo Search
CTRW	Continuous Time Random Work

EOM	Equation of Motion
fBm	Fractional Brownian Motion
fGn	Fractional Gaussian Noise
FARIMA	Fractional Autoregressive Integrated Moving Average
FC	Fractional Calculus
FIGARCH	Fractional Integral Generalized Autoregressive Conditional Heteroscedasticity
FLOM	Fractional Lower-Order Moments
FOCV	Fractional-Order Calculus of Variation
FODA	Fractional-Order Data Analytics
FOEL	Fractional-Order Euler–Lagrange
FOT	Fractional-Order Thinking
GARMA	Gegenbauer Autoregressive Moving Average
GD	Gradient Descent
GDM	Gradient Descent Momentum
GEV	Generalized Extreme Value
IMP	Internal Model Principle
IPL	Inverse Power Law
ISE	Integral Squared Error
LGD	Long Range Dependence
LTI	Linear Time Invariant
MAD	Modeling, Analysis and Design
ML	Machine Learning
MLL	Mittag–Leffler Law
MNIST	Modified National Institute of Standards and Technology Database
NAGD	Nesterov Accelerated Gradient Descent
NDVI	Normalized Difference Vegetation Index
NILT	Numerical Inverse Laplace Transform
NN	Neural Networks
PA	Precision Agriculture
PDF	Probability Density Function
PID	Proportional, Integral, Derivative
PSO	Particle Swarm Optimization
RBF	Randomized Radial Basis Function (RBF) Networks
RGB	Red, Green, Blue
RMSE	Root Mean Squared Error
RVFL	Random Vector Functional Link
RW-FNN	Feed-Forward Networks with Random Weights
SCN	Stochastic Configuration Network
SGD	Stochastic Gradient Descent
SLFNNs	Single-Layer Feed-Forward Neural Networks
UAVs	Unmanned Aerial Vehicles
USDA	United States Department of Agriculture
wGn	White Gaussian Noise

Appendix A. SCN Codes

The Matlab and Python codes can be found at https://github.com/niuhaoyu16/StochasticConfigurationNetwork (accessed on 2 February 2021).

References

1. Vinagre, B.M.; Chen, Y. Lecture notes on fractional calculus applications in automatic control and robotics. In Proceedings of the 41st IEEE CDC Tutorial Workshop, Las Vegas, NV, USA, 9 December 2002; pp. 1–310.
2. Valério, D.; Machado, J.; Kiryakova, V. Some Pioneers of the Applications of Fractional Calculus. *Fract. Calc. Appl. Anal.* **2014**, *17*, 552–578. [CrossRef]
3. Abel, N. Solution of a Couple of Problems by Means of Definite Integrals. *Mag. Naturvidenskaberne* **1823**, 2, 2.
4. Podlubny, I.; Magin, R.L.; Trymorush, I. Niels Henrik Abel and the Birth of Fractional Calculus. *Fract. Calc. Appl. Anal.* **2017**, *20*, 1068–1075. [CrossRef]
5. Ross, B. The Development of Fractional Calculus 1695–1900. *Hist. Math.* **1977**, *4*, 75–89. [CrossRef]

6. Tarasov, V.E. *Fractional Dynamics: Applications of Fractional Calculus to Dynamics of Particles, Fields and Media*; Springer Science & Business Media: Berlin, Germany, 2011.
7. Klafter, J.; Lim, S.; Metzler, R. *Fractional Dynamics: Recent Advances*; World Scientific: Singapore, 2012.
8. Pramukkul, P.; Svenkeson, A.; Grigolini, P.; Bologna, M.; West, B. Complexity and the Fractional Calculus. *Adv. Math. Phys.* **2013**, *2013*, 498789. [CrossRef]
9. Chen, D.; Xue, D.; Chen, Y. More optimal image processing by fractional order differentiation and fractional order partial differential equations. In Proceedings of the International Symposium on Fractional PDEs, Newport, RI, USA, 3–5 June 2013.
10. Chen, D.; Sun, S.; Zhang, C.; Chen, Y.; Xue, D. Fractional-order TV-L 2 Model for Image Denoising. *Cent. Eur. J. Phys.* **2013**, *11*, 1414–1422. [CrossRef]
11. Yang, Q.; Chen, D.; Zhao, T.; Chen, Y. Fractional Calculus in Image Processing: A Review. *Fract. Calc. Appl. Anal.* **2016**, *19*, 1222–1249. [CrossRef]
12. Seshadri, V.; West, B.J. Fractal dimensionality of Lévy processes. *Proc. Natl. Acad. Sci. USA* **1982**, *79*, 4501. [CrossRef] [PubMed]
13. Metzler, R.; Klafter, J. The Random Walk's Guide to Anomalous Diffusion: A Fractional Dynamics Approach. *Phys. Rep.* **2000**, *339*, 1–77. [CrossRef]
14. Metzler, R.; Glöckle, W.G.; Nonnenmacher, T.F. Fractional Model Equation for Anomalous Diffusion. *Phys. A Stat. Mech. Appl.* **1994**, *211*, 13–24. [CrossRef]
15. Sheng, H.; Chen, Y.; Qiu, T. *Fractional Processes and Fractional-Order Signal Processing: Techniques and Applications*; Springer Science & Business Media: Berlin, Germany, 2011.
16. Mandelbrot, B.B.; Wallis, J.R. Robustness of the Rescaled Range R/S in the Measurement of Noncyclic Long Run Statistical Dependence. *Water Resour. Res.* **1969**, *5*, 967–988. [CrossRef]
17. Geweke, J.; Porter-Hudak, S. The Estimation and Application of Long Memory Time Series Models. *J. Time Ser. Anal.* **1983**, *4*, 221–238. [CrossRef]
18. Liu, K.; Chen, Y.; Zhang, X. An Evaluation of ARFIMA (Autoregressive Fractional Integral Moving Average) Programs. *Axioms* **2017**, *6*, 16. [CrossRef]
19. Montroll, E.W.; Weiss, G.H. Random Walks on Lattices. II. *J. Math. Phys.* **1965**, *6*, 167–181. [CrossRef]
20. Liakos, K.G.; Busato, P.; Moshou, D.; Pearson, S.; Bochtis, D. Machine Learning in Agriculture: A Review. *Sensors* **2018**, *18*, 2674. [CrossRef] [PubMed]
21. Nesterov, Y. A Method for Unconstrained Convex Minimization Problem with the Rate of Convergence O $(1/k^2)$. *Doklady an Ussr* **1983**, *269*, 543–547.
22. Montroll, E.W.; West, B.J. On An Enriched Collection of Stochastic Processes. *Fluct. Phenom.* **1979**, *66*, 61.
23. Francis, B.A.; Wonham, W.M. The Internal Model Principle of Control Theory. *Automatica* **1976**, *12*, 457–465. [CrossRef]
24. Zadeh, L.A. Fuzzy Logic. *Computer* **1988**, *21*, 83–93. [CrossRef]
25. Unser, M.; Blu, T. Fractional Splines and Wavelets. *SIAM Rev.* **2000**, *42*, 43–67. [CrossRef]
26. Samoradnitsky, G. *Stable Non-Gaussian Random Processes: Stochastic Models with Infinite Variance*; Routledge: Oxford, UK, 2017.
27. Crovella, M.E.; Bestavros, A. Self-similarity in World Wide Web Traffic: Evidence and Possible Causes. *IEEE/ACM Trans. Netw.* **1997**, *5*, 835–846. [CrossRef]
28. Burnecki, K.; Weron, A. Levy Stable Processes. From Stationary to Self-similar Dynamics and Back. An Application to Finance. *Acta Phys. Pol. Ser. B* **2004**, *35*, 1343–1358.
29. Pesquet-Popescu, B.; Pesquet, J.C. Synthesis of Bidimensional α-stable Models with Long-range Dependence. *Signal Process.* **2002**, *82*, 1927–1940. [CrossRef]
30. Hartley, T.T.; Lorenzo, C.F. Fractional-order System Identification Based on Continuous Order-distributions. *Signal Process.* **2003**, *83*, 2287–2300. [CrossRef]
31. Wolpert, R.L.; Taqqu, M.S. Fractional Ornstein–Uhlenbeck Lévy Processes and the Telecom Process: Upstairs and Downstairs. *Signal Process.* **2005**, *85*, 1523–1545. [CrossRef]
32. Bahg, G.; Evans, D.G.; Galdo, M.; Turner, B.M. Gaussian process linking functions for mind, brain, and behavior. *Proc. Natl. Acad. Sci. USA* **2020**, *117*, 29398–29406. [CrossRef]
33. West, B.J.; Geneston, E.L.; Grigolini, P. Maximizing Information Exchange between Complex Networks. *Phys. Rep.* **2008**, *468*, 1–99. [CrossRef]
34. West, B.J. Sir Isaac Newton Stranger in a Strange Land. *Entropy* **2020**, *22*, 1204. [CrossRef] [PubMed]
35. Csete, M.; Doyle, J. Bow Ties, Metabolism and Disease. *Trends Biotechnol.* **2004**, *22*, 446–450. [CrossRef]
36. Zhao, J.; Yu, H.; Luo, J.H.; Cao, Z.W.; Li, Y.X. Hierarchical Modularity of Nested Bow-ties in Metabolic Networks. *BMC Bioinform.* **2006**, *7*, 1–16. [CrossRef] [PubMed]
37. Doyle, J. Universal Laws and Architectures. Available online: http://www.ieeecss-oll.org/lecture/universal-laws-and-architectures. (accessed on 2 February 2021).
38. Doyle, J.C.; Csete, M. Architecture, Constraints, and Behavior. *Proc. Natl. Acad. Sci. USA* **2011**, *108*, 15624–15630. [CrossRef]
39. Sheng, H.; Chen, Y.Q.; Qiu, T. Heavy-tailed Distribution and Local Long Memory in Time Series of Molecular Motion on the Cell Membrane. *Fluct. Noise Lett.* **2011**, *10*, 93–119. [CrossRef]
40. Graves, T.; Gramacy, R.; Watkins, N.; Franzke, C. A Brief History of Long Memory: Hurst, Mandelbrot and the Road to ARFIMA, 1951–1980. *Entropy* **2017**, *19*, 437. [CrossRef]

41. West, B.J.; Grigolini, P. *Complex Webs: Anticipating the Improbable*; Cambridge University Press: Cambridge, UK, 2010.
42. Barabási, A.L.; Albert, R. Emergence of Scaling in Random Networks. *Science* **1999**, *286*, 509–512. [CrossRef]
43. Sun, W.; Li, Y.; Li, C.; Chen, Y. Convergence Speed of a Fractional Order Consensus Algorithm over Undirected Scale-free Networks. *Asian J. Control* **2011**, *13*, 936–946. [CrossRef]
44. Li, M. Modeling Autocorrelation Functions of Long-range Dependent Teletraffic Series Based on Optimal Approximation in Hilbert Space—A Further Study. *Appl. Math. Model.* **2007**, *31*, 625–631. [CrossRef]
45. Zhao, Z.; Guo, Q.; Li, C. A Fractional Model for the Allometric Scaling Laws. *Open Appl. Math. J.* **2008**, *2*, 26–30. [CrossRef]
46. Sun, H.; Chen, Y.; Chen, W. Random-order Fractional Differential Equation Models. *Signal Process.* **2011**, *91*, 525–530. [CrossRef]
47. Kello, C.T.; Brown, G.D.; Ferrer-i Cancho, R.; Holden, J.G.; Linkenkaer-Hansen, K.; Rhodes, T.; Van Orden, G.C. Scaling Laws in Cognitive Sciences. *Trends Cogn. Sci.* **2010**, *14*, 223–232. [CrossRef] [PubMed]
48. Gorenflo, R.; Mainardi, F. Fractional Calculus and Stable Probability Distributions. *Arch. Mech.* **1998**, *50*, 377–388.
49. Mainardi, F. The Fundamental Solutions for the Fractional Diffusion-wave Equation. *Appl. Math. Lett.* **1996**, *9*, 23–28. [CrossRef]
50. Luchko, Y.; Mainardi, F.; Povstenko, Y. Propagation Speed of the Maximum of the Fundamental Solution to the Fractional Diffusion–wave Equation. *Comput. Math. Appl.* **2013**, *66*, 774–784. [CrossRef]
51. Luchko, Y.; Mainardi, F. Some Properties of the Fundamental Solution to the Signalling Problem for the Fractional Diffusion-wave Equation. *Open Phys.* **2013**, *11*, 666–675. [CrossRef]
52. Luchko, Y.; Mainardi, F. Cauchy and Signaling Problems for the Time-fractional Diffusion-wave Equation. *J. Vib. Acoust.* **2014**, *136*, 050904. [CrossRef]
53. Li, Z.; Liu, L.; Dehghan, S.; Chen, Y.; Xue, D. A Review and Evaluation of Numerical Tools for Fractional Calculus and Fractional Order Controls. *Int. J. Control* **2017**, *90*, 1165–1181. [CrossRef]
54. Asmussen, S. Steady-state Properties of GI/G/1. In *Applied Probability and Queues*; Springer: New York, NY, USA, 2003; pp. 266–301.
55. Bernardi, M.; Petrella, L. Interconnected Risk Contributions: A Heavy-tail Approach to Analyze US Financial Sectors. *J. Risk Financ. Manag.* **2015**, *8*, 198–226. [CrossRef]
56. Ahn, S.; Kim, J.H.; Ramaswami, V. A New Class of Models for Heavy Tailed Distributions in Finance and Insurance Risk. *Insur. Math. Econ.* **2012**, *51*, 43–52. [CrossRef]
57. Resnick, S.I. *Heavy-tail Phenomena: Probabilistic and Statistical Modeling*; Springer Science & Business Media: Berlin, Germany, 2007.
58. Rolski, T.; Schmidli, H.; Schmidt, V.; Teugels, J.L. *Stochastic Processes for Insurance and Finance*; John Wiley & Sons: Hoboken, NJ, USA, 2009; Volume 505.
59. Foss, S.; Korshunov, D.; Zachary, S. *An Introduction to Heavy-Tailed and Subexponential Distributions*; Springer: Berlin, Germany, 2011; Volume 6.
60. Niu, H.; Chen, Y.; Chen, Y. Fractional-order extreme learning machine with Mittag-Leffler distribution. In Proceedings of ASME 2019 International Design Engineering Technical Conferences and Computers and Information in Engineering Conference, Anaheim, CA, USA, 18–21 August 2019.
61. Hariya, Y.; Kurihara, T.; Shindo, T.; Jin'no, K. Lévy flight PSO. In Proceedings of the IEEE Congress on Evolutionary Computation (CEC), Sendai, Japan, 25–28 May 2015.
62. Yang, X.S. *Nature-Inspired Metaheuristic Algorithms*; Luniver Press: London, UK, 2010.
63. Yang, X.S.; Deb, S. Engineering Optimisation by Cuckoo Search. *Int. J. Math. Model. Numer. Optim.* **2010**, *1*, 330–343. [CrossRef]
64. Haubold, H.J.; Mathai, A.M.; Saxena, R.K. Mittag-Leffler Functions and Their Applications. *J. Appl. Math.* **2011**, *2011*, 298628. [CrossRef]
65. Jayakumar, K. Mittag-Leffler Process. *Math. Comput. Model.* **2003**, *37*, 1427–1434. [CrossRef]
66. Wei, J.; Chen, Y.; Yu, Y.; Chen, Y. Improving cuckoo search algorithm with Mittag-Leffler distribution. In Proceedings of the ASME 2019 International Design Engineering Technical Conferences and Computers and Information in Engineering Conference, Anaheim, CA, USA, 18–21 August 2019.
67. Rinne, H. *The Weibull Distribution: A Handbook*; CRC Press: Boca Raton, FL, USA, 2008.
68. Johnson, N.L.; Kotz, S.; Balakrishnan, N. *Continuous Univariate Distributions*; John Wiley & Sons, Ltd.: Hoboken, NJ, USA, 1995.
69. Feller, W. *An Introduction to Probability Theory and Its Application Vol II*; John Wiley & Sons: Hoboken, NJ, USA, 1971.
70. Liu, T.; Zhang, P.; Dai, W.S.; Xie, M. An Intermediate Distribution Between Gaussian and Cauchy Distributions. *Phys. A Stat. Mech. Appl.* **2012**, *391*, 5411–5421. [CrossRef]
71. Bahat, D.; Rabinovitch, A.; Frid, V. *Tensile Fracturing in Rocks*; Springer: Berlin, Germany, 2005.
72. Geerolf, F. A Theory of Pareto Distributions. Available online: https://fgeerolf.com/geerolf-pareto.pdf (accessed on 2 February 2021).
73. Mandelbrot, B. The Pareto-Levy Law and the Distribution of Income. *Int. Econ. Rev.* **1960**, *1*, 79–106. [CrossRef]
74. Levy, M.; Solomon, S. New Evidence for the Power-law Distribution of Wealth. *Phys. A Stat. Mech. Appl.* **1997**, *242*, 90–94. [CrossRef]
75. Lu, J.; Ding, J. Mixed-Distribution-Based Robust Stochastic Configuration Networks for Prediction Interval Construction. *IEEE Trans. Ind. Inform.* **2019**, *16*, 5099–5109. [CrossRef]
76. Spiegel, M.R.; Schiller, J.J.; Srinivasan, R. *Probability and Statistics*; McGraw-Hill: New York, NY, USA, 2013.

77. Embrechts, P.; Klüppelberg, C.; Mikosch, T. *Modelling Extremal Events: For Insurance and Finance*; Springer Science & Business Media: Berlin, Germany, 2013; Volume 33.

78. Novak, S.Y. *Extreme Value Methods with Applications to Finance*; CRC Press: Boca Raton, FL, USA, 2011.

79. De Haan, L.; Ferreira, A. *Extreme Value Theory: An Introduction*; Springer Science & Business Media: Berlin, Germany, 2007.

80. Bottou, L.; Bousquet, O. The Tradeoffs of Large Scale Learning. *Adv. Neural Inf. Process. Syst.* **2007**, *20*, 161–168.

81. Bottou, L. Large-scale machine learning with stochastic gradient descent. In *Proceedings of the COMPSTAT*; Springer: Berlin, Germany, 2010; pp. 177–186.

82. Simsekli, U.; Sagun, L.; Gurbuzbalaban, M. A tail-index analysis of stochastic gradient noise in deep neural networks. *arXiv* **2019**, arXiv:1901.06053.

83. Yanovsky, V.; Chechkin, A.; Schertzer, D.; Tur, A. Lévy Anomalous Diffusion and Fractional Fokker–Planck Equation. *Phys. A Stat. Mech. Appl.* **2000**, *282*, 13–34. [CrossRef]

84. Viswanathan, G.M.; Afanasyev, V.; Buldyrev, S.; Murphy, E.; Prince, P.; Stanley, H.E. Lévy Flight Search Patterns of Wandering Albatrosses. *Nature* **1996**, *381*, 413–415. [CrossRef]

85. Hilbert, M.; López, P. The World's Technological Capacity to Store, Communicate, and Compute Information. *Science* **2011**, *332*, 60–65. [CrossRef] [PubMed]

86. Ward, J.S.; Barker, A. Undefined by data: A survey of big data definitions. *arXiv* **2013**, arXiv:1309.5821.

87. Reinsel, D.; Gantz, J.; Rydning, J. Data Age 2025: The Evolution of Data to Life-critical. *Don't Focus Big Data* **2017**, *2*, 2–24.

88. Firican, G. The 10 Vs of Big Data. 2017. Available online: https://tdwi.org/articles/2017/02/08/10-vs-of-big-data.aspx (accessed on 2 February 2021).

89. Nakahira, Y.; Liu, Q.; Sejnowski, T.J.; Doyle, J.C. Diversity-enabled sweet spots in layered architectures and speed-accuracy trade-offs in sensorimotor control. *arXiv* **2019**, arXiv:1909.08601.

90. Arabas, J.; Opara, K. Population Diversity of Non-elitist Evolutionary Algorithms in the Exploration Phase. *IEEE Trans. Evol. Comput.* **2019**, *24*, 1050–1062. [CrossRef]

91. Ko, M.; Stark, B.; Barbadillo, M.; Chen, Y. An Evaluation of Three Approaches Using Hurst Estimation to Differentiate Between Normal and Abnormal HRV. In Proceedings of the ASME 2015 International Design Engineering Technical Conferences and Computers and Information in Engineering Conference, Boston, MA, USA, 2–5 August 2015.

92. Li, N.; Cruz, J.; Chien, C.S.; Sojoudi, S.; Recht, B.; Stone, D.; Csete, M.; Bahmiller, D.; Doyle, J.C. Robust Efficiency and Actuator Saturation Explain Healthy Heart Rate Control and Variability. *Proc. Natl. Acad. Sci. USA* **2014**, *111*, E3476–E3485. [CrossRef] [PubMed]

93. Hutton, E.L. *Xunzi: The Complete Text*; Princeton University Press: Princeton, NJ, USA, 2014.

94. Boyer, C.B. *The History of the Calculus and Its Conceptual Development: (The Concepts of the Calculus)*; Courier Corporation: Chelmsford, MA, USA, 1959.

95. Bardi, J.S. *The Calculus Wars: Newton, Leibniz, and the Greatest Mathematical Clash of All Time*; Hachette UK: Paris, France, 2009.

96. Tanner, R.I.; Walters, K. *Rheology: An Historical Perspective*; Elsevier: Amsterdam, The Netherlands, 1998.

97. Chen, Y.; Sun, R.; Zhou, A. An Improved Hurst Parameter Estimator Based on Fractional Fourier Transform. *Telecommun. Syst.* **2010**, *43*, 197–206. [CrossRef]

98. Sheng, H.; Sun, H.; Chen, Y.; Qiu, T. Synthesis of Multifractional Gaussian Noises Based on Variable-order Fractional Operators. *Signal Process.* **2011**, *91*, 1645–1650. [CrossRef]

99. Sun, R.; Chen, Y.; Zaveri, N.; Zhou, A. Local analysis of long range dependence based on fractional Fourier transform. In Proceedings of the IEEE Mountain Workshop on Adaptive and Learning Systems, Logan, UT, USA, 24–26 July 2006; pp. 13–18.

100. Pipiras, V.; Taqqu, M.S. *Long-Range Dependence and Self-Similarity*; Cambridge University Press: Cambridge, UK, 2017; Volume 45.

101. Samorodnitsky, G. Long Range Dependence. Available online: https://onlinelibrary.wiley.com/doi/abs/10.1002/9781118445112.stat04569 (accessed on 2 February 2021).

102. Gubner, J.A. *Probability and Random Processes for Electrical and Computer Engineers*; Cambridge University Press: Cambridge, UK, 2006.

103. Clegg, R.G. A practical guide to measuring the Hurst parameter. *arXiv* **2006**, arXiv:math/0610756.

104. Decreusefond, L. Stochastic Analysis of the Fractional Brownian Motion. *Potential Anal.* **1999**, *10*, 177–214. [CrossRef]

105. Koutsoyiannis, D. The Hurst Phenomenon and Fractional Gaussian Noise Made Easy. *Hydrol. Sci. J.* **2002**, *47*, 573–595. [CrossRef]

106. Mandelbrot, B.B.; Van Ness, J.W. Fractional Brownian Motions, Fractional Noises and Applications. *SIAM Rev.* **1968**, *10*, 422–437. [CrossRef]

107. Ortigueira, M.D.; Batista, A.G. On the Relation between the Fractional Brownian Motion and the Fractional Derivatives. *Phys. Lett. A* **2008**, *372*, 958–968. [CrossRef]

108. Chen, Y.; Sun, R.; Zhou, A. An overview of fractional order signal processing (FOSP) techniques. In Proceedings of the International Design Engineering Technical Conferences and Computers and Information in Engineering Conference, Las Vegas, NV, USA, 4–7 September 2007.

109. Liu, K.; Domański, P.D.; Chen, Y. Control performance assessment with fractional lower order moments. In Proceedings of the 2020 7th International Conference on Control, Decision and Information Technologies (CoDIT), Prague, Czech Republic, 29 June–2 July 2020.

110. Cottone, G.; Di Paola, M. On the Use of Fractional Calculus for the Probabilistic Characterization of Random Variables. *Probabilistic Eng. Mech.* **2009**, *24*, 321–330. [CrossRef]
111. Cottone, G.; Di Paola, M.; Metzler, R. Fractional Calculus Approach to the Statistical Characterization of Random Variables and Vectors. *Phys. A Stat. Mech. Appl.* **2010**, *389*, 909–920. [CrossRef]
112. Ma, X.; Nikias, C.L. Joint Estimation of Time Delay and Frequency Delay in Impulsive Noise Using Fractional Lower Order Statistics. *IEEE Trans. Signal Process.* **1996**, *44*, 2669–2687.
113. RongHua, F. Modeling and Application of Theory Based on Time Series ARMA. *Sci. Technol. Inf.* **2012**, *2012*, 153.
114. Shalalfeh, L.; Bogdan, P.; Jonckheere, E. Fractional Dynamics of PMU Data. *IEEE Trans. Smart Grid* **2020**. [CrossRef]
115. Harmantzis, F. Heavy network traffic modeling and simulation using stable FARIMA processes. In Proceedings of the 19th International Teletraffic Congress (ITC19), Beijing, China, 29 August–2 September 2005.
116. Sheng, H.; Chen, Y. FARIMA with Stable Innovations Model of Great Salt Lake Elevation Time Series. *Signal Process.* **2011**, *91*, 553–561. [CrossRef]
117. Li, Q.; Tricaud, C.; Sun, R.; Chen, Y. Great Salt Lake surface level forecasting using FIGARCH model. In Proceedings of the International Design Engineering Technical Conferences and Computers and Information in Engineering Conference, Las Vegas, NV, USA, 4–7 September 2007; pp. 1361–1370.
118. Brockwell, P.J.; Davis, R.A.; Fienberg, S.E. *Time Series: Theory and Methods*; Springer Science & Business Media: Berlin, Germany, 1991.
119. Boutahar, M.; Dufrénot, G.; Péguin-Feissolle, A. A Simple Fractionally Integrated Model with a Time-varying Long Memory Parameter d_t. *Comput. Econ.* **2008**, *31*, 225–241. [CrossRef]
120. Gray, H.L.; Zhang, N.F.; Woodward, W.A. On Generalized Fractional Processes. *J. Time Ser. Anal.* **1989**, *10*, 233–257. [CrossRef]
121. Woodward, W.A.; Cheng, Q.C.; Gray, H.L. A k-factor GARMA Long-memory Model. *J. Time Ser. Anal.* **1998**, *19*, 485–504. [CrossRef]
122. West, B.J. *Fractional Calculus View of Complexity: Tomorrow's Science*; CRC Press: Boca Raton, FL, USA, 2016.
123. Zaslavsky, G.M.; Sagdeev, R.; Usikov, D.; Chernikov, A. *Weak Chaos and Quasi-Regular Patterns*; Cambridge University Press: Cambridge, UK, 1992.
124. Hilfer, R.; Anton, L. Fractional Master Equations and Fractal Time Random Walks. *Phys. Rev. E* **1995**, *51*, R848. [CrossRef]
125. Gorenflo, R.; Mainardi, F.; Vivoli, A. Continuous-time Random Walk and Parametric Subordination in Fractional Diffusion. *Chaos Solitons Fractals* **2007**, *34*, 87–103. [CrossRef]
126. Niu, H.; Hollenbeck, D.; Zhao, T.; Wang, D.; Chen, Y. Evapotranspiration Estimation with Small UAVs in Precision Agriculture. *Sensors* **2020**, *20*, 6427. [CrossRef]
127. Díaz-Varela, R.; de la Rosa, R.; León, L.; Zarco-Tejada, P. High-resolution Airborne UAV Imagery to Assess Olive Tree Crown Parameters Using 3D Photo Reconstruction: Application in Breeding Trials. *Remote Sens.* **2015**, *7*, 4213–4232. [CrossRef]
128. Gonzalez-Dugo, V.; Goldhamer, D.; Zarco-Tejada, P.J.; Fereres, E. Improving the Precision of Irrigation in a Pistachio Farm Using an Unmanned Airborne Thermal System. *Irrig. Sci.* **2015**, *33*, 43–52. [CrossRef]
129. Swain, K.C.; Thomson, S.J.; Jayasuriya, H.P. Adoption of an Unmanned Helicopter for Low-altitude Remote Sensing to Estimate Yield and Total Biomass of a Rice Crop. *Trans. ASABE* **2010**, *53*, 21–27. [CrossRef]
130. Zarco-Tejada, P.J.; González-Dugo, V.; Williams, L.; Suárez, L.; Berni, J.A.; Goldhamer, D.; Fereres, E. A PRI-based Water Stress Index Combining Structural and Chlorophyll Effects: Assessment Using Diurnal Narrow-band Airborne Imagery and the CWSI Thermal Index. *Remote Sens. Environ.* **2013**, *138*, 38–50. [CrossRef]
131. Niu, H.; Zhao, T.; Wang, D.; Chen, Y. Estimating evapotranspiration with UAVs in agriculture: A review. In Proceedings of the ASABE Annual International Meeting, Boston, MA, USA, 7–10 July 2019.
132. Niu, H.; Zhao, T.; Wang, D.; Chen, Y. A UAV resolution and waveband aware path planning for onion irrigation treatments inference. In Proceedings of the 2019 International Conference on Unmanned Aircraft Systems (ICUAS), Atlanta, GA, USA, 11–14 June 2019; pp. 808–812.
133. Niu, H.; Wang, D.; Chen, Y. Estimating crop coefficients using linear and deep stochastic configuration networks models and UAV-based Normalized Difference Vegetation Index (NDVI). In Proceedings of the International Conference on Unmanned Aircraft Systems (ICUAS), Athens, Greece, 1–4 September 2020.
134. Niu, H.; Wang, D.; Chen, Y. Estimating actual crop evapotranspiration using deep stochastic configuration networks model and UAV-based crop coefficients in a pomegranate orchard. In Proceedings of the Autonomous Air and Ground Sensing Systems for Agricultural Optimization and Phenotyping V. International Society for Optics and Photonics, 27 April–8 May 2020, held online.
135. Che, Y.; Wang, Q.; Xie, Z.; Zhou, L.; Li, S.; Hui, F.; Wang, X.; Li, B.; Ma, Y. Estimation of Maize Plant Height and Leaf Area Index Dynamic Using Unmanned Aerial Vehicle with Oblique and Nadir Photography. *Ann. Bot.* **2020**, *126*, 765–773. [CrossRef]
136. Deng, R.; Jiang, Y.; Tao, M.; Huang, X.; Bangura, K.; Liu, C.; Lin, J.; Qi, L. Deep Learning-based Automatic Detection of Productive Tillers in Rice. *Comput. Electron. Agric.* **2020**, *177*, 105703. [CrossRef]
137. Zhao, T.; Chen, Y.; Ray, A.; Doll, D. Quantifying almond water stress using unmanned aerial vehicles (UAVs): Correlation of stem water potential and higher order moments of non-normalized canopy distribution. In Proceedings of the ASME 2017 International Design Engineering Technical Conferences and Computers and Information in Engineering Conference, Cleveland, OH, USA, 6–9 August 2017.

138. Zhao, T.; Niu, H.; de la Rosa, E.; Doll, D.; Wang, D.; Chen, Y. Tree canopy differentiation using instance-aware semantic segmentation. In Proceedings of the 2018 ASABE Annual International Meeting, Detroit, MI, USA, 29 July–1 August 2018.

139. Géron, A. *Hands-on Machine Learning with Scikit-Learn, Keras, and TensorFlow: Concepts, Tools, and Techniques to Build Intelligent Systems*; O'Reilly Media: Sevastopol, CA, USA, 2019.

140. Mitchell, T.M. *Machine Learning*; McGraw-Hill: New York, NY, USA, 1997.

141. Polyak, B.T. Some Methods of Speeding up the Convergence of Iteration Methods. *USSR Comput. Math. Math. Phys.* **1964**, *4*, 1–17. [CrossRef]

142. Duchi, J.; Hazan, E.; Singer, Y. Adaptive Subgradient Methods for Online Learning and Stochastic Optimization. *J. Mach. Learn. Res.* **2011**, *12*, 2121–2159.

143. Hinton, G.; Tieleman, T. Slide 29 in Lecture 6. 2012. Available online: http://www.cs.toronto.edu/~tijmen/csc321/slides/lecture_slides_lec6.pdf (accessed on 2 February 2021).

144. Kingma, D.P.; Ba, J. Adam: A method for stochastic optimization. *arXiv* **2014**, arXiv:1412.6980.

145. Zeng, C.; Chen, Y. Optimal Random Search, Fractional Dynamics and Fractional Calculus. *Fract. Calc. Appl. Anal.* **2014**, *17*, 321–332. [CrossRef]

146. Wei, J.; Yu, Y.; Wang, S. Parameter Estimation for Noisy Chaotic Systems Based on an Improved Particle Swarm Optimization Algorithm. *J. Appl. Anal. Comput.* **2015**, *5*, 232–242.

147. Wei, J.; Yu, Y.; Cai, D. Identification of Uncertain Incommensurate Fractional-order Chaotic Systems Using an Improved Quantum-behaved Particle Swarm Optimization Algorithm. *J. Comput. Nonlinear Dyn.* **2018**, *13*, 051004. [CrossRef]

148. Wei, J.; Chen, Y.; Yu, Y.; Chen, Y. Optimal Randomness in Swarm-based Search. *Mathematics* **2019**, *7*, 828. [CrossRef]

149. Wei, J.; Yu, Y. A Novel Cuckoo Search Algorithm under Adaptive Parameter Control for Global Numerical Optimization. *Soft Comput.* **2019**, *24*, 4917–4940. [CrossRef]

150. Wei, J.; Yu, Y. An adaptive cuckoo search algorithm with optional external archive for global numerical optimization. In Proceedings of the International Conference on Fractional Differentiation and its Applications (ICFDA), Amman, Jordan, 16–18 July 2018.

151. Wilson, A.C.; Recht, B.; Jordan, M.I. A Lyapunov analysis of momentum methods in optimization. *arXiv* **2016**, arXiv:1611.02635.

152. Feynman, R.P. The Principle of Least Action in Quantum Mechanics. In *Feynman's Thesis—A New Approach to Quantum Theory*; World Scientific: Singapore, 2005; pp. 1–69.

153. Hamilton, S.W.R. *On a General Method in Dynamics*; Richard Taylor, 1834. Available online: http://www.kurims.kyoto-u.ac.jp/EMIS/classics/Hamilton/GenMeth.pdf (accessed on 2 February 2021).

154. Hawking, S.W. The Path-integral Approach to Quantum Gravity. In *General Relativity*; World Scientific: Singapore, 1979.

155. Kerrigan, E. What the Machine Should Learn about Models for Control. 2020. Available online: https://www.ifac2020.org/program/workshops/machine-learning-meets-model-based-control (accessed on 2 February 2021).

156. Vinnicombe, G. *Uncertainty and Feedback: H∞ Loop-Shaping and the ν-Gap Metric*; World Scientific: Singapore, 2001.

157. Viola, J.; Chen, Y.; Wang, J. Information-based model discrimination for digital twin behavioral matching. In Proceedings of the International Conference on Industrial Artificial Intelligence (IAI), Shenyang, China, 23–25 October 2020; pp. 1–6.

158. Kashima, K.; Yamamoto, Y. System Theory for Numerical Analysis. *Automatica* **2007**, *43*, 1156–1164. [CrossRef]

159. An, W.; Wang, H.; Sun, Q.; Xu, J.; Dai, Q.; Zhang, L. A PID controller approach for stochastic optimization of deep networks. In Proceedings of the IEEE Conference on Computer Vision and Pattern Recognition, Salt Lake City, UT, USA, 18–22 June 2018; pp. 8522–8531.

160. Fan, Y.; Koellermeier, J. Accelerating the Convergence of the Moment Method for the Boltzmann Equation Using Filters. *J. Sci. Comput.* **2020**, *84*, 1–28. [CrossRef]

161. Kuhlman, K.L. Review of Inverse Laplace Transform Algorithms for Laplace-space Numerical Approaches. *Numer. Algorithms* **2013**, *63*, 339–355. [CrossRef]

162. Xue, D.; Chen, Y. *Solving Applied Mathematical Problems with MATLAB*; CRC Press: Boca Raton, FL, USA, 2009.

163. Wang, D.; Li, M. Stochastic Configuration Networks: Fundamentals and Algorithms. *IEEE Trans. Cybern.* **2017**, *47*, 3466–3479. [CrossRef] [PubMed]

164. Bebis, G.; Georgiopoulos, M. Feed-forward Neural Networks. *IEEE Potentials* **1994**, *13*, 27–31. [CrossRef]

165. Broomhead, D.; Lowe, D. Multivariable Functional Interpolation and Adaptive Networks. *Complex Syst.* **1988**, *2*, 321–355.

166. Pao, Y.H.; Takefuji, Y. Functional-link Net Computing: Theory, System Architecture, and Functionalities. *Computer* **1992**, *25*, 76–79. [CrossRef]

167. Wang, D.; Li, M. Deep stochastic configuration networks with universal approximation property. In Proceedings of the International Joint Conference on Neural Networks (IJCNN), Rio de Janeiro, Brazil, 8–13 July 2018; pp. 1–8.

168. Li, M.; Wang, D. 2-D Stochastic Configuration Networks for Image Data Analytics. *IEEE Trans. Cybern.* **2019**, *51*, 359–372. [CrossRef] [PubMed]

169. Huang, C.; Huang, Q.; Wang, D. Stochastic Configuration Networks Based Adaptive Storage Replica Management for Power Big Data Processing. *IEEE Trans. Ind. Inf.* **2019**, *16*, 373–383. [CrossRef]

170. Scardapane, S.; Wang, D. Randomness in Neural Networks: An Overview. *Wiley Interdiscip. Rev. Data Min. Knowl. Discov.* **2017**, *7*, e1200. [CrossRef]

171. Wei, J. Research on Swarm Intelligence Optimization Algorithms and Their Applications to Parameter Identification of Fractional-Order Systems. Ph.D. Thesis, Beijing Jiaotong University, Beijing, China, 2020.
172. Chen, Y. Fundamental Principles for Fractional Order Gradient Methods. Ph.D. Thesis, University of Science and Technology of China, Hefei, China, 2020.
173. Tyukin, I.Y.; Prokhorov, D.V. Feasibility of random basis function approximators for modeling and control. In Proceedings of the IEEE Control Applications, (CCA) & Intelligent Control, (ISIC), St. Petersburg, Russia, 8–10 July 2009.
174. Nagaraj, S. Optimization and learning with nonlocal calculus. *arXiv* **2020**, arXiv:2012.07013.
175. Tarasov, V.E. Fractional Vector Calculus and Fractional Maxwell's Equations. *Ann. Phys.* **2008**, *323*, 2756–2778. [CrossRef]
176. Ortigueira, M.; Machado, J. On Fractional Vectorial Calculus. *Bull. Pol. Acad. Sci. Tech. Sci.* **2018**, *66*, 389–402.
177. Feliu-Faba, J.; Fan, Y.; Ying, L. Meta-learning Pseudo-differential Operators with Deep Neural Networks. *J. Comput. Phys.* **2020**, *408*, 109309. [CrossRef]
178. Hall, D.L. *Dao De Jing: A Philosophical Translation*; Ballantine Books: New York, NY, USA, 2010.

Article

Skellam Type Processes of Order k and Beyond

Neha Gupta [1], Arun Kumar [1] and Nikolai Leonenko [2,*

[1] Department of Mathematics, Indian Institute of Technology Ropar, Rupnagar, Punjab 140001, India;
    2017maz0002@iitrpr.ac.in (N.G.); arun.kumar@iitrpr.ac.in (A.K.)
[2] Cardiff School of Mathematics, Cardiff University, Senghennydd Road, Cardiff CF24 4AG, UK
* Correspondence: LeonenkoN@cardiff.ac.uk

Received: 31 August 2020; Accepted: 19 October 2020; Published: 22 October 2020

Abstract: In this article, we introduce the Skellam process of order k and its running average. We also discuss the time-changed Skellam process of order k. In particular, we discuss the space-fractional Skellam process and tempered space-fractional Skellam process via time changes in Skellam process by independent stable subordinator and tempered stable subordinator, respectively. We derive the marginal probabilities, Lévy measures, governing difference-differential equations of the introduced processes. Our results generalize the Skellam process and running average of Poisson process in several directions.

Keywords: Skellam process; subordination; Lévy measure; Poisson process of order k; running average

1. Introduction

The Skellam distribution is obtained by taking the difference between two independent Poisson distributed random variables, which was introduced for the case of different intensities λ_1, λ_2 by (see [1]) and for equal means in [2]. For large values of $\lambda_1 + \lambda_2$, the distribution can be approximated by the normal distribution and if λ_2 is very close to 0, then the distribution tends to a Poisson distribution with intensity λ_1. Similarly, if λ_1 tends to 0, the distribution tends to a Poisson distribution with non-positive integer values. The Skellam random variable is infinitely divisible, since it is the difference of two infinitely divisible random variables (see Proposition 2.1 in [3]). Therefore, one can define a continuous time Lévy process for Skellam distribution, which is called Skellam process.

The Skellam process is an integer valued Lévy process and it can also be obtained by taking the difference of two independent Poisson processes. Its marginal probability mass function (pmf) involves the modified Bessel function of the first kind. The Skellam process has various applications in different areas, such as to model the intensity difference of pixels in cameras (see [4]) and for modeling the difference of the number of goals of two competing teams in a football game [5]. The model based on the difference of two point processes is proposed in (see [6–9]).

Recently, the time-fractional Skellam process has been studied in [10], which is obtained by time-changing the Skellam process with an inverse stable subordinator. Further, they provided the application of time-fractional Skellam process in modeling of arrivals of jumps in high frequency trading data. It is shown that the inter-arrival times between the positive and negative jumps follow a Mittag–Leffler distribution rather then the exponential distribution. Similar observations are observed in the case of Danish fire insurance data (see [11]). Buchak and Sakhno, in [12], have also proposed the governing equations for time-fractional Skellam processes. Recently, [13] introduced time-changed Poisson process of order k, which is obtained by time changing the Poisson process of order k (see [14]) by general subordinators.

In this paper, we introduce Skellam process of order k and its running average. We also discuss the time-changed Skellam process of order k. In particular, we discuss space-fractional Skellam process and tempered space-fractional Skellam process via time changes in Skellam process by independent stable subordinator and tempered stable subordinator, respectively. We obtain closed form expressions for the marginal distributions of the considered processes and other important properties. Skellam process is used to model the difference between the number of goals between two teams in a football match. At the beginning, both teams have scores 0 each and at time t the team 1 score is $N_1(t)$, which is the cumulative sum of arrivals (goals) of size 1 until time t with exponential inter-arrival times. Similarly for team 2, the score is $N_2(t)$ at time t. The difference between the number of goals can be modeled using $N_1(t) - N_2(t)$ at time t. Similarly, the Skellam process of order k can be used to model the difference between the number of points scored by two competing teams in a basketball match where $k = 3$. Note that, in a basketball game, a free throw is count as one point, any basket from a shot taken from inside the three-point line counts for two points and any basket from a shot taken from outside the three-point line is considered as three points. Thus, a jump in the score of any team may be of size one, two, or three. Hence, a Skellam process of order 3 can be used to model the difference between the points scored.

In [10], it is shown that the fractional Skellam process is a better model then the Skellam process for modeling the arrivals of the up and down jumps for the tick-by-tick financial data. Equivalently, it is shown that the Mittag–Leffler distribution is a better model than the exponential distribution for the inter-arrival times between the up and down jumps. However, it is evident from Figure 3 of [10] that the fractional Skellam process is also not perfectly fitting the arrivals of positive and negative jumps. We hope that a more flexible class of processes like time-changed Skellam process of order k (see Section 6) and the introduced tempered space-fractional Skellam process (see Section 7) would be better model for arrivals of jumps. Additionally, see [8] for applications of integer-valued Lévy processes in financial econometrics. Moreover, distributions of order k are interesting for reliability theory [15]. The Fisher dispersion index is a widely used measure for quantifying the departure of any univariate count distribution from the equi-dispersed Poisson model [16–18]. The introduced processes in this article can be useful in modeling of over-dispersed and under-dispersed data. Further, in (49), we present probabilistic solutions of some fractional equations.

The remainder of this paper proceeds, as follows: in Section 2, we introduce all the relevant definitions and results. We also derive the Lévy density for space- and tempered space-fractional Poisson processes. In Section 3, we introduce and study running average of Poisson process of order k. Section 4 is dedicated to Skellam process of order k. Section 5 deals with running average of Skellam process of order k. In Section 6, we discuss the time-changed Skellam process of order k. In Section 7, we determine the marginal pmf, governing equations for marginal pmf, Lévy densities, and moment generating functions for space-fractional Skellam process and tempered space-fractional Skellam process.

2. Preliminaries

In this section, we collect relevant definitions and some results on Skellam process, subordinators, space-fractional Poisson process, and tempered space-fractional Poisson process. These results will be used to define the space-fractional Skellam processes and tempered space-fractional Skellam processes.

2.1. Skellam Process

In this section, we revisit the Skellam process and also provide a characterization of it. Let $S(t)$ be a Skellam process, such that

$$S(t) = N_1(t) - N_2(t), \ t \geq 0,$$

where $N_1(t)$ and $N_2(t)$ are two independent homogeneous Poisson processes with intensity $\lambda_1 > 0$ and $\lambda_2 > 0$, respectively. The Skellam process is defined in [8] and the distribution has been introduced and studied in [1], see also [2]. This process is only symmetric when $\lambda_1 = \lambda_2$. The pmf $s_k(t) = \mathbb{P}(S(t) = k)$ of $S(t)$ is given by (see e.g., [1,10])

$$s_k(t) = e^{-t(\lambda_1 + \lambda_2)} \left(\frac{\lambda_1}{\lambda_2} \right)^{k/2} I_{|k|}(2t\sqrt{\lambda_1 \lambda_2}), \quad k \in \mathbb{Z}, \tag{1}$$

where I_k is modified Bessel function of first kind (see [19], p. 375),

$$I_k(z) = \sum_{n=0}^{\infty} \frac{(z/2)^{2n+k}}{n!(n+k)!}. \tag{2}$$

The pmf $s_k(t)$ satisfies the following differential difference equation (see [10])

$$\frac{d}{dt} s_k(t) = \lambda_1(s_{k-1}(t) - s_k(t)) - \lambda_2(s_k(t) - s_{k+1}(t)), \quad k \in \mathbb{Z}, \tag{3}$$

with initial conditions $s_0(0) = 1$ and $s_k(0) = 0$, $k \neq 0$. For a real-valued Lévy process $Z(t)$ the characteristic function admits the form

$$\mathbb{E}(e^{iuZ(t)}) = e^{t\psi_Z(u)}, \tag{4}$$

where the function ψ_Z is called characteristic exponent and it admits the following Lévy-Khintchine representation (see [20])

$$\psi_Z(u) = iau - bu^2 + \int_{\mathbb{R} \setminus \{0\}} (e^{iux} - 1 - iux\mathbf{1}_{\{|x| \leq 1\}}) \pi_Z(dx). \tag{5}$$

Here, $a \in \mathbb{R}$, $b \geq 0$ and π_Z is a Lévy measure. If $\pi_Z(dx) = v_Z(x)dx$ for some function v_Z, then v_Z is called the Lévy density of the process Z. The Skellam process is a Lévy process, its Lévy density v_S is a linear combination of two Dirac delta functions, $v_S(y) = \lambda_1 \delta_1(y) + \lambda_2 \delta_{-1}(y)$ and the corresponding characteristic exponent is given by

$$\psi_{S(1)}(u) = \int_{-\infty}^{\infty} (1 - e^{-uy}) v_S(y) dy.$$

The moment generating function (mgf) of Skellam process is

$$\mathbb{E}[e^{\theta S(t)}] = e^{-t(\lambda_1 + \lambda_2 - \lambda_1 e^{\theta} - \lambda_2 e^{-\theta})}, \quad \theta \in \mathbb{R}. \tag{6}$$

With the help of mgf, one can easily find the moments of Skellam process. In the next result, we give a characterization of Skellam process, which is not available in literature as per our knowledge. For a function h, we write $h(\delta) = o(\delta)$ if $\lim_{\delta \to 0} h(\delta)/\delta = 0$.

Theorem 1. *Suppose that an arrival process has the independent and stationary increments and it also satisfies the following incremental condition, then the process is Skellam.*

$$\mathbb{P}(S(t + \delta) = m | S(t) = n) = \begin{cases} \lambda_1 \delta + o(\delta), & m > n, \ m = n + 1; \\ \lambda_2 \delta + o(\delta), & m < n, \ m = n - 1; \\ 1 - \lambda_1 \delta - \lambda_2 \delta + o(\delta), & m = n; \\ o(\delta) & \text{otherwise.} \end{cases}$$

Proof. Consider the interval [0,t], which is discretized with n sub-intervals of size δ each, such that $n\delta = t$. For $k \geq 0$, we have

$$\mathbb{P}(S(t) = k) = \sum_{m=0}^{\left[\frac{n-k}{2}\right]} \frac{n!}{m!(m+k)!(n-2m-k)!} (\lambda_1\delta)^{m+k}(\lambda_2\delta)^m (1 - \lambda_1\delta - \lambda_2\delta)^{n-2m-k} + o(\delta)$$

$$= \sum_{m=0}^{\left[\frac{n-k}{2}\right]} \frac{n!}{m!(m+k)!(n-2m-k)!} \left(\frac{\lambda_1 t}{n}\right)^{m+k} \left(\frac{\lambda_2 t}{n}\right)^m \left(1 - \frac{\lambda_1 t}{n} - \frac{\lambda_2 t}{n}\right)^{n-2m-k} + o(\delta)$$

$$= \sum_{m=0}^{\left[\frac{n-k}{2}\right]} \frac{(\lambda_1 t)^{m+k}(\lambda_2 t)^m}{m!(m+k)!} \frac{n!}{(n-2m-k)! n^{2m+k}} \left(1 - \frac{\lambda_1 t}{n} - \frac{\lambda_2 t}{n}\right)^{n-2m-k} + o(\delta)$$

$$= e^{-(\lambda_1+\lambda_2)t} \sum_{m=0}^{\infty} \frac{(\lambda_1 t)^{m+k}(\lambda_2 t)^m}{m!(m+k)!},$$

by taking $n \to \infty$. The result follows now by using the definition of modified Bessel function of first kind I_k. Similarly, it can be proved for $k < 0$. $\square$

2.2. Poisson Process of Order k

In this section, we recall the definition and some important properties of Poisson process of order k (PPoK). Kostadinova and Minkova (see [14]) introduced and studied the PPoK. Let $x_1, x_2, \cdots, x_k$ be non-negative integers and $\zeta_k = x_1 + x_2 + \cdots + x_k$, $\Pi_k! = x_1! x_2! \ldots x_k!$ and

$$\Omega(k,n) = \{X = (x_1, x_2, \ldots, x_k) | x_1 + 2x_2 + \cdots + kx_k = n\}. \tag{7}$$

Additionally, let $\{N^k(t)\}_{t\geq 0}$, represent the PPoK with rate parameter λt, then probability mass function (pmf) is given by

$$p_n^{N^k}(t) = \mathbb{P}(N^k(t) = n) = \sum_{X=\Omega(k,n)} e^{-k\lambda t} \frac{(\lambda t)^{\zeta_k}}{\Pi_k!}. \tag{8}$$

The pmf of $N^k(t)$ satisfies the following differential-difference equations (see [14])

$$\frac{d}{dt} p_n^{N^k}(t) = -k\lambda p_n^{N^k}(t) + \lambda \sum_{j=1}^{n \wedge k} p_{n-j}^{N^k}(t), \quad n = 1, 2, \ldots$$

$$\frac{d}{dt} p_0^{N^k}(t) = -k\lambda p_0^{N^k}(t), \tag{9}$$

with initial condition $p_0^{N^k}(0) = 1$ and $p_n^{N^k}(0) = 0$ and $n \wedge k = \min\{k,n\}$. The characteristic function of PPoK $N^k(t)$

$$\phi_{N^k(t)}(u) = \mathbb{E}[e^{iuN^k(t)}] = e^{-\lambda t(k-\sum_{j=1}^{k} e^{iuj})}, \tag{10}$$

where $i = \sqrt{-1}$. The process PPoK is Lévy, so it is infinite divisible i.e. $\phi_{N^k(t)}(u) = (\phi_{N^k(1)}(u))^t$. The Lévy density for PPoK is easy to derive and it is given by

$$\nu_{N^k}(x) = \lambda \sum_{j=1}^{k} \delta_j(x),$$

where δ_j is the Dirac delta function concentrated at j. The transition probability of the PPoK $\{N^k(t)\}_{t \geq 0}$ is also given by Kostadinova and Minkova [14],

$$\mathbb{P}(N^k(t+\delta) = m | N^k(t) = n) = \begin{cases} 1 - k\lambda\delta, & m = n; \\ \lambda\delta & m = n+i, i = 1, 2, \ldots, k; \\ 0 & \text{otherwise.} \end{cases} \tag{11}$$

The probability generating function (pgf) $G^{N^k}(s,t)$ is given by (see [14])

$$G^{N^k}(s,t) = e^{-\lambda t(k - \sum_{j=1}^{k} s^j)}. \tag{12}$$

The mean, variance and covariance function of the PPoK are given by

$$\mathbb{E}[N^k(t)] = \frac{k(k+1)}{2}\lambda t;$$

$$\text{Var}[N^k(t)] = \frac{k(k+1)(2k+1)}{6}\lambda t;$$

$$\text{Cov}[N^k(t), N^k(s)] = \frac{k(k+1)(2k+1)}{6}\lambda(t \wedge s). \tag{13}$$

2.3. Subordinators

Let $D_f(t)$ be a real valued Lévy process with non-decreasing sample paths and its Laplace transform has the form

$$\mathbb{E}[e^{-sD_f(t)}] = e^{-tf(s)},$$

where

$$f(s) = bs + \int_0^\infty (1 - e^{xs})\pi(dx), \quad s > 0, \ b \geq 0,$$

is the integral representation of Bernstein functions (see [21]). The Bernstein functions are C^∞, non-negative and such that $(-1)^m \frac{d^m}{dx^m} f(x) \leq 0$ for $m \geq 1$ [21]. Here, π denote the non-negative Lévy measure on the positive half line, such that

$$\int_0^\infty (x \wedge 1)\pi(dx) < \infty, \quad \pi([0, \infty)) = \infty,$$

and b is the drift coefficient. The right continuous inverse $E_f(t) = \inf\{u \geq 0 : D_f(u) > t\}$ is the inverse and first exist time of $D_f(t)$, which is non-Markovian with non-stationary and non-independent increments. Next, we analyze some special cases of Lévy subordinators with drift coefficient b = 0,

$$f(s) = \begin{cases} p\log(1 + \frac{s}{\alpha}), \ p > 0, \ \alpha > 0, & \text{(gamma subordinator);} \\ (s+\mu)^\alpha - \mu^\alpha, \ \mu > 0, \ 0 < \alpha < 1, & \text{(tempered } \alpha\text{-stable subordinator);} \\ \delta(\sqrt{2s+\gamma^2} - \gamma), \ \gamma > 0, \ \delta > 0, & \text{(inverse Gaussian subordinator);} \\ s^\alpha, \ 0 < \alpha < 1, & \text{(} \alpha\text{-stable subordinator).} \end{cases} \tag{14}$$

It is worth noting that, among the subordinators given in (14), all of the integer order moments of α-stable subordinators are infinite and others subordinators have all finite moments.

2.4. The Space-Fractional Poisson Process

In this section, we discuss the main properties of a space-fractional Poisson process (SFPP). We also provide the Lévy density for SFPP, which is not discussed in the literature. The SFPP $N_\alpha(t)$ was introduced by (see [22]), as follows

$$N_\alpha(t) = \begin{cases} N(D_\alpha(t)), \ t \geq 0, & 0 < \alpha < 1, \\ N(t), \ t \geq 0, & \alpha = 1, \end{cases} \tag{15}$$

where $D_\alpha(t)$ is an α-stable subordinator, which is independent of the homogeneous Poisson process $N(t)$.

The probability generating function (pgf) of this process is

$$G^{N_\alpha}(s,t) = \mathbb{E}[s^{N_\alpha(t)}] = e^{-\lambda^\alpha(1-s)^\alpha t}, \ |s| \leq 1, \ \alpha \in (0,1). \tag{16}$$

The pmf of SFPP is

$$P^\alpha(k,t) = \mathbb{P}\{N_\alpha(t) = k\} = \frac{(-1)^k}{k!} \sum_{r=0}^{\infty} \frac{(-\lambda^\alpha)^r t^r}{r!} \frac{\Gamma(r\alpha + 1)}{\Gamma(r\alpha - k + 1)}$$

$$= \frac{(-1)^k}{k!} {}_1\psi_1 \left[\begin{matrix} (1,\alpha); \\ (1-k,\alpha); \end{matrix} (-\lambda^\alpha t) \right], \tag{17}$$

where $_h\psi_i(z)$ is the Fox Wright function (see formula (1.11.14) in [23]). It was shown in [22] that the pmf of the SFPP satisfies the following fractional differential-difference equations

$$\frac{d}{dt}P^\alpha(k,t) = -\lambda^\alpha(1-B)^\alpha P^\alpha(k,t), \ \alpha \in (0,1], \ k = 1, 2, \ldots \tag{18}$$

$$\frac{d}{dt}P^\alpha(0,t) = -\lambda^\alpha P^\alpha(0,t), \tag{19}$$

with initial conditions

$$P^\alpha(k,0) = \delta_{k,0}, \tag{20}$$

where $\delta_{k,0}$ is the Kronecker delta function, given by

$$\delta_{k,0} = \begin{cases} 0, & k \geq 1, \\ 1, & k = 0. \end{cases} \tag{21}$$

The fractional difference operator

$$(1-B)^\alpha = \sum_{j=0}^{\infty} \binom{\alpha}{j}(-1)^j B^j \tag{22}$$

is defined in [24], where B is the backward shift operator. The characteristic function of SFPP is

$$\mathbb{E}[e^{iuN_\alpha(t)}] = e^{-\lambda^\alpha(1-e^{iu})^\alpha t}. \tag{23}$$

Proposition 1. *The Lévy density $\nu_{N_\alpha}(x)$ of SFPP is given by*

$$\nu_{N_\alpha}(x) = \lambda^\alpha \sum_{n=1}^{\infty} (-1)^{n+1} \binom{\alpha}{n} \delta_n(x). \tag{24}$$

Proof. We use Lévy-Khintchine formula (see [20]),

$$\int_{\mathbb{R}\backslash\{0\}} (e^{iux} - 1)\lambda^{\alpha} \sum_{n=1}^{\infty} (-1)^{n+1} \binom{\alpha}{n} \delta_n(x) dx$$

$$= \lambda^{\alpha} \left[\sum_{n=1}^{\infty} (-1)^{n+1} \binom{\alpha}{n} e^{iun} + \sum_{n=0}^{\infty} (-1)^n \binom{\alpha}{n} - 1 \right]$$

$$= \lambda^{\alpha} \sum_{n=0}^{\infty} (-1)^{n+1} \binom{\alpha}{n} e^{iun} = -\lambda^{\alpha} (1 - e^{iu})^{\alpha},$$

which is the characteristic exponent of SFPP from Equation (23). □

2.5. Tempered Space-Fractional Poisson Process

The tempered space-fractional Poisson process (TSFPP) can be obtained by subordinating the homogeneous Poisson process $N(t)$ with the independent tempered stable subordinator $D_{\alpha,\mu}(t)$ (see [25])

$$N_{\alpha,\mu}(t) = N(D_{\alpha,\mu}(t)), \ \alpha \in (0,1), \ \mu > 0. \tag{25}$$

This process has finite integer order moments due to the tempered α-stable subordinator. The pmf of TSFPP is given by (see [25])

$$P^{\alpha,\mu}(k,t) = (-1)^k e^{t\mu^{\alpha}} \sum_{m=0}^{\infty} \mu^m \sum_{r=0}^{\infty} \frac{(-t)^r}{r!} \lambda^{\alpha r - m} \binom{\alpha r}{m} \binom{\alpha r - m}{k}$$

$$= e^{t\mu^{\alpha}} \frac{(-1)^k}{k!} \sum_{m=0}^{\infty} \frac{\mu^m \lambda^{-m}}{m!} \, _1\psi_1 \left[\begin{matrix} (1,\alpha); \\ (1-k-m,\alpha); \end{matrix} (-\lambda^{\alpha} t) \right], \ k = 0,1,\ldots. \tag{26}$$

The governing difference-differential equation is given by

$$\frac{d}{dt} P^{\alpha,\mu}(k,t) = -((\mu + \lambda(1-B))^{\alpha} - \mu^{\alpha}) P^{\alpha,\mu}(k,t), \ k > 0. \tag{27}$$

The characteristic function of TSFPP,

$$\mathbb{E}[e^{iuN_{\alpha,\mu}(t)}] = e^{-t((\mu+\lambda(1-e^{iu}))^{\alpha}-\mu^{\alpha})}. \tag{28}$$

While using a standard conditioning argument, the mean and variance of TSFPP are given by

$$\mathbb{E}[N_{\alpha,\mu}(t)] = \lambda\alpha\mu^{\alpha-1}t, \ \text{Var}[N_{\alpha,\mu}(t))] = \lambda\alpha\mu^{\alpha-1}t + \lambda^2\alpha(1-\alpha)\mu^{\alpha-2}t. \tag{29}$$

Proposition 2. *The Lévy density $\nu_{N_{\alpha,\mu}}(x)$ of TSFPP is*

$$\nu_{N_{\alpha,\mu}}(x) = \sum_{n=1}^{\infty} \mu^{\alpha-n} \binom{\alpha}{n} \lambda^n \sum_{l=1}^{n} \binom{n}{l} (-1)^{l+1} \delta_l(x), \ \mu > 0. \tag{30}$$

Proof. Using (28), the characteristic exponent of TSFPP is given by $\psi_{N_{\alpha,\mu}}(u) = -((\mu + \lambda(1 - e^{iu}))^{\alpha} - \mu^{\alpha})$. We find the Lévy density with the help of Lévy-Khintchine formula (see [20]),

$$\int_{\mathbb{R}\setminus\{0\}} (e^{iux} - 1) \sum_{n=1}^{\infty} \mu^{\alpha-n} \binom{\alpha}{n} \lambda^n \sum_{l=1}^{n} \binom{n}{l} (-1)^{l+1} \delta_l(x) dx$$

$$= \sum_{n=1}^{\infty} \mu^{\alpha-n} \binom{\alpha}{n} \lambda^n \left(\sum_{l=1}^{n} \binom{n}{l} (-1)^{l+1} e^{iux} - \sum_{l=1}^{n} \binom{n}{l} (-1)^{l+1} \right)$$

$$= \sum_{n=0}^{\infty} \mu^{\alpha-n} \binom{\alpha}{n} \lambda^n \sum_{l=0}^{n} \binom{n}{l} (-1)^{l+1} \delta_l(x) - \mu^{\alpha}$$

$$= -((\mu + \lambda(1 - e^{iu}))^{\alpha} - \mu^{\alpha}),$$

hence proved. □

Definition 1. *A stochastic process $X(t)$ is over-dispersed, equi-dispersed or under-dispersed [18], if the Fisher index of dispersion, given by (see e.g., [17])*

$$\text{FI}[X(t)] = \frac{\text{Var}[X(t)]}{\mathbb{E}[X(t)]}$$

is more than 1, equal to 1, or smaller than 1, respectively, for all $t > 0$.

Remark 1. *Using (29), we have $\text{FI}[N_{\alpha,\mu}(t)] = 1 + \frac{\lambda(1-\alpha)}{\mu} > 1$, i.e. TSFPP $N_{\alpha,\mu}(t)$ is over-dispersed.*

3. Running Average of PPoK

In this section, we first introduced the running average of PPoK and their main properties. These results will be used further to discuss the running average of SPoK.

Definition 2 (Running average of PPoK). *We define the running average process $N_A^k(t)$, $t \geq 0$ by taking time-scaled integral of the path of the PPoK (see [26]),*

$$N_A^k(t) = \frac{1}{t} \int_0^t N^k(s) ds. \tag{31}$$

We can write the differential equation with initial condition $N_A^k(0) = 0$,

$$\frac{d}{dt}(N_A^k(t)) = \frac{1}{t} N^k(t) - \frac{1}{t^2} \int_0^t N^k(s) ds.$$

Which shows that it has continuous sample paths of bounded total variation. We explored the compound Poisson representation and distribution properties of running average of PPoK. The characteristic of $N_A^k(t)$ is obtained using the Lemma 1 of [26]. We recall Lemma 1 from [26] for ease of reference.

Lemma 1. *If X_t is a Lévy process and Y_t its Riemann integral is defined by*

$$Y_t = \int_0^t X_s ds,$$

then the characteristic functions of Y satisfies

$$\phi_{Y(t)}(u) = \mathbb{E}[e^{iuY(t)}] = e^{t\left(\int_0^1 \log \phi_{X(1)}(tuz)dz\right)}, \ u \in \mathbb{R}. \tag{32}$$

Proposition 3. *The characteristic function of $N_A^k(t)$ is given by*

$$\phi_{N_A^k(t)}(u) = e^{-t\lambda\left(k-\sum_{j=1}^{k}\frac{(e^{iuj}-1)}{iuj}\right)}. \tag{33}$$

Proof. Using the Equation (10), we have

$$\int_0^1 \log \phi_{N^k(1)}(tuz)dz = -\lambda\left(k - \sum_{j=1}^{k}\frac{(e^{ituzj}-1)}{ituj}\right).$$

Using (32) and (31), we have

$$\phi_{N_A^k(t)}(u) = e^{t\left(\int_0^1 \log \phi_{N^k(1)}(uz)dz\right)} = e^{-t\lambda\left(k-\sum_{j=1}^{k}\frac{(e^{iuj}-1)}{iuj}\right)}.$$

$\square$

Proposition 4. *The running average process has a compound Poisson representation, such that*

$$Y(t) = \sum_{i=1}^{N(t)} X_i, \tag{34}$$

where $X_i = 1, 2, \ldots$ are independent, identically distributed (iid) copies of X random variables, independent of $N(t)$ and $N(t)$ is a Poisson process with intensity $k\lambda$. Subsequently,

$$Y(t) \stackrel{law}{=} N_A^k(t).$$

Further, the random variable X has the following pdf

$$f_X(x) = \sum_{i=1}^{k} p_{V_i}(x)f_{U_i}(x) = \frac{1}{k}\sum_{i=1}^{k} f_{U_i}(x), \tag{35}$$

where V_i follows discrete uniform distribution over $(0, k)$ and U_i follows continuous uniform distribution over $(0, i)$, $i = 1, 2, \ldots, k$.

Proof. The pdf of U_i is $f_{U_i}(x) = \frac{1}{i}$, $0 \leq x \leq i$. Using (45), the characteristic function of X is given by

$$\phi_X(u) = \frac{1}{k}\sum_{j=1}^{k}\frac{(e^{iuj}-1)}{iuj}.$$

For fixed t, the characteristic function of $Y(t)$ is

$$\phi_{Y(t)}(u) = e^{-k\lambda t(1-\phi_X(u))} = e^{-t\lambda\left(k-\sum_{j=1}^{k}\frac{(e^{iuj}-1)}{iuj}\right)}, \tag{36}$$

which is equal to the characteristic function of PPoK that is given in (33). Hence, by the uniqueness of characteristic function, the result follows. $\square$

Using the definition

$$m_r = \mathbb{E}[X^r] = (-i)^r\frac{d^r\phi_X(u)}{du^r}, \tag{37}$$

the first two moments for random variable X given in Proposition (4) are $m_1 = \frac{(k+1)}{4}$ and $m_2 = \frac{1}{18}[(k+1)(2k+1)]$. Further, using the mean, variance, and covariance of compound Poisson process, we have

$$\mathbb{E}[N_A^k(t)] = \mathbb{E}[N(t)]\mathbb{E}[X] = \frac{k(k+1)}{4}\lambda t;$$

$$\text{Var}[N_A^k(t)] = \mathbb{E}[N(t)]\mathbb{E}[X^2] = \frac{1}{18}[k(k+1)(2k+1)]\lambda t;$$

$$\text{Cov}[N_A^k(t), N_A^k(s)] = \mathbb{E}[N_A^k(t), N_A^k(s)] - \mathbb{E}[N_A^k(t)]\mathbb{E}[N_A^k(s)]$$

$$= \mathbb{E}[N_A^k(s)]\mathbb{E}[N_A^k(t-s)] - \mathbb{E}[N_A^k(s)^2] - \mathbb{E}[N_A^k(t)]\mathbb{E}[N_A^k(s)]$$

$$= \frac{1}{18}[k(k+1)(2k+1)]\lambda s - \frac{k^2(k+1)^2}{16}\lambda^2 s^2, \quad s < t.$$

Corollary 1. *Putting $k = 1$, the running average of PPoK $N_A^k(t)$ reduces to the running average of standard Poisson process $N_A(t)$ (see Appendix in [26]).*

Corollary 2. *The mean and variance of PPoK and running average of PPoK satisfy, $\mathbb{E}[N_A^k(t)]/\mathbb{E}[N^k(t)] = \frac{1}{2}$ and $\text{Var}[N_A^k(t)]/\text{Var}[N^k(t)] = \frac{1}{3}$.*

Remark 2. *The Fisher index of dispersion for running average of PPoK $N_A^k(t)$ is given by $\text{FI}[N_A^k(t)] = \frac{2}{9}(2k+1)$. If $k = 1$ the process is under-dispersed and for $k > 1$ it is over-dispersed.*

Next we discuss the long-range dependence (LRD) property of running average of PPoK. We recall the definition of LRD for a non-stationary process.

Definition 3 (Long range dependence (LRD)). *Let $X(t)$ be a stochastic process that has a correlation function for $s \geq t$ for fixed s, that satisfies,*

$$c_1(s)t^{-d} \leq \text{Cor}(X(t), X(s)) \leq c_2(s)t^{-d},$$

for large t, $d > 0$, $c_1(s) > 0$ and $c_2(s) > 0$. For the particular case when $c_1(s) = c_2(s) = c(s)$, the above equation reduced to

$$\lim_{t \to \infty} \frac{\text{Cor}(X(t), X(s))}{t^{-d}} = c(s).$$

We say that, if $d \in (0, 1)$, then $X(t)$ has the LRD property and if $d \in (1, 2)$ it has short-range dependence (SRD) property [27].

Proposition 5. *The running average of PPoK has LRD property.*

Proof. Let $0 \leq s < t < \infty$, then the correlation function for running average of PPoK $N_A^k(t)$ is

$$\text{Cor}[N_A^k(t), N_A^k(s)] = \frac{(8(2k+1) - 9(k+1)k\lambda s)\, s^{1/2}t^{-1/2}}{8(2k+1)}.$$

Subsequently, for $d = 1/2$, it follows

$$\lim_{t \to \infty} \frac{\text{Cor}[N_A^k(t), N_A^k(s)]}{t^{-d}} = \frac{(8(2k+1) - 9(k+1)k\lambda s)\, s^{1/2}}{8(2k+1)} = c(s).$$

□

4. Skellam Process of Order k (SPoK)

In this section, we introduce and study the Skellam process of order k (SPoK).

Definition 4 (SPoK). *Let $N_1^k(t)$ and $N_2^k(t)$ be two independent PPoK with intensities $\lambda_1 > 0$ and $\lambda_2 > 0$. The stochastic process*

$$S^k(t) = N_1^k(t) - N_2^k(t)$$

is called a Skellam process of order k (SPoK).

Proposition 6. *The marginal distribution $R_m(t) = \mathbb{P}(S^k(t) = m)$ of SPoK $S^k(t)$ is given by*

$$R_m(t) = e^{-kt(\lambda_1+\lambda_2)} \left(\frac{\lambda_1}{\lambda_2}\right)^{m/2} I_{|m|}(2tk\sqrt{\lambda_1\lambda_2}), \ m \in \mathbb{Z}. \tag{38}$$

Proof. For $m \geq 0$, using the pmf of PPoK that is given in (8), it follows

$$R_m(t) = \sum_{n=0}^{\infty} \mathbb{P}(N_1^k(t) = n + m)\mathbb{P}(N_2^k(t) = n)\mathbb{I}_{m \geq 0}$$

$$= \sum_{n=0}^{\infty} \left(\sum_{X=\Omega(k,n+m)} e^{-k\lambda_1 t} \frac{(\lambda_1 t)^{\zeta_k}}{\Pi_k!} \right) \left(\sum_{X=\Omega(k,n)} e^{-k\lambda_2 t} \frac{(\lambda_2 t)^{\zeta_k}}{\Pi_k!} \right).$$

Setting $x_i = n_i$ and $n = x + \sum_{i=1}^{k}(i-1)n_i$, we have

$$R_m(t) = e^{-kt(\lambda_1+\lambda_2)} \sum_{x=0}^{\infty} \frac{(\lambda_2 t)^x}{x!} \frac{(\lambda_1 t)^{m+x}}{(m+x)!} \left(\sum_{n_1+n_2+\ldots+n_k=m+x} \binom{m+x}{n_1!n_2!\ldots n_k!} \right) \left(\sum_{n_1+n_2+\ldots+n_k=x} \binom{x}{n_1!n_2!\ldots n_k!} \right)$$

$$= e^{-kt(\lambda_1+\lambda_2)} \sum_{x=0}^{\infty} \frac{(\lambda_2 t)^x}{x!} \frac{(\lambda_1 t)^{m+x}}{(m+x)!} k^{m+x} k^x,$$

using the multinomial theorem and modified Bessel function given in (2). Similarly, it follows for $m < 0$. □

Proposition 7. *The Lévy density for SPoK is*

$$\nu_{S^k}(x) = \lambda_1 \sum_{j=1}^{k} \delta_j(x) + \lambda_2 \sum_{j=1}^{k} \delta_{-j}(x).$$

Proof. The proof follows by using the independence of two PPoK used in the definition of SPoK. □

Remark 3. *Using* (12), *the pgf of SPoK is given by*

$$G^{S^k}(s,t) = \sum_{m=-\infty}^{\infty} s^m R_m(t) = e^{-t\left(k(\lambda_1+\lambda_2)-\lambda_1\sum_{j=1}^{k}s^j-\lambda_2\sum_{j=1}^{k}s^{-j}\right)}. \tag{39}$$

Further, the characteristic function of SPoK is given by

$$\phi_{S^k(t)}(u) = e^{-t[k(\lambda_1+\lambda_2)-\lambda_1\sum_{j=1}^{k}e^{iju}-\lambda_2\sum_{j=1}^{k}e^{-iju}]}. \tag{40}$$

SPoK as a Pure Birth and Death Process

In this section, we provide the transition probabilities of SPoK at time $t + \delta$, given that we started at time t. Over such a short interval of length $\delta \to 0$, it is nearly impossible to observe more than k event; in fact, the probability to see more than k event is $o(\delta)$.

Proposition 8. *The transition probabilities of SPoK are given by*

$$\mathbb{P}(S^k(t+\delta) = m | S^k(t) = n) = \begin{cases} \lambda_1\delta + o(\delta), & m > n, \ m = n+i, i = 1, 2, \ldots, k; \\ \lambda_2\delta + o(\delta), & m < n, \ m = n-i, i = 1, 2, \ldots, k; \\ 1 - k\lambda_1\delta - k\lambda_2\delta + o(\delta), & m = n; \\ o(\delta) & \text{otherwise.} \end{cases} \quad (41)$$

Basically, at most k events can occur in a very small interval of time δ. Additionally, even though the probability for more than k event is non-zero, it is negligible.

Proof. Note that $S^k(t) = N_1^k(t) - N_2^k(t)$. We call $N_1^k(t)$ as the first process and $N_2^k(t)$ as the second process. For $i = 1, 2, \cdots, k$, we have

$$\mathbb{P}(S^k(t+\delta) = n+i | S^k(t) = n) = \sum_{j=1}^{k-i} \mathbb{P}(\text{the first process has i+j arrivals and the second process has j arrivals})$$

$$+ \mathbb{P}(\text{the first process has i arrivals and the second process has 0 arrivals}) + o(\delta)$$

$$= \sum_{j=0}^{k-i} (\lambda_1\delta + o(\delta)) \times (\lambda_2\delta + o(\delta)) + (\lambda_1\delta + o(\delta)) \times (1 - k\lambda_2\delta + o(\delta)) + o(\delta)$$

$$= \lambda_1\delta + o(\delta).$$

Similarly, for $i = 1, 2, \cdots, k$, we have

$$\mathbb{P}(S^k(t+\delta) = n-i | S^k(t) = n) = \sum_{j=1}^{k-i} \mathbb{P}(\text{the first process has j arrivals and the second process has i+j arrivals})$$

$$+ \mathbb{P}(\text{the first process has 0 arrivals and the second process has i arrivals}) + o(\delta)$$

$$= \sum_{j=0}^{k-i} (\lambda_1\delta + o(\delta)) \times (\lambda_2\delta + o(\delta)) + (1 - k\lambda_1\delta + o(\delta)) \times (\lambda_2\delta + o(\delta)) + o(\delta)$$

$$= \lambda_2\delta + o(\delta).$$

Further,

$$\mathbb{P}(S^k(t+\delta) = n | S^k(t) = n) = \sum_{j=1}^{k} \mathbb{P}(\text{the first process has j arrivals and the second process has j arrivals})$$

$$+ \mathbb{P}(\text{the first process has 0 arrivals and the second process has 0 arrivals}) + o(\delta)$$

$$= \sum_{j=0}^{k} (\lambda_1\delta + o(\delta)) \times (\lambda_2\delta + o(\delta)) + (1 - k\lambda_1\delta + o(\delta)) \times (1 - k\lambda_2\delta + o(\delta)) + o(\delta)$$

$$= 1 - k\lambda_1\delta - k\lambda_2\delta + o(\delta).$$

□

Remark 4. *The pmf $R_m(t)$ of SPoK satisfies the following difference differential equation*

$$\frac{d}{dt}R_m(t) = -k(\lambda_1 + \lambda_2)R_m(t) + \lambda_1 \sum_{j=1}^{k} R_{m-j}(t) + \lambda_2 \sum_{j=1}^{k} R_{m+j}(t)$$

$$= -\lambda_1 \sum_{j=1}^{k}(1 - B^j)R_m - \lambda_2 \sum_{j=1}^{k}(1 - F^j)R_m(t), \quad m \in \mathbb{Z},$$

with initial condition $R_0(0) = 1$ and $R_m(0) = 0$ for $m \neq 0$. Let B be the backward shift operator defined in (22) and F be the forward shift operator defined by $F^j X(t) = X(t+j)$, such that $(1-F)^\alpha = \sum_{j=0}^{\infty} \binom{\alpha}{j} F^j$. Multiplying by s^m and summing for all m in (42), we obtain the following differential equation for the pgf

$$\frac{d}{dt}G^{S^k}(s,t) = \left(-k(\lambda_1 + \lambda_2) + \lambda_1 \sum_{j=1}^{k} s^j + \lambda_2 \sum_{j=1}^{k} s^{-j}\right) G^{S^k}(s,t).$$

The mean, variance and covariance of SPoK can be easily calculated by using the pgf,

$$\mathbb{E}[S^k(t)] = \frac{k(k+1)}{2}(\lambda_1 - \lambda_2)t;$$

$$\text{Var}[S^k(t)] = \frac{1}{6}\left[k(k+1)(2k+1)\right](\lambda_1 + \lambda_2)t;$$

$$\text{Cov}[S^k(t), S^k(s)] = \frac{1}{6}\left[k(k+1)(2k+1)\right](\lambda_1 + \lambda_2)s, \quad s < t.$$

Remark 5. *For the SPoK, when $\lambda_1 > \lambda_2$, $\text{Var}[S^k(t)] - \mathbb{E}[S^k(t)] = \frac{k(k+1)}{3}[(k-1)\lambda_1 + (k+2)\lambda_2] > 0$, which implies that $\text{FI}[S^k(t)] > 1$ and hence $S^k(t)$ exhibits over-dispersion. For $\lambda_1 < \lambda_2$, the process is under-dispersed.*

Next, we show the LRD property for SPoK.

Proposition 9. *The SPoK has LRD property defined in Definition 3.*

Proof. The correlation function of SPoK satisfies

$$\lim_{t\to\infty} \frac{\text{Cor}(S^k(t), S^k(s))}{t^{-d}} = \frac{s^{1/2}t^{-1/2}}{t^{-1/2}} = c(s).$$

Hence, SPoK exhibits the LRD property. □

5. Running Average of SPoK

In this section, we introduce and study the new stochastic Lévy process, which is the running average of SPoK.

Definition 5. *The following stochastic process defined by taking the time-scaled integral of the path of the SPoK,*

$$S_A^k(t) = \frac{1}{t}\int_0^t S^k(s)ds, \tag{42}$$

is called the running average of SPoK.

Next, we provide the compound Poisson representation of running average of SPoK.

Proposition 10. *The characteristic function $\phi_{S_A^k(t)}(u) = \mathbb{E}[e^{iuS_A^k(t)}]$ of $S_A^k(t)$ is given by*

$$\phi_{S_A^k(t)}(u) = e^{-kt\left\{\lambda_1\left(1 - \frac{1}{k}\sum_{j=1}^{k}\frac{(e^{iuj}-1)}{iuj}\right) + \lambda_2\left(1 - \frac{1}{k}\sum_{j=1}^{k}\frac{(1-e^{-iuj})}{iuj}\right)\right\}}, \quad u \in \mathbb{R}. \tag{43}$$

Proof. By using the Lemma 3.1 to Equation (40) after scaling by $1/t$. □

Remark 6. *It is easily observable that Equation (43) has removable singularity at $u = 0$. To remove that singularity, we can define $\phi_{S_A^k(t)}(0) = 1$.*

Proposition 11. *Let $Y(t)$ be a compound Poisson process*

$$Y(t) = \sum_{n=1}^{N(t)} J_n, \tag{44}$$

where $N(t)$ is a Poisson process with rate parameter $k(\lambda_1 + \lambda_2) > 0$ and $\{J_n\}_{n \geq 1}$ are iid random variables with mixed double uniform distribution function p_j, which are independent of $N(t)$. Subsequently,

$$Y(t) \stackrel{law}{=} S_A^k(t).$$

Proof. Rearranging the $\phi_{S_A^k(t)}(u)$,

$$\phi_{S_A^k(t)}(u) = e^{(\lambda_1 + \lambda_2)kt\left(\frac{\lambda_1}{\lambda_1 + \lambda_2}\frac{1}{k}\sum_{j=1}^{k}\frac{(e^{iuj}-1)}{iuj} + \frac{\lambda_2}{\lambda_1 + \lambda_2}\frac{1}{k}\sum_{j=1}^{k}\frac{(1-e^{-iuj})}{iuj} - 1\right)}$$

The random variable J_1 being a mixed double uniformly distributed has density

$$p_{J_1}(x) = \sum_{i=1}^{k} p_{V_i}(x) f_{U_i}(x) = \frac{1}{k}\sum_{i=1}^{k} f_{U_i}(x), \tag{45}$$

where V_j follows discrete uniform distribution over $(0, k)$ with pmf $p_{V_j}(x) = \mathbb{P}(V_j = x) = \frac{1}{k}$, $j = 1, 2, \ldots k$, and U_i be doubly uniform distributed random variables with density

$$f_{U_i}(x) = (1 - w)1_{[-i,0]}(x) + w1_{[0,i]}(x), \quad -i \leq x \leq i.$$

Further, $0 < w < 1$ is a weight parameter and $1(\cdot)$ is the indicator function. Here, we obtained the characteristic of J_1 using the Fourier transform of (45),

$$\phi_{J_1}(u) = \frac{\lambda_1}{\lambda_1 + \lambda_2}\frac{1}{k}\sum_{j=1}^{k}\frac{(e^{iuj}-1)}{iuj} + \frac{\lambda_2}{\lambda_1 + \lambda_2}\frac{1}{k}\sum_{j=1}^{k}\frac{(1-e^{-iuj})}{iuj}.$$

The characteristic function of $Y(t)$ is

$$\phi_{Y(t)}(u) = e^{-kt(\lambda_1+\lambda_2)t(1-\phi_{J_1}(u))}, \tag{46}$$

putting the characteristic function $\phi_{J_1}(u)$ in the above expression yields the characteristic function of $S_A^k(t)$, which completes the proof. □

Remark 7. *The q-th order moments of J_1 can be calculated using (37) and also using Taylor series expansion of the characteristic $\phi_{J_1}(u)$, around 0, such that*

$$\frac{(e^{iuj}-1)}{iuj} = 1 + \sum_{r=1}^{\infty}\frac{(iuj)^r}{(r+1)!} \quad \& \quad \frac{(1-e^{-iuj})}{iuj} = 1 + \sum_{r=1}^{\infty}\frac{(-iuj)^r}{(r+1)!}.$$

We have $m_1 = \frac{(k+1)(\lambda_1-\lambda_2)}{4(\lambda_1+\lambda_2)}$ and $m_2 = \frac{1}{18}[(k+1)(2k+1)]$. Further, the mean, variance, and covariance of running average of SPoK are

$$\mathbb{E}[S_A^k(t)] = \mathbb{E}[N(t)]\mathbb{E}[J_1] = \frac{k(k+1)}{4}(\lambda_1 - \lambda_2)t$$

$$\text{Var}[S_A^k(t)] = \mathbb{E}[N(t)]\mathbb{E}[J_1^2] = \frac{1}{18}[k(k+1)(2k+1)](\lambda_1 + \lambda_2)t$$

$$\text{Cov}[S_A^k(t), S_A^k(s)] = \frac{1}{18}[k(k+1)(2k+1)](\lambda_1 - \lambda_2)s - \frac{k^2(k+1)^2}{16}(\lambda_1 - \lambda_2)^2 s^2.$$

Corollary 3. *For $\lambda_2 = 0$ the running average of SPoK is same as the running average of PPoK, i.e.,*

$$\phi_{S_A^k(t)}(u) = \phi_{N_A^k(t)}(u).$$

Corollary 4. *For $k = 1$ this process behave like the running average of Skellam process.*

Corollary 5. *The ratio of mean and variance of SPoK and running average of SPoK are 1/2 and 1/3, respectively.*

Remark 8. *For running average of SPoK, when $\lambda_1 > \lambda_2$ and $k > 1$, the process is over-dispersed. Otherwise, it exhibits under-dispersion.*

6. Time-Changed Skellam Process of Order k

We consider time-changed SPoK, which can be obtained by subordinating SPoK $S^k(t)$ with the independent Lévy subordinator $D_f(t)$ satisfying $\mathbb{E}[D_f(t)]^c < \infty$ for all $c > 0$. The time-changed SPoK is defined by

$$Z_f(t) = S^k(D_f(t)), \quad t \geq 0.$$

Note that the stable subordinator does not satisfy the condition $\mathbb{E}[D_f(t)]^c < \infty$. The mgf of time-changed SPoK $Z_f(t)$ is given by

$$\mathbb{E}[e^{\theta Z_f(t)}] = e^{-tf(k(\lambda_1+\lambda_2) - \lambda_1\sum_{j=1}^k e^{\theta j} - \lambda_2\sum_{j=1}^k e^{-\theta j})}.$$

Theorem 2. *The pmf $H_f(t) = \mathbb{P}(Z_f(t) = m)$ of time-changed SPoK is given by*

$$H_f(t) = \sum_{x=\max(0,-m)}^{\infty} \frac{(k\lambda_1)^{m+x}(k\lambda_2)^x}{(m+x)!x!}\mathbb{E}[e^{-k(\lambda_1+\lambda_2)D_f(t)}D_f^{2m+x}(t)], \quad m \in \mathbb{Z}. \tag{47}$$

Proof. Let $h_f(x,t)$ be the probability density function of Lévy subordinator. Using conditional argument

$$\begin{aligned}
H_f(t) &= \int_0^{\infty} R_m(y)h_f(y,t)dy \\
&= \int_0^{\infty} e^{-ky(\lambda_1+\lambda_2)}\left(\frac{\lambda_1}{\lambda_2}\right)^{m/2} I_{|m|}(2yk\sqrt{\lambda_1\lambda_2})h_f(y,t)dy \\
&= \sum_{x=\max(0,-m)}^{\infty} \frac{(k\lambda_1)^{m+x}(k\lambda_2)^x}{(m+x)!x!} \int_0^{\infty} e^{-k(\lambda_1+\lambda_2)y}y^{2m+x}h_f(y,t)dy \\
&= \sum_{x=\max(0,-m)}^{\infty} \frac{(k\lambda_1)^{m+x}(k\lambda_2)^x}{(m+x)!x!}\mathbb{E}[e^{-k(\lambda_1+\lambda_2)D_f(t)}D_f^{2m+x}(t)].
\end{aligned}$$

□

The mean and covariance of time changed SPoK are given by,

$$\mathbb{E}[Z_f(t)] = \frac{k(k+1)}{2}(\lambda_1 - \lambda_2)\mathbb{E}[D_f(t)].$$

$$\text{Cov}[Z_f(t), Z_f(s)] = \frac{1}{6}[k(k+1)(2k+1)](\lambda_1 + \lambda_2))\mathbb{E}[D_f(s)] + \frac{k^2(k+1)^2}{4}(\lambda_1 - \lambda_2)^2\text{Var}[D_f(s)].$$

7. Space Fractional Skellam Process and Tempered Space Fractional Skellam Process

In this section, we introduce time-changed Skellam processes where time-change are stable subordinator and tempered stable subordinator. These processes give the space-fractional version of the Skellam process similar to the time-fractional version of the Skellam process introduced in [10].

7.1. The Space-Fractional Skellam Process

In this section, we introduce space-fractional Skellam processes (SFSP). Further, for introduced processes, we study main results, such as state probabilities and governing difference-differential equations of marginal pmf.

Definition 6 (SFSP). *Let $N_1(t)$ and $N_2(t)$ be two independent homogeneous Poison processes with intensities $\lambda_1 > 0$ and $\lambda_2 > 0$,, respectively. Let $D_{\alpha_1}(t)$ and $D_{\alpha_2}(t)$ be two independent stable subordinators with indices $\alpha_1 \in (0,1)$ and $\alpha_2 \in (0,1)$, respectively. These subordinators are independent of the Poisson processes $N_1(t)$ and $N_2(t)$. The subordinated stochastic process*

$$S_{\alpha_1,\alpha_2}(t) = N_1(D_{\alpha_1}(t)) - N_2(D_{\alpha_2}(t))$$

is called a SFSP.

Next, we derive the mgf of SFSP. We use the expression for marginal (pmf) of SFPP that is given in (17) to obtain the marginal pmf of SFSP.

$$M_\theta(t) = \mathbb{E}[e^{\theta S_{\alpha_1,\alpha_2}(t)}] = \mathbb{E}[e^{\theta(N_1(D_{\alpha_1}(t)) - N_2(D_{\alpha_2}(t)))}] = e^{-t[\lambda_1^{\alpha_1}(1-e^\theta)^{\alpha_1} + \lambda_2^{\alpha_2}(1-e^{-\theta})^{\alpha_2}]}, \quad \theta \in \mathbb{R}.$$

In the next result, we obtain the state probabilities of the SFSP.

Theorem 3. *The pmf $H_k(t) = \mathbb{P}(S_{\alpha_1,\alpha_2}(t) = k)$ of SFSP is given by*

$$H_k(t) = \sum_{n=0}^\infty \frac{(-1)^k}{n!(n+k)!}\left({}_1\psi_1\left[\begin{matrix}(1,\alpha_1); \\ (1-n-k,\alpha_1);\end{matrix}(-\lambda_1^{\alpha_1}t)\right]\right)\left({}_1\psi_1\left[\begin{matrix}(1,\alpha_2); \\ (1-n,\alpha_2);\end{matrix}(-\lambda_2^{\alpha_2}t)\right]\right)\mathbb{I}_{k\geq0}$$

$$+ \sum_{n=0}^\infty \frac{(-1)^{|k|}}{n!(n+|k|)!}\left({}_1\psi_1\left[\begin{matrix}(1,\alpha_1); \\ (1-n,\alpha_1);\end{matrix}(-\lambda_1^{\alpha_1}t)\right]\right)\left({}_1\psi_1\left[\begin{matrix}(1,\alpha_2); \\ (1-n-|k|,\alpha_2);\end{matrix}(-\lambda_2^{\alpha_2}t)\right]\right)\mathbb{I}_{k<0} \quad (48)$$

for $k \in \mathbb{Z}$.

Proof. Note that $N_1(D_{\alpha_1}(t))$ and $N_2(D_{\alpha_2}(t))$ are independent, hence

$$\mathbb{P}(S_{\alpha_1,\alpha_2}(t) = k) = \sum_{n=0}^\infty \mathbb{P}(N_1(D_{\alpha_1}(t)) = n+k)\mathbb{P}(N_2(D_{\alpha_2}(t)) = n)\mathbb{I}_{k\geq0}$$

$$+ \sum_{n=0}^\infty \mathbb{P}(N_1(D_{\alpha_1}(t)) = n)\mathbb{P}(N_2(D_{\alpha_2}(t)) = n+|k|)\mathbb{I}_{k<0}.$$

Using (17), the result follows. □

In the next theorem, we discuss the governing differential-difference equation of the marginal pmf of SFSP.

Theorem 4. *The marginal distribution $H_k(t) = \mathbb{P}(S_{\alpha_1,\alpha_2}(t) = k)$ of SFSP satisfies the following differential difference equations*

$$\frac{d}{dt}H_k(t) = -\lambda_1^{\alpha_1}(1-B)^{\alpha_1}H_k(t) - \lambda_2^{\alpha_2}(1-F)^{\alpha_2}H_k(t), \quad k \in \mathbb{Z} \tag{49}$$

$$\frac{d}{dt}H_0(t) = -\lambda_1^{\alpha_1}H_0(t) - \lambda_2^{\alpha_2}H_1(t), \tag{50}$$

with initial conditions $H_0(0) = 1$ and $H_k(0) = 0$ for $k \neq 0$.

Proof. The proof follows by using pgf. $\square$

Remark 9. *The mgf of the SFSP solves the differential equation*

$$\frac{dM_\theta(t)}{dt} = -M_\theta(t)(\lambda_1^{\alpha_1}(1-e^\theta)^{\alpha_1} + \lambda_2^{\alpha_2}(1-e^{-\theta})^{\alpha_2}). \tag{51}$$

Proposition 12. *The Lévy density $\nu_{S_{\alpha_1,\alpha_2}}(x)$ of SFSP is given by*

$$\nu_{S_{\alpha_1,\alpha_2}}(x) = \lambda_1^{\alpha_1}\sum_{n_1=1}^\infty (-1)^{n_1+1}\binom{\alpha_1}{n_1}\delta_{n_1}(x) + \lambda_2^{\alpha_2}\sum_{n_2=1}^\infty (-1)^{n_2+1}\binom{\alpha_2}{n_2}\delta_{-n_2}(x).$$

Proof. Substituting the Lévy density $\nu_{N_{\alpha_1}}(x)$ and $\nu_{N_{\alpha_2}}(x)$ of $N_1(D_{\alpha_1}(t))$ and $N_2(D_{\alpha_2}(t))$, respectively, from the Equation (24), we obtain

$$\nu_{S_{\alpha_1,\alpha_2}}(x) = \nu_{N_{\alpha_1}}(x) + \nu_{N_{\alpha_2}}(x),$$

which gives the desired result. $\square$

7.2. Tempered Space-Fractional Skellam Process (TSFSP)

In this section, we present the tempered space-fractional Skellam process (TSFSP). We discuss the corresponding fractional difference-differential equations, marginal pmfs, and moments of this process.

Definition 7 (TSFSP). *The TSFSP is obtained by taking the difference of two independent tempered space fractional Poisson processes. Let $D_{\alpha_1,\mu_1}(t)$, $D_{\alpha_2,\mu_2}(t)$ be two independent TSS (see [28]) and $N_1(t)$, $N_2(t)$ be two independent Poisson processes that are independent of TSS. Subsequently, the stochastic process*

$$S_{\alpha_1,\alpha_2}^{\mu_1,\mu_2}(t) = N_1(D_{\alpha_1,\mu_1}(t) - N_2(D_{\alpha_2,\mu_2}(t))$$

is called the TSFSP.

Theorem 5. *The PMF $H_k^{\mu_1,\mu_2}(t) = \mathbb{P}(S_{\alpha_1,\alpha_2}^{\mu_1,\mu_2}(t) = k)$ is given by*

$$H_k^{\mu_1,\mu_2}(t) = \sum_{n=0}^\infty \frac{(-1)^k}{n!(n+k)!}e^{t(\mu_1^{\alpha_1}+\mu_1^{\alpha_1})}\left(\sum_{m=0}^\infty \frac{\mu_1^m\lambda_1^{-m}}{m!}{}_1\psi_1\left[\begin{matrix}(1,\alpha_1);\\(1-n-k-m,\alpha_1);\end{matrix}(-\lambda_1^{\alpha_1}t)\right]\right) \times$$

$$\left(\sum_{l=0}^\infty \frac{\mu_2^l\lambda_2^{-l}}{l!}{}_1\psi_1\left[\begin{matrix}(1,\alpha_2);\\(1-l-k,\alpha_2);\end{matrix}(-\lambda_2^{\alpha_2}t)\right]\right) \tag{52}$$

when $k \geq 0$ and similarly for $k < 0$,

$$H_k^{\mu_1,\mu_2}(t) = \sum_{n=0}^{\infty} \frac{(-1)^{|k|}}{n!(n+|k|)!} e^{t(\mu_1^{\alpha_1}+\mu_1^{\alpha_1})} \left(\sum_{m=0}^{\infty} \frac{\mu_1^m \lambda_1^{-m}}{m!} {}_1\psi_1 \left[\begin{matrix} (1,\alpha_1); \\ (1-n-m,\alpha_1); \end{matrix} (-\lambda_1^{\alpha_1} t) \right] \right) \times$$
$$\left(\sum_{l=0}^{\infty} \frac{\mu_2^l \lambda_2^{-l}}{l!} {}_1\psi_1 \left[\begin{matrix} (1,\alpha_2); \\ (1-l-n-|k|,\alpha_2); \end{matrix} (-\lambda_2^{\alpha_2} t) \right] \right). \tag{53}$$

Proof. Because $N_1(D_{\alpha_1,\mu_1}(t))$ and $N_2(D_{\alpha_2,\mu_2}(t))$ are independent,

$$\mathbb{P}\left(S_{\alpha_1,\alpha_2}^{\mu_1,\mu_2}(t) = k \right) = \sum_{n=0}^{\infty} \mathbb{P}(N_1(D_{\alpha_1,\mu_1}(t)) = n+k)\mathbb{P}(N_2(D_{\alpha_2,\mu_2}(t)) = n)\mathbb{I}_{k \geq 0}$$
$$+ \sum_{n=0}^{\infty} \mathbb{P}(N_1(D_{\alpha_1,\mu_1}(t)) = n)\mathbb{P}(N_2(D_{\alpha_2,\mu_2}(t)) = n+|k|)\mathbb{I}_{k<0},$$

which gives the marginal pmf of TSFPP using (26). □

Remark 10. *We use this expression to calculate the marginal distribution of TSFSP. The mgf is obtained using the conditioning argument. Let $f_{\alpha,\mu}(x,t)$ be the density function of $D_{\alpha,\mu}(t)$. Subsequently,*

$$\mathbb{E}[e^{\theta N(D_{\alpha,\mu}(t))}] = \int_0^{\infty} \mathbb{E}[e^{\theta N(u)}] f_{\alpha,\mu}(u,t)du = e^{-t\{(\lambda(1-e^{\theta})+\mu)^{\alpha}-\mu^{\alpha}\}}. \tag{54}$$

Using (54), the mgf of TSFSP is

$$\mathbb{E}[e^{\theta S_{\alpha_1,\alpha_2}^{\mu_1,\mu_2}(t)}] = \mathbb{E}\left[e^{\theta N_1(D_{\alpha_1,\mu_1}(t))}\right] \mathbb{E}\left[e^{-\theta N_2(D_{\alpha_2,\mu_2}(t))}\right] = e^{-t[\{(\lambda_1(1-e^{\theta})+\mu_1)^{\alpha_1}-\mu_1^{\alpha_1}\}+\{(\lambda_2(1-e^{-\theta})+\mu_2)^{\alpha_2}-\mu_2^{\alpha_2}\}]}.$$

Remark 11. *We have $\mathbb{E}[S_{\alpha_1,\alpha_2}^{\mu_1,\mu_2}(t)] = t(\lambda_1\alpha_1\mu_1^{\alpha_1-1} - \lambda_2\alpha_2\mu_2^{\alpha_2-1})$. Further, the covariance of TSFSP can be obtained by using (29) and*

$$\mathrm{Cov}\left[S_{\alpha_1,\alpha_2}^{\mu_1,\mu_2}(t), S_{\alpha_1,\alpha_2}^{\mu_1,\mu_2}(s) \right] = \mathrm{Cov}[N_1(D_{\alpha_1,\mu_1}(t)), N_1(D_{\alpha_1,\mu_1}(s))] + \mathrm{Cov}[N_2(D_{\alpha_2,\mu_2}(t)), N_2(D_{\alpha_2,\mu_2}(s))]$$
$$= \mathrm{Var}(N_1(D_{\alpha_1,\mu_1}(\min(t,s)))) + \mathrm{Var}(N_2(D_{\alpha_2,\mu_2}(\min(t,s)))).$$

Proposition 13. *The Lévy density $\nu_{S_{\alpha_1,\alpha_2}^{\mu_1,\mu_2}}(x)$ of TSFSP is given by*

$$\nu_{S_{\alpha_1,\alpha_2}^{\mu_1,\mu_2}}(x) = \sum_{n_1=1}^{\infty} \mu_1^{\alpha_1-n_1} \binom{\alpha_1}{n_1} \lambda_1^{n_1} \sum_{l_1=1}^{n_1} \binom{n_1}{l_1} (-1)^{l_1+1} \delta_{l_1}(x)$$
$$+ \sum_{n_2=1}^{\infty} \mu_2^{\alpha_2-n_2} \binom{\alpha_2}{n_2} \lambda_2^{n_2} \sum_{l_2=1}^{n_2} \binom{n_2}{l_2} (-1)^{l_2+1} \delta_{l_2}(x), \quad \mu_1, \mu_2 > 0.$$

Proof. By adding Lévy density $\nu_{N_{\alpha_1,\mu_1}}(x)$ and $\nu_{N_{\alpha_2,\mu_2}}(x)$ of $N_1(D_{\alpha_1,\mu_1}(t))$ and $N_2(D_{\alpha_2,\mu_2}(t))$, respectively, from Equation (30), which leads to

$$\nu_{S_{\alpha_1,\alpha_2}^{\mu_1,\mu_2}}(x) = \nu_{N_{\alpha_1,\mu_1}}(x) + \nu_{N_{\alpha_2,\mu_2}}(x).$$

□

7.3. Simulation of SFSP and TSFSP

We present the algorithm to simulate the sample trajectories for SFSP and TSFSP. We use *Python 3.7* and its libraries *Numpy* and *Matplotlib* for the simulation purpose. It is worth mentioning that Python is an open source and freely available software.

Simulation of SFSP: fix the values of the parameters α_1, α_2, λ_1 and λ_2;

> **Step-1:** generate independent and uniformly distributed random vectors U, V of size 1000 each in the interval $[0,1]$;
>
> **Step-2:** generate the increments of the α_1-stable subordinator $D_{\alpha_1}(t)$ (see [29]) with pdf $f_{\alpha_1}(x,t)$, while using the relationship $D_{\alpha_1}(t+dt) - D_{\alpha_1}(t) \overset{d}{=} D_{\alpha_1}(dt) \overset{d}{=} (dt)^{\frac{1}{\alpha_1}} D_{\alpha_1}(1)$, where
>
> $$D_{\alpha_1}(1) = \frac{\sin(\alpha_1 \pi U)[\sin((1-\alpha_1)\pi U)]^{1/\alpha_1 - 1}}{[\sin(\pi U)]^{1/\alpha_1} |\log V|^{1/\alpha_1 - 1}};$$
>
> **Step-3:** generate the increments of Poisson distributed rvs $N_1(D_{\alpha_1}(dt))$ with parameter $\lambda_1(dt)^{1/\alpha_1} D_{\alpha_1}(1)$;
>
> **Step-4:** cumulative sum of increments gives the space fractional Poisson process $N_1(D_{\alpha_1}(t))$ sample trajectories; and,
>
> **Step-5:** similarly generate $N_2(D_{\alpha_2}(t))$ and subtract these to obtain the SFSP $S_{\alpha_1,\alpha_2}(t)$.

We next present the algorithm for generating the sample trajectories of TSFSP.

Simulation of TSFSP: fix the values of the parameters α_1, α_2, λ_1, λ_2, μ_1 and μ_2.

Use the first two steps of previous algorithm for generating the increments of α-stable subordinator $D_{\alpha_1}(t)$.

> **Step-3:** for generating the increments of TSS $D_{\alpha_1,\mu_1}(t)$ with pdf $f_{\alpha_1,\mu_1}(x,t)$, we use the following steps, called "acceptance-rejection method";
>
> **(a)** generate the stable random variable $D_{\alpha_1}(dt)$;
> **(b)** generate uniform $(0,1)$ rv W (independent from D_{α_1});
> **(c)** if $W \leq e^{-\mu_1 D_{\alpha_1}(dt)}$, then $D_{\alpha_1,\mu_1}(dt) = D_{\alpha_1}(dt)$ ("accept"); otherwise, go back to (a) ("reject").
> Note that, here we used that $f_{\alpha_1,\mu_1}(x,t) = e^{-\mu_1 x + \mu_1^{\alpha_1} t} f_{\alpha_1}(x,t)$, which implies $\frac{f_{\alpha_1,\mu_1}(x,t)(x,dt)}{c f_{\alpha_1}(x,dt)} = e^{-\mu_1 x}$ for $c = e^{\mu_1^{\alpha_1} dt}$ and the ratio is bounded between 0 and 1;
>
> **Step-4:** generate Poisson distributed rv $N(D_{\alpha_1,\mu_1}(dt))$ with parameter $\lambda_1 D_{\alpha_1,\mu_1}(dt)$
>
> **Step-5:** cumulative sum of increments gives the tempered space fractional Poisson process $N_1(D_{\alpha_1,\mu_1}(t))$ sample trajectories; and,
>
> **Step-6:** similarly generate $N_2(D_{\alpha_2,\mu_2}(t))$, then take difference of these to get the sample paths of the TSFSP$S_{\alpha_1,\alpha_2}^{\mu_1,\mu_2}(t)$.

The tail probability of α-stable subordinator behaves asymptotically as (see e.g., [30])

$$\mathbb{P}(D_\alpha(t) > x) \sim \frac{t}{\Gamma(1-\alpha)} x^{-\alpha}, \text{ as } x \to \infty.$$

For $\alpha_1 = 0.6$ and $\alpha_2 = 0.9$ and fixed t, it is more probable that the value of the rv $D_{\alpha_1}(t)$ is higher than the rv $D_{\alpha_2}(t)$. Thus, for same intensity parameter λ for Poisson process the process $N(D_{\alpha_1}(t))$ will have generally more arrivals than the process $N(D_{\alpha_2}(t))$ until time t. This is evident from the trajectories of the SFSP in Figure 1, because the trajectories are biased towards positive side. The TSFPP is a finite mean process, however SFPP is an infinite mean process and hence SFSP paths are expected to have large jumps, since there could be a large number of arrivals in any interval.

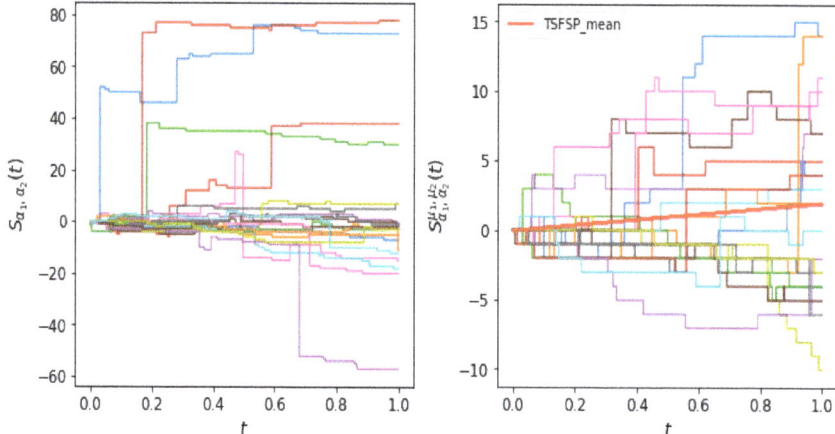

Figure 1. The left hand figure shows the sample trajectories of SFSP with parameters $\alpha_1 = 0.6$, $\alpha_2 = 0.9$, $\lambda_1 = 6$ and $\lambda_2 = 10$. The sample trajectories of TSFSP are shown in the right figure with parameters $\alpha_1 = 0.6$, $\alpha_2 = 0.9$, $\lambda_1 = 6$, $\lambda_2 = 10$, $\mu_1 = 0.2$ and $\mu_2 = 0.5$.

Author Contributions: Conceptualization, N.G., A.K. and N.L.; Methodology, A.K. and N.L.; Simulation, N.G.; Writing-Original Draft Preparation, N.G., A.K. and N.L.; Writing-Review & Editing, N.G., A.K. and N.L. All authors have read and agreed to the published version of the manuscript.

Funding: N.G. would like to thank Council of Scientific and Industrial Research (CSIR), India for supporting her research under the fellowship award number 09/1005(0021)2018-EMR-I. Further, A.K. would like to express his gratitude to Science and Engineering Research Board (SERB), India for the financial support under the MATRICS research grant MTR/2019/000286.

Acknowledgments: N.G. would like to thank Council of Scientific and Industrial Research(CSIR), India, for the award of a research fellowship.

Conflicts of Interest: The authors declare no conflicts of interest.

References

1. Skellam, J.G. The frequency distribution of the difference between two Poisson variables belonging to different populations. *J. Roy. Statist. Soc. Ser. A* **1946**, *109*, 296. [CrossRef]

2. Irwin, J.O. The frequency distribution of the difference between two independent variates following the same Poisson distribution. *J. Roy. Statist. Soc. Ser. A* **1937**, *100*, 415–416. [CrossRef]

3. Steutel, F.W.; Van Harn, K. *Infinite Divisibility of Probability Distributions on the Real Line*; Marcel Dekker: New York, NY, USA, 2004.

4. Hwang, Y.; Kim, J.; Kweon, I. Sensor noise modeling using the Skellam distribution: Application to the color edge detection. In Proceedings of the 2007 IEEE Conference on Computer Vision and Pattern Recognition, Minneapolis, MN, USA, 17–22 June 2007; pp. 1–8.

5. Karlis, D.; Ntzoufras, I. Bayesian modeling of football outcomes: Using the Skellam's distribution for the goal difference. *IMA J. Manag. Math.* **2008**, *20*, 133–145. [CrossRef]

6. Bacry, E.; Delattre, M.; Hoffman, M.; Muzy, J. Modeling microstructure noise with mutually exciting point processes. *Quant. Finance* **2013**, *13*, 65–77. [CrossRef]

7. Bacry, E.; Delattre, M.; Hoffman, M.; Muzy, J. Some limit theorems for Hawkes processes and applications to financial statistics. *Stoch. Proc. Appl.* **2013**, *123*, 2475–2499. [CrossRef]

8. Barndorff-Nielsen, O.E.; Pollard, D.; Shephard, N. Integer-valued Lévy processes and low latency financial econometrics. *Quant. Financ.* **2011**, *12*, 587–605. [CrossRef]

9. Carr, P. Semi-static hedging of barrier options under Poisson jumps. *Int. J. Theor. Appl. Financ.* **2011**, *14*, 1091–1111. [CrossRef]

10. Kerss, A.; Leonenko, N.N; Sikorskii, A. Fractional Skellam processes with applications to finance. *Fract. Calc. Appl. Anal.* **2014**, *17*, 532–551. [CrossRef]

11. Kumar, A., Leonenko, N.N.; Pichler, A. Fractional risk process in insurance. *Math. Financ. Econ.* **2019**, *529*, 121–539.

12. Buchak, K.V.; Sakhno, L.M. On the governing equations for Poisson and Skellam processes time-changed by inverse subordinators. *Theory Probab. Math. Statist.* **2018**, *98*, 87–99. [CrossRef]

13. Sengar, A.S.; Maheshwari, A.; Upadhye, N.S. Time-changed Poisson processes of order *k*. *Stoch. Anal. Appl.* **2020**, *38*, 124–148. [CrossRef]

14. Kostadinova, K.Y.; Minkova, L.D. On the Poisson –process of order *k*. *Pliska Stud. Math. Bulgar.* **2012**, *22*, 117–128.

15. Philippou, A.N. Distributions and Fibonacci polynomials of order *k*, longest runs, and reliability of consecutive *k*-out-of-*n*:F systems. In *Fibonacci Numbers and Their Applications*; Philippou, A.N., Ed.; D. Reidel: Dordrecht, The Netherlands, 1986; pp. 203–227.

16. Fisher, R.A. The effects of methods of ascertainment upon the estimation of frequencies. *Ann. Eugen.* **1934**, *6*, 13–15. [CrossRef]

17. Cox, D.R.; Lewis, P.A.W. *The Statistical Analysis of Series of Events*; Wiley: New York, NY, USA, 1966.

18. Beghin, L.; Macci, C. Fractional discrete processes: Compound and mixed Poisson representations. *J. Appl. Probab.* **2014**, *51*, 9–36. [CrossRef]

19. Abramowitz, M.; Stegun, I.A. *Handbook of Mathematical Functions*; Dover: New York, NY, USA, 1974.

20. Sato, K-I. *Lévy Processes and Infinitely Divisible Distributions*; Cambridge University Press: Cambridge, UK, 1999.

21. Schilling, R.L.; Song, R.; Vondracek, Z. *Bernstein Functions: Theory and Applications*; De Gruyter: Berlin, Germany, 2012.

22. Orsingher, E.; Polito, F. The space-fractional Poisson process. *Statist. Probab. Lett.* **2012**, *82*, 852–858. [CrossRef]

23. Kilbas; A.; Srivastava, H.; Trujillo, J. *Theory and Applications of Fractional Differential Equations*; Elsevier: Amsterdam, The Netherlands, 2006.

24. Beran, J. *Statistics for Long-Memory Processes*; Chapman & Hall: New York, NY, USA, 1994.

25. Gupta, N.; Kumar, A.; Leonenko, N. Tempered Fractional Poisson Processes and Fractional Equations with Z-Transform. *Stoch. Anal. Appl.* **2020**, *38*, 939–957. [CrossRef]

26. Xia, W. On the distribution of running average of Skellam process. *Int. J. Pure Appl. Math.* **2018**, *119*, 461–473.

27. Maheshwari, A.; Vellaisamy, P. On the long-range dependence of fractional Poisson and negative binomial processes. *J. Appl. Probab.* **2016**, *53*, 989–1000. [CrossRef]

28. Rosiński, J. Tempering stable processes. *Stochastic. Process. Appl.* **2007**, *117*, 677–707. [CrossRef]

29. Cahoy, D.; Uchaikin, V.; Woyczynski, A. Parameter estimation from fractional Poisson process. *J. Statist. Plann. Inference* **2013**, *140*, 3106–3120. [CrossRef]

30. Samorodnitsky, G.; Taqqu, M.S. *Stable Non-Gaussian Random Processes*; Chapman and Hall: Boca Raton, FL, USA, 1994.

Publisher's Note: MDPI stays neutral with regard to jurisdictional claims in published maps and institutional affiliations.

MDPI

Article

A Simplified Fractional Order PID Controller's Optimal Tuning: A Case Study on a PMSM Speed Servo

Weijia Zheng [1], Ying Luo [2,*], YangQuan Chen [3] and Xiaohong Wang [4]

1 School of Mechatronic Engineering and Automation, Foshan University, 33 Guangyun Road, Foshan 528225, China; z.wj08@mail.scut.edu.cn
2 Department of Mechanical Science and Engineering, Huazhong University of Science and Technology, 1037 Luoyu Road, Wuhan 430074, China
3 School of Engineering, University of California, Merced, 5200 North Lake Road, Merced, CA 95340, USA; ychen53@ucmerced.edu
4 School of Automation Science and Engineering, South China University of Technology, 381 Wushan Road, Guangzhou 510641, China; xhwang@scut.edu.cn
* Correspondence: ying.luo@hust.edu.cn

Abstract: A simplified fractional order PID (FOPID) controller is proposed by the suitable definition of the parameter relation with the optimized changeable coefficient. The number of the pending controller parameters is reduced, but all the proportional, integral, and derivative components are kept. The estimation model of the optimal relation coefficient between the controller parameters is established, according to which the optimal FOPID controller parameters can be calculated analytically. A case study is provided, focusing on the practical application of the simplified FOPID controller to a permanent magnet synchronous motor (PMSM) speed servo. The dynamic performance of the simplified FOPID control system is tested by motor speed control simulation and experiments. Comparisons are performed between the control systems using the proposed method and those using some other existing methods. According to the simulation and experimental results, the simplified FOPID control system achieves the optimal dynamic performance. Therefore, the validity of the proposed controller structure and tuning method is demonstrated.

Keywords: fractional order PID control; PMSM; frequency-domain control design; optimal tuning

Citation: Zheng, W.; Luo, Y.; Chen, Y.; Wang, X. A Simplified Fractional Order PID Controller'sOptimal Tuning: A Case Study on a PMSM Speed Servo. *Entropy* **2021**, *23*, 130. https://doi.org/10.3390/e23020130

Received: 27 November 2020
Accepted: 14 January 2021
Published: 20 January 2021

Publisher's Note: MDPI stays neutral with regard to jurisdictional claims in published maps and institutional affiliations.

1. Introduction

Recently, fractional calculus has attracted increasing interest in various fields of science and engineering [1–4]. Fractional calculus is a generalization of the traditional integral and differential operators from integer order to real number order [5–8]. Thus, it has a larger feasible scope and greater flexibility in the system modeling and controller design methodology than the classical integer order one [9–11]. Fractional control has aroused theoretical and practical interest in the control community. Different kinds of fractional order controllers and tuning methods have been introduced and studied [12–14].

The fractional order proportional-integral-derivative (FOPID) controller has the tunable integral and differential orders, creating the possibility to provide better control performance [15]. However, the design of the FOPID controller is also more difficult. Generally, the tuning methods of the FOPID controller can mainly be divided into the analytic design methods and the optimization methods. The classic frequency-domain method is a typical analytic design method for the FOPI/D controller. Applying this method, three equations can be derived from three frequency-domain specifications [16], according to which the controller parameters can be calculated. However, with only three specifications, this method may not be directly used to design the FOPID controller with five degrees of freedom. On the other hand, the optimization design methods are based on iterative optimization [17,18]. Applying the optimization methods, the FOPID controller parameters are obtained by optimizing an objective function characterizing the performance of the control system, under the

197

constraints corresponding to specific design requirements, such as the system stability and sensitivity [19]. Thus, an optimal FOPID controller can be obtained using the optimization method, but the optimization process requires sufficient time and computing capability.

In our previous work, an analytic design method was proposed for the FOPID controller, according to the linear relation between the controller parameters [20]. On this basis, an improved FOPID controller is proposed in this paper, building the nonlinear relation between the integral gain K_i and derivative gain K_d, with a changeable coefficient. The optimal coefficient is modeled using the numerical fitting method, based on its optimal distribution with regard to the plant model characteristics and design specifications. With the estimated model, the parameters of the optimal FOPID controller can be calculated analytically according to the design specifications. Compared with our previous work, the improved FOPID controller proposed in this paper can be applied to a larger scope of plant models and design specifications because a more sophisticated relation between the controller parameters is adopted.

A case study of the proposed controller on the PMSM speed control is provided. The robustness to the gain variations, step response performance, and anti-load disturbance performance of the FOPID control system are tested by simulations and experiments. Comparisons are performed between the control systems using the proposed controller and those using some existing FOPID controllers. The advantages of the proposed method are demonstrated by simulation and experimental results.

The contributions of this paper mainly include: (1) The relations among the FOPID controller parameters being reasonably defined with a changeable coefficient, obtaining a simplified FOPID controller structure, but a complete P&I&D tunability. (2) The estimation model of the optimal relation coefficient between the controller parameters is built, realizing the optimal estimation of the fractional orders and the subsequent analytical calculation of the remaining parameters of the controller.

The paper is organized as follows: The simplified FOPID controller and the corresponding tuning method are proposed in Section 2. The estimation model of the optimal relation coefficient is discussed and established in Section 3. In Sections 4 and 5, the application of the improved FOPID controller to the PMSM speed control is studied. The robustness and dynamic performance of the control system using the simplified FOPID controller are verified by simulations and experiments. The conclusion is presented in Section 6.

2. Simplified FOPID Controller

The FOPID controller can be represented as (1),

$$C(s) = K_p \left(1 + \frac{K_i}{s^\lambda} + K_d s^\mu \right),$$ (1)

where K_p, K_i, and K_d represent the gains of the proportional, integral, and derivative components, respectively; λ and μ are the real number orders with $0 < \lambda < 2$ and $0 < \mu < 2$.

The typical unit negative feedback control system can be represented as Figure 1, where $G(s)$ and $C(s)$ are the plant and controller, respectively, and n_r and n are the reference and output signals, respectively. The classic frequency-domain method depends on three specifications, i.e., the gain crossover frequency ω_c, the phase margin φ_m, and the slope of the phase at ω_c [21], yielding,

$$|C(j\omega_c)G(j\omega_c)| = 1,$$ (2)

$$\mathrm{Arg}[C(j\omega_c)] + \mathrm{Arg}[G(j\omega_c)] = -\pi + \varphi_m,$$ (3)

$$\left. \frac{d[\mathrm{Arg}[C(j\omega)G(j\omega)]]}{d\omega} \right|_{\omega=\omega_c} = 0,$$ (4)

Therefore, the parameters of the FOPI or FOPD controllers can be calculated according to these specifications. However, five pending parameters of the FOPID controller cannot be solved according to only three equations.

To solve this problem, a relation between K_i and K_d is proposed as (5),

$$K_d = \frac{1}{aK_i}, \tag{5}$$

where a is a changeable coefficient. The dynamic characteristics of the FOPID controller, e.g., the overshoot and oscillation of the step response, are affected by the fractional orders λ and μ. Taking advantage of a simple assumption [22], a relation between λ and μ is proposed as (6),

$$\lambda = \mu. \tag{6}$$

Thus, the FOPID controller is converted into a simplified form,

$$C(s) = K_p\left(1 + \frac{K_i}{s^\lambda} + \frac{1}{aK_i}s^\lambda\right). \tag{7}$$

The amplitude and phase of the simplified FOPID controller can be obtained,

$$|C(j\omega)| = K_p\sqrt{P(\omega)^2 + Q(\omega)^2}, \tag{8}$$

$$\mathrm{Arg}[C(j\omega)] = \arctan\left(\frac{Q(\omega)}{P(\omega)}\right), \tag{9}$$

where:

$$P(\omega) = 1 + K_i\omega^{-\lambda}\cos\left(\frac{\pi}{2}\lambda\right) + \frac{1}{aK_i}\omega^\lambda\cos\left(\frac{\pi}{2}\lambda\right), \tag{10}$$

$$Q(\omega) = \frac{1}{aK_i}\omega^\lambda\sin\left(\frac{\pi}{2}\lambda\right) - K_i\omega^{-\lambda}\sin\left(\frac{\pi}{2}\lambda\right). \tag{11}$$

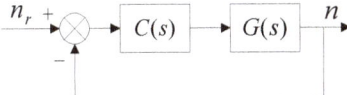

Figure 1. The closed-loop control system.

If ω_c and φ_m are given as the design specifications, substituting (9) into (3) yields,

$$\arctan\left(\frac{Q(\omega_c)}{P(\omega_c)}\right) + \mathrm{Arg}[G(j\omega_c)] = -\pi + \varphi_m. \tag{12}$$

Assuming that the coefficient a has been determined, denoting T as $\tan(-\pi + \varphi_m - \mathrm{Arg}[G(j\omega_c)])$, an equation relating K_i and λ can be obtained,

$$s_1 K_i{}^2 + s_0 K_i - \frac{1}{a} = 0, \tag{13}$$

where:

$$s_1 = \frac{T\omega_c^{-\lambda}\cos\left(\frac{\pi}{2}\lambda\right) + \omega_c^{-\lambda}\sin\left(\frac{\pi}{2}\lambda\right)}{\omega_c^\lambda\sin\left(\frac{\pi}{2}\lambda\right) - T\omega_c^\lambda\cos\left(\frac{\pi}{2}\lambda\right)}, \tag{14}$$

$$s_0 = \frac{T}{\omega_c^\lambda\sin\left(\frac{\pi}{2}\lambda\right) - T\omega_c^\lambda\cos\left(\frac{\pi}{2}\lambda\right)}. \tag{15}$$

Substituting (9) into (4), another equation about K_i and λ is obtained,

$$\frac{\lambda\omega_c^{\lambda-1}}{aK_i}\sin\left(\frac{\pi}{2}\lambda\right) + \frac{2\lambda}{a\omega_c}\sin(\lambda\pi) + \frac{M}{\omega_c^{2\lambda}}K_i{}^2 + \frac{M\omega_c^{2\lambda}}{aK_i{}^2} + \frac{2M}{a}\cos(\lambda\pi)$$
$$+ \frac{2M\omega_c^\lambda}{aK_i}\cos\left(\frac{\pi}{2}\lambda\right) + \lambda\omega_c^{\lambda-1}K_i\sin\left(\frac{\pi}{2}\lambda\right) + \frac{2M}{\omega_c^\lambda}\cos\left(\frac{\pi}{2}\lambda\right) + M = 0. \tag{16}$$

where:

$$M = \frac{d[\text{Arg}[G(j\omega)]]}{d\omega}\bigg|_{\omega=\omega_c}. \tag{17}$$

The integral gain K_i and order λ can be calculated by solving (13) and (16), and then, the proportional gain K_p can also be calculated by solving (2). Thus, if a is determined, all the parameters of the simplified FOPID controller can be calculated according to the design specifications.

3. Estimation Model Establishment

According to the proposed tuning method, the coefficient a should be determined before the calculation of the FOPID controller parameters. Thus, in order to improve the control performance, the distribution of the optimal a should be studied. In this paper, we concentrate on the third-order plant model described by (18),

$$G(s) = \frac{K}{s^3 + \tau_1 s^2 + \tau_2 s}, \tag{18}$$

where K, τ_1, and τ_2 are the parameters of the plant. The estimation model of a is established in the hyperspace defined by the ranges of the plant model parameters (τ_1, τ_2) and the design specifications (ω_c, φ_m). The ranges of τ_1 and τ_2 are determined according to the parameters of the plant models in actual applications, while those of ω_c and φ_m are determined according to the design requirements. In this paper, the range of τ_1 is set from 90 to 180 and that of τ_2 is set from 6000 to 11,000. The range of the gain crossover frequency ω_c is set from 35 rad/s to 70 rad/s, and that of the phase margin φ_m is set from 30° to 60°, covering the design requirements of a class of motion control systems [23].

3.1. Optimal Samples' Collection

Several values of τ_1 and τ_2 are uniformly selected from their ranges, respectively, obtaining ($\tau_{1,1}$, $\tau_{1,2}$, ..., $\tau_{1,m}$) and ($\tau_{2,1}$, $\tau_{2,2}$, ..., $\tau_{2,n}$). Since the plant model gain K has no influence on the estimation of a, it is given a fixed value. Thus, several test models can be established by combining the values of τ_1 and τ_2,

$$G_{i,j}(s) = \frac{K}{s^3 + \tau_{1,i} s^2 + \tau_{2,j} s}, \tag{19}$$

where $i = 1, 2, ..., m, j = 1, 2, ..., n$. Similarly, several values of ω_c ($\omega_{c,1}, \omega_{c,2}, ..., \omega_{c,p}$) and φ_m ($\varphi_{m,1}, \varphi_{m,2}, ..., \varphi_{m,q}$) are selected from their ranges to be the given design specifications.

The integral of time and absolute error (ITAE) is adopted as the loss function to evaluate the dynamic performance of the control system,

$$J = \int_0^\infty t|e(t)|dt, \tag{20}$$

where $e(t)$ represents the error between the reference and output signals.

The optimal sample of a for each test model (τ_1, τ_2) and design index (ω_c, φ_m) is collected following the steps shown in Figure 2. An accuracy threshold σ is set for the search of the optimal a. If the value resolution of the obtained a is smaller than σ, this value is considered to be the optimum; otherwise, another loop of search needs to be performed in a smaller range of a. For example, as shown in Figure 3, if the kth value of a, a_k, is the current optimal value, but its resolution is larger than σ, namely $a_{k+1} - a_k > \sigma$, then a new range of a will be created as (a_{k-1}, a_{k+1}), in which a new optimum will be obtained.

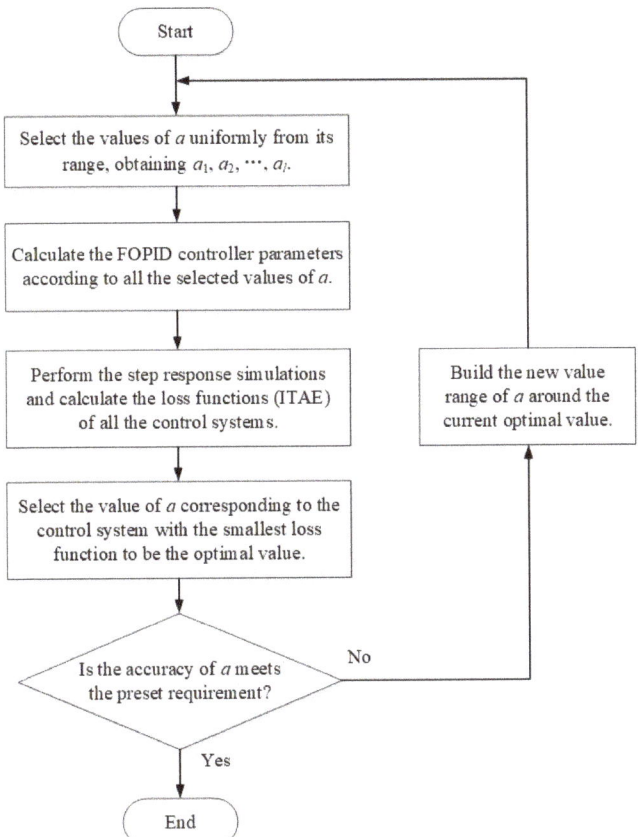

Figure 2. The determining process of the optimal *a*.

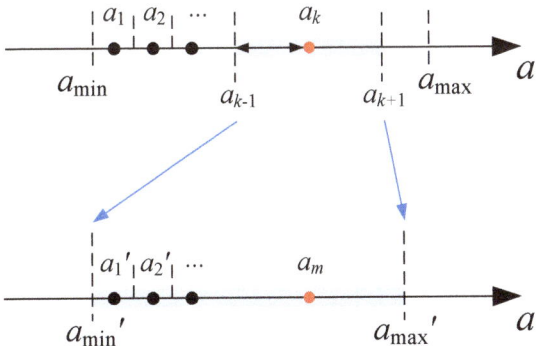

Figure 3. The construction of the new range of *a*.

According to the model parameter ranges, several values of τ_1: 90, 100, ..., 180, and τ_2: 6000, 6200, ..., 11,000, are selected to generate the test models. Similarly, several values of ω_c: 34 rad/s, 36 rad/s, ..., 70 rad/s and φ_m: 30°, 32°, ..., 60° are selected to be the design specifications. The initial range of *a* is from 0.001 to 500. The accuracy threshold σ is 0.001. Thus, following the steps shown in Figure 2, the optimal values of *a* corresponding to all the test models and design specifications are collected.

3.2. Estimation Model Establishment

Given the design specifications (ω_c, φ_m), an optimal FOPID controller can be designed for a plant model $G(s)$, according to an optimal value of a, which depends on the plant model characteristics (τ_1, τ_2) and design specifications (ω_c, φ_m). The estimation model is established to approximate the distribution law of the optimal a.

Firstly, the distribution of the optimal a for a single plant model with regard to ω_c and φ_m is studied. Taking the test model $G_{2,5}(s)$ ($\tau_1 = 100$, $\tau_2 = 6800$) as an example, the optimal values of a corresponding to different given crossover frequencies ω_c and a fixed phase margin φ_m ($\varphi_m = 30°$) are selected and plotted as the ω_c–a relation curve in Figure 4. According to Figure 4, the distribution of the optimal a can be approximated as a curve.

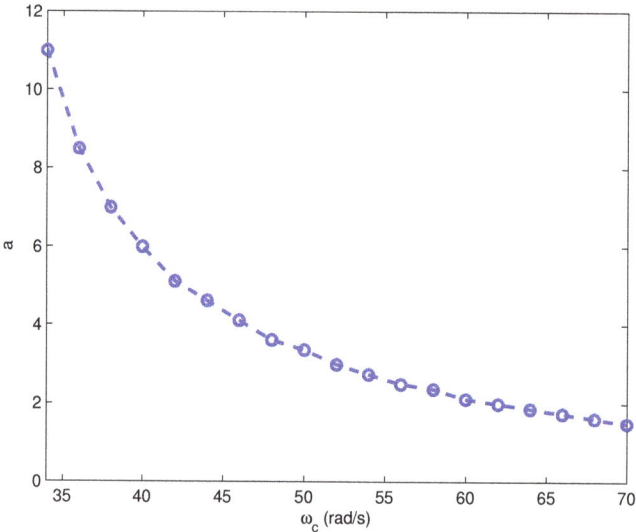

Figure 4. The ω_c–a relation curve with φ_m fixed to be $30°$.

The ω_c–a relation curves of different φ_m for test model $G_{2,5}(s)$ are plotted in Figure 5. It can be seen that the ω_c–a relation curves corresponding to different φ_m are close to each other. Thus, an assumption is adopted to simplify the analysis, i.e., the difference between the ω_c–a relation curves corresponding to different φ_m can be ignored. Therefore, for the same plant model, the optimal value of a is assumed to be only determined by ω_c.

Adopting the simplifying assumption, an estimation model needs to be built for the mean values of the optimal a. The ω_c–mean a relation corresponding to $G_{2,5}(s)$ is plotted as data spots in Figure 6.

It can be seen that the mean a values with regard to ω_c obey an obvious distribution law, which can be described by an exponential function,

$$a = A(\tau_1, \tau_2)e^{B(\tau_1, \tau_2)\omega_c}, \tag{21}$$

where A and B are the coefficients determined by the model parameters τ_1 and τ_2. The values of A and B can be obtained using the numerical fitting methods. The fitting function is plotted as the red curve in Figure 6. Fitting the ω_c–mean a relations of all the plant models, the values of A and B corresponding to different plant models: $A_{i,j}$ and $B_{i,j}$, are obtained, where the subscript i corresponds to that of $\tau_{1,i}$ and the subscript j corresponds to that of $\tau_{2,j}$, $i = 1, 2, ..., m, j = 1, 2, ..., n$.

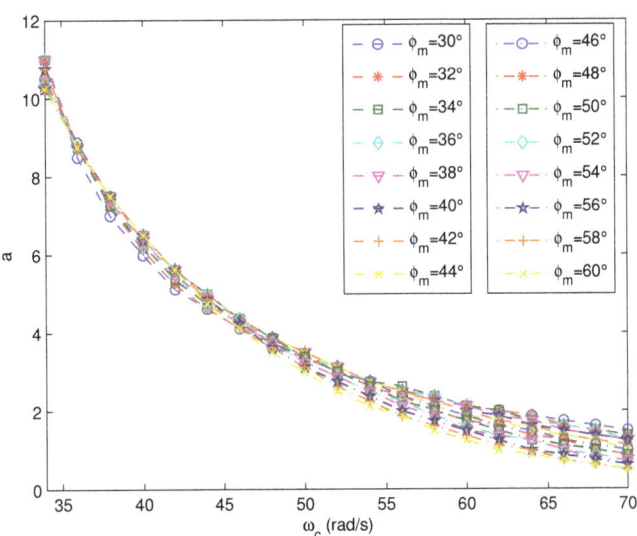

Figure 5. The ω_c–a relation curves correspond to different φ_m.

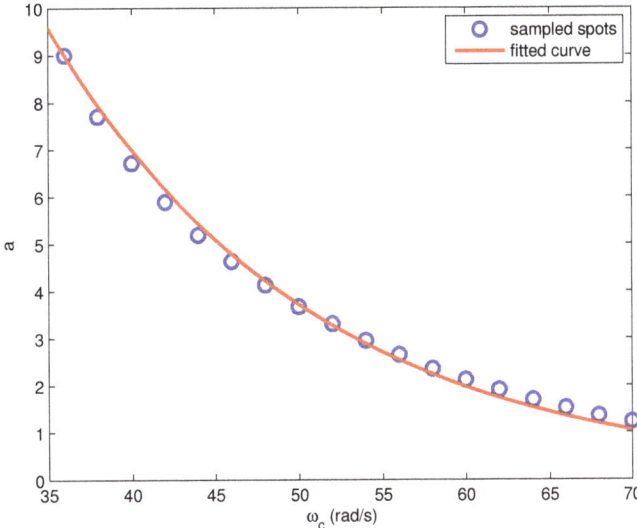

Figure 6. The ω_c–mean a relation and fitting curve of the test model $G_{2,5}(s)$.

Secondly, the relation between the coefficient A and the model parameters (τ_1, τ_2) is studied. Taking τ_2/τ_1 as the abscissa and the corresponding coefficient A as the ordinate, the distribution of A with regard to τ_2/τ_1 is plotted in Figure 7. As can be seen, the distribution of A with regard to τ_2/τ_1 can be approximated as a curve.

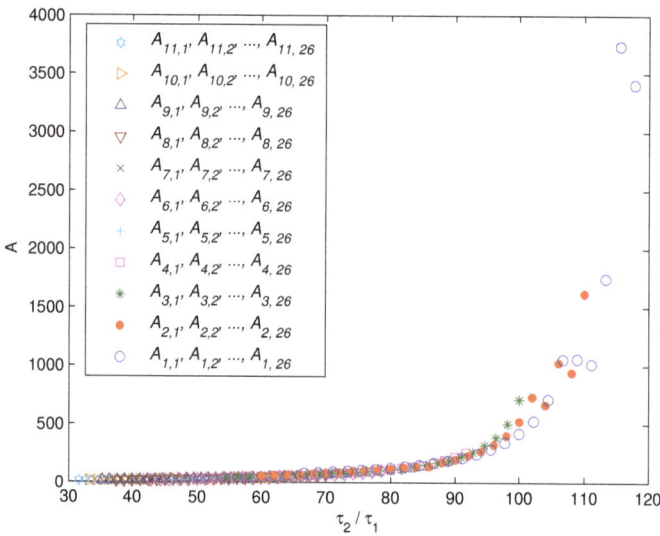

Figure 7. The distribution of A with regard to τ_2/τ_1.

The τ_2/τ_1–A relation is plotted again in Figure 8, without distinguishing the data spots corresponding to different plant models. According to the distribution of the data spots, the τ_2/τ_1–A relation can be fitted by a model with two exponential functions,

$$A(\tau_1, \tau_2) = Me^{P\frac{\tau_2}{\tau_1}} + Ne^{Q\frac{\tau_2}{\tau_1}}, \tag{22}$$

where M, N, P, and Q are the model coefficients, which can be obtained using numerical fitting methods. The fitting function is plotted as the red curve in Figure 8.

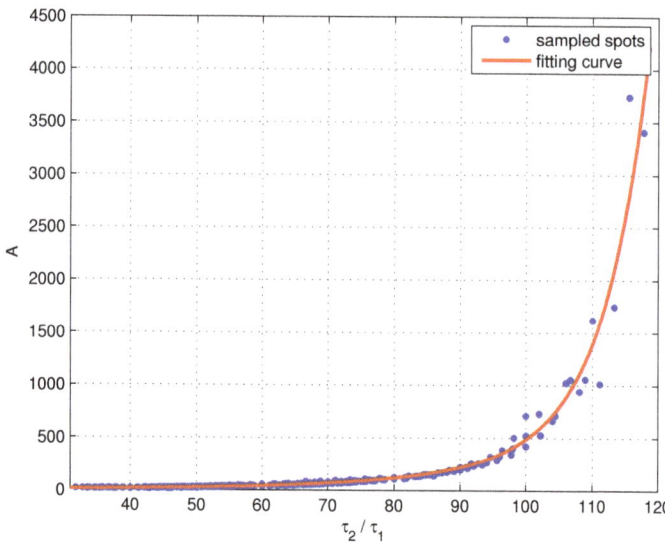

Figure 8. The τ_2/τ_1–A relation and the fitting curve.

Thirdly, the three-dimensional distribution of coefficient B with regard to τ_1 and τ_2 is plotted in Figure 9.

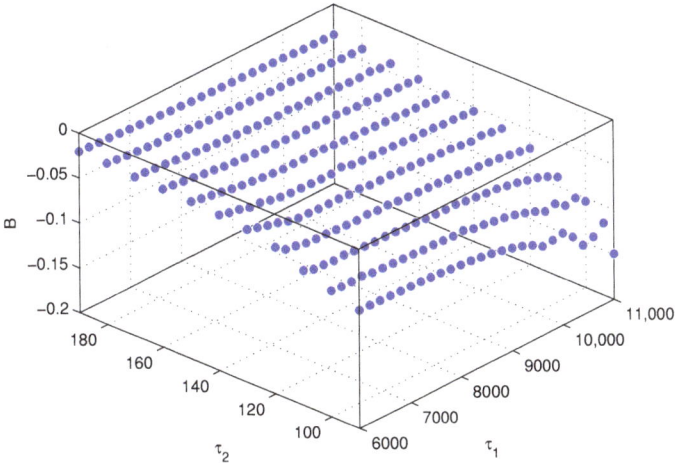

Figure 9. The distribution of B with regard to τ_1 and τ_2.

Taking τ_1 and τ_2 as the independent variables, the (τ_1, τ_2)–B relation can be fitted by a cubic polynomial function,

$$B(\tau_1, \tau_2) = p_{00} + p_{10}\tau_2 + p_{01}\tau_1 + p_{20}\tau_2^2 + p_{11}\tau_2\tau_1 + p_{02}\tau_1^2 + p_{30}\tau_2^3 + p_{21}\tau_2^2\tau_1 + p_{12}\tau_2\tau_1^2, \quad (23)$$

where p_{00}, p_{10}, p_{01}, p_{20}, p_{11}, p_{02}, p_{30}, p_{21}, and p_{12} are the model coefficients, which can be obtained using the numerical fitting methods. Therefore, all the coefficients of the estimation model are obtained.

4. Simulation Study

4.1. Feasible Region Study

The design flexibility of the proposed FOPID controller can be verified by studying the feasible regions of the design specifications. The feasible region of the design specifications includes the (ω_c, φ_m) combinations, according to which the reasonable FOPID controller can be obtained by solving (2)–(4). To demonstrate the advantage of the proposed method, the feasible region of the simplified FOPID controller is compared with those of the FOPI and IOPID controllers.

Taking the test model $G_{1,26}(s)$ ($\tau_1 = 90$, $\tau_2 = 11{,}000$) as an example, the feasible regions of the FOPI, IOPID, and FOPID controllers are plotted in Figures 10–12, respectively, where the feasible design specifications are marked in blue. According to Figure 10, if the design specifications are in the region where both ω_c and φ_m are large, we are unable to design an FOPI controller to satisfy (2)–(4) simultaneously. Similarly, according to Figure 11, we are unable to design an IOPID controller if both ω_c and φ_m are small. In contrast, according to Figure 12, the feasible region of the FOPID controller covers the entire region of the design specifications. Therefore, the proposed FOPID controller achieves more design options and flexibility than the FOPI and IOPID controllers.

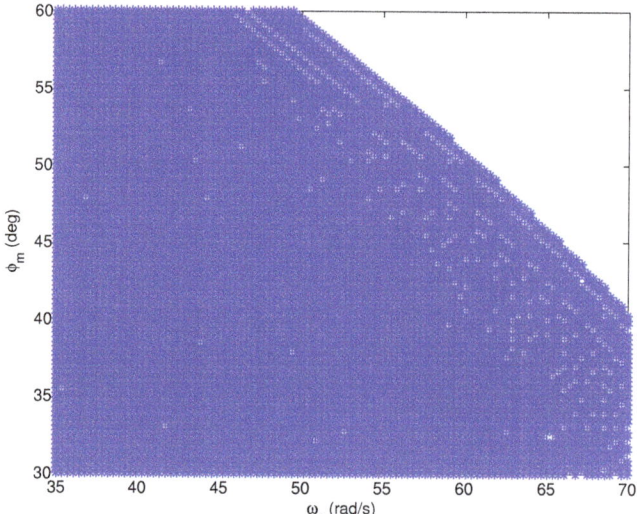

Figure 10. The feasible region of the FOPI controller.

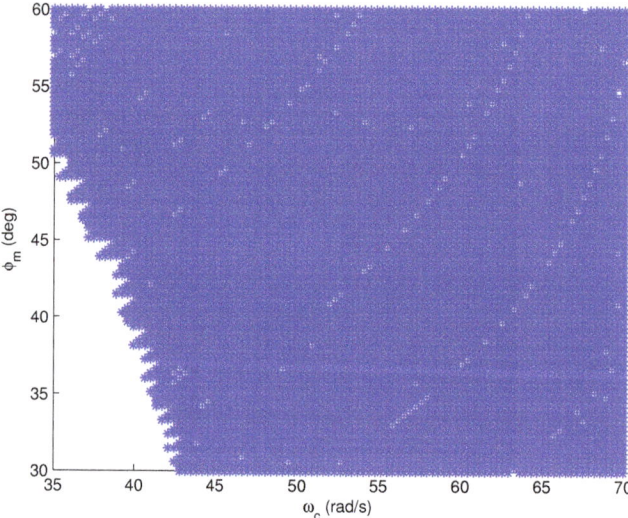

Figure 11. The feasible region of the IOPID controller.

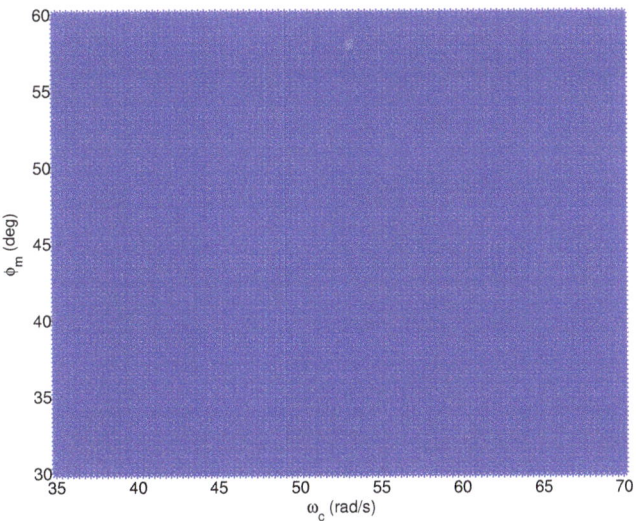

Figure 12. The feasible region of the FOPID controller.

4.2. PMSM Speed Servo Plant

The proposed estimation model and tuning method are applied to design the FOPID controllers for a class of PMSM speed servo systems. Applying the $d - q$ coordinates and the field-oriented control scheme, the dynamic characteristics of a PMSM can be described by the following equations,

$$u_q = Ri_q + L_q \frac{di_q}{dt} + C_e n, \tag{24}$$

$$\frac{GD^2}{375} \frac{dn}{dt} = C_m i_q - T_L, \tag{25}$$

where u_q and i_q are the q-axis voltage and current, respectively, R is the stator resistance, L_q is the q-axis stator inductance, C_e is the induced voltage constant, n is the motor speed in revolutions per minute (RPM), C_m is the torque constant, T_L is the load disturbance torque, and GD^2 is the flywheel inertia.

In the PMSM servo system, the q-axis voltage is often supplied by the pulse-width modulation (PWM) inverter, whose dynamic characteristics can be approximated by a first-order filter with time constant T_s. Adopting a PI controller as the feedback controller of the q-axis current,

$$C_i(s) = K_s (1 + \frac{1}{T_s s}), \tag{26}$$

the q-axis voltage can be obtained as:

$$u_q(s) = \frac{K_s}{T_s s}(i_{qr}(s) - i_q(s)), \tag{27}$$

where i_{qr} is the q-axis reference current. Thus, according to (24), (25), and (27), the transfer function of the PMSM speed servo plant (from i_{qr} to n) can be represented as:

$$G(s) = \frac{\frac{K_s}{C_e T_m T_s T_l}}{s^3 + \frac{1}{T_l}s^2 + \frac{K_s K_1}{R T_s T_l}s}, \tag{28}$$

where T_l is the electromagnetic time constant, $T_l = L/R$, and T_m is the electromechanical time constant, $T_m = GD^2R/(375C_eC_m)$. The transfer function of the PMSM speed servo plant model used in this paper is described as:

$$G(s) = \frac{47,979.257}{s^3 + 127.38s^2 + 9995.678s}.$$ (29)

4.3. Gain Robustness Study

Taking the PMSM speed servo as the plant model, setting the design specifications as $\omega_c = 40$ rad/s and $\varphi_m = 55°$, the optimal coefficient a is estimated as 9.968. Thus, the FOPID controller is obtained,

$$C_1(s) = 8.032\left(1 + \frac{13.207}{s^{0.983}} + 0.0076s^{0.983}\right).$$ (30)

The open-loop Bode diagram of the PMSM servo system using the FOPID controller is shown in Figure 13. It can be seen that the magnitude and phase characteristics of the control system satisfy the design specifications. The phase characteristic has zero slope at ω_c. Thus, the systems with gain variations will have similar phase margins as the nominal system.

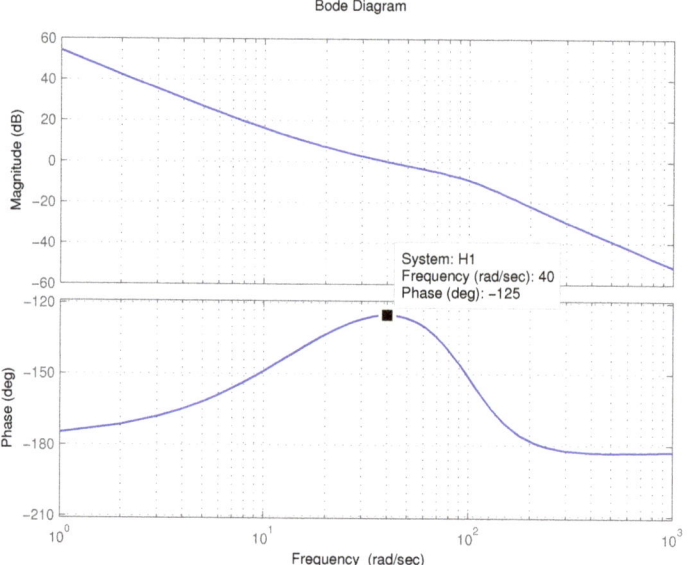

Figure 13. The open-loop Bode diagram of the control system.

The step response is performed to test the overshoots of the control systems with gain variations. The nominal gain of the plant is multiplied by 120% and 80% to simulate the gain variations. The step responses of the nominal system and those with gain variations are shown in Figure 14.

It can be seen that the responses of the control systems with gain variations have similar overshoots, satisfying the robustness requirement.

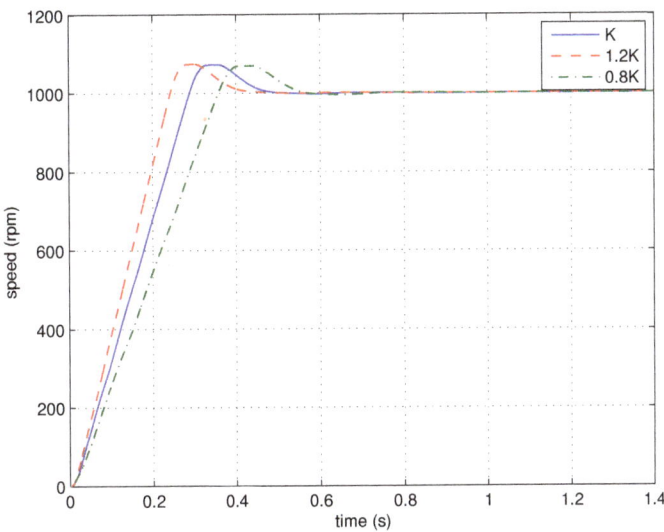

Figure 14. The step responses of the simplified FOPID control systems with different loop-gains (simulation).

4.4. Comparisons with Some Existing Methods

An optimization-based tuning method was proposed in [24], with the sensitivity and complementary sensitivity functions introduced as the constraints. Applying this method, an optimal FOPID controller is designed for the PMSM speed control system,

$$C_3(s) = 8.896\left(1 + \frac{29.815}{s^{1.299}} + 0.0685s^{0.403}\right). \tag{31}$$

The gain crossover frequency of the obtained control system is $\omega_c = 51.6$ rad/s, and the phase margin is $\varphi_m = 50°$. According to these design specifications, the optimal coefficient a is estimated as 5.047, and the FOPID controller is obtained,

$$C_4(s) = 10.451\left(1 + \frac{21.017}{s^{0.991}} + 0.0094s^{0.991}\right). \tag{32}$$

The step response simulation is performed, using the optimal FOPID controller $C_3(s)$ (denoted as opt-FOPID) and the proposed FOPID controller $C_4(s)$ (denoted as a-FOPID) as the speed controllers, respectively. To guarantee a fair comparison, the two systems are made to have similar rising times. The response curves and the performance indexes are shown in Figure 15 and Table 1, respectively.

The load disturbance response simulation is also performed to test the anti-load disturbance performance of the control systems. The response curves and performance indexes are shown in Figure 16 and Table 2, respectively.

Table 1. The step response performance indexes of the control systems using the optimal (opt)-FOPID and a-FOPID (simulation).

Control System	Settling Time (s)	Overshoot (%)
opt-FOPID	0.313	19.49
a-FOPID	0.255	21.46

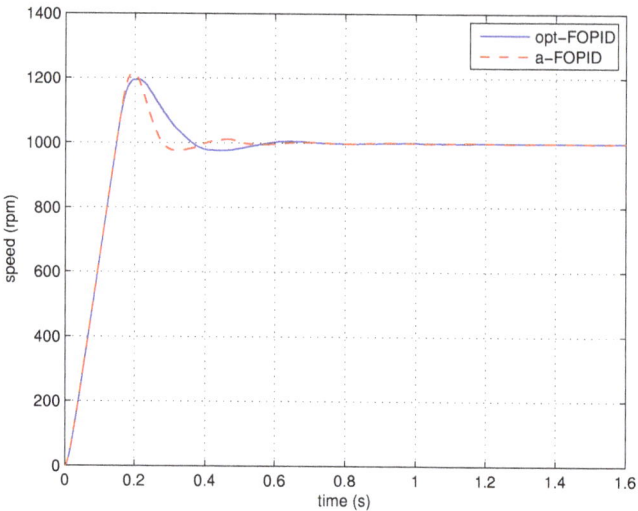

Figure 15. The step responses of the control systems using the opt-FOPID and a-FOPID (simulation).

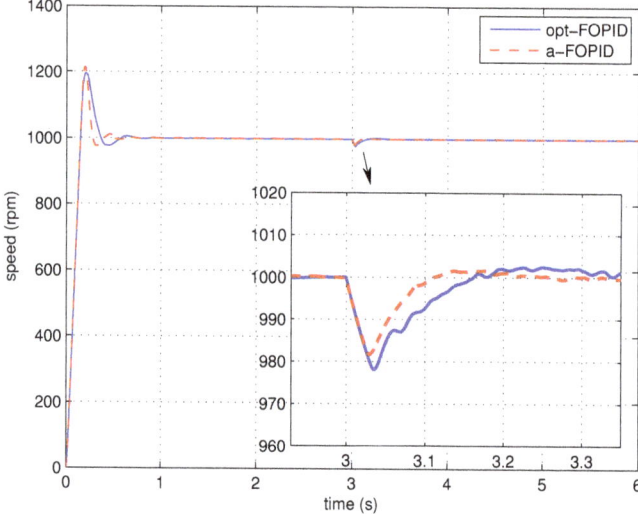

Figure 16. The load disturbance responses of the control systems using the opt-FOPID and a-FOPID (simulation).

Table 2. The anti-load disturbance performance indexes of the control systems using the opt-FOPID and a-FOPID (simulation).

Control System	Recovery Time (s)	Dynamic Speed Drop (%)
opt-FOPID	0.080	2.19
a-FOPID	0.055	1.83

According to Figure 15 and Table 1, the responses of two systems have similar over-shoots, but the system using the a-FOPID has a shorter settling time. Therefore, the system using the a-FOPID achieves better step response performance. According to Figure 16 and Table 2, the response of the system using the a-FOPID has a smaller speed drop and a shorter recovery time. Therefore, the system using the a-FOPID achieves better anti-load disturbance performance.

A Bode shaping-based tuning method for the FOPID controller is proposed in [25]. Applying this method, a FOPID controller is designed for the PMSM control system,

$$C_5(s) = 7.532\left(1 + \frac{49.843}{s^{1.27}} + 0.0604s^{0.556}\right). \tag{33}$$

The gain crossover frequency of the obtained control system is $\omega_c = 41.5$ rad/s, and the phase margin is $\varphi_m = 55.7°$. According to these design specifications, the optimal coefficient a is estimated as 9.128, and the FOPID controller is obtained,

$$C_6(s) = 8.362\left(1 + \frac{13.628}{s^{0.986}} + 0.008s^{0.986}\right). \tag{34}$$

Step response simulation is performed, using the Bode shaping-based FOPID controller $C_5(s)$ (denoted as BS-FOPID) and the proposed FOPID controller $C_6(s)$ (denoted as a-FOPID) as the speed controllers, respectively. The response curves and the performance indexes are shown in Figure 17 and Table 3, respectively. The load disturbance response simulation is also performed. The response curves and performance indexes are shown in Figure 18 and Table 4, respectively.

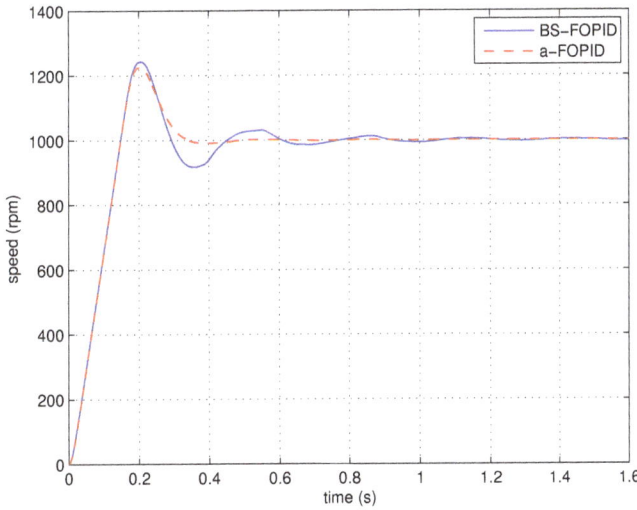

Figure 17. The step responses of the control systems using the Bode shaping-based (BS)-FOPID and a-FOPID (simulation).

Table 3. The step response performance indexes of the control systems using the BS-FOPID and a-FOPID (simulation).

Control System	Settling Time (s)	Overshoot (%)
BS-FOPID	0.408	24.24
a-FOPID	0.292	22.44

According to Figure 17 and Table 3, the response of the system using the a-FOPID has a smaller oscillation and a shorter settling time. Therefore, the system using the a-FOPID achieves better step response performance. According to Figure 18 and Table 4, the two responses have a similar speed drop and recovery time, but the response of the system using the a-FOPID has a smaller oscillation. Therefore, the system using the a-FOPID achieves better anti-load disturbance performance.

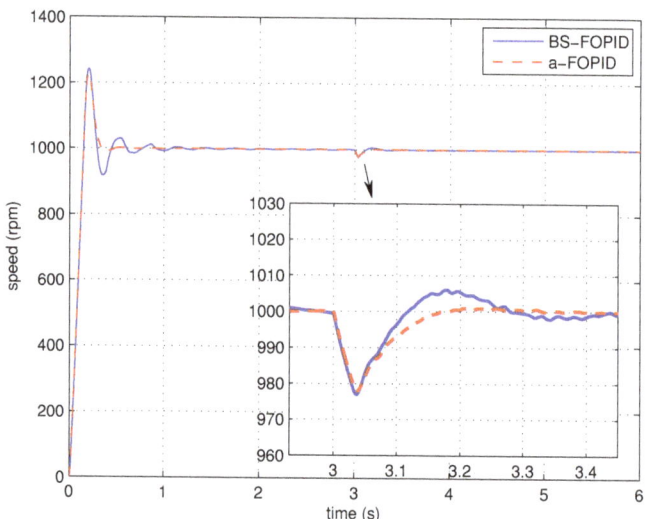

Figure 18. The load disturbance responses of the control systems using the BS-FOPID and a-FOPID (simulation).

Table 4. The anti-load disturbance performance indexes of the control systems using the BS-FOPID and a-FOPID (simulation).

Control System	Recovery Time (s)	Dynamic Speed Drop (%)
BS-FOPID	0.075	2.30
a-FOPID	0.082	2.22

5. Experimental Study

Figure 19 shows the PMSM speed control platform used in this paper. The PMSM is the model Sanyo-P10B18200BXS PMSM. In the experiments, the fractional order operator s^r is realized by applying the impulse invariant discretization method [26].

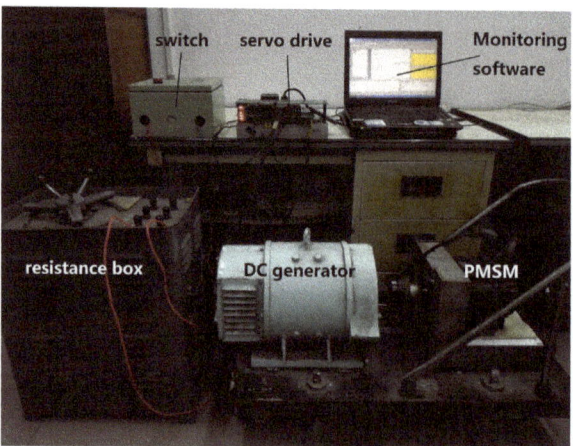

Figure 19. The PMSM speed control platform.

5.1. Gain Robustness Study

Step response experiments are performed to test the gain robustness of the control system using the proposed FOPID controller. The proportional gain of the FOPID controller is multiplied by 120% and 80% to simulate the gain variations. The step responses of the nominal system and those with gain variations are shown in Figure 20.

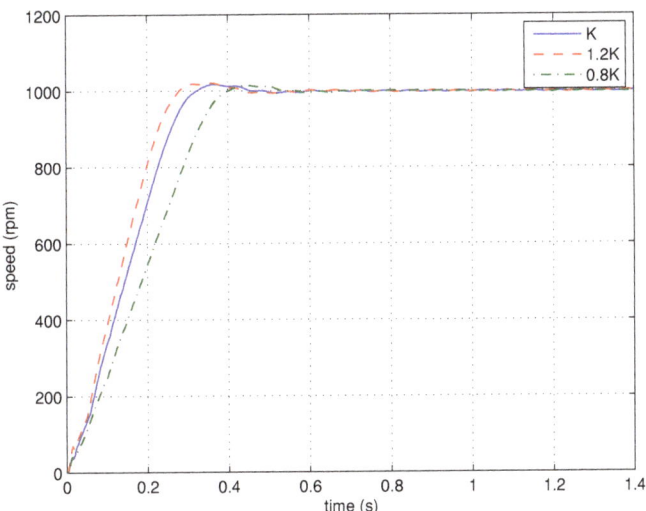

Figure 20. The step responses of the simplified FOPID control systems with different loop-gains (experiment).

According to Figure 20, similar to the simulation result, the responses of the control systems with gain variations have similar overshoots, satisfying the robustness requirement.

5.2. Comparisons with Some Existing Methods

Step response experiments are performed, using the optimal FOPID controller $C_3(s)$ (opt-FOPID) and the proposed FOPID controller $C_4(s)$ (a-FOPID) as the speed controllers, respectively. The response curves and the performance indexes are shown in Figure 21 and Table 5, respectively. The load disturbance response simulation is also performed to test the anti-load disturbance performance of the control systems. The response curves and performance indexes are shown in Figure 22 and Table 6, respectively.

According to Figure 21 and Table 5, similar to the simulation result, the responses of the two systems have similar overshoots, but the response of the system using the a-FOPID has a shorter settling time. Therefore, the system using the a-FOPID achieves better step response performance. According to Figure 22 and Table 6, the responses of two systems have similar speed drops, but the response of the system using the a-FOPID has a shorter recovery time. Therefore, the system using the a-FOPID achieves better anti-load disturbance performance.

Table 5. The step response performance indexes of the control systems using the opt-FOPID and a-FOPID (experiment).

Control System	Settling Time (s)	Overshoot (%)
opt-FOPID	0.325	23.61
a-FOPID	0.273	21.91

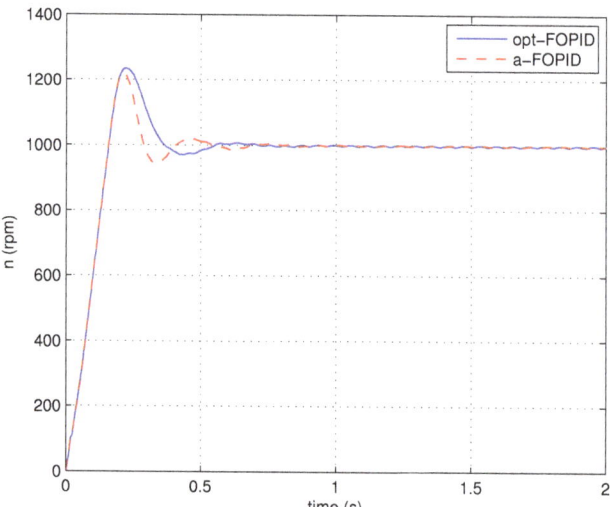

Figure 21. The step responses of the control systems using the opt-FOPID and a-FOPID (experiment).

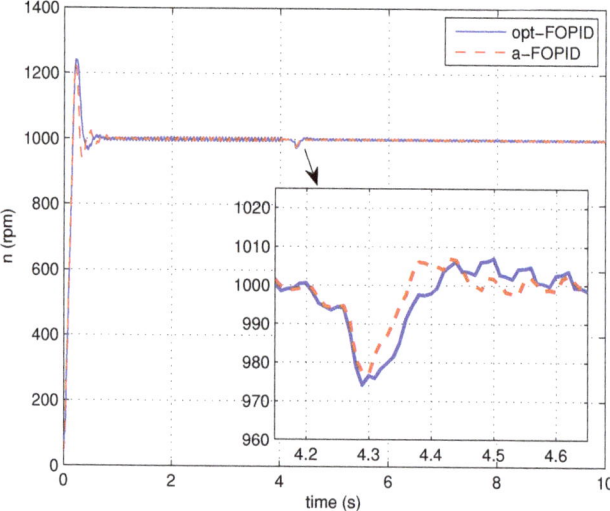

Figure 22. The load disturbance responses of the control systems using the opt-FOPID and a-FOPID (experiment).

Table 6. The anti-load disturbance performance indexes of the control systems using the opt-FOPID and a-FOPID (experiment).

Control System	Recovery Time (s)	Dynamic Speed Drop (%)
opt-FOPID	0.255	2.55
a-FOPID	0.195	2.30

Step response experiments are performed, using the Bode shaping-based FOPID controller $C_5(s)$ (BS-FOPID) and the simplified FOPID controller $C_6(s)$ (a-FOPID) as the speed controllers, respectively. The response curves and the performance indexes are shown in

Figure 23 and Table 7, respectively. The load disturbance response simulation is also performed to test the anti-load disturbance performance of the control systems. The response curves and performance indexes are shown in Figure 24 and Table 8, respectively.

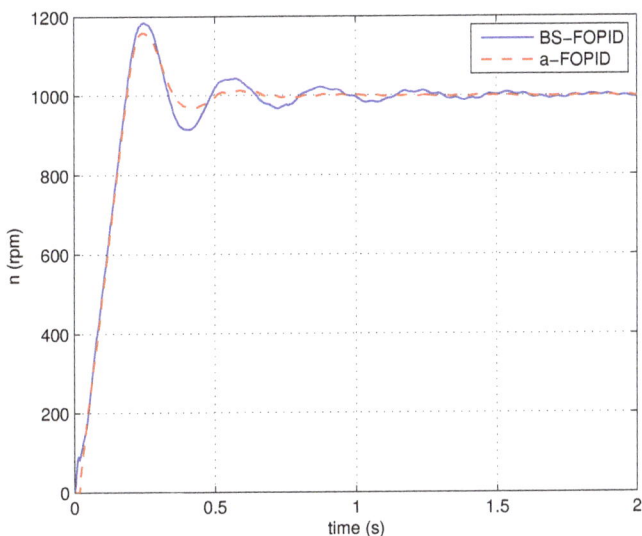

Figure 23. The step responses of the control systems using the BS-FOPID and a-FOPID (experiment).

Table 7. The step response performance indexes of the control systems using the BS-FOPID and a-FOPID (experiment).

Control System	Settling Time (s)	Overshoot (%)
BS-FOPID	0.452	18.45
a-FOPID	0.324	15.89

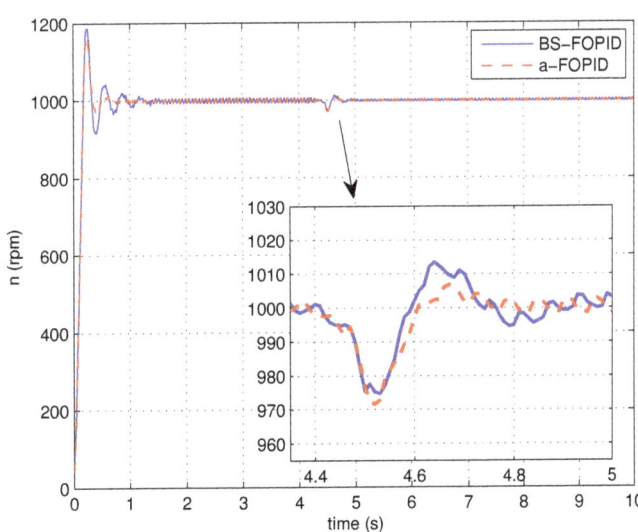

Figure 24. The load disturbance responses of the control systems using the BS-FOPID and a-FOPID (experiment).

Table 8. The anti-load disturbance performance indexes of the control systems using the BS-FOPID and a-FOPID (experiment).

Control System	Recovery Time (s)	Dynamic Speed Drop (%)
BS-FOPID	0.265	2.52
a-FOPID	0.236	2.83

According to Figure 23 and Table 7, the response of the system using the a-FOPID has a smaller overshoot and a shorter settling time. Therefore, the system using the a-FOPID achieves better step response performance. According to Figure 24 and Table 8, the speed drops and recovery time of two responses are close to each other, but the response of the system using the a-FOPID has smaller oscillation. Therefore, the system using the a-FOPID achieves better anti-load disturbance performance. From the simulation and experimental results, the simplified FOPID controller achieves flexible tuning capability, sufficient robustness to gain variations, and the optimal step response performance.

6. Conclusions

A simplified FOPID controller is proposed by building the relations between the controller parameters. An estimation model for the optimal relation coefficient a is built for a class of third-order models, according to which the optimal FOPID controller controllers can be obtained analytically. An actual application of the proposed controller and tuning method on the PMSM speed servo is studied by simulation and experiments, verifying the robustness and dynamic performance of the simplified FOPID control system. The advantages of the proposed method are demonstrated by the comparisons with some other existing methods. Some issues may be studied in the future works, such as improving the relation between the fractional orders and applying the simplified FOPID controller to other classes of plants.

Author Contributions: Methodology, Y.C. and W.Z.; validation, W.Z. and Y.L.; resources, X.W.; writing, W.Z.; supervision, Y.L.; funding acquisition, W.Z., Y.L. and X.W. All authors have read and agreed to the published version of the manuscript.

Funding: This research was funded by the National Natural Science Foundation of China grant number 51975234 and 61803087, the Natural Science Foundation of Guangdong, China grant number 2019A1515110180, the Projects of Guangdong Provincial Department of Education grant number 2017KQNCX215, 2018KTSCX237 and 2019KZDZX1034, the Science and Technology Planning Project of Guangdong, China grant number 2016B090911003, and the Science and Technology Program of Guangzhou, China grant number 201902010066. The APC was funded by 2019A1515110180.

Institutional Review Board Statement: Not applicable.

Informed Consent Statement: Not applicable.

Data Availability Statement: The data presented in this study are available on request from the corresponding author.

Conflicts of Interest: The authors declare no conflict of interest.

References

1. Bhrawy, A.; Zaky, M. Numerical simulation for two-dimensional variable-order fractional nonlinear cable equation. *Nonlinear Dyn.* **2015**, *80*, 101–106. [CrossRef]
2. Duarte-Mermoud, M.; Aguila-Camacho, N.; Gallegos, J. Using general quadratic lyapunov functions to prove lyapunov uniform stability for fractional order systems. *Commun. Nonlinear Sci.* **2015**, *22*, 650–659. [CrossRef]
3. Velmurugan, G.; Rakkiyappan, J.C.R. Finite-time synchronization of fractional-order memristor-based neural networks with time delays. *Neural Netw.* **2016**, *73*, 36–46. [CrossRef]
4. Zhang, L.; Hu, X.; Wang, Z.; Sun, F.; Dorrell, D.G. A review of supercapacitor modeling, estimation, and applications: A control/management perspective. *Renew. Sustain. Energy Rev.* **2018**, *81*, 1868–1878. [CrossRef]

5. Li, Z.; Liu, L.; Dehghan, S.; Chen, Y.; Xue, D. A review and evaluation of numerical tools for fractional calculus and fractional order controls. *Int. J. Control* **2017**, *90*, 1165–1181. [CrossRef]

6. Huang, J.; Chen, Y.; Li, H.; Shi, X. Fractional order modeling of human operator behavior with second order controlled plant and experiment research. *IEEE/CAA J. Autom. Sin.* **2016**, *3*, 271–280.

7. Hartley, T.T.; Lorenzo, C.F. Dynamics and control of initialized fractional-order systems. *Nonlinear Dyn.* **2002**, *29*, 201–233. [CrossRef]

8. Sabatier, J.; Lanusse, P.; Melchior, P.; Oustaloup, A. Fractional order differentiation and robust control design: CRONE, H-infinity and motion control. In *Intelligent Systems, Control and Automation: Science and Engineering*; Springer: Dordrecht, The Netherlands, 2015.

9. Malek, H.; Dadras, S.; Chen, Y. Fractional order equivalent series resistance modeling of electrolytic capacitor and fractional order failure prediction with application to predictive maintenance. *IET Power Electron.* **2016**, *9*, 1608–1613. [CrossRef]

10. Zheng, W.; Luo, Y.; Chen, Y.; Pi, Y. Fractional-order modeling of permanent magnet synchronous motor speed servo system. *J. Vib. Control* **2016**, *22*, 2255–2280. [CrossRef]

11. Tian, J.; Xiong, R.; Yu, Q. Fractional-order model-based incremental capacity analysis for degradation state recognition of lithium-ion batteries. *IEEE Trans. Ind. Electron.* **2019**, *66*, 1576–1584. [CrossRef]

12. Xu, Y.; Zhou, J.; Xue, X.; Fu, W.; Zhu, W.; Li, C. An adaptively fast fuzzy fractional order PID control for pumped storage hydro unit using improved gravitational search algorithm. *Energy Convers. Manag.* **2016**, *111*, 67–78. [CrossRef]

13. Keyser, R.D.; Muresan, C.; Ionescu, C. A novel auto-tuning method for fractional order PI/PD controllers. *ISA Trans.* **2016**, *6*, 268–275. [CrossRef] [PubMed]

14. Li, C.; Zhang, N.; Lai, X.; Zhou, J.; Xu, Y. Design of a fractional-order PID controller for a pumped storage unit using a gravitational search algorithm based on the cauchy and gaussian mutation. *Inform. Sci.* **2017**, *396*, 162–181. [CrossRef]

15. Shah, P.; Agashe, S. Review of fractional PID controller. *Mechatronics* **2016**, *38*, 29–41. [CrossRef]

16. Malek, H.; Luo, Y.; Chen, Y. Identification and tuning fractional order proportional integral controllers for time delayed systems with a fractional pole. *Mechatronics* **2013**, *23*, 746–754. [CrossRef]

17. Zamani, M.; Karimi-Ghartemani, M.; Sadati, N. Design of a fractional order PID controller for an avr using particle swarm optimization. *Control Eng. Pract.* **2009**, *17*, 1380–1387. [CrossRef]

18. Biswas, A.; Das, S.; Abraham, A. Design of fractional-order $PI^\lambda D^\mu$ controllers with an improved differential evolution. *Eng. Appl. Artif. Intell.* **2009**, *22*, 343–350. [CrossRef]

19. Zheng, W.; Luo, Y.; Wang, X.; Pi, Y.; Chen, Y. Fractional order $PI^\lambda D^\mu$ controller design for satisfying time and frequency domain specifications simultaneously. *ISA Trans.* **2017**, *84*, 212–222. [CrossRef]

20. Zheng, W.; Luo, Y.; Pi, Y.; Chen, Y. Improved frequency-domain design method for the fractional order proportional-integral-derivative controller optimal design: A case study of permanent magnet synchronous motor speed control. *IET Control Theory Appl.* **2018**, *12*, 2478–2487. [CrossRef]

21. Luo, Y.; Chen, Y. Stabilizing and robust FOPI controller synthesis for first order plus time delay systems. *Automatica* **2011**, *48*, 2040–2045.

22. Chevalier, A.; Francis, C.; Copot, C.; Ionescu, C.M.; De Keyser, R. Fractional-order PID design: Towards transition from state-of-art to state-of-use. *ISA Trans.* **2019**, *84*, 178–186. [CrossRef] [PubMed]

23. Ruan, Y.; Yang, Y.; Chen, B. Dynamic performance index of control system. In *Control System of Electric Drives–Motion Control Systems*, 5th ed.; Yang, G.; Ed., China Machine Press: Beijing, China, 2016; pp. 62–64. (In Chinese)

24. Mercader, P.; Banos, A.; Vilanova, R. Robust proportional-integral-derivative design for processes with interval parametric uncertainty. *IET Control Theory Appl.* **2017**, *11*, 1016–1023. [CrossRef]

25. Saidi, B.; Amairi, M.; Najar, S.; Aoun, M. Bode shaping-based design methods of a fractional order PID controller for uncertain systems. *Nonlinear Dyn.* **2015**, *80*, 1817–1838. [CrossRef]

26. Impulse Response Invariant Discretization of Fractional Order Integrators/Differentiators. Available online: http://www.mathworks.com/matlabcentral/fileexchange/21342-impulse-response-invariant-discretization-of-fractional-order-integrators-differentiators (accessed on 6 September 2020).

MDPI

Article

Statistical Assessment of Discrimination Capabilities of a Fractional Calculus Based Image Watermarking System for Gaussian Watermarks

Mario Gonzalez-Lee [1,*,†], Hector Vazquez-Leal [2,3,†], Luis J. Morales-Mendoza [1,†], Mariko Nakano-Miyatake [4,†], Hector Perez-Meana [4,†] and Juan R. Laguna-Camacho [5,†]

1   Facultad de Ingeniería en Electrónica y Comunicaciones, Universidad Veracruzana, Av. Venustiano Carranza S/N, Poza Rica Veracruz C.P. 93390, Mexico; javmorales@uv.mx
2   Facultad de Instrumentación Electrónica, Universidad Veracruzana Lomas del Estadio S/N, Xalapa Veracruz C.P. 91090, Mexico; hvazquez@uv.mx
3   Consejo Veracruzano de Investigación Científica y Desarrollo Tecnológico (COVEICYDET), Av. Rafael Murillo Vidal No. 1735, Cuauhtemoc, Xalapa Veracruz C.P. 91069, Mexico
4   Seccion de Estudios de Posgrado e Investigacion, Instituto Politécnico Nacional, Av. Santa Ana No. 1000, Del. Coyacan, Ciudad de Mexico C.P. 04440, Mexico; mnakano@ipn.mx (M.N.-M.); hmperezm@ipn.mx (H.P.-M.)
5   Facultad de Ingeniería en Mecánica Electrica, Universidad Veracruzana, Av. Venustiano Carranza S/N, Poza Rica Veracruz C.P. 93390, Mexico; jlaguna@uv.mx
*   Correspondence: mgonzalez01@uv.mx
†   These authors contributed equally to this work.

Abstract: In this paper, we explore the advantages of a fractional calculus based watermarking system for detecting Gaussian watermarks. To reach this goal, we selected a typical watermarking scheme and replaced the detection equation set by another set of equations derived from fractional calculus principles; then, we carried out a statistical assessment of the performance of both schemes by analyzing the Receiver Operating Characteristic (ROC) curve and the False Positive Percentage (FPP) when they are used to detect Gaussian watermarks. The results show that the ROC of a fractional equation based scheme has 48.3% more Area Under the Curve (AUC) and a False Positives Percentage median of 0.2% whilst the selected typical watermarking scheme has 3%. In addition, the experimental results suggest that the target applications of fractional schemes for detecting Gaussian watermarks are as a semi-fragile image watermarking systems robust to Gaussian noise.

Keywords: fractional calculus; Gaussian watermarks; statistical assessment; false positive rate; semi-fragile watermarking system

Citation: Gonzalez-Lee, M.; Vazquez-Leal, H.; Morales-Mendoza, L.J.; Nakano-Miyatake, M.; Perez-Meana, H.; Laguna-Camacho, J.R. Statistical Assessment of Discrimination Capabilities of a Fractional Calculus Based Image Watermarking System for Gaussian Watermarks. *Entropy* 2021, 23, 255. https://doi.org/10.3390/e23020255

Received: 18 December 2020
Accepted: 13 January 2021
Published: 23 February 2021

1. Introduction

Digital watermarking has gained popularity in the past few decades as a copyright enforcement tool. It is an active research field that includes applications such as data authentication and data indexing among other practical applications [1–3]. The scenario of copyright enforcement is as follows: the copyright holder wants to exploit some digital media, so he embeds a watermark under the premise that, in case of an unauthorized person exploiting the media, the copyright holder would be able to demonstrate in court that his watermark was embedded in the media and hence he owns all rights to the media.

A watermarking system embeds a signal, called the watermark, into another signal known as the cover; a cover might be digital media such as an image, audio, video, or other digital media. Most of the proposed watermarking systems generate a pseudo-random signal (the watermark) using a user's key and then embeds this watermark into the cover; conversely, the watermarking system is able to detect the watermark or even retrieve it from the watermarked cover. If watermark samples are in the set $\{-1,1\}$, then the watermark is called binary; sometimes, designers let the watermark be a pseudo-random sequence with Gaussian distribution, this kind or watermark is called a Gaussian Watermark.

A watermarking system can have two types of errors during its attempt to detect a watermark:

Error type I: The system failed to find a watermark; this is called a False Negative (FN).

Error type II: The system found a given watermark even when either no watermark or another watermark was embedded; this is called False Positive (FP).

A FP is considered flawed that must be avoided because this might lead to a legal dispute on the copyrights of the digital media. For this reason, systems that exhibit a high FPP are impractical and thus excluded from literature. Usually, a watermarking system has a negligible FPP for detecting binary watermarks; conversely, some systems might have a high FPP when detecting Gaussian watermarks.

To clarify this issue, consider the following example: Figure 1 (left) shows an image watermarked with a Gaussian watermark. Only one watermark was embedded; however, the system detects several watermarks as if they were actually embedded as shown in Figure 1 (right). Assuming that a court acknowledges as the copyright owner any individual who claims the rights to some digital media granted, he can prove that the watermarking system detects his watermark within the media. Under these conditions, an attacker would have to search for a watermark that produces a positive detection and could then claim ownership of the media, causing a legal dispute.

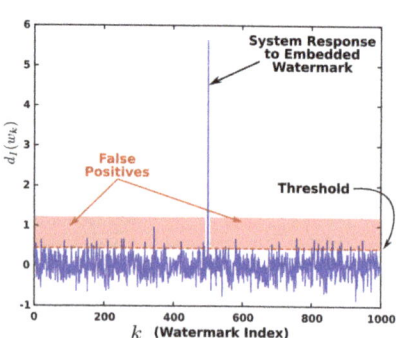

Figure 1. Faulty detection of a Gaussian watermark due to False Positives. (**left**) Watermarked image. $g = 5$, $PSNR = 34.14$ dB. (**right**) The systems verify the presence of several watermarks; the cases that fall in the red zone are False Positives.

To help to mitigate this issue, we proposed in a previous paper to replace the detection equations of watermarking systems to reduce the FPP. Although results were interesting and seem promising, our tests were not conclusive due to the low number of images in the database used in the experiments; thus, the purpose of this paper is to fill the remaining gaps in our previous proposal by analyzing the cases we left unexplored using a bigger image database. In this paper, we put our early proposal on a firmer basis, we:

- Assessed statistically meaningful results by extending the data set up to 10,000 images.
- Carried out a statistical analysis to compare the FPP of the original watermarking scheme versus the corresponding version with detection equations derived from fractional calculus.
- Evaluated the quality of both schemes as a watermark detector by comparing their ROC curves.

- Examined the successful detection rate after performing a number of signal processing operations on the watermarked images to define robustness of the system and recommend target applications of fractional detector equations.
- Complemented our previous study about binary watermarks with this study about Gaussian watermarks.

With these results as a basis, we expect designers of watermarking systems to take advantage of Gaussian watermarks when appropriate to meet their design goals. At the moment, it is difficult to detect watermarks using simple equations, so we look forward to providing an alternative to reuse previously proposed schemes by using fractional calculus based equations.

The usage scenarios for such schemes include:

- The system designer wants to enhance the discriminative power of a system already proposed.
- The watermark is some information that closely holds the Gaussian distribution.
- The complexity of the watermarking system has to be low.

Another scenario will be discussed later.

The rest of the paper is organized as follows: in Section 2, we review the background of the analyzed watermarking scheme. A discussion about related works is presented in Section 3. Section 4 presents a Fractional Scheme for watermarking. In Section 5, we discuss the materials and methods of analysis used to carry out the experiments; next, in Section 6, we present the experimental results; then, in Section 7, we discuss the experimental results and present the conclusions, and finally the references are in the last section.

2. The Watermarking Model

Before continuing with the background fundamentals, let us define the terminology used in the remainder of this paper. One often refers to different watermarks, so we will call the set of different watermarks $\mathbb{W}$; w_k is the k-th watermark of the set $\mathbb{W}$ and $w_k[i]$ denotes the i-th sample of the k-th watermark. The set of images that serve as covers is $\mathbb{X}$; similarly, x_k denotes the k-th image and $x_k[i]$ is the i-th sample of the k-th. $y_k[i]$ is the i-th sample of the k-th watermarked image. Note that, although we are focusing on images, we will use one index for the sake of simplicity, so consider $i = (r, c)$ a coordinate pair of the image.

A simple model approach to watermarking is to make analogies to the field of the theory of communications. In this context, we assume that the watermark is transmitted through a communications channel as pictured in Figure 2. The model has the following variables: the cover which is a signal used as host for the watermark; a user's key as input for a pseudo-random number generator, and the embedding gain which is related to the watermark's energy. In an ideal scenario, the cover does not distort the watermark; however, in practice, this can not be achieved, so the effects of the cover on a watermark are modeled as the distortion caused by the channel. Attacks to the watermark are modeled as noise. An attack is a signal processing operation performed on the watermarked with the goal of making the watermark undetectable by the watermarking system.

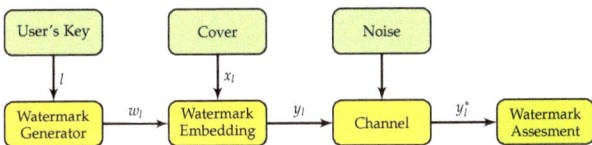

Figure 2. General model of watermarking as a communication process.

2.1. Watermark Embedding

There are two basic rules for embedding watermarks: the additive rule and the multiplicative rule. We will focus on the additive embedding rule since it is widely used in most related works.

The watermarking system embeds the watermark w_l into the cover x_l producing the watermarked signal y_l as shown in Figure 3. This scheme uses the additive rule defined as:

$$y_l = x_l + g w_l, \tag{1}$$

where y_l is the watermarked signal and g is the embedding gain.

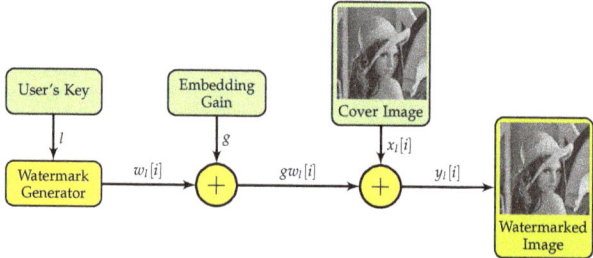

Figure 3. Watermark embedding scheme.

2.2. Watermark Detection

A typical watermark system assesses the presence of the watermark by computing two statistics: a decision variable which is a measurement of the presence of the watermark within the watermarked image, and a threshold that helps to decide if the watermark is present or absent. If the decision variable is greater than or equal to the threshold, then the watermark was detected; otherwise, the watermark is absent as shown in Figure 4.

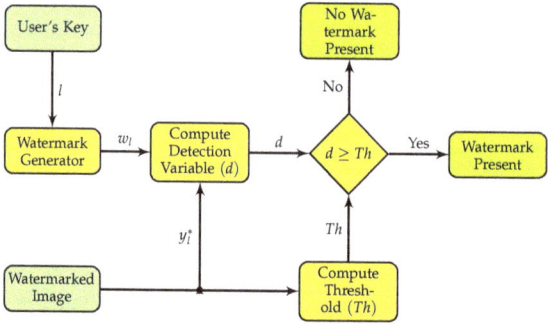

Figure 4. Block diagram of the watermark detection process.

Most watermarking systems have detected watermarks using the cross-correlation formula since the early works on watermarking; an example is the highly influential paper by Cox et al. [3]. The watermarking system uses the received and possible noisy watermarked media (y_l^*) for detecting the watermark; first, it computes a decision variable $d_I(w_l)$ as follows:

$$d_I(w_l) = \frac{1}{N} \sum_{i=1}^{N} w_l[i] y_l^*[i]; \tag{2}$$

Next, the system compares $d_I(w_l)$ to a threshold ($Th_I(w_l)$) and, if $d_I(w_l) \geq Th_I(w_l)$, then the detection is positive; the threshold is computed using the following equation [4]:

$$Th_I(w_l) = 3.3\sqrt{2\frac{\sigma^2}{N}} \qquad (3)$$

where σ^2 is the variance of y_l^*.

Many state-of-the-art algorithms use (1)–(3) for inserting and detecting the watermark as discussed later in this paper. We will call (2) and (3), the integer equation set—hence, the subscript I of the detection variable set. A watermarking scheme based on Equations (1)–(3) is shown in Figure 5.

The decision of the system for the integer equation set is computed as:

$$D_I(w_l) = \begin{cases} 1 & d_I(w_l) \geq Th_I(w_l) \\ 0 & \text{Otherwise} \end{cases}, \qquad (4)$$

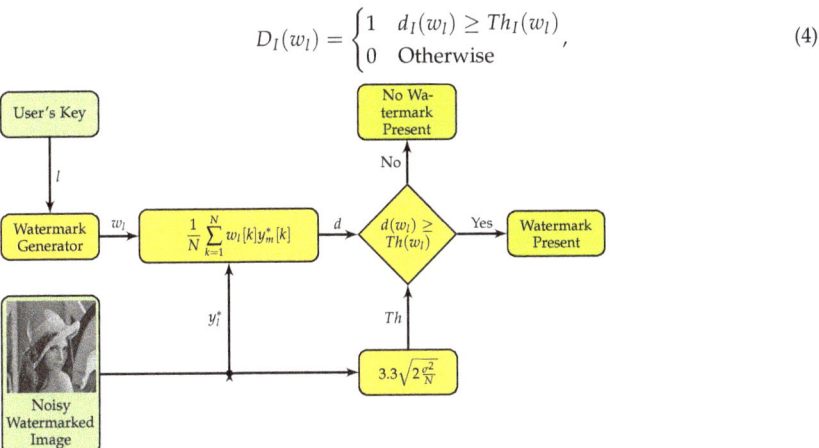

Figure 5. Integer watermark detecting scheme.

Many works use (1) and (2) to embed and detect watermarks respectively as discussed in next section.

3. Works Related to Watermarking Based on Fractional Calculus

On the other hand, Fractional Calculus (FC) has gained attention in recent years; for example, Refs. [5–8] are good references that cover the basics on FC ranging from introductory to advanced FC theory. Many scientists used it for modeling several physical phenomena with applications to engineering; for example, in [9–11], the authors present applications of FC to the analysis of control systems. In [12–14], the authors present applications to Digital Filters design. In [15,16], the authors discuss an approach to linear systems analysis for both continuous and discrete cases. Researchers already started to develop FC applications to watermarking; related works exhibit a tendency to adapt (1) and (2) for working with fractional calculus based approaches.

Some authors use a fractional derivative for watermarking since there is a relationship between the order of the derivative and the resulting function; this relationship is difficult to establish. For example, the authors of [17] use the Grünwald–Letnikov fractional operator for computing a pseudo-random sine function, allowing two fractional orders α and β to act as keys. The authors claim that this scheme is robust toward occlusion attack; however, this is the only test they reported. The work [18] is similar to [17]. The main difference between those works is that authors of [18] use the fractional Cauchy formula for the sine function. Authors report that the system is robust; nevertheless, their results are supported by the test in just one image lacking evidence for confirming the system's reliability.

Other authors use the Fractional Fourier Transform (FrFT) for watermarking since there is a strong dependency between the orders and the resulting coefficient set of the FrFT, a dependency that seems random. The algorithm proposed in [19] uses the FrFT coefficients as the embedding domain. The authors report good results; however, they present just a case of study. A similar approach is presented in [20]. This approach also uses the fractional orders as the secret keys. The watermark is detected using standard cross-correlation. The authors claim that the system is robust toward JPEG compression, noise addition, and image manipulation operations such as median filtering, Gaussian smoothing, and sharpening filtering. Another work that uses the FrFT is [21]; its authors affirm that their proposal is robust to geometrical transform, filtering, and histogram stretching; however, they carried out too few experiments. In [22], the authors present an approach based on the FrFT with a random modification to the phase. The resulting system is more similar to a digital signature based system than to a typical watermarking system. This system is robust against cropping, salt and pepper noise addition, uniform noise addition, Gaussian noise addition, noise addition in both the amplitude and the phase, JPEG compression, and histogram equalization operations. Another idea presented in [23] is to generate a watermark in the FrFT domain and embed it into an image also in the FrFT domain using the additive rule. The authors used the cross-correlation for detecting the watermark. This scheme is robust toward occlusion attack, which is the only attack reported by the authors.

The Random Fractional Fourier Transform (RFrFT) is a variation of the FrFT; it has the same properties of FrFT but has the advantage that the spectrum is random and exhibits a high embedding capacity and robustness for watermarking applications. An RFrFT application to watermarking is presented in [24]. This system computes the RFrFT with a given random phase; then, it divides the transformed image into blocks and computes their fractal dimension; next, it selects a set of those blocks and uses the highest amplitude in each block for watermark embedding using Amplitude Shift Keying (ASK). The watermark extraction is accomplished by reversing previous steps. The system computes the Mean Square Error to measure the robustness using both the extracted and the real watermark. They tested their system by performing three attacks: noise addition, cropping, and JPEG compression.

Another fractional calculus based transform, the Discrete Fractional Random Transform (DFRNT), was used in [25]. This work is similar to [24]; first, the system computes the DFRNT; then, it divides the signal into blocks and selects a set of blocks randomly; next, it selects the highest amplitudes for watermark embedding using Phase Shift Keying (PSK). The authors report that their proposal is robust against Gaussian noise addition, cropping, and low pass filtering; however, they present too few tests.

One more fractional based transform is the Fractional Dual-Tree Complex Wavelet Transform (FrDT-WT); the FrDT-WT is used to find the wavelet transform in the Fourier domain resulting in a mathematical description of the multiresolution properties. The work presented in [26] and exploits that the randomness of the FrDT-WT coefficients depends on the fractional order, also using a biometric pattern to further enhance the security. The main idea is to build two biometric images; then, use the SURF algorithm to compute the robust matching point vectors; next, use these vectors to compute the keys for building a chaotic map. The watermark extraction uses both the original and the watermarked images. The authors report that their system is robust. The attacks covered in the test include average filtering, median filtering, Gaussian noise addition, salt and pepper noise addition, JPEG compression, SPIHT compression, row-column deletion, resizing, cropping, rotation, histogram equalization, contrast adjustment, and sharpen attacks; however, there were only six images used for the test; furthermore, the reported results correspond to their best case.

Another work is [27] that is almost the same as the system presented in [26]. The main difference between these works is that Ref. [27] uses the Redundant Fractional Wavelet Transform (RFrWT) due to a problem with the discrete FrDT-WT related to the use of decimators.

The authors of [28] present an interesting idea; unlike most watermarking schemes, their system does not embed a watermark into a host image, but they use Visual Cryptography (VC) and a Visual Secret Sharing Scheme. The system constructs two shares that convey a secret message in the following way: the encoder divides the host image into blocks; then, it selects a set of blocks and computes the FrFT using orders α and β; next, it computes the Singular Value Decomposition (SVD) of the transformed blocks and uses the first value of the resulting SVD for computing the master share according to the standard rules of secret sharing schemes. The authors report that their scheme resists various signal processing operations such as JPEG compression, average filtering, median filtering, blurring filtering, sharpening filtering, Gaussian noise addition, contrast adjustment, gamma correction, histogram equalization, resizing, rotation, and geometrical distortion.

All of these works have in common the use of (1) and (2) to embed and detect watermarks; from this perspective, we can say that the overall difference among them is the use of some transform coefficient set for watermarking. In other words, they use already proposed equations, and the novelty of these works rely on the use of a different embedding domain. This leads to incrementing the complexity of the watermarking system and other problems related to the multiple definitions of fractional operators proposed until now.

On the other hand, the authors of [29] analyzed the watermarking systems proposed in [17] through [28] and observed that they use (1) and (2), so they proposed a new improved equation set to substitute (2) and (3). They showed that this modification increases the system's robustness, so the watermarking system designer might prefer to use fractional equations as a reliable solution for copyright enforcement; however, they limited their study to the case where the watermark is binary, and they added that they would skip the case of Gaussian watermarks since the system based on (2) and (3) was not reliable in this case, so a fair comparison to their proposed equation set was not possible in the context of the experiments carried out to test their proposed scheme.

The authors of [30] explored the case of Gaussian watermarks, and their results suggest that the scheme proposed in [29] reduces the False Positive Percentage; however, they limited the benchmark corpus to 20 images from the standard image set.

The case of the fractional scheme proposed in [30] for the Gaussian watermarks case needs a deeper study; for this reason, we accomplished this study where we explore the behavior of the fractional scheme for detecting Gaussian watermarks; we looked for confirming that the fractional scheme proposed in [29] reduces the false positive percentage of the detector when Gaussian watermarks were embedded, and, by reaching this goal, we confirmed that the fractional scheme is reliable for watermarking applications when Gaussian watermarks are used; thus, the novelty of this paper is to generalize the results presented in [29,30]. The main advantage of the proposed scheme is that it avoids the problems related to the use of fractional transforms found in previous works, keeping the complexity almost the same, however, for detecting Gaussian watermarks.

4. Fractional Calculus Approach to Watermark Detection

The detection variable derived from FC principles proposed in [29] is:

$$d_F(w_l) = -\text{Im}\left[\frac{3}{4}\frac{1}{N}\sqrt{\left(\sum_{i=1}^{N} y_l^*[i]w_l[i]\right)^2 - \frac{2}{3}\sigma^2 N[2NH-1]} - \epsilon\right], \tag{5}$$

where $\text{Im}[\cdot]$ is the imaginary part operator, and the threshold is:

$$Th_F(w_l) = k_p \sigma^2 \sqrt{\frac{H}{\epsilon}}, \tag{6}$$

where:

$$\epsilon = \frac{3}{4}\frac{1}{N}\sqrt{\frac{2}{3}\sigma^2 N[2NH-1]}, \tag{7}$$

with $H = \ln(\sqrt{2\pi\sigma^2 e})$; σ^2 is the variance of y_l^*. We call (5) and (6) the Fractional equation set. A fractional scheme based on (5) and (6) is shown in Figure 6.

The decision of the system for the fractional equation set is:

$$D_F(w_l) = \begin{cases} 1 & d_F(w_l) \geq Th_F(w_l) \\ 0 & \text{otherwise} \end{cases}.$$

(8)

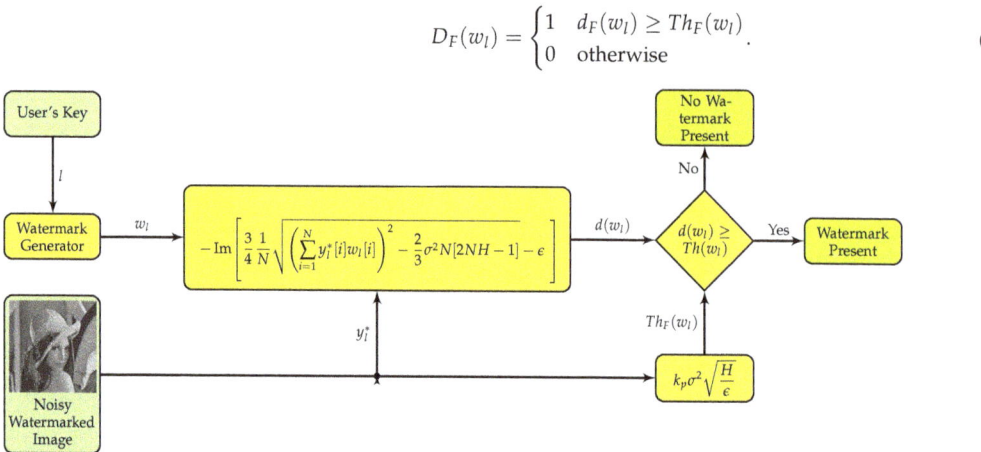

Figure 6. Fractional watermark detecting scheme.

If we use (5) and (6) for detecting Gaussian watermarks, we get the result shown in Figure 7, which clearly has improved detection characteristics since it has no false positives. We are looking for confirming that the scheme in Figure 6 is more reliable than the scheme in Figure 5. This follows the strategy stated early in this paper about reusing the algorithm in Figure 5 by replacing detection equations with fractional calculus based equations resulting in a possibly improved algorithm and then verifying the effectiveness of this strategy. For this reason, there is a lack of comparison to related works as a means of control to the experiments. In other words, a fair experiment in the purpose's context of this paper is to compare the original algorithm versus the same algorithm with fractional equations and assess its improvement. Thus, the only control needed is the original algorithm and, as a result, the outcomes of the experiments are reliable.

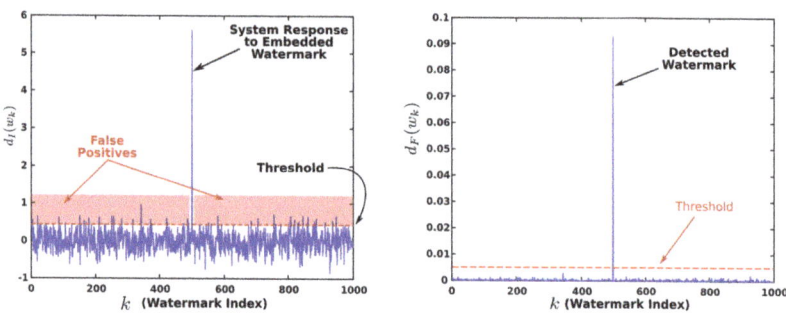

Figure 7. Detection of the watermark. (**left**) using (2) and (3); (**right**) using (5) and (6), and note the lack of False Positives. The cases that fall in the red zone are False Positives.

5. Materials and Methods

To carry out experiments, we used 10,000 images of the BOWS database as the set $\mathbb{X}$; each image of this set is grayscale with size 512×512 pixels and their luminance values are in the range $[0, 255]$.

We used the embedding scheme shown in Figure 3 for watermarking each image in the set $\mathbb{X}$. In addition, the embedding gain was fixed for all cases to the value $g = 5$; this setting leads to a Peak Signal-to-Noise Ratio (PSNR) mean of 34.21 dB for the entire set giving a fair balance between robustness and imperceptibility of the watermark. The embedded watermark w_l was selected at random from the watermark set $\mathbb{W}$ for each image; each watermark in the set was equally probable.

The goal of the tests was to assess the capacity of watermarking schemes shown in Figures 5 and 6 to reduce the false positives by computing the FPP and the ROC curve.

For the first test, we computed the FP of each image. To achieve this, each image in the set $\mathbb{X}$ was watermarked; then, the system tried to detect all watermarks in $\mathbb{W}$ within a single image; next, all FP were identified and counted using (4) and (8). The false positives computing process is summarized in Procedure 1, and it is described in Figure 8 (left). False positives gave us an insight about the reliability of both the integer and the fractional schemes.

For the sake of simplicity, and without loss of generality, we indicated $D(w_l)$ instead of $D_I(w_l)$ or $D_F(w_l)$ in all procedures since the same steps were followed for both schemes.

We selected the Receiver Operating Characteristic (ROC) since it is regarded as an objective measure to evaluate performance of a decision technique. Thus, as a second test, we computed the ROC as follows: each image in the set $\mathbb{X}$ was watermarked using watermark w_l; then, we computed $d_I(w_l)$, $d_F(w_l)$, $D_I(w_l)$, $D_F(w_l)$. Those values and the corresponding ground truth values of $D_I(w_l)$, $D_F(w_l)$ were recorded. The data were used to derive a Generalized Lineal Model for estimating the ROC for both the integer and the fractional schemes. Data collecting steps were summarized in Procedure 2 and further explained in Figure 8 (right). The ROC curves were used to evaluate the integer and the fractional to clarify which of them is more reliable.

With a last test, we examined the robustness of the fractional scheme. To achieve this goal, each image in the set $\mathbb{X}$ was watermarked to get the set of watermarked images $\mathbb{Y}$, and then an attack was carried out on the watermarked images; next, we added up the cases where the embedded watermark was detected to compute the percentage of detected watermark cases. The process of computing the detection rate is summarized in Procedure 3. The percentage of detected watermarks after the watermarked image was attacked suggested the target applications of the fractional scheme based on its robustness.

Procedure 1 Procedure to record measures for False positives.

Require: Image set $\mathbb{X}$, Watermark set $\mathbb{W}$.

1: Open log file for writing.

2: **for** Each image $x_k \in \mathbb{X}$ **do**

3:   Select randomly a watermark w_l from the watermark set $\mathbb{W}$.

4:   Partition set $\mathbb{W}$ into two subsets $\mathbb{W}_e$ and $\mathbb{W}_n$ that hold $\mathbb{W}_e \cap \mathbb{W}_n = \varnothing$ and $\mathbb{W}_e \cup \mathbb{W}_n = \mathbb{W}$, $\mathbb{W}_e = \{w_l\}$.

5:   Embed the watermark w_l into Image x_k

6:   **for** Each watermark w_m in $\mathbb{W}_n$ **do**

7:     **for** Each watermarking scheme **do**

8:       Compute $R(w_m)$

9:       **if** $R(w_m)==1$ **then**

10:          $FP = FP + 1$

11:       **end if**

12:     **end for**

13:   **end for**

14:   Record FP a of a current image in log file.

15: **end for**

16: **return** Log file.

Procedure 2 Procedure to record measures for getting the ROC curve.

Require: Image set $\mathbb{X}$, Watermark set $\mathbb{W}$.

1: Open log file for writing.

2: **for** Each image $x_k \in \mathbb{X}$ **do**

3:    Select a random watermark w_l from the watermark set $\mathbb{W}$.

4:    Embed the watermark w_l into Image x_k.

5:    Compute $d(w_l)$ and $D(w_l)$.

6:    Record $d(w_l)$, $D(w_l)$, and ground truth values in log file.

7:    Get the set of indexes of true negatives, these indexes form a set $\mathbb{T}$.

8:    Draw at random an index m from set $\mathbb{T}$.

9:    Record $d(w_m)$, $R(w_m)$, and ground truth values.

10:    Get the set of indexes false positives, these indexes form a set $\mathbb{P}$.

11:    **for** Each index k in $\mathbb{P}$ **do**

12:        Record the values of $d(w_m)$, $R(w_m)$, and ground truth values in log file (Record all the False Positives).

13:    **end for**

14: **end for**

15: Close log file.

16: **return** Log file.

Procedure 3 Procedure to record Detection Rate.

Require: Image set $\mathbb{X}$, Watermark set $\mathbb{W}$.

1: Open log file for writing.

2: **for** Each image $x_k \in \mathbb{X}$ **do**

3:    Select randomly a watermark w_l from the watermark set $\mathbb{W}$.

4:    Embed the watermark w_l into Image x_k

5:    Perform an attack on the watermarked image.

6:    Compute $D(w_l)$ using the attacked image.

7:    **if** $D(w_l)==1$ **then**

8:        $D = D + 1$

9:    **end if**

10: **end for**

11: Compute detection rate ($D_r = \frac{D}{10,000}$)

12: Record D_r in log file.

13: **return** Log file.

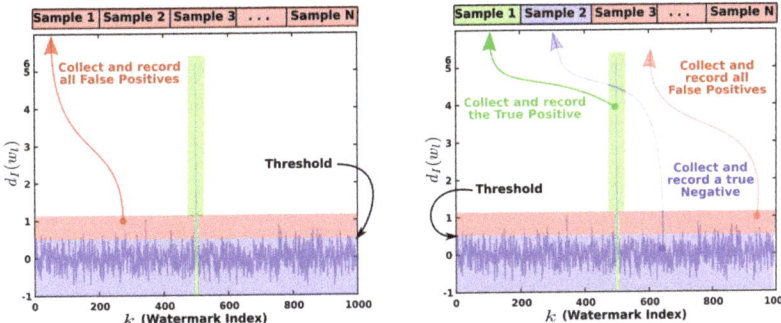

Figure 8. Collecting data. (**left**), all the system's responses that cross the threshold when no watermark was embedded are collected since these are False Positives. These data fall in the red zone. (**right**) We collected the true positive (data in the green zone), a single true negative (data in the blue zone), and all false positives for each image in the set $\mathbb{Y}$ (data in the red zone).

6. Experimental Results

As a first test, we computed the false positive percentages for both the integer and the fractional schemes and build a boxplot. Figure 9 (left) shows that the false positives for the integer scheme span from 0.1% to 16.15%; in contrast, the fractional scheme has very low percentages of false positives and the range of values concentrates around 0.2%.

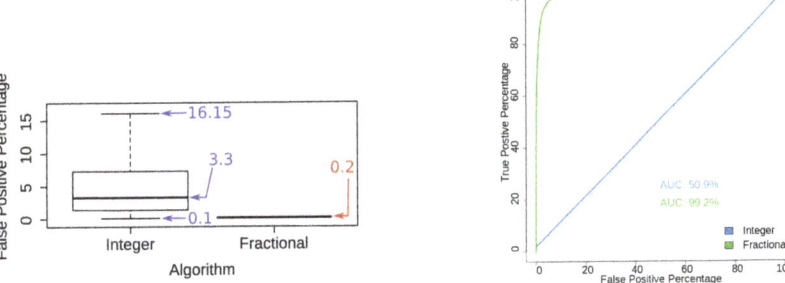

Figure 9. Statistical assessment of discrimination characteristics. (**left**) value ranges of false positives for both the Integer and Fractional watermarking schemes. No outliers are drawn for the sake of clearness; (**right**) comparison of the ROC curves of the integer and fractional schemes.

In our second test, we evaluated the quality of both schemes; we used the data we collected to draw the ROC curve shown in Figure 9 (right). As a result of this test, we found that the ROC of the integer scheme has an Area Under the Curve (AUC) of 50.9%, whereas the fractional scheme obtained an AUC of 99.2% for the same test.

The last test consisted of examining the successful detection rate after attacks; this was accomplished by attacking each watermarked image in the set $\mathbb{Y}$. The attacks performed were: average filtering, median filtering, Gaussian noise addition, speckle noise addition, salt and pepper noise addition, JPEG compression, cropping, removing random rows and columns, substituting random rows and columns, and scaling. The corresponding figures are in Appendix A for the sake of readability of this section.

A bar plot of the percentage of detected watermarks after the image set $\mathbb{Y}$ filtered using an average filter is shown in Figure A1. The attack was repeated for window sizes of 3×3, 5×5, and 7×7. The resulting bar plot shows that the percentage of successful watermark detection after the attack is about 6% and became lower as the window size increases.

We performed a similar attack; this time, a median filter was used to filter the set $\mathbb{Y}$. Results shown in Figure A2 reveal a similarity to the results reached for the average filtering attack; this time, detection percentages are lower than 10%.

The next test is comprised of adding Gaussian noise to each watermarked image in the set $\mathbb{Y}$ and then we tried to detect the watermark. We constructed a bar plot showing the percentages of detected watermark for various noise variances. Figure A3 shows that the scheme is robust to Gaussian noise. This figure might look suspicious because it looks atypical; thus, to discard that the Gaussian noise triggers false positives and this causes a high detection rate, we inspected some cases and present an example in Figure 10.

 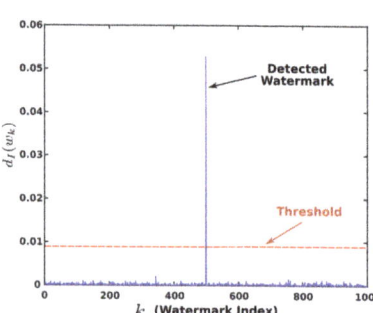

Figure 10. Detection of the watermark. (**left**) noisy watermarked imaged; noise variance was 0.05. (**right**) the corresponding evaluation of (5).

The following test consists of adding salt and pepper noise and then detecting the watermark. Results depicted in Figure A4 show that the fractional scheme is robust up to noise densities of 20%, the detection rate drops for noise densities higher than 20%.

We carried out the next test by adding speckle noise before trying to detect the watermark. Results in Figure A5 show that the fractional scheme is robust up to noise variances of 0.2. After this limit, the detection rate drops.

A very common scenario is to compress images using the JPEG standard, so the next test comprised watermarking the image and compressing the image with the JPEG standard, and then detecting the watermark. Figure A6 shows that the fractional scheme is robust to JPEG compression up to a quality factor of 90% and then the detection percentage starts to decline.

Another common signal processing operation is the cropping attack. Figure A7 shows results when the watermarked image is cropped. This figure shows that the fractional scheme is robust up to 20% of cropped pixels.

The next test selects t rows and t columns at random, and then removes these rows and columns from the watermarked image; the resulting image is smaller than the original watermarked image, so the image is then scaled to match the size of the original watermarked image. The watermark was then detected, and the results are shown in Figure A8. Results show that the fractional scheme is not robust since it exhibits a detection rate around 8% for removing 10 rows and columns.

Another test, similar to the previous one, selects t rows and t columns at random, and then substitutes the selected rows and columns with the adjacent row or column of the same image. The watermark was detected, and results are shown in Figure A9. Results show that the fractional scheme is robust up to substituting 100 rows and columns.

Finally, we carried out a scaling attack; the watermarked image was scaled to make it smaller and then the image was restored back to its original size. The results are shown in Figure A10; this figure shows that the fractional scheme is robust up to 90%. In other words, we shrank the image to 90% of its original size and then restored to the original size before we tried to detect the watermark.

7. Conclusions

In this study, we compared the FPP of the original watermarking scheme versus the corresponding version with detection equations derived from fractional calculus; evaluated the quality of both schemes as a watermark detector by comparing their ROC curves, and examined the successful detection rate after attacking the watermarked images to define robustness of the system. We performed several tests that allowed us to conclude the following facts:

The False Positives percentage is much lower for the fractional scheme than the corresponding percentages of the integer scheme. According to Figure 9 (left), the FPP spans from 0.2% to 16.2% for the integer scheme whilst the FPP concentrates around 0.2% for the fractional scheme. This means it is more likely to get a 0.2% FPP when using a fractional scheme and also the FP rate will be lower for this fractional scheme than the corresponding results for an integer scheme.

Results show that the fractional scheme is a reliable method for detecting Gaussian watermarks according to Figure 9 (right); the fractional scheme has a significant advantage compared to the integer scheme since the AUC is higher for the fractional case ($AUC = 99.20\%$ versus $AUC = 50.90\%$); this means that the fractional scheme has higher discriminative power compared to the integer scheme.

In addition, the experimental results in Figures A1–A10 show that this system is fragile to all attacks presented in Section 6, except for the case of the Gaussian noise addition attack, this is because the noise is added in the same manner as the watermarks are; thus, the systems treats Gaussian noise as a watermark. The target applications of such a scheme include cases where the watermarks should not survive attacks; an example of practical application of a fractional scheme is for authenticating information.

Since the target application might be authenticating information, it will be convenient to propose another value of k_p; this value should be higher than that used in this study since this will help to reduce the detection rate after attacks. Additional usage scenarios include: The system designer wants to enhance the discriminative power of a system already proposed, the watermark is some information that closely holds the Gaussian distribution, and the complexity of the watermarking system has to be low.

The results provide designers of watermarking systems with an alternative to take advantage of Gaussian watermarks when appropriate to meet their design goals. Thus, the proposed strategy is an alternative to reuse previously proposed schemes by using fractional calculus based equations.

The results obtained in this study complement the study in [29] since the case of Gaussian watermarks was left unexplored, so this paper provides the designer of watermark systems with a more logical insight of the potential and practical applications of a fractional watermark detector.

The characteristics to discriminate between patterns with Gaussian statistical distribution suggest that the fractional equations might be used in pattern recognition applications where samples have a Gaussian distribution.

Author Contributions: Conceptualization, M.G.-L. and H.V.-L.; Methodology, M.G.-L. and H.V.-L.; Software, L.J.M.-M. and J.R.L.-C.; Validation, M.N.-M. and H.P.-M. All authors have read and agreed to the published version of the manuscript.

Funding: This research received no external funding.

Data Availability Statement: Not applicable.

Acknowledgments: The authors thank the Universidad Veracruzana for their support for this work. In addition, the authors thank the students Ivan de Gaona-Marquez, Marco A. Salas-Moreno, and Flavio C. Garcia-Salas who helped during the development of this work.

Conflicts of Interest: The authors declare no conflict of interest.

Appendix A

This appendix has complementary experimental data that the reader might need to check out closer.

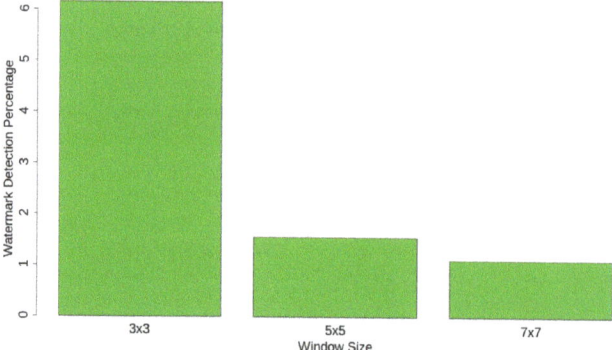

Figure A1. Percentage of successful watermark detection after an average filter attack for various window sizes.

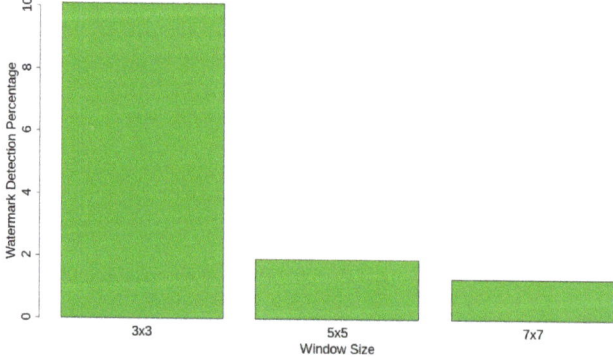

Figure A2. Percentage of successful watermark detection after a median filter attack for various window sizes.

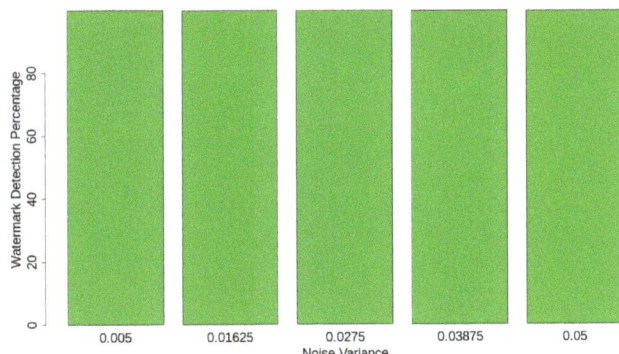

Figure A3. Percentage of successful watermark detection when Gaussian noise is added to the watermarked image. The mean of the noise was zero and the horizontal axis corresponds to the variance of the noise.

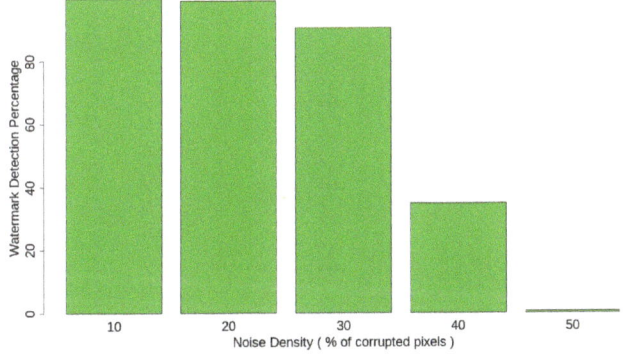

Figure A4. Percentage of successful watermark detection when salt and pepper noise is added to the watermarked image. Horizontal axis show the noise density.

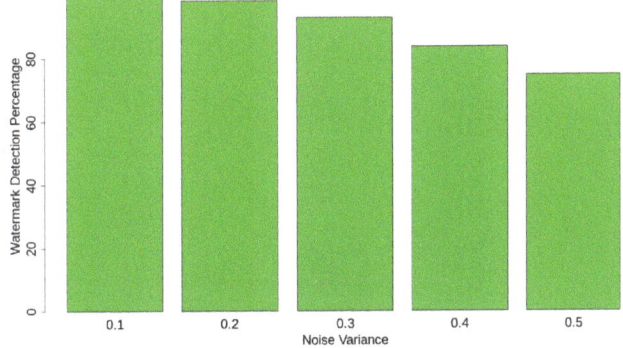

Figure A5. Percentage of successful watermark detection when speckle noise is added to the watermarked image. Horizontal axis show the noise variance.

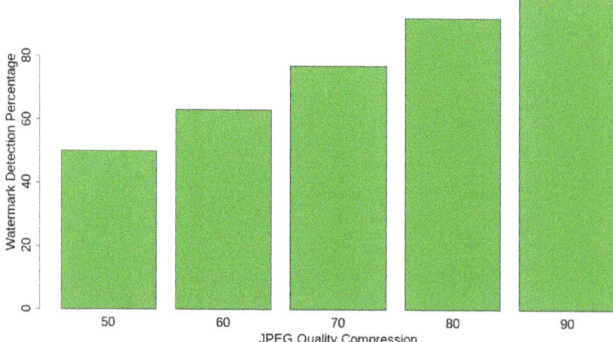

Figure A6. Percentage of successful watermark detection after the watermarked image was compressed using the JPEG standard. The horizontal axis corresponds to the JPEG compression quality factor.

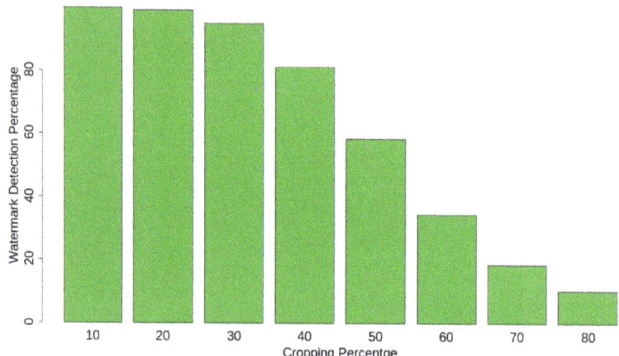

Figure A7. Percentage of successful watermark detection after cropping. This figure spans various cropping percentages.

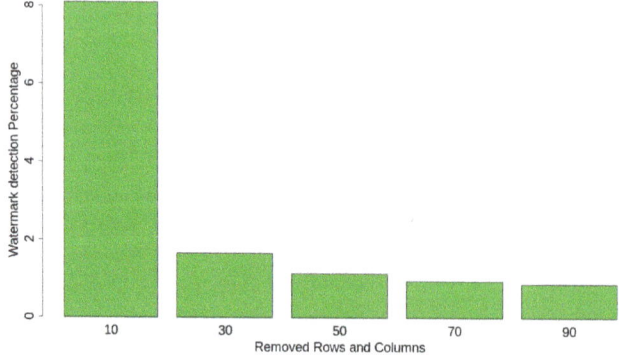

Figure A8. Percentage of successful watermark detection after a number of rows and columns were removed from the watermarked image. The horizontal axis shows the number of rows and columns removed.

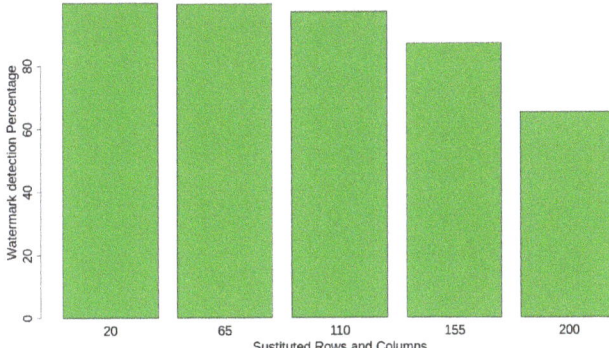

Figure A9. Percentage of successful watermark detection after a number of rows and columns of the watermarked image were substituted with another row or column. The horizontal axis shows the number of rows and columns replaced.

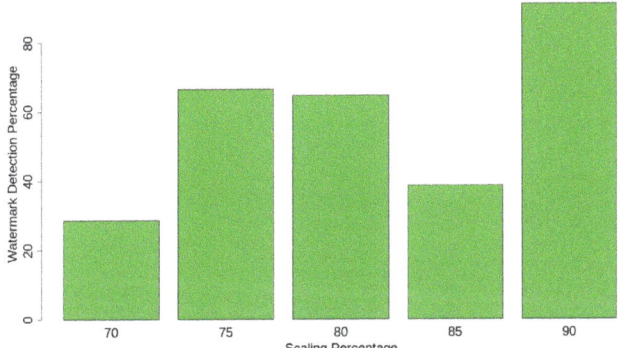

Figure A10. Percentage of successful watermark detection when the watermarked image is scaled. The horizontal axis shows the scaling factor.

References

1. Langelaar, G.C.; Setyawan, I.; Lagendijk, R.L. Watermarking digital image and video data. A state-of-the-art overview. *IEEE Signal Process. Mag.* **2000**, *17*, 20–46. [CrossRef]
2. Gonzalez-Lee, M.; Morales-Mendoza, L.J.; Gonzalez-Lee, E. Detección óptima de marcas de agua digitales. In *Detección Óptima de Marcas de agua Digitales: Fundamentos y Aplicaciones (Spanish Edition)*; Editorial Académica Española: Riga, Latvia, 2012; pp. 33–54.
3. Cox, I.J.; Kilian, J.; Leighton, F.T.; Shamoon, T. Secure spread spectrum watermarking for multimedia. *IEEE Trans. Image Process.* **1997**, *6*, 1673–1687. [CrossRef] [PubMed]
4. Piva, A.; Barni, M.; Bartolini, F.; Cappellini, V. Threshold selection for correlation-based watermark detection. In Proceedings of the COST254 Workshop on Intelligent Communications, L'Aquila, Italy, 4–6 June 1998; pp. 67–72.
5. Baleanu, D.; Diethelm, K.; Scalas, E.; Trujillo, J.J. Preliminaries. In *Fractional Calculus Models and Numerical Methods*; World Scientific: Singapore, 2012; pp.1–40.
6. Podlubny, I. Fractional derivatives and integrals. In *Fractional Differential Equations*; Academic Press: Cambridge, MA, USA, 1998; pp. 41–117.
7. Oldham, K.B.; Spanier, J. Fractional derivatives and integrals: Definitions and Equivalences. In *The Fractional Calculus*; Academic Press: Cambridge, MA, USA, 1974; pp. 46–60.
8. Debnath, L. A brief historical introduction to fractional calculus. *Int. J. Math. Educ. Sci. Technol.* **2006**, *35*, 487–501. [CrossRef]
9. Yin, C.; Huang, X.; Dadras, S.; Cheng, Y.; Cao, J.; Malek, H.; Mei, J. Design of optimal lighting control strategy based on multi-variable fracional-order extremum seeking method. *Inf. Sci.* **2018**, *465*, 38–60. [CrossRef]

10. Machado, J.A.T.; Silva, M.F.; Barbosa, R.S.; Jesus, I.S.; Reis, C.M.; Marcos, M.G.; Galhano, A.F. Some applications of fractional calculus in engineering. *Math. Probl. Eng.* **2010**, *2010*, 639801.
11. Yin, C.; Huang, X.; Chen, Y.; Dadras, S.; Zhong, S.; Cheng, Y. Fractional-Order exponential switching technique to enhance sliding model control. *Appl. Math. Model.* **2017**, *44*, 705–726. [CrossRef]
12. Tseng, C. Design of fractional order digital FIR differentiators. *IEEE Signal Process. Lett.* **2001**, *8*, 77–79. [CrossRef]
13. Ferdi, Y. Fractional order calculus-based filters for biomedical signal processing. In Proceedings of the 2011 1st Middle East Conference on Biomedical Engineering, Sharjah, UAEnited Arab Emirates, 21–24 February 2011; pp. 73–76.
14. Vainio, O.; Lehto, R.; Saramaki, T. Fractional order FIR differentiator with optimum noise attenuation. In Proceedings of the IEEE Instrumentation and Measurement Technology Conference, Warsaw, Poland, 1–3 May 2007; pp. 1–4.
15. Ortigueira, M.D. Introduction to fractional linear systems, part 1: Continuous-time case. *IEE Proc. Vis. Image Signal Process.* **2000**, *147*, 62–70. [CrossRef]
16. Ortigueira, M.D. Introduction to fractional linear systems, part 2: Discrete-time case. *IEEE Signal Process. Lett.* **2000**, *147*, 71–78. [CrossRef]
17. Huading, J.; Yifei, P. Application and numerical implementation of fractional calculus to digital watermark. In Proceedings of the 2006 8th international Conference on Signal Processing, Beijing, China, 16–20 November 2006.
18. Miao, Q.G.; Shi, C.; Wang, W. A novel image digital watermark algorithm with weighted fractional calculus based on wavelet coefficients. In Proceedings of the 2010 International Conference on Computer Application and System Modeling (ICCASM 2010), Taiyuan, China, 22–24 October 2010; pp. 610–613.
19. Taba, M.T. The fractional Fourier transform and its application to digital watermarking. In Proceedings of the 8th International Workshop Systems, Signal Process. and their Applications (WoSSPA), Algiers, Algeria, 12–15 May 2013; pp. 262–266.
20. Lang, J.; Zhang, Z.G. Blind digital watermarking method in the fractional Fourier transform domain. *Opt. Lasers Eng.* **2014**, *53*, 112–121. [CrossRef]
21. Djurovic, I.; Stankovic, S.; Pitas, I. Digital watermarking in the fractional Fourier transformation domain. *J. Netw. Comput. Appl.* **2001**, *24*, 167–173. [CrossRef]
22. Guo, Q.; Liu, Z.; Liucora, S. Image watermarking algorithm based on fractional Fourier transform and random phase encoding. *Opt. Commun.* **2011**, *284*, 3918–3923. [CrossRef]
23. Nishchal, N.K. Optical image watermarking using fractional Fourier transform. *J. Opt.* **2009**, *38*, 22–28. [CrossRef]
24. Guo, Q.; Guo, J.; Liu, Z.; Liu, S. An adaptive watermarking using fractal dimension based on random fractional Fourier transform. *Opt. Laser Technol.* **2012**, *44*, 124–129. [CrossRef]
25. Guo, J.; Liu, Z.; Liu, S. Watermarking based on discrete fractional random transform. *Opt. Commun.* **2007**, *272*, 344–348. [CrossRef]
26. Bhatnagar, G.; Wu, Q.J. Biometrics inspired watermarking based on a fractional dual tree complex wavelet transform. *Future Gener. Comput. Syst.* **2013**, *29*, 182–195. [CrossRef]
27. Bhatnagar, G.; Wu, Q.J. A new logo watermarking based on redundant fractional wavelet transform. *Math. Comput. Model.* **2013**, *58*, 204–218. [CrossRef]
28. Rawat, S.; Raman, B. A blind watermarking algorithm based on fractional Fourier transform and visual cryptography. *Signal Process.* **2012**, *92*, 1480–1491. [CrossRef]
29. Gonzalez-Lee, M.; Vazquez-Leal, H.; Gomez-Aguilar, J.F.; Morales-Mendoza, L.J.; Jimenez-Fernandez, V.M.; Laguna-Camacho, R.; Calderon-Ramon, C.M. Exploring the Cross-correlation as a Means for Detecting Digital Watermarks In addition, its Reformulation Into The Fractional Calculus Framework. *IEEE Access* **2018**, *6*, 71699–71718. [CrossRef]
30. Gonzalez-Lee, M.; Calderon-Ramon, C.M.; Morales-Mendoza, L.J.; Escalante-Martinez, J.E.; De Gaona-Marquez, I.; Salas-Moreno, M.A.; Hernandez-Cadenas, L.; Vazquez-Bautista, R.F. Decreasing false positive detection of Gaussian watermarks by means of fractional calculus principles. In Proceedings of the 2019 IEEE International Conference on Engineering Veracruz, ICEV 2019, Xalapa, Veracruz, Mexico, 14–17 October 2019.

MDPI

Article

Variable-Order Fractional Models for Wall-Bounded Turbulent Flows

Fangying Song [1] and George Em Karniadakis [2,*]

1   College of Mathematics and Computer Science, Fuzhou University, Fuzhou 350108, China; fysong@stu.xmu.edu.cn

2   Division of Applied Mathematics, School of Engineering, Brown University, Providence, RI 02912, USA

*   Correspondence: george_karniadakis@brown.edu

Abstract: Modeling of wall-bounded turbulent flows is still an open problem in classical physics, with relatively slow progress in the last few decades beyond the log law, which only describes the intermediate region in wall-bounded turbulence, i.e., 30–50 y+ to 0.1–0.2 R+ in a pipe of radius R. Here, we propose a fundamentally new approach based on fractional calculus to model the entire mean velocity profile from the wall to the centerline of the pipe. Specifically, we represent the Reynolds stresses with a non-local fractional derivative of variable-order that decays with the distance from the wall. Surprisingly, we find that this variable fractional order has a universal form for all Reynolds numbers and for three different flow types, i.e., channel flow, Couette flow, and pipe flow. We first use existing databases from direct numerical simulations (DNSs) to lean the variable-order function and subsequently we test it against other DNS data and experimental measurements, including the Princeton superpipe experiments. Taken together, our findings reveal the continuous change in rate of turbulent diffusion from the wall as well as the strong nonlocality of turbulent interactions that intensify away from the wall. Moreover, we propose alternative formulations, including a divergence variable fractional (two-sided) model for turbulent flows. The total shear stress is represented by a two-sided symmetric variable fractional derivative. The numerical results show that this formulation can lead to smooth fractional-order profiles in the whole domain. This new model improves the one-sided model, which is considered in the half domain (wall to centerline) only. We use a finite difference method for solving the inverse problem, but we also introduce the fractional physics-informed neural network (fPINN) for solving the inverse and forward problems much more efficiently. In addition to the aforementioned fully-developed flows, we model turbulent boundary layers and discuss how the streamwise variation affects the universal curve.

Keywords: fractional conservations laws; variable fractional model; turbulent flows; fractional PINN; physics-informed learning

Citation: Song, F.; Karniadakis, G.E. ariable-Order Fractional Models for Wall-Bounded Turbulent Flows. *Entropy* **2021**, *23*, 782. https://doi.org/10.3390/e23060782

Academic Editors: Bruce J. West and José A. Tenreiro Machado

Received: 14 May 2021
Accepted: 15 June 2021
Published: 20 June 2021

1. Introduction

Reynolds [1] was the first to statistically describe turbulence by decomposing the instantaneous velocity vector into an average field and its fluctuation. Upon substitution into the Navier–Stokes equations and averaging, assuming quasi-stationarity, a new modified equation emerged for the average velocity that includes an additional term, namely, the averaged dissipation tensor leading to the turbulence-closure problem [2]. Addressing the closure complexity has been a century-long pursuit, starting with the seminal work of Prandtl [3], who proposed a simplified mixing length model analogous with Fick's law of local diffusion. Interestingly, at about the same time, Richardson [4], in an attempt to unify turbulent diffusion with molecular diffusion, combined geophysical measurements with Brownian motion to produce the famous scaling law on turbulent pair diffusivity. While ingenious, both approaches assume implicitly locality in turbulent interactions, which limits the universality of the derived correlations—an open standing question for over a century. As stated by Kraichnan [5], Prandtl's approach is valid only when the

237

spatial scale of inhomogeneity of the mean field is large compared to the mixing length. This assumption is clearly violated in most turbulent flows, e.g., in Reynolds' pipe flow, where the turbulent eddies are of the size of the pipe radius. This has motivated research in nonlocal constitutive equations of turbulence, and Prandltl, in subsequent work [6], developed a turbulent shear-layer model in an attempt to introduce non-locality in his approach. Kraichnan [5] pioneered such non-local approximations and, based on his work, more recently generalized versions of the second Prandtl non-local model were proposed in the literature [7].

Fractional calculus is an effective tool to solve complex problems with nonlocality and scale-free self-similar processes as well as non-Gaussian statistics. Lévy statistics lead to anomalous diffusion [8] and can effectively model turbulent intermittency [9]. Hence, it is possible that turbulent eddy diffusion could be accurately modeled by fractional Reynolds stresses [10]. Based on physical arguments, in order to represent nonlocality and intermittency, Chen [11] proposed a fractional Laplacian as a model for representing the Reynolds stress with a fixed fractional exponent $\alpha = 2/3$. More recently, starting with the Boltzmann equation, Epps et al. [12] rigorously derived the fractional Navier–Stokes equations by replacing the Maxwell–Boltzmann distribution with the more general Levy α-stable distribution; see a recent extension of this work in [13]. For $\alpha = 2$, the new equations revert to the standard Navier–Stokes equations, while for $\alpha = 1$, we obtain the logarithmic velocity profile known as the law of the wall [14]. The work of Epps et al. [12] laid a new framework for turbulence modeling that may lead to new fundamental understanding of turbulence, but it is only valid in an open domain and thus ignores the important issue of nonlocal boundary conditions encountered in defining fractional Laplacians in bounded domains [15].

The work we include here incorporates our first paper [16] published in the archives, and is a significant extension. We also refer to the work of [17], who modeled the total shear stress directly in wall units by formulating a one-sided variable-order model using the Caputo fractional derivative for Couette flow [17] and in ongoing work on transitional and turbulent boundary layers. For the case of Couette flow, universality was found. We note that directly formulating the problem in wall units does not require modeling of any additional coefficients, unlike the formulation in the present study.

The remainder of this paper is organized as follows: Since the small-scale components can be described as an anomalous diffusion [11], we introduce the variable-order fractional calculus in the next section. Then, we formulate the inverse optimization problem corresponding to the governing equations. We present the fractional differential equations to model different turbulent flows (e.g., channel flow, Couette flow, and pipe flow) in Section 2. The inverse problem is solved by a finite difference (FD) method to obtain the fractional order. Moreover, we introduce the fractional physics-informed neural network for solving the inverse problem to find the variable-orders. In Section 3, we present the numerical results that show that the universal fractional-order profiles of the channel and pipe flow as a function of the distance from the wall, a unique capability enabled by fractional calculus. In particular, we discovered that this fractional-order function is universal for all Reynolds numbers and for different geometries. Finally, we provide a short summary in Section 4.

2. Variable-Order Fractional Models for Turbulent Flows

The first fractional model for the Reynolds averaged Navier–Stokes equations was developed by Chen [11], who proposed a fractional Laplacian to model the Reynolds stresses and to account for intermittency [18,19] as follows:

$$\frac{\partial U}{\partial t} + U \cdot \nabla U = -\frac{1}{\rho} \nabla P + \nu_0 \Delta U - \gamma (-\Delta)^{1/3} U, \tag{1}$$

where U is the average velocity and γ is the turbulent diffusion coefficient. Hence, the effective fractional order in this model is fixed at $\alpha = 2/3$. This value is consistent with the Richardson superdiffusion scaling for homogeneous turbulence that leads to a t^3 scaling for the mean square displacement, but it is not valid for wall-bounded turbulence where anisotropy and the distance from the wall determine the effective rate of turbulent diffusion. Defining a fractional Laplacian in multiple dimensions and in bounded domains is still an open issue in fractional calculus and extending it to variable orders is challenging [15]. However, other somewhat equivalent definitions based on tempered fractional calculus [20] may lead to satisfactory nonlocal representations as well; specifically, in a Boltzmannian framework, Samiee et al. [13] developed a tempered fractional subgrid-scale model to capture high-order structures at the inertial and dissipative ranges. As Richardson first noted, the velocity field in the atmosphere shares a number of properties with the Weierstrass function, i.e., it appears to be continuous but non-differentiable, and this provides a strong case for fractional modeling of turbulence in the atmosphere but also in wall-bounded flows in engineering applications.

In this section, we present a variable-order fractional model for turbulent flows. We firstly consider a one-sided model for channel and pipe flows. Furthermore, we formulate an inverse problem for the fractional order $\alpha(y)$. We present a finite difference method and design a physics-informed neural network (PINN) to obtain the fractional order. Finally, we propose a divergence variable fractional (two-sided) model for turbulent flows.

2.1. Turbulent Channel Flow and Pipe Flow

2.1.1. One-Sided Fractional Derivative Modeling

For wall-bounded turbulence, the effective rate of diffusion varies with distance from the wall. Hence, we exploit the power of fractional calculus that allows variable fractional order, and we propose a variable-order fractional differential equation for modeling the Reynolds stresses, i.e., $\alpha(y)$, where y is the distance from the wall. In particular, we consider fully developed turbulent flows with one-dimensional (dimensionless) averaged velocity $U(y) = u/V$ (where V is the characteristic velocity), including channel flows and pipe flows for which we apply a unified fractional modeling approach. Specifically, assuming that the flow direction is along x and y is the wall-normal direction (distance from the wall), we consider the variable fractional model (VFM-I) in the normalized interval [0, 1]:

$$\text{(VFM-I)} \quad \frac{\partial}{\partial y}\left(\nu_0 \frac{\partial U}{\partial y} - \overline{u'v'}\right) = \nu(y) D_y^{\alpha(y)} U = f, \ \forall y \in \Lambda = (0,1], \tag{2}$$

with $\alpha(0) = 1$, $0 \leq \alpha(y) \leq 1$, D_y^α is the (Caputo) fractional derivative, $f = -\frac{1}{\rho}\partial P/\partial x$ is a constant pressure gradient, $U(y)$ is the mean velocity we want to model, and ν_0 is the kinematic viscosity. The Caputo derivative is defined as:

$$D_y^\alpha U(y) = \frac{1}{\Gamma(1-\alpha)} \int_0^y (y-\tau)^{-\alpha} U'(\tau) d\tau,$$

and it is identical to the Riemann–Liouville left-sided derivative because $U(0) = 0$. Interestingly, we can obtain the scalar coefficient $\nu(y)$ (we refer to it as turbulent diffusivity, although it does not have the correct units) explicitly in terms of the fractional order $\alpha(y)$ from:

$$\nu(y) = f\Gamma(2 - \alpha(y)) Re_\tau^{-\alpha(y)} V/u_\tau, \tag{3}$$

where $Re_\tau = u_\tau R/\nu_0$ is the friction Reynolds number, R is the radius of the pipe (or the half channel width), and u_τ is the wall friction velocity $u_\tau = \sqrt{\tau_w/\rho}$, where $\tau_w = \mu\partial U/\partial y|_{y=0}$ is the wall shear stress with μ being the dynamic viscosity.

We discuss an alternative model, where the variable fractional order $\alpha(y)$ is between one and two instead of the VFM-I we presented, where $0 < \alpha(y) \leq 1$; this model is analogous to VFM-I and is defined by:

$$\text{(VFM-II)} \quad \frac{\partial}{\partial y}\left(\nu_0 \frac{\partial U}{\partial y} - \overline{u'v'}\right) = \nu(y) D_y^{\alpha(y)} U = f, \ \forall y \in \Lambda = (0,1], \tag{4}$$

with $\alpha(0) = 2$, and the variable-order $1 \leq \alpha(y) \leq 2$ is an unknown function to be determined by the data. The scalar coefficient $\nu(y)$ can also be computed from a similar formula as before, i.e.,

$$\nu(y) = \lim_{y_0 \to \frac{1}{Re_\tau}} \frac{f}{D_y^{\alpha(y)}(U|_{y_0})}. \tag{5}$$

2.1.2. Numerical Method

We assume that we know the mean velocity $U(y)$ (also $U^+(y^+)$) from the DNS data or experimental results. The VFM-I can be written in the form:

$$\nu(y) D_y^{\alpha(y)} U = f, \tag{6}$$

where $f = -\frac{1}{\rho} \partial P / \partial x$. Since the fractional order $\alpha(y)$ is unknown in Equation (6), we need to solve a nonlinear problem to obtain $\alpha(y)$. Alternatively, we consider the following optimization problem: given U and f, find the $\alpha(y)$ that satisfies

$$J(\alpha(y)) = \inf_{\alpha(y) \in S} \|\nu(y) D_y^{\alpha(y)} U - f\|^2, \tag{7}$$

where, $S(\Lambda) := \{0 \leq a(y) \leq 1, a(y) \in C^0(\Lambda)\}$. If $\alpha^*(y)$ satisfies Equation (6), then we obtain $J(\alpha^*(y)) \equiv 0$.

Next, we present a numerical method for solving the optimization problem (7). The fractional derivative is discretized with the finite difference method. Then, the fractional order $\alpha(y)$ can be solved point-by-point; for each point $y_n = n\Delta y, \Delta y = 1/N, n = 1, 2, \cdots, N$, we calculate the fractional derivative $D_y^{\alpha(y_n)} U^n$ with the DNS data using the finite difference method [21]

$$D_y^{\alpha(y_n)} U^n = \frac{1}{\Gamma(2 - \alpha(y_n))} \sum_{j=0}^{n} b_j^n \frac{U^{n+1-j} - U^{n-j}}{\Delta y^{\alpha(y_n)}}, \tag{8}$$

where $b_j^n := (j+1)^{\alpha(y_n)} - j^{\alpha(y_n)}$ and $U^n = U(y_n)$. The discrete optimization problem can now be written as

$$J_N(\alpha(y)) = \inf_{\alpha(y) \in S} \sum_{n=1}^{N} \left| \nu(y_n) D_y^{\alpha(y_n)} U^n - f(y_n) \right|^2 \Delta y. \tag{9}$$

Finally, we formulate the fractional physics-informed neural network (fPINN) for the inverse problems of the proposed turbulence model; see Figure 1.

The aim of the inverse problem is to estimate the fractional order $\alpha(y)$ given the mean velocity profile U in the DNS data. We approximate the variable fractional order $\alpha(y)$ by a multi-layer feedforward neural network $\alpha_{NN}(y; \theta = \{W_j, b_j\}_{j=1}^l)$, where θ are a collection of parameters of the NN. The locations y are the input of the NN, and the output U is computed by a recursive formula $Y^j = \sigma(W_j Y^{j-1} + b_j)$ with the initial value $Y^0 = y$. The weight matrix between the $(j-1)$th and jth layers has the dimension $W_j \in \mathbb{R}^{n_j \times n_{j-1}}$, and the bias vector b_j in the jth layer. The column vectors $Y^{j-1} \in \mathbb{R}^{n_{j-1} \times 1}$ and $Y^j \in \mathbb{R}^{n_j \times 1}$ denote the input and output of the jth layer, respectively. The input vector Y^{j-1} is first subject to a linear transformation and then an element-wise nonlinear function $\sigma(\cdot)$, which is called the activation function. The NN consists of one input layer ($j = 0$), $l - 1$ hidden layers ($j = 1, 2, \cdots, l-1$), and one output layer ($j = l$). The depth of the NN is l, and the width of the jth layer is n_j. To determine the parameters θ, we minimize the following loss function with respect to θ

$$\mathcal{L}(\theta) = \frac{1}{N_t} \sum_{i=1}^{N_t} \left(D_y^{\alpha_{NN}(y_i; \theta)} U(y_i) - 1 \right)^2 + (\alpha_{NN}(0; \theta) - 1)^2, y_i \in (0, 1]. \tag{10}$$

The first term on the right-hand side is the equation residual, and the second term is the constraint on the fractional order at the wall, i.e., $\alpha(0) = 1$. We select N_t training points, $\{y_i\}_{i=1}^{N_t}$, to enforce the equation residual on them to be zero. The fractional derivative is evaluated using the finite difference method (8). We optimize the loss function with respect to θ, employing a stochastic gradient descent, Adam, written in TensorFlow. Finally, we estimate the variable fractional order using $\alpha_{NN}(y; \theta)$.

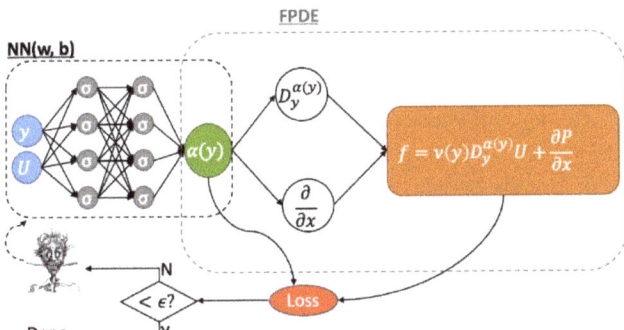

Figure 1. Basic structure of fPINN in 1D for the inverse fractional-order problem. The left uninformed DNN processes data to predict the fractional order, which also has to satisfy the correct physics of turbulence for the channel fully developed flow, represented by the right informed DNN induced by the fractional governing equation.

2.2. Two-Sided Turbulent Channel Flow

2.2.1. Fractional Modeling in Divergence Form

We consider the Reynolds averaged momentum equation for incompressible fully developed channel flow; the governing equation is as follows

$$\frac{\partial}{\partial y}\left(\nu_0 \frac{\partial U}{\partial y} - \overline{u'v'}\right) + \frac{1}{\rho}\frac{\partial P}{\partial x} = 0, \ y \in (0,2), \tag{11}$$

where ρ is the density; and P and U are the mean pressure and velocity, respectively. The process of Reynolds averaging introduces the unclosed Reynolds stress, $\tau_{ij} = -\rho\overline{u'v'}$. The total shear stress on the wall is τ_w. Integrating the above equation from wall to an arbitrary position in wall-wise y, we obtain a new formula as follows

$$\nu_0 \frac{\partial U}{\partial y} - \overline{u'v'} = \tau_w/\rho - \frac{1}{\rho}\frac{\partial P}{\partial x}y. \tag{12}$$

We assume the dimensionless wall shear τ_w and pressure gradient $\frac{\partial P}{\partial x} = C$ are constants. Additionally, we introduce a symmetric divergence variable fractional model for approximating the total shear stress,

$$\text{(DVFM)} \quad \nu_0 \frac{\partial U}{\partial y} + \overline{u'v'} = \nu(y)D_{|y|}^{\alpha(y)}U = 1 - y, \tag{13}$$

with the boundary conditions $\alpha(0) = \alpha(2) = 1$, where the fractional derivative is defined as follows

$$D_{|y|}^{\alpha(y)}U = \frac{1}{2}(D_y^{\alpha(y)}U + {}_yD^{\alpha(y)}U), \tag{14}$$

and $D_y^{\alpha(y)}$ and $_yD^{\alpha(y)}U$ are left and right Caputo derivatives, respectively. The definitions are given as follows

$$\text{Left Caputo derivative: } D_y^\alpha U(y) = \frac{1}{\Gamma(1-\alpha)}\int_0^y (y-\tau)^{-\alpha}U'(\tau)d\tau,$$

and

$$\text{Right Caputo derivative: } _yD^\alpha U(y) = -\frac{1}{\Gamma(1-\alpha)}\int_y^2 (\tau-y)^{-\alpha}U'(\tau)d\tau,$$

and it is identical to the Riemann–Liouville derivatives because $U(0) = 0$ and $U(2) = 0$. We also propose the eddy viscosity in the fractional momentum equation, and the explicit formula is as follows

$$\nu(y) = \Gamma(2-\alpha(y))Re_\tau^{-\alpha(y)}, \tag{15}$$

where $Re_\tau = u_\tau R/\nu_0$ is the friction Reynolds number, R is the radius of the pipe (or the half channel width), and u_τ is the wall friction velocity, $u_\tau = \sqrt{\tau_w/\rho}$, where $\tau_w = \mu \partial U/\partial y|_{y=0}$ is the wall shear stress with μ being the dynamic viscosity.

2.2.2. Numerical Method

We assume that we know the mean velocity $U(y)$ (also $U^+(y^+)$) from the DNS data or experimental results. Since the fractional order $\alpha(y)$ is unknown in Equation (13), we need to solve a nonlinear problem to obtain $\alpha(y)$. Alternatively, we consider the following optimization problem: given U and f, find $\alpha(y)$ that satisfies

$$J(\alpha(y)) = \inf_{\alpha(y)\in S}\|\nu(y)D_{|y|}^{\alpha(y)}U - f\|^2, \tag{16}$$

where $f = 1 - y$ and $S(\Lambda) := \{0 \le a(y) \le 1, a(y) \in C^0(\Lambda)\}$. If $\alpha^*(y)$ satisfies Equation (13), then we obtain $J(\alpha^*(y)) \equiv 0$.

Next, we present a numerical method for solving the optimization problem (16). The fractional derivative is discretized with the finite difference (FD) method. Then, the fractional order $\alpha(y)$ can be solved point-by-point; for each point $y_n = n\Delta y$, $\Delta y = 1/N$, $n = 1, 2, \cdots, N$, we calculate the fractional derivatives $D_{|y|}^{\alpha(y_n)}U^n$ with the DNS data using the finite difference method [21]

$$\text{Left: } D_y^{\alpha(y_n)}U^n = \frac{1}{\Gamma(2-\alpha(y_n))}\sum_{j=0}^n b_j^n \frac{U^{n+1-j}-U^{n-j}}{\Delta y^{\alpha(y_n)}}, \tag{17}$$

and

$$\text{Right: } _yD^{\alpha(y_n)}U^n = -\frac{1}{\Gamma(2-\alpha(y_n))}\sum_{j=0}^{N-n+1} c_j^n \frac{U^{N-j}-U^{N-j-1}}{\Delta y^{\alpha(y_n)}}, \tag{18}$$

where $b_j^n := (j+1)^{\alpha(y_n)} - j^{\alpha(y_n)}$, $c_j^n = b_j^n$ and $U^n = U(y_n)$.

The discretized optimization problem can be now written as

$$J_N(\alpha(y)) = \inf_{\alpha(y)\in S}\sum_{n=1}^N |\nu(y_n)D_{|y|}^{\alpha(y_n)}U^n - f(y_n)|^2\Delta y. \tag{19}$$

Here, we use $N \approx Re_\tau$ points to solve the above optimization for the channel flow at a given Reynolds number Re_τ.

Alternatively, we propose the fractional fPINN for solving the inverse DVFM with the loss function

$$\mathcal{L}(\boldsymbol{\theta}) = \sum_{n=1}^{N_t} |\nu(y_n)D_{|y|}^{\alpha_{NN}(y_n;\boldsymbol{\theta})}U^n - f(y_n)|^2 + |\alpha_{NN}(0;\boldsymbol{\theta}) - 1|^2 + |\alpha_{NN}(2;\boldsymbol{\theta}) - 1|^2. \tag{20}$$

2.3. Turbulent Boundary Layer and Couette Flow

For a boundary layer and Couette flow with zero pressure gradient, the mean two-dimensional continuity and stream-wise momentum reduce to

$$\frac{\partial UU}{\partial x} + \frac{\partial VU}{\partial y} = \frac{\partial}{\partial y}(\nu_0 \frac{\partial U}{\partial y} - \overline{u'v'}). \tag{21}$$

If we assume that the convective effects are small near the wall for the boundary layer problem, then the above equation reduces to

$$\frac{\partial}{\partial y}(\nu_0 \frac{\partial U}{\partial y} - \overline{uv}) = 0. \tag{22}$$

Here, U is viewed as a function of y due to $\frac{\partial U}{\partial x} = 0$. Since the two plates are infinitely long for the Couette flow, the flow properties cannot change with x and all partial derivatives with respect to x vanish. Flow motion only occurs in the x direction, and thus, $V = 0$. After simplifying the RANS equations, the turbulent Couette flow is governed by Equation (22) too.

Further integrating the above equation provides

$$\nu_0 \frac{\partial U}{\partial y} - \overline{u'v'} = C, \tag{23}$$

where C is a constant and $\overline{uv} = 0$ at the wall, while $\nu \frac{\partial U}{\partial y}$ is simply the wall shear stress τ_w/ρ. Then, we have the following equation

$$(\text{TCM}) \quad \nu(y)D_y^{\alpha(y)}U = \frac{\tau_w}{\rho},$$

with $\alpha(0) = 1, 0 < \alpha \le 1$, D_y^{α} is the (Caputo) fractional derivative, and $\nu(y)$ is the eddy viscosity defined as

$$\nu(y) = \Gamma(2 - \alpha(y))Re_\tau^{-\alpha(y)}.$$

Numerical Method

We solve the fractional order $\alpha(y)$ for the turbulent boundary layer problem and Couette flow using fPINN (Figure 1) with the loss function

$$
\begin{aligned}
\mathcal{L}(\theta) &= \sum_{k=0}^{N_t} (\nu(y_k)D_y^{\alpha_{NN}(y_k;\theta)} - \frac{\tau_w}{\rho})^2 + (\alpha_{NN}(0;\theta) - 1)^2 \\
&= \sum_{k=1}^{N_t} (Re_\tau^{-\alpha_{NN}(y_k;\theta)} \sum_{j=0}^{k} \frac{b_j^k}{\Delta y^{\alpha_{NN}(y_k;\theta)}}(U^{k+1-j} - U^{k-j}) - \frac{\tau_w}{\rho})^2 + (\alpha_{NN}(0;\theta) - 1)^2,
\end{aligned}
$$

where U is the DNS data. It changes with Re_θ for the boundary layer problem, so there is (implicit) x dependence as well.

3. Numerical Results

In this section, we present the results for the turbulent channel, pipe, Couette, and boundary layer flows.

3.1. Channel Flow

3.1.1. Numerical Results of the One-Sided Models

We first consider turbulent channel flow for which DNS data are available up to $Re_\tau = 5200$ [22]. Here, we use the FD scheme with $N \approx Re_\tau$ points to solve the aforementioned inverse problem for the channel flow at a given Reynolds number Re_τ. Solving for

$\alpha(y)$, which uniquely determines the Reynolds stresses, Figure 2a depicts the profiles of the fractional order $\alpha(y)$ for different Re_τ as a function of the non-dimensional distance from the wall $y \in [0,1]$. We see a strong dependence of $\alpha(y)$ on Re_τ; however, if we re-plot all data in terms of the viscous wall units, i.e., $y^+ = yu_\tau/\nu_0$ we see a collapse of all results into a single universal curve, as shown in Figure 2b. Moreover, we employ the empirical Spalding formula [23] for $U^+ = u/u_\tau$ in order to extend the results up to high $Re_\tau = 10^6$, and again we obtain a similar universal scaling with the exception of low Re_τ for which the Spalding formula is known to be somewhat inaccurate. We fit the fractional order using these numerical results to obtain the fractional order $\alpha(y^+)$ in wall units as follows

$$\alpha^*(y^+) = \frac{1 - \phi(y^+)}{2} + \frac{\phi(y^+) + 1}{2} a(y^+), \tag{24}$$

where $\phi(y^+) = \tanh(\ln(y^+/9.5)/1.049)$ and $a(y^+) = 1/(b + \kappa|\ln(y^+)|^{0.9})$ with $b = 0.855, \kappa = 0.301$ are constants. This is a remarkable result as it goes beyond the logarithmic profile and seamlessly connects the viscous sublayer with the buffer zone, the logarithmic profile, and the wake region. Although at first it appears to be a perfect fitting exercise, it has important consequences due to the nonlocal interpretation of the fractional derivative involved, i.e., it shows that nonlocality is stronger away from the wall and at high Reynolds numbers. Using the same data for $U(y)$, we show that the alternative model VFM-II with $1 \leq \alpha(y) \leq 2$ also leads to the same type of universality (Figure 3). However, unlike the aforementioned VFM-I, we are unable to obtain an explicit formula for $\nu(y)$, relating it to the Reynolds number as in the first model (i.e., $\alpha(y) \in (0,1]$); instead, we can compute it numerically from the DNS data of turbulent channel flow. As shown in Section 3, this alternative fractional model also exhibits a universal scaling if plotted in terms of wall units, with the lowest value of $\alpha(10^5+) \approx 1.3$.

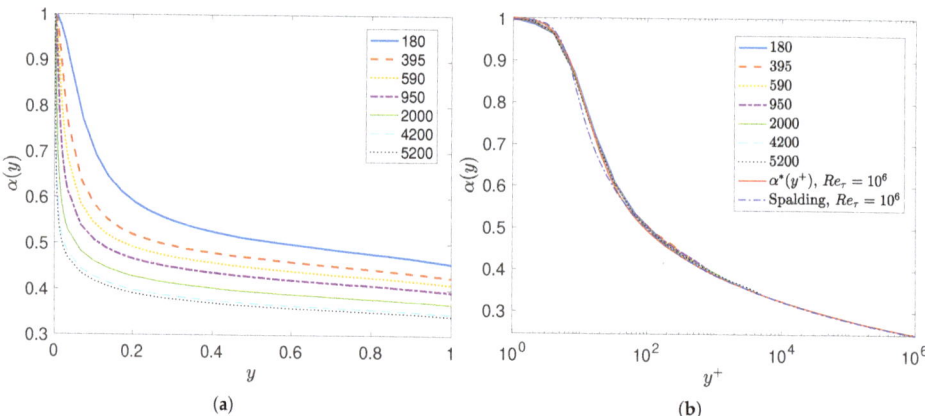

Figure 2. Channel flow modeled with VFM-I: Learning the fractional variable order $\alpha(y)$ using DNS databases at $Re_\tau = 180$ to 5200: (**a**) profiles of the fractional order $\alpha(y)$; (**b**) rescaled fractional order $\alpha(y^+)$ in viscous units.

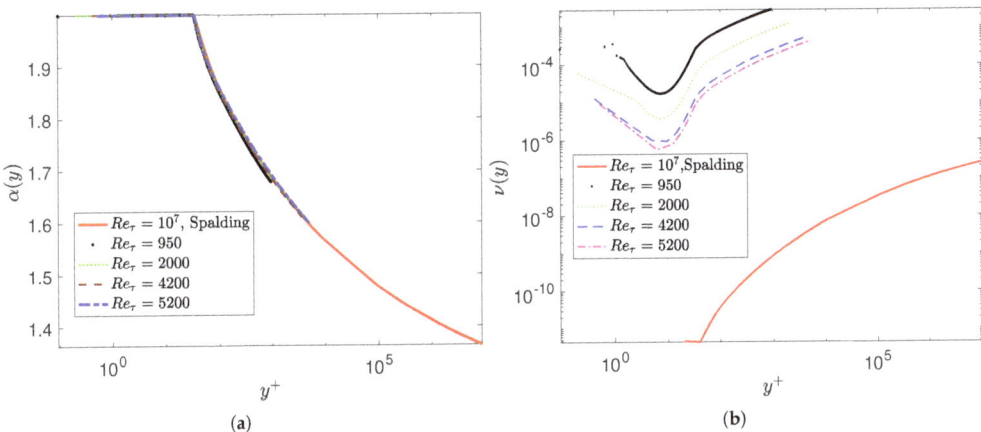

Figure 3. Alternative fractional modeled with VFM-II with $1 \leq \alpha(y) \leq 2$. The numerical fractional orders are computed based on DNS data for turbulent channel flow at $Re_\tau = 950, 2000, 4200, 5200$: (**a**) plots of the fractional orders $\alpha(y^+)$ in wall units; (**b**) corresponding eddy viscosity coefficients.

To evaluate the predictability of the universal scaling, we now solve the forward Equation (2) to obtain $U(y)$ at $Re_\tau = [4200, 6000, 8600]$, which are cases not used in the training of the model for $\alpha(y^+)$. The results presented in Figures 4 and 5 are in good agreement with DNS and experimental data. We also include the turbulent channel flow results obtained by nested LES [24]. Figures 4 and 5 show that the mean velocity profiles predicted by VFM-I exhibit the correct behavior throughout the channel for Reynolds numbers up to $Re_\tau = 8600$, including the correct slope in the logarithmic layer, and agree with DNS and experimental data in the wake region for all $Re_\tau = [4200, 6000, 8600]$.

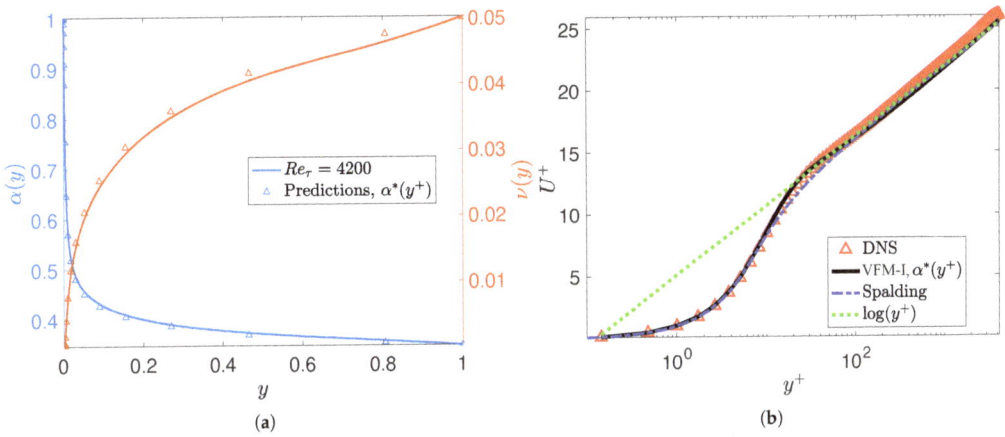

Figure 4. VFM-I: Model predictions for the turbulent channel flow at $Re_\tau = 4200$: (**a**) the solid line ($-$) represents the numerical solution of the optimization problem and the triangle symbols ($\triangle$) represent Equation (24). The blue line represents the fractional order $\alpha(y)$ and the red line is the eddy viscosity coefficient. This Reynolds number $Re_\tau = 4200$ is not included in the training of the model; (**b**) mean velocity obtained by VFM-I corresponding to the fractional order $\alpha^*(y^+)$ from the left plot.

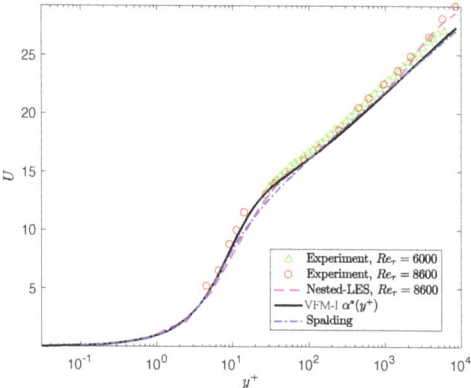

Figure 5. VFM-I: Profiles of the mean velocity for turbulent channel flow at $Re_\tau = 6000, 8600$: the triangle symbol ($\triangle$) represents experimental data from [25], the circle symbol (o) represents experimental data from [26], the solid line ($-$) represents the VFM-I profile, and the dashed line ($--$) represents the LES results [24].

We used fPINN to investigate the turbulent channel flows. We used different training points for investigating the convergence using DNS data at $Re_\tau = 2000$. Figure 6 shows the training results with uniform training points in the interval for $N_t = 500, 1000, 2000$. Figure 7 shows the training results with log-uniform training points in wall units scaling for $N_t = 10, 20, 40, 80$. The corresponding loss histories are listed in Table 1. Figure 7 presents the comparison profiles between the training sets. We can observe that the results trained by the log-uniform are smoother than the uniform training points near the wall.

Table 1. VFM-I: The history of the loss function with different training data sets for $Re_\tau = 2000$. Log represents the log-uniform training points set.

Itr	$N_t = 500$	$N_t = 1000$	$N_t = 2000$	Log, $N_t = 10$	$N_t = 20$	$N_t = 40$	$N_t = 80$
0	6.08×10^{-1}	4.70×10^{-1}	6.57×10^{-1}	6.92×10^{-1}	7.01×10^{-1}	6.74×10^{-1}	6.90×10^{-1}
5000	1.04×10^{-4}	8.11×10^{-5}	9.12×10^{-5}	4.72×10^{-1}	5.94×10^{-5}	5.08×10^{-5}	4.61×10^{-5}
10,000	1.79×10^{-5}	1.32×10^{-5}	1.21×10^{-5}	8.27×10^{-6}	8.93×10^{-6}	1.08×10^{-5}	9.51×10^{-6}
20,000	3.34×10^{-6}	1.75×10^{-5}	1.40×10^{-6}	9.26×10^{-7}	4.68×10^{-7}	2.84×10^{-6}	2.77×10^{-6}
30,000	2.41×10^{-6}	7.31×10^{-7}	7.12×10^{-7}	4.41×10^{-7}	2.05×10^{-7}	1.55×10^{-6}	2.08×10^{-6}

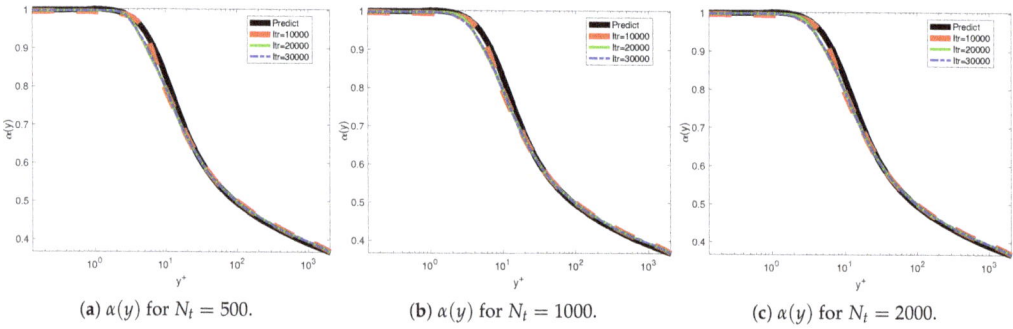

(**a**) $\alpha(y)$ for $N_t = 500$.　　(**b**) $\alpha(y)$ for $N_t = 1000$.　　(**c**) $\alpha(y)$ for $N_t = 2000$.

Figure 6. VFM-I: The fractional order obtained from fPINN and from the universal formula derived using point-by-point minimization ("Predict", Equation (24)). The training results for the uniform training sets at iteration steps *Itr* = 10,000, 20,000, 30,000: (**a**) for $N_t = 500$; (**b**) for $N_t = 1000$; (**c**) for $N_t = 2000$.

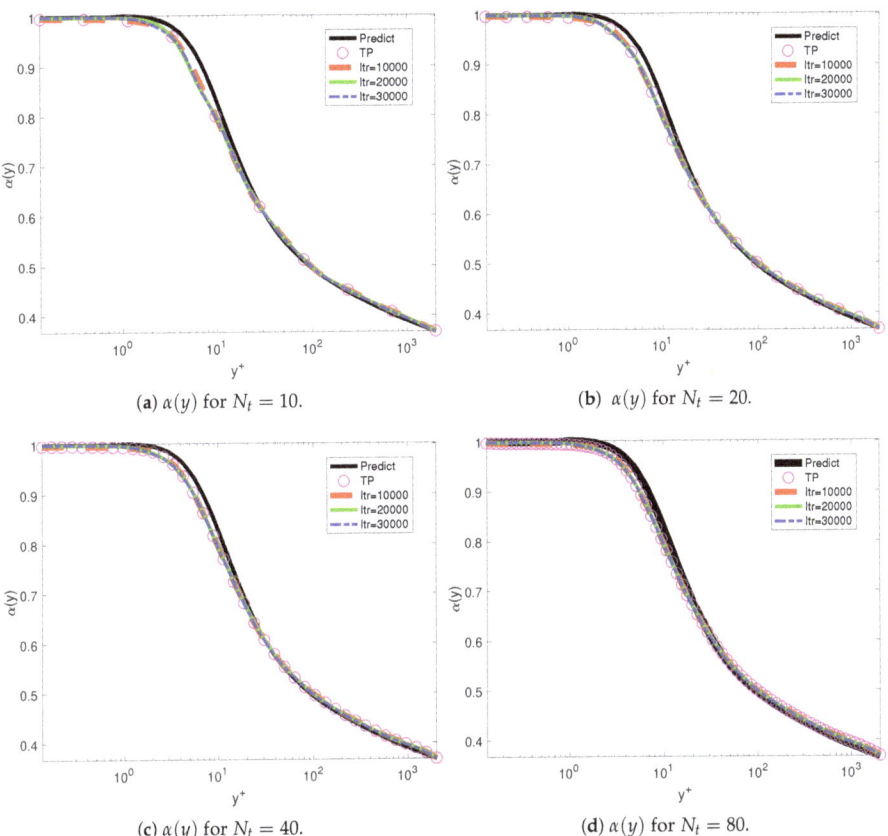

Figure 7. VFM-I: Fractional order for uniform training sets at iteration steps Itr = 10,000, 20,000, 30,000 for different $N_t = 10, 20, 30, 40$. "Predict" presents the profiles from Equation (1). The friction Reynolds number $Re_\tau = 2000$. TP, the distribution of the log-uniform training points.

Next, we test the accuracy of the forward problem and the loss function error with the training fractional order predicted by log-uniform training points $N_t = 20$. We solve the fractional equation as follows:

$$\nu(y) D_y^{\alpha(y)} U = f, \ \forall y \in (0,1], \tag{25}$$

with $U(0) = 0$, and the fractional order is obtained by training fPINN with $N_t = 20$ and Equation (24). The corresponding loss functional error is defined as follows

$$\mathcal{L}(\boldsymbol{\theta}) = \sum_{k=1}^{N_t} \left(Re_\tau^{-\alpha(y_k)} \sum_{j=0}^{k} \frac{b_j^k}{\Delta y^{\alpha(y_k)}} \left(U^{k+1-j}(\boldsymbol{\theta}) - U^{k-j}(\boldsymbol{\theta}) \right) - f_k \right)^2 + (U(0; \boldsymbol{\theta}))^2.$$

Figure 8 plots the pointwise error of the mean velocity and the loss function for $Re_\tau = 4000$ and 5000.

Finally, we use the simplified one-dimensional equation

$$\frac{\partial}{\partial y} \left(\tau_{uv} - R_{uv} \right) = \nu(y) D_y^{\alpha(y)} U = \frac{\partial P}{\partial x}, \ y \in (0,1), \tag{26}$$

where the R_{uv} denotes the Reynolds stress $R_{uv} = \overline{u'v'}$, τ_{uv} denotes the viscous shear stress $\tau_{uv} = v_0 \partial U / \partial y$, and U is the mean velocity, which is the solution to the above fractional Equation (26). Then, we obtain the Reynolds stresses by integration,

$$-R_{uv} = \int_y^1 v(s) D_s^{\alpha(s)} U ds - \tau_{uv}. \tag{27}$$

We can compare the predicted Reynolds stresses to their counterparts, R_D from DNS data for turbulent channel flow, and using the corresponding viscous shear stress denoted by $\tau_D = \mu \partial U_D / \partial y$, where U_D denotes the mean velocity from the DNS database. In Figure 9, we plot the predicted and DNS profiles for Reynolds numbers $Re_\tau = 4000, 5200$ and the corresponding pointwise error. We can observe that they are all in very good agreement. The numerical results of the mean velocities and shear stresses for all Reynolds number Re_τ match very well with the DNS data; here, we only show the high Reynolds number cases due to space limitations.

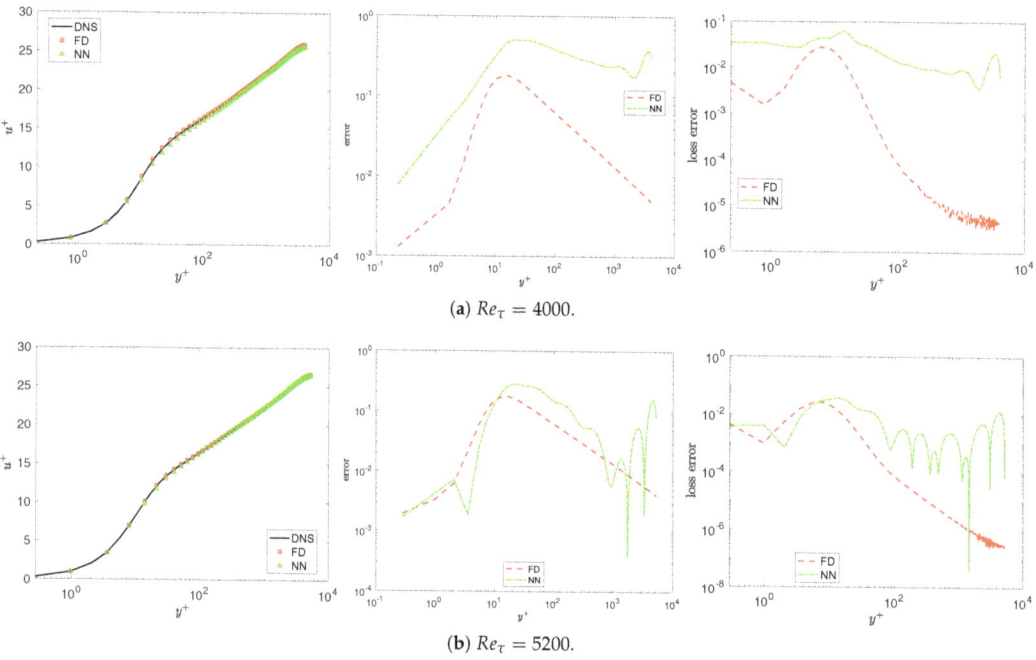

(a) $Re_\tau = 4000$.

(b) $Re_\tau = 5200$.

Figure 8. VFM-I: The mean velocity (**left**) for different Reynolds numbers, the pointwise errors of the mean velocity between predictor and DNS data (**middle**), and the loss function (**right**). FD, the fractional order solved by the finite difference method; NN, the results from the neural network.

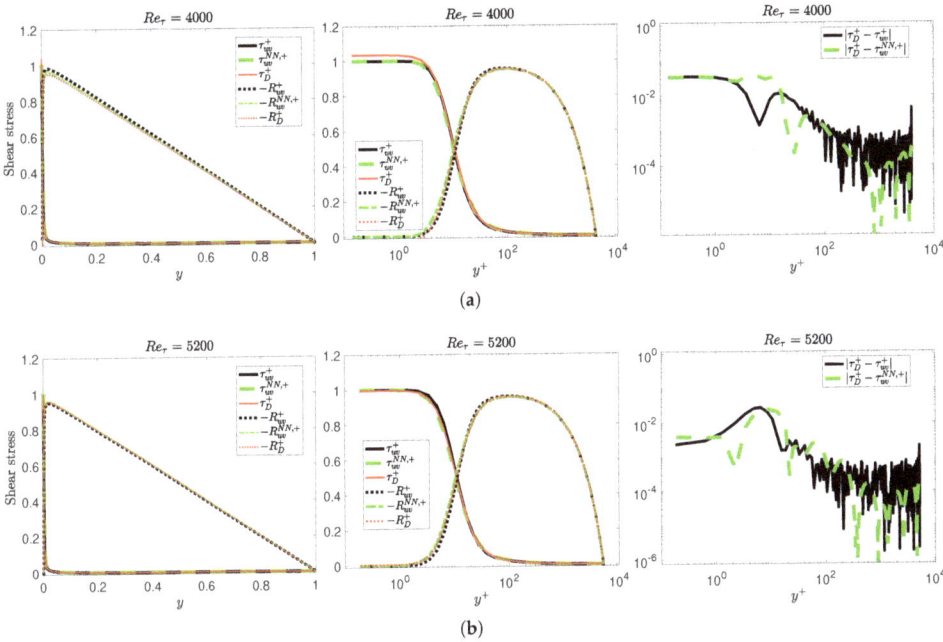

Figure 9. VFM-I: Accurate prediction of the shear stress at (**a**,**b**) $Re_\tau = 4000$, 5200 in outer units and wall units: (**left**) outer scaling; (**middle**) wall units scaling; (**right**) pointwise error of the wall shear stress. Here, τ_{uv} denotes the wall shear stress for the fractional order predicted by the finite difference (FD) method, τ_{uv}^{NN} denotes the wall shear stress predicted by the NN, and τ_D is the corresponding profile from DNS data. $-R_{uv}$ denotes the Reynolds shear stress predicted by Equation (24), $-R_{uv}$ denotes the wall shear stress predicted by the NN, and $-R_D$ is the corresponding profile from DNS data.

3.1.2. Numerical Results of the Two-Sided Models

In this subsection, we focus on the two-sided models. Solving for $\alpha(y)$, which uniquely determines the total shear stresses, Figure 10 plots the profiles of the fractional order $\alpha(y)$ for different Re_τ as a function of the non-dimensional distance between the two walls $y \in [0, 2]$. We see a strong dependence of $\alpha(y)$ of Re_τ, which is the same conclusion as for the previous variable fractional model. Furthermore, we re-plot all data in terms of the viscous wall units, i.e., $y^+ = yRe_\tau$, and we see an approximate collapse of all results into a single universal curve in the half-plane excluding the wake region (i.e., near the centerline), as shown in Figure 10.

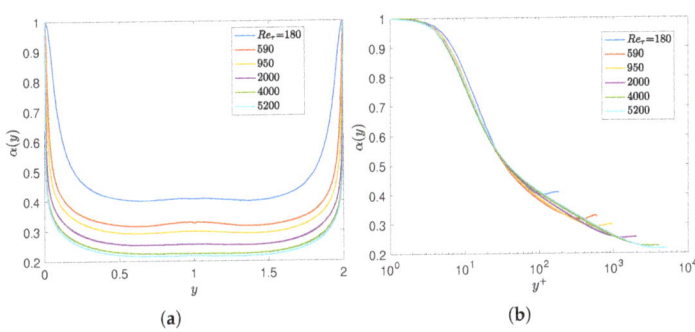

Figure 10. Learning the fractional variable-order $\alpha(y)$ using DNS data bases at $Re_\tau = 180$ to 5200: (**a**) profiles of the fractional order $\alpha(y)$; (**b**) rescaled fractional order $\alpha(y^+)$ in viscous wall units.

Next, we test the accuracy of the forward problem with the fractional order provided by the inverse optimization problem (19). We solve the divergence variable fractional equation as follows

$$-D\left(\nu(y)D_{|y|}^{\alpha(y)}U\right) = 1, \forall y \in (0,2),$$ (28)

with $U(0) = U(2) = 0$. Figure 11 plots the solutions (left) of the above equation and the pointwise error (right) of the mean velocity in each subfigure for several Re_τ. We can observe that this model predicts the mean velocity well. Moreover, it can obtain a smooth mean velocity profile in the whole domain along the wall-wise direction.

We also use fPINN (20) to solve the inverse problem to obtain the variable order $\alpha(y)$. The two results from the two different methods (i.e., FD and fPINN) agree well for all Reynolds numbers.

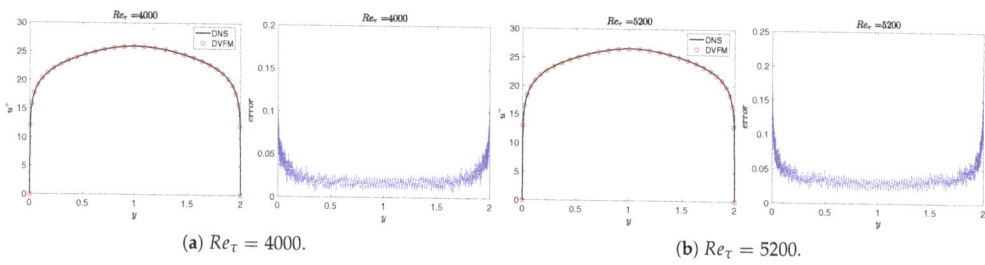

(a) $Re_\tau = 4000$.   (b) $Re_\tau = 5200$.

Figure 11. The mean velocity (**left**) and the pointwise difference between the numerical solution and the DNS data (**right**) in each sub-figure.

3.2. Turbulent Pipe Flow

In this subsection, we consider turbulent pipe flow and again test the universal variable fraction order $\alpha(y^+)$ against DNS and experimental data. First, we examine the highest Reynolds number available from the superpipe experiment [27,28] at $Re_\tau = 5 \times 10^5$, estimated at $Re_R \approx 3.525 \times 10^7$ based on the pipe radius R. As the experimental data were only available for $y^+ > 10,000$, we synthesized an entire profile from the pipe wall to centerline using multifidelity Gaussian process regression (M-GPR) [29] as follows: we considered as high fidelity data the superpipe data in the outer region together with the highest DNS data for channel flow at $Re_\tau = 5200$. We then employed the Spalding curve to provide the low-fidelity data and, using M-GPR, we constructed the final profile as shown in Figure 12a. Having this profile and the VFM-I model transformed in polar coordinates, we can then solve the inverse problem and obtain a new variable fractional order $\alpha(y^+)$. Figure 13a shows that the variable fractional order we obtain for this problem is identical to the function defined by Equation (24). This finding further confirms the universality of the variable fractional order even at very high Reynolds numbers. Having validated the accuracy of the variable fractional order, we can now solve the forward fractional differential problem to obtain predictions of the entire velocity profiles from $Re_\tau = 10^5$ to $Re_\tau = 5 \times 10^5$. Figure 12b plots the results, showing that there is excellent agreement with all available data from the superpipe experiment. Figure 13b plots the mean velocity profiles from the DNS data base [30] at low Reynolds numbers, the corresponding VFM predictions, and the Spalding profile. The universal defect law for pipe flows is not valid for the low Reynolds number range, and this is also in agreement with [27], who argued that the lowest Re_τ for universality is approximately 5000.

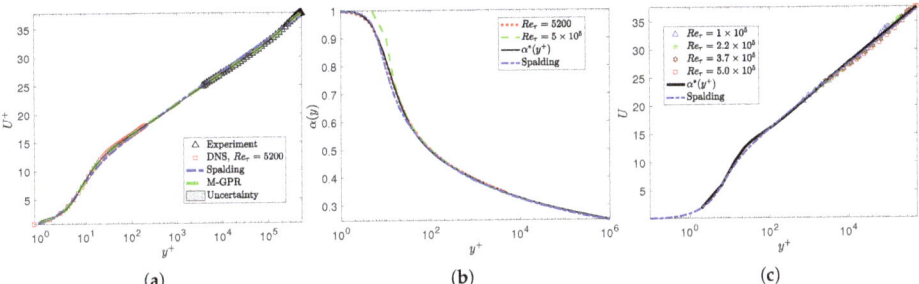

Figure 12. Predictions of the mean velocity profile for the superpipe flow from $Re_\tau = 1 \times 10^5$ to 5×10^5: (**a**) velocity profile reconstructed from the experimental data ($\triangle$, [28]), DNS data at $Re_\tau = 5200$ ($\square$, [22]), and the Spalding profile (blue line [23]) using multifidelity Gaussian process regression (M-GPR); (**b**) "- -", fractional order with the M-GPR profile at $Re_\tau = 5 \times 10^5$; "-", the profile of Equation (24); and '$\cdot\cdot$', the corresponding Spalding profile; (**c**) velocity profiles solving the forward fractional model and the Spalding curve against the experimental data.

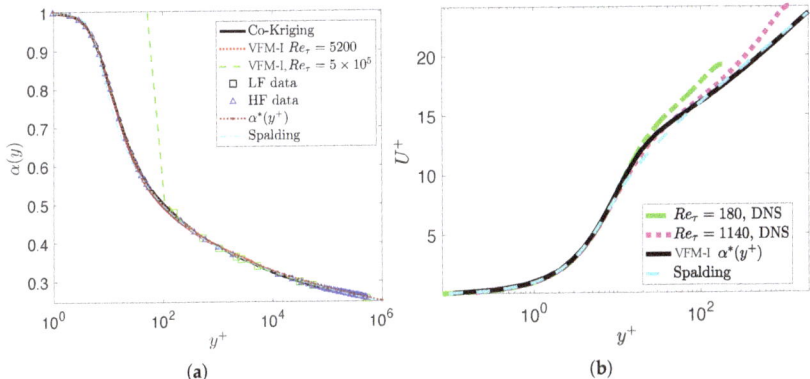

Figure 13. VFM-I for turbulent pipe flow: (**a**) "$\cdot\cdot$", VFM-I model with the channel flow DNS data at $Re_\tau = 5200$; "- -", VFM-I model with the M-GPR profile at $Re_\tau = 5 \times 10^5$; "-", the profile of Equation (24); and '$\cdot\cdot$', the corresponding Spalding profile; (**b**) '$\cdot$-' and '$\cdot\cdot$' plot the DNS data at $Re_\tau = 180$ and $Re_\tau = 1140$; '-' the VFM-I model at $Re_\tau = 2000$ and the corresponding Spalding profile.

3.3. Turbulent Couette Flow

In reference [12], the authors proposed the double-log profile to predict the mean velocity for the Couette flow as follows

$$U(y) = \frac{1}{2} - \frac{1}{2}\frac{\ln\left((d+y)/(d+1-y)\right)}{\ln\left(d/(d+1)\right)}, \tag{29}$$

where d is a small number ($d \ll 1$) that represents a viscous sublayer or roughness height. The non-dimensional boundary conditions are $U(0) = 0$ and $U(1) = 1$.

Here, we consider the predictions from the universal scaling fractional order $\alpha^*(y^+)$, and we also compare it against the double-log profile. The variable fractional order $\alpha^*(y^+)$ is between zero and one in our turbulence model. So, we work in the half-plane $y \in [0, 0.5]$ (see the dashed square in Figure 14a). We then obtain the results in the other half of the domain with $U(y) = 1 - U(1 - y), y \in (0.5, 1]$. Figure 14 shows the mean velocity profiles predicted using (29) and the mean velocity, which is predicted by the variable fractional order $\alpha^*(y^+)$. We can observe that the variable fractional model is in agreement with the experiment data as well as the double-log profile. However, the double-log profile is unable to capture the correct mean velocity near the wall. We also tested the profiles for low Reynolds number $Re_\tau = 52$, where the numerical data were obtained from reference [31]. For the double-log profile, we could not find a suitable parameter d to obtain a good fit for the low $Re_\tau = 52$. Finally, we show the comparisons between the TCM predicted mean velocities and DNS data at $Re_\tau = 250$ obtained from reference [32]. Figure 15 shows that the fractional predictions are correct almost everywhere, especially near the wall regions for high Reynolds numbers.

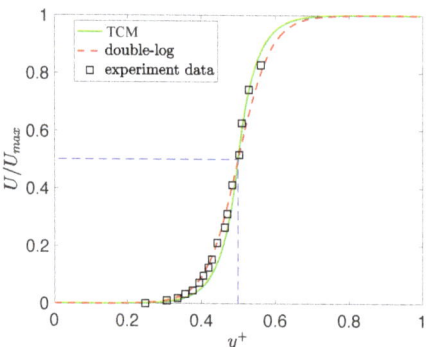

Figure 14. Turbulent Couette flow—numerical results for $Re = 16{,}500$: "-", TCM predictions at $Re_\tau = 1650$; "- -", best fit of the double-log profile in Equation (29) with $d = 1.06 \times 10^{-5}$; "□", experimental data from [33].

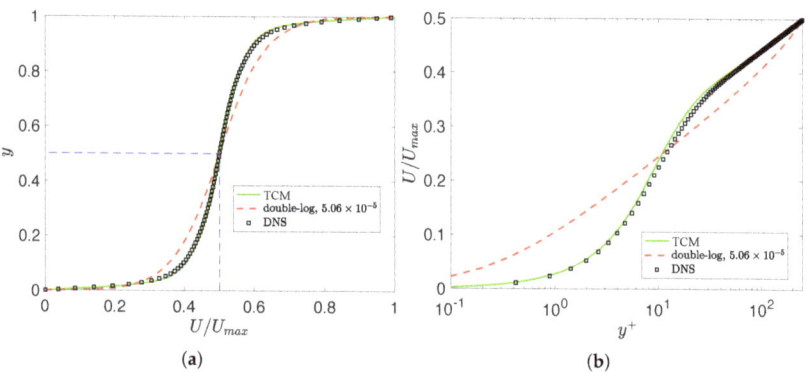

Figure 15. Turbulent Couette flow at $Re_\tau = 250$: (**a**) "-", TCM predictions; "- -", best fit of the double-log profile in Equation (29) with $d = 1.06 \times 10^{-5}$; "□", DNS data from [32]; (**b**) wall units scaling for the mean velocity profiles.

3.4. Turbulent Boundary Layer Flow

In this subsection. we focus on the boundary layer problem. We use data available from the KTH turbulence group from the turbulent boundary layer DNS [34,35]. We first investigate the correlations between Re_θ (x-variable) and Re_τ (y-variable); Figure 16 shows the downstream variations in the friction Reynolds number Re_τ, and unlike the channel flow, here, Re_τ is a function of the streamwise distance x.

Then, we test if the mean velocity of the boundary layer problem exhibits any universality as the channel and pipe flow. We solve the forward boundary layer problem with the fractional order predicted by Equation (24) (i.e., the formula is the same as the channel flow case) including the wake region. Figure 17 presents the mean velocity profiles from the DNS [34] and fractional modeling near the wall for several Re_θ from 670 to 4060, with the corresponding Re_τ varying from 252 to 1200. We observe that the mean velocities are different in the wake region for different Re_τ. Figure 18 plots the wake region, which is between δ_{99}^+ and the error $E = 1\%$. We define this error as the difference in the mean velocity between the DNS data and the fractional model as follows:

$$E = \frac{U - U_f}{U_\infty},\tag{30}$$

where U is the DNS data and U_f presents the numerical results from the fractional model.

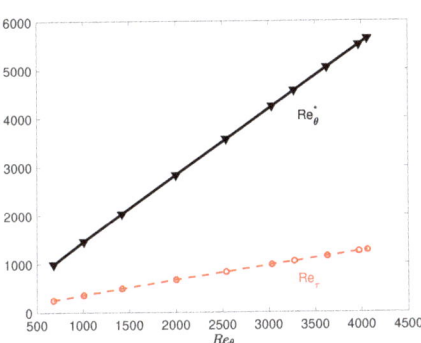

Figure 16. The relation between the friction Reynolds number Re_τ and Re_θ.

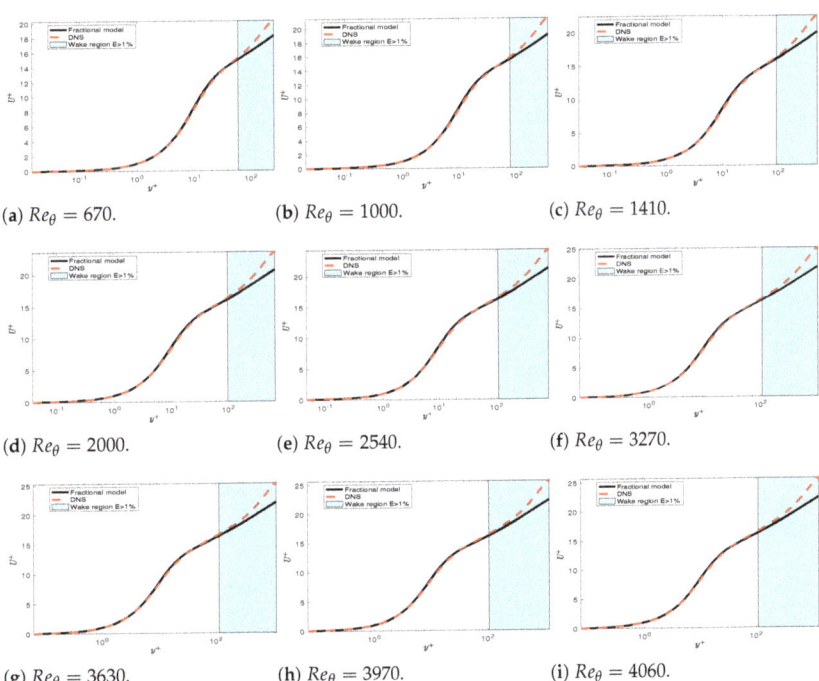

(**a**) $Re_\theta = 670$.　　　　(**b**) $Re_\theta = 1000$.　　　　(**c**) $Re_\theta = 1410$.

(**d**) $Re_\theta = 2000$.　　　　(**e**) $Re_\theta = 2540$.　　　　(**f**) $Re_\theta = 3270$.

(**g**) $Re_\theta = 3630$.　　　　(**h**) $Re_\theta = 3970$.　　　　(**i**) $Re_\theta = 4060$.

Figure 17. TCM: Boundary layer mean velocity profiles from the DNS and fractional modeling near the wall and in the wake region for several Re_θ from 670 to 4060.

Since the mean velocity does not exhibit universality in the wake region, we solve the fPINNs to investigate the variations in the fractional order in the wake region. In Figure 19, we plot the fractional order inferred by fPINN based on the DNS data for $Re_\theta = 670$ to

4060. We can observe that the fractional order varies for different Re_θ in the wake region. Then, we train the fractional order in the wake region selecting the data set $Re_\theta = 670$ to 4060 but excluding $Re_\theta = 2000$. In Figure 20, we present the factional order in the 2D plane for the x-axis and y^+-axis. Finally, we solve the fractional turbulent boundary layer model with the fractional orders presented in Figure 20. The comparison between the numerical results and the DNS data set is presented in Figure 21.

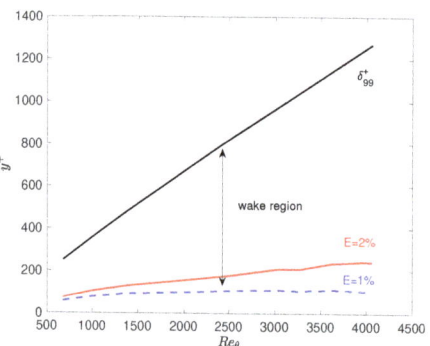

Figure 18. Downstream variations in δ_{99}^+, and the error $E = 2\%$ and $E = 1\%$. The lower bounds of the wake region are denoted by the blue curve with $E = 1\%$ and the red curve with $E = 2\%$ (see Equation (30)).

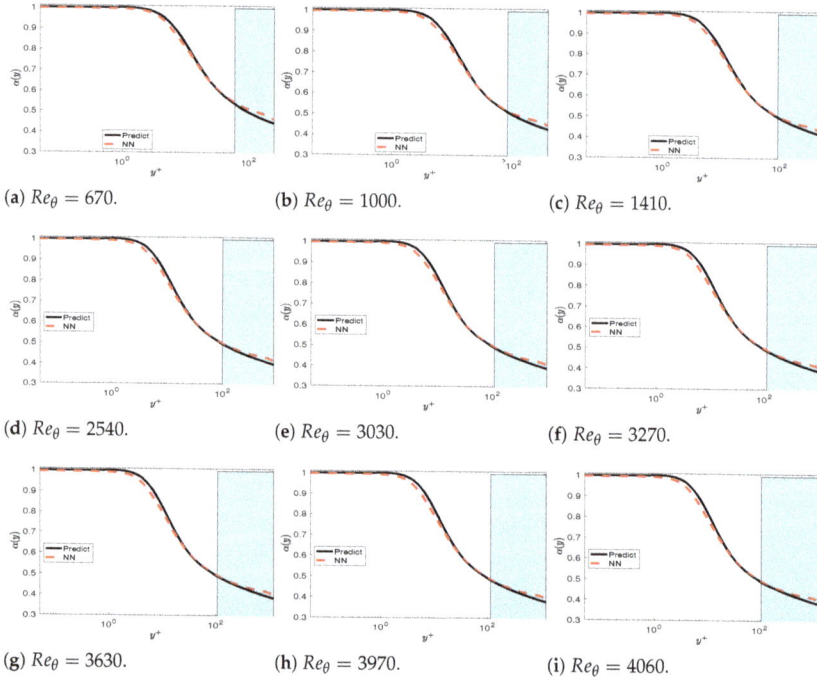

(a) $Re_\theta = 670$.

(b) $Re_\theta = 1000$.

(c) $Re_\theta = 1410$.

(d) $Re_\theta = 2540$.

(e) $Re_\theta = 3030$.

(f) $Re_\theta = 3270$.

(g) $Re_\theta = 3630$.

(h) $Re_\theta = 3970$.

(i) $Re_\theta = 4060$.

Figure 19. TCM: The fractional order $\alpha(y)$ learning from a neural network (NN) near the wall and in the wake region for several Re_θ from 670 to 4060, and the corresponding Re_τ from 252 to 1200. We can observe that the fractional order is different for different Re_θ in the wake region. The black line represents the reference fractional order predicted by channel flows; the red curve represents the NN results for different Re_θ.

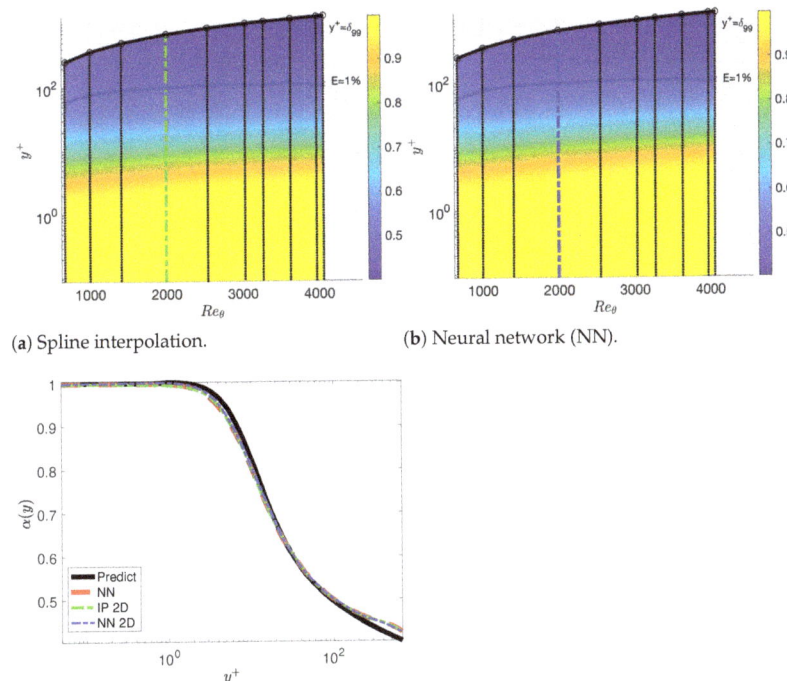

(**a**) Spline interpolation.　　　　(**b**) Neural network (NN).

(**c**) Extract alone the line.

Figure 20. We train the fractional order in the wake region and near the wall selecting the data set $Re_\theta = 670$ to 4060, excluding $Re_\theta = 2000$; the training region is $(Re_\theta, y^+) \in [670, 4060] \times [0, 1200]$. The training data set is represented as black dots: (**a**) we use spline interpolation (IP) in 2D; (**b**) the fractional order is trained by a neural network with 2 hidden layers and 20 neurons in each hidden layer. (**c**) The black line represents the reference fractional order predicted by channel flows; the red curve represents the fPINN results at $Re_\theta = 2000$; the green line plots the interpolation results IP2D along the green line in (**a**); the blue curve presents the NN along the blue line in (**b**).

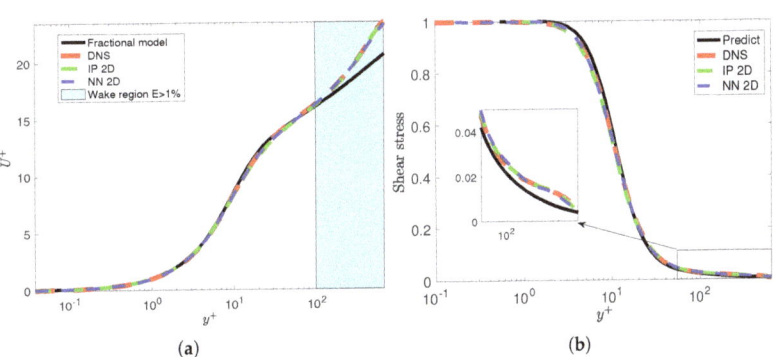

(**a**)　　　　(**b**)

Figure 21. We solve the fractional turbulent boundary layer model with the fractional orders represented in Figure 20c at $Re_\theta = 2000$. (**a**) The mean velocity; (**b**) the viscous shear stress. The black line represents the reference fractional order predicted by channel flows; the red curve represents the NN1D results at $Re_\theta = 2000$; the green line plots the interpolation results IP2D for Re_θ; the blue curve represents the NN for $Re_\theta = 2000$.

4. Summary

We proposed multiple fractional models for wall-bounded turbulent flows in benchmark cases where the mean flow is either one-dimensional (channel, pipe, and Couette flows) or two-dimensional (boundary layer). The main idea is to employ a variable-order fractional gradient that depends on the distance from the wall, starting with an integer order at the wall. The computational problem we addressed is the discovery of the fractional variable-order profile given DNS or experimental data for the mean velocity profile. To this end, we formulated an inverse problem for the fractional order as a function of the distance from the wall, and we solved it using a finite difference method point-by-point and through a new fractional physics-informed neural network (fPINN) that encodes the physics of turbulence expressed via the fractional derivative of variable order. The fractional order is a function of the distance from the wall, a unique capability enabled by fractional calculus. We discovered that this fractional order function is universal for all Reynolds numbers and for different geometries.

The main contributions of this work are: (1) new fractional turbulent models with variable order are presented to model the total shear stress of RANS; (2) two solution methods for the non-trivial inverse problem, a FD method, and a fPINN for obtaining the fractional order function; (3) a universal fractional order profile was discovered for the channel and pipe flows that allowed us to accurately predict the fractional order for the boundary layer flows.

Author Contributions: Conceptualization, F.S. and G.E.K.; methodology, F.S.; software, F.S.; validation, G.E.K.; formal analysis, F.S.; investigation, F.S.; writing—original draft preparation, F.S.; writing—review and editing, G.E.K.; supervision, G.E.K.; funding acquisition, G.E.K. All authors have read and agreed to the published version of the manuscript.

Funding: This research was funded by Artificial Intelligence Research Associate (AIRA) program of the Defense Advance Research Projects Agency (DARPA) and the NSF of China (grant No. 11901100).

Institutional Review Board Statement: Not applicable.

Informed Consent Statement: Not applicable.

Data Availability Statement: Not applicable.

Acknowledgments: G.E.K. acknowledges support by Artificial Intelligence Research Associate (AIRA) program of the Defense Advance Research Projects Agency (DARPA). F.S. was also supported by the NSF of China (grant No. 11901100).

Conflicts of Interest: The authors declare no conflict of interest.

References

1. Reynolds, O. On the dynamical theory of incompressible viscous fluids and the determination of the criterion. *Philos. Trans. R. Soc. Lond. A* **1895**, *186*, 123–164.
2. Pope, S.B. *Turbulent Flows*; Cambridge University Press: Cambridge, MA, USA, 2000.
3. Prandtl, L. Bericht uber Untersuchungen zur ausgebildeten Turbulenz. *Z. Angew. Math. Mech.* **1925**, *5*, 136–139. [CrossRef]
4. Richardson, L.F. Atmospheric diffusion shown on a distance-neighbour graph. *R. Soc. Lond.* **1926**, *110*, 709–737.
5. Kraichnan, R.H. Direct-Interaction Approximation for Shear and Thermally Driven Turbulence. *Phys. Fluids* **1964**, *7*, 1048–1062. [CrossRef]
6. Prandtl, L. Bemerkungen zur Theorie der freien Turbulenz. *ZAMM-J. Appl. Math. Mech. Angew. Math. Mech.* **1942**, *22*, 241–243. [CrossRef]
7. Egolf, P.W.; Hutter, K. A nonlocal zero-Equation turbulence model and a deficit-power law of the wall with a dynamical critical phenomenon. *Proc. Appl. Math. Mech.* **2016**, *16*, 889–890. [CrossRef]
8. Gorenflo, R.; Mainardi, F. Fractional calculus and continuous-time finance III: The diffusion limit. Mathematical finnance. *Trends Math. Birkhäuser* **2001**, *2001*, 171–180.
9. Shlesinger, M.; West, B.; Klafter, J. Lévy dynamics of enhanced diffusion: Application to turbulence. *Phys. Rev. Lett.* **1987**, *58*, 1100. [CrossRef]
10. Egolf, P.W.; Hutter, K. Fractional Turbulence Models. In *Progress in Turbulence VII*; Springer: Berlin/Heidelberg, Germany, 2017; pp. 123–131.

11. Chen, W. A speculative study of 2/3-order fractional Laplacian modeling of turbulence: Some thoughts and conjectures. *Chaos Interdisc. J. Nonlinear Sci.* **2006**, *16*, 023126. [CrossRef]
12. Epps, B.P.; Cushman-Roisin, B. Turbulence modeling via the fractional Laplacian. *arXiv* **2018**, arXiv:1803.05286.
13. Samiee, M.; Akhavan-Safaei, A.; Zayernouri, M. A fractional subgrid-scale model for turbulent flows: Theoretical formulation and a priori study. *Phys. Fluids* **2020**, *32*, 055102. [CrossRef]
14. George, W.K. Is there a universal log law for turbulent wall-bounded flows? *Philos. Trans. R. Soc. Lond. A Math. Phys. Eng. Sci.* **2007**, *365*, 789–806. [CrossRef]
15. Lischke, A.; Pang, G.; Gulian, M.; Song, F.; Glusa, C.; Zheng, X.; Mao, Z.; Cai, W.; Meerschaert, M.M.; Ainsworth, M.; et al. What is the fractional Laplacian? A comparative review with new results. *J. Comput. Phys.* **2020**, *404*, 109009. [CrossRef]
16. Song, F.; Karniadakis, G.E. A universal fractional model of wall-turbulence. *arXiv* **2018**, arXiv:1808.10276.
17. Mehta, P.P.; Pang, G.; Song, F.; Karniadakis, G.E. Discovering a universal variable-order fractional model for turbulent Couette flow using a physics-informed neural network. *Fract. Calc. Appl. Anal.* **2019**, *22*, 1675–1688. [CrossRef]
18. Marusic, I.; McKeon, B.; Monkewitz, P.; Nagib, H.; Smits, A.; Sreenivasan, K. Wall-bounded turbulent flows at high Reynolds numbers: Recent advances and key issues. *Phys. Fluids* **2010**, *22*, 065103. [CrossRef]
19. Johnson, P.L.; Meneveau, C. Turbulence intermittency in a multiple-time-scale Navier-Stokes-based reduced model. *Phys. Rev. Fluids* **2017**, *2*, 072601. [CrossRef]
20. Sabzikar, F.; Meerschaert, M.M.; Chen, J. Tempered fractional calculus. *J. Comput. Phys.* **2015**, *293*, 14–28. [CrossRef]
21. Lin, Y.; Xu, C. Finite difference/spectral approximations for the time-fractional diffusion equation. *J. Comput. Phys.* **2007**, *225*, 1533–1552. [CrossRef]
22. Lee, M.; Moser, R.D. Direct numerical simulation of turbulent channel flow up to $Re_\tau = 5200$. *J. Fluid Mech.* **2015**, *774*, 395–415. [CrossRef]
23. White, F.M. *Viscous Fluid Flow*; MacGraw-Hill: New York, NY, USA, 1991; pp. 335–393.
24. Tang, Y. A Nested-LES Approach for Computation of High-Reynolds Number, Equilibrium and Non-Equilibrium Turbulent Wall-Bounded Flows. Ph.D. Thesis, University of Michigan, Ann Arbor, MI, USA, 2016.
25. Schultz, M.P.; Flack, K.A. Reynolds-number scaling of turbulent channel flow. *Phys. Fluids* **2013**, *25*, 025104. [CrossRef]
26. Comte-Bellot, G. Turbulent Flow between Two Parallel Walls. Ph.D. Thesis, University of Grenoble, Grenoble, France, 1963.
27. Zagarola, M.V.; Smits, A.J. Mean-flow scaling of turbulent pipe flow. *J. Fluid Mech.* **1998**, *373*, 33–79. [CrossRef]
28. McKeon, B.J.; Smits, A.J. Static pressure correction in high Reynolds number fully developed turbulent pipe flow. *Meas. Sci. Technol.* **2002**, *13*, 1608. [CrossRef]
29. Kennedy, M.C.; O'Hagan, A. Predicting the output from a complex computer code when fast approximations are available. *Biometrika* **2000**, *87*, 1–13. [CrossRef]
30. Wu, X.; Moin, P. A direct numerical simulation study on the mean velocity characteristics in turbulent pipe flow. *J. Fluid Mech.* **2008**, *608*, 81–112. [CrossRef]
31. Liu, C.H. Turbulent plane Couette flow and scalar transport at low Reynolds number. *J. Heat Transf.* **2003**, *125*, 988–998. [CrossRef]
32. Avsarkisov, V.; Hoyas, S.; Oberlack, M.; García-Galache, J. Turbulent plane Couette flow at moderately high Reynolds number. *J. Fluid Mech.* **2014**, *751*. [CrossRef]
33. Robertson, J.M.; Johnson, H.F. Turbulence structure in plane Couette flow. *J. Eng. Mech. Div.* **1970**, *96*, 1171–1182. [CrossRef]
34. Schlatter, P.; Örlü, R. Assessment of direct numerical simulation data of turbulent boundary layers. *J. Fluid Mech.* **2010**, *659*, 116. [CrossRef]
35. Schlatter, P.; Li, Q.; Brethouwer, G.; Johansson, A.V.; Henningson, D.S. Simulations of spatially evolving turbulent boundary layers up to $Re_\theta = 4300$. *Int. J. Heat Fluid Flow* **2010**, *31*, 251–261. [CrossRef]

Review

Applications of Distributed-Order Fractional Operators: A Review

Wei Ding, Sansit Patnaik, Sai Sidhardh and Fabio Semperlotti *

Ray W. Herrick Laboratories, School of Mechanical Engineering, Purdue University,
West Lafayette, IN 47907, USA; ding242@purdue.edu (W.D.); spatnai@purdue.edu (S.P.);
ssidhard@purdue.edu (S.S.)
* Correspondence: fsemperl@purdue.edu

Abstract: Distributed-order fractional calculus (DOFC) is a rapidly emerging branch of the broader area of fractional calculus that has important and far-reaching applications for the modeling of complex systems. DOFC generalizes the intrinsic multiscale nature of constant and variable-order fractional operators opening significant opportunities to model systems whose behavior stems from the complex interplay and superposition of nonlocal and memory effects occurring over a multitude of scales. In recent years, a significant amount of studies focusing on mathematical aspects and real-world applications of DOFC have been produced. However, a systematic review of the available literature and of the state-of-the-art of DOFC as it pertains, specifically, to real-world applications is still lacking. This review article is intended to provide the reader a road map to understand the early development of DOFC and the progressive evolution and application to the modeling of complex real-world problems. The review starts by offering a brief introduction to the mathematics of DOFC, including analytical and numerical methods, and it continues providing an extensive overview of the applications of DOFC to fields like viscoelasticity, transport processes, and control theory that have seen most of the research activity to date.

Keywords: fractional calculus; distributed-order operators; viscoelasticity; transport processes; control theory

Citation: Ding, W.; Patnaik, S.; Sidhardh, S.; Semperlotti, F. Applications of Distributed-Order Fractional Operators: A Review. *Entropy* **2021**, *23*, 110. https://doi.org/10.3390/e23010110

Received: 25 December 2020
Accepted: 11 January 2021
Published: 15 January 2021

Content

1. Introduction

Fractional calculus (FC) was first introduced as a mathematical generalization of integer-order integration and differentiation. Started in 1695 from a discussion between Leibniz and de L'Hôpital about the possible interpretation of the operator d^n/dx^n when $n = 1/2$ [1], FC has been the object of studies for more than 300 years. In the early years, research mostly focused on mathematical aspects of the fractional-order operators; their physical interpretations and potential applications followed much later. Likely, the first application of FC can be traced back to Abel in 1826. Abel [2] applied FC to formulate an integral equation describing a tautochrone problem. Following Abel's study, the integral representation of FC started gaining increasing attention in the mathematics community. Early works mostly focused on the development of analytical formulations to solve selected mathematical problems. The most immediate result of this rapidly growing interest in FC was the expansion of the possible definitions of a fractional operator including, but not limited to, the integral representation (Liouville, Riemann, and Hadamard) and the convergent series representation (Grünwald and Letnikov). While these early studies had pointed out the intriguing role that FC can play when modeling complex processes in physical systems, the bulk of the early research kept focusing on the development of the mathematical framework [3] and on the integration of these operators into ordinary and partial differential equations [4]. It was only in the second half of the twentieth century that the concept of FC started percolating to fields other than mathematics. An area of application that has seen a remarkably rapid growth is that involving the modeling of complex physical phenomena. Unlike integer-order operators, the intrinsic multiscale nature of fractional operators enabled a very unique and effective approach to model historically challenging physical processes involving, as an example, nonlocality or memory effects. Indeed, many of the early applications of FC to physical modeling included viscoelastic effects [5–12], nonlocal behavior [8,12–24], anomalous and hybrid transport [9–11,24–30], fractal media [12,31–35], and even control theory [36–39]. The interested reader is referred to the work in [40] for a detailed account of the birth and evolution of fractional calculus.

For more than a century, the study of fractional calculus focused on operators accepting a constant and single-valued order; we will refer to these operators as constant-order operators in order to differentiate them from the distributed (but constant) order operators that will be introduced below. Despite constant-order operators being considerably more general than their integer-order counterpart, the constant and single-valued nature of the order still limits its ability to accurately capture certain complex phenomena whose underlying physics could either evolve in time or emerge as the result of the interplay of multiple orders. In relatively recent years, this observation led to the formulation of two remarkable and unique forms of FC operators, namely, the distributed-order and the variable-order operators. The latter definition accounts for operators whose order can be a function of either dependent (e.g., state variables of the system) or independent (e.g., space or time) variables and can change value following the evolution of the system. While this review does not focus on this class of operators, the interested reader is referred to the works in [41,42] for a detailed overview of the mathematical aspects and applications of variable-order operators.

Before proceeding further, we clarify the different acronyms that will be used in this review in order to refer to the different types of fractional-order operators. The single constant-order operators are denoted as "CO" operators, the distributed-order operators (with constant order distribution) are denoted as "DO" operators, and the variable-order operators are denoted as "VO" operators. While VO operators can certainly be single or distributed in nature, with the acronym "VO" we specifically refer to single variable-order operators. Distributed-variable-order operators, which will be introduced later, are denoted as "DVO" operators.

The distributed-order definition of the operator allows considering a superposition of orders and accounting for, as an example, physical phenomena such as memory effects in composite materials [43] or multi-scale effects [44]. A typical example that illustrates the capabilities of this class of operators is the mechanical behavior of viscoelastic materials having spatially varying properties [45]. Distributed-order fractional calculus presents a natural generalization of constant-order fractional calculus (COFC) by integrating the fractional kernel of CO operators over an extended range of orders. Given that the fundamental kernel of a CO operator is retained in the DO operator, DO operators inherit the fundamental properties of COFC, such as the ability to model nonlocality and memory effects, and further extend them to multiple coexisting orders. This latter argument can be interpreted as a superposition of the behavior captured by individual CO operators using different orders within a given range.

The original concept of distributed-order fractional calculus (DOFC) can be traced back to the seminal studies by Caputo on dissipative elastodynamics [46–48]. In these studies, a generalization of the viscoelastic stress–strain constitutive laws, by employing a parallel sequence of fractional-order derivatives, was undertaken. Initially, the author dubbed this operator as the "*mean fractional-order derivative*". A couple of decades later, Caputo [49] formalized the original proposition into the concept of DO derivative and also explored possible solutions to differential equations employing DO derivatives. Later, detailed investigations on the properties of DO operators, and on the properties and solution techniques of DO differential equations (DODE) were conducted in [45,50,51]. Following these pioneering studies on the mathematics of DO operators, in the 1990s and early 2000s, the interest in this topic went beyond the mathematical community and started percolating into several branches of engineering and physics. To date, we estimate that a total of approximately 300 papers have been published in the general area of DOFC. This estimate includes both journal and conference publications spanning a variety of fields including, but not limited to, theoretical and applied mathematics, analytical and numerical methods, viscoelasticity, transport processes, and control theory. A detailed time history and a quantitative assessment of the scientific studies produced in the general area of DOFC are provided in Figure 1.

Given the substantial critical mass reached by this field to date, and in view of the drastic acceleration of the research on DOFC observed in recent years, the time is ripe to assess the state of the field not only in terms of the mathematical formulation, but from the perspective of practical applications. In this review, we will provide a comprehensive discussion of the different fields of application and possible opportunities offered by DOFC to model complex physical problems. We expect that this review would serve as a starting point for the reader interested in approaching this fascinating field. Engineering, physics, chemistry, biology, and finance are only some of the communities that should find several points of interest and material for further consideration in this work.

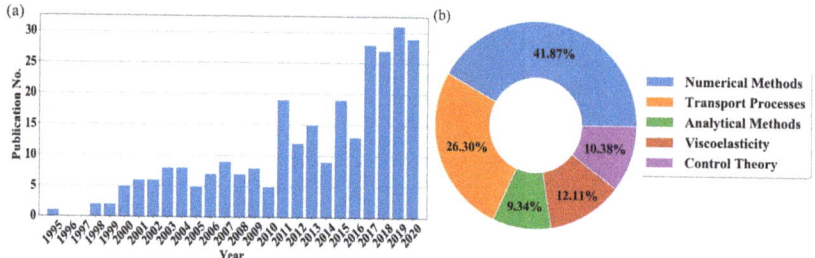

Figure 1. (**a**) Histogram chart showing the historical evolution of scientific publications per year starting from 1995. Note that the first study on distributed-order fractional calculus (DOFC) was published in 1966 by Caputo [46]. Approximately five studies were produced until 1995, which was taken as the starting year for the histogram. (**b**) Pie chart showing the distribution of publications per field. The data used in this figure were collected from Google Scholar.

The remainder of this paper is organized as follows. Section 2 focuses on providing an overview of the main mathematical concepts including basic definitions and properties of DOFC. The section also covers analytical and numerical methods for the calculation of DO operators and for the solution of DODEs. Section 3 briefly discusses the relevance of DO operators with respect to the modeling of complex physical processes. The remaining sections provide a review of the applications of DOFC to real-world problems including viscoelastic systems, transport processes, and control theory.

2. Mathematical Background

We begin this review by providing a brief summary of the basic definitions and properties of DO operators. Further, we will discuss the properties of differential equations with DO operators, and provide a brief overview of the corresponding analytical and numerical simulation techniques. We highlight here that, unless otherwise mentioned, the DO operator is defined on the basis of a general fractional-order derivative denoted by $_c^\square D_t^\alpha$, evaluated with respect to a generalized independent variable t. We emphasize that the notation t used in this section must not be interpreted necessarily as time. Note that c denotes the lower terminal of the fractional derivative. The fractional derivative $_c^\square D_t^\alpha$ can accept different definitions, although the most common for DO operators are those provided by Riemann–Liouville $_c^{RL} D_t^\alpha$ and by Caputo $_c^C D_t^\alpha$ [45]. Finally, also for the sake of brevity, we shall provide only the definitions corresponding to the left-handed fractional derivatives (the right-handed DO derivatives being an immediate extension).

2.1. Definitions and Properties

From a mathematical perspective, DO derivatives are defined as an integration of either the constant-order or the variable-order fractional derivatives with respect to the non-integer order of differentiation [48–51]. Two approaches to the definition of DO derivatives have been explored [45]. First, the so-called *direct approach* treats the order as a variable so that the DO derivative is defined as [45,49]

$$_{\alpha_1,\alpha_2}\mathcal{D}_{c,t}^\alpha(f(t),\kappa(\alpha),\alpha) = \int_{\alpha_1}^{\alpha_2} \kappa(\alpha)_c^\square D_t^\alpha f(t) \, \mathrm{d}\alpha \tag{1}$$

where the integrand $\kappa(\alpha)_c^\square D_t^\alpha f(t)$ undergoes integration with respect to the independent variable α, that is, the fractional order within the interval $\alpha \in [\alpha_1, \alpha_2]$. $\kappa(\alpha)$ is denominated as the order-weighting/strength function, or simply the strength function. The second approach, referred to as the indirect approach, treats the order as a function of a different independent variable x leading to the following definition [45],

$$_{x_1,x_2}\mathcal{D}_{c,t}^{\alpha(x)}(f(t),\kappa(\alpha),x) = \int_{x_1}^{x_2} \kappa(x)_c^\square D_t^{\alpha(x)} f(t)\, dx \tag{2}$$

where $x \in [x_1, x_2]$ is the interval of integration. Similar to $\kappa(\alpha)$, $\kappa(x)$ is also an order strength distribution [45]. The strength function ($\kappa(\alpha)$ or $\kappa(x)$) determines the contribution of each individual CO derivative to the overall DO derivative. As an example, a constant value of the strength function $\kappa(\alpha) = \kappa_0$ would mean the all the CO derivatives contribute equally to the final DO derivative [49]. The specific choice of this strength function depends on the underlying physics of the problem to be modeled and could be defined as either a continuous or a discrete function of the order α (direct approach) or the independent variable x (indirect approach). This latter comment is further clarified in the following section by using practical examples.

To better illustrate the above concepts, we present a numerical demonstration of the DO derivatives evaluated for two representative functions of the variable t: (1) a sinusoidal function $f(t) = \sin \pi t$ in Figure 2 and (2) a step function $f(t) = \mathcal{H}(t-1)$ in Figure 3, where $\mathcal{H}$ is the Heaviside function. In Figures 2a and 3a, the strength function is chosen to be $\kappa(\alpha) = 1$, such that it is constant and continuous. In the Figures 2b and 3b, a discontinuous strength function $\kappa(\alpha) = \Sigma_{\alpha_j \in \{0.5,0.7,0.9\}} \tau_0^\alpha \delta(\alpha - \alpha_j)$, where τ_0 is a positive constant. In generating the above results, we employed the Caputo definition of the fractional derivatives with terminals $(-\infty, t]$. The CO Caputo fractional derivative of the two different functions to an order $\alpha \in (0,1)$ is [52]:

$$_{-\infty}^C D_t^\alpha (\sin \pi t) = \pi^\alpha \sin\left(\frac{\pi\,(2t + \alpha)}{2}\right) \tag{3a}$$

$$_{-\infty}^C D_t^\alpha (\mathcal{H}(t - 1)) = \mathcal{H}(t - 1)\left[\frac{(t-1)^{-\alpha}}{\Gamma(1-\alpha)}\right] \tag{3b}$$

The above CO derivatives are also provided in the Figures 2 and 3 to facilitate comparison with the DO derivatives. Note that above expressions for the different CO derivatives identically reduce to their respective first-order (integer) derivatives for the choice of $\alpha = 1$.

As evident from the Figures 2 and 3, the DO derivatives can be perceived as the weighted sum of individual CO derivatives over the specified range of fractional-order α. Particularly for $\kappa(\alpha) = 1$, as evident from Figures 2a and 3a, the DO derivative is the linear sum of the CO derivatives with fractional-order α spanning the range $[\alpha_1, \alpha_2]$. This concept is further illustrated by the examples in Figures 2b and 3b. In these figures, the DO derivatives evaluated for $\tau_0 = 1$ are the sum of the individual CO derivatives. In contrast, for $\tau_0 = 2$ wherein the strength function is also a function of the order α, we observe a weighted contribution of the different CO derivatives to the DO derivative. The above discussion also explains the shift in the phase of the harmonic function in Figure 2a. More specifically, the phase shift in the DO derivative with respect to the original signal is caused due to the contribution of a phase difference of $\pi\alpha/2$ (see Equation (3a)) by each CO derivative. The effect of the strength function on the amplitude, without changes in the phase, is illustrated in Figure 2b. Similarly, for the case of the Heaviside step function in Figure 3, different decaying characteristics can be obtained by varying the definitions of the strength function $\kappa(\alpha)$ and its support $[\alpha_1, \alpha_2]$. Interesting applications to viscoelasticity based on this observation will be discussed in Section 4.

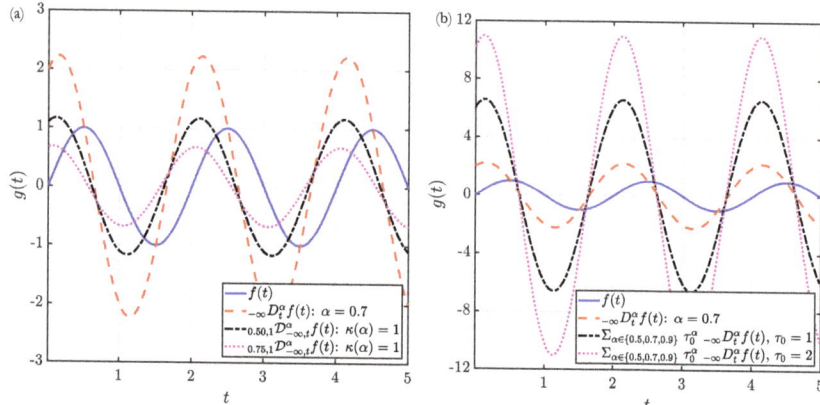

Figure 2. DO derivative of a harmonic function $f(t) = \sin \pi t$ derived following the definitions given in Equation (1). The plot shows the behavior of the derivative for (**a**) continuous and (**b**) discrete strength functions.

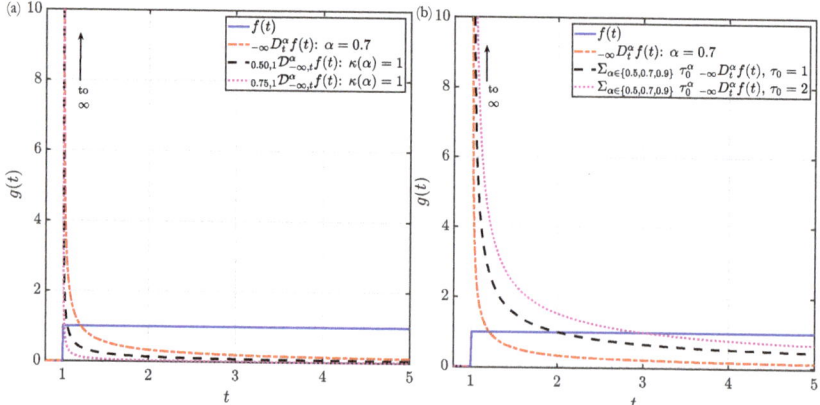

Figure 3. DO derivative of the Heaviside function $f(t) = \mathcal{H}(t-1)$ derived following the definitions given in Equation (1). The plot shows the behavior of the derivative for (**a**) continuous and (**b**) discrete strength functions.

Lorenzo and Hartley [45] also extended the definitions of DO derivatives by allowing for the order distribution to be a function of different variables (such as, for example, space, time, or external loads). This extension introduced the concept of distributed-variable-order (DVO) operator. Following this extension, the direct and indirect approaches to the definition of DO operators can be reformulated as

$$_{\alpha_1,\alpha_2}\mathcal{D}_{c,t}^{\alpha(t)}(f(t),\kappa(\alpha),\alpha) = \int_{\alpha_1}^{\alpha_2} \kappa(\alpha)_c^{\square}D_t^{\alpha(t)}f(t)\,d\alpha \qquad (4\text{a})$$

$$_{x_1,x_2}\mathcal{D}_{c,t}^{\alpha(x,t)}(f(t),\kappa(\alpha),x) = \int_{x_1}^{x_2} \kappa(x)_c^{\square}D_t^{\alpha(x,t)}f(t)\,dx \qquad (4\text{b})$$

Although providing a very general form of the operator that can capture both multifractal (DO) and evolutionary (VO) behavior, the application of these operators has been rather limited. To date, most applications of DVO operators have been in the area of complex viscoelastic materials (see Section 4.3).

2.2. Distributed-Order Differential Equations

The present section is intended to briefly introduce the concept of differential equations based on DO operators. Clearly, the concept of DODEs is fairly extensive in itself and the reader is referred to the works in [53,54] for a detailed discussion on the different forms of DODEs and the corresponding solution techniques. Here, we simply introduce the general concept of DODE in order to facilitate the understanding of the discussion on applications presented in the remainder of the paper. Consider the following DODE [49],

$$_{0,m}\mathcal{D}_{0,t}^{\alpha}(\kappa(\alpha), u(t), \alpha) = f(t) \tag{5}$$

for $m \in \mathbb{N}$. Note that a discrete distribution function $\kappa(\alpha) = \sum_{j=1}^{n} b_j \delta(\alpha - \alpha_j)$ reduces the above equation to following multi-term fractional-order differential equation,

$$\sum_{j=1}^{n} b_j \, _0^{\square}D_t^{\alpha_j} u(t) = f(t) \tag{6}$$

At the same time, a continuous distribution $\kappa(\alpha) = \mathbb{C}[0, m]$ can be perceived as a limiting case of the multi-term definition provided above when $n \to \infty$ [49]. While Equation (5) is an example of linear DODE, a nonlinear DODE can be given as [55]

$$\int_{m_1}^{m_2} \kappa(\alpha) F\left(_0^{\square}D_t^{\alpha}u(t)\right) d\alpha = f(t, u(t)) \tag{7}$$

where $F\left(_0^{\square}D_t^{\alpha}u(t)\right)$ is a nonlinear function in the primary variable $u(t)$ including its fractional derivatives.

For the linear DODE in Equation (5), some common assumptions are employed to ensure that the problem is well-posed, that is, the solution is both bounded and convergent [55,56]:

Hypothesis 1. κ *is absolutely integrable on the interval* $[\alpha_1, \alpha_2]$ *and satisfies the following inequality,*

$$\int_{\alpha_1}^{\alpha_2} \kappa(\alpha)s^{\alpha} d\alpha \neq 0, \quad for \ Re(s) > 0 \tag{8}$$

Hypothesis 2. $f \in \mathbb{L}^1[0, \infty)$, *where* $\mathbb{L}^1$ *is the Lebesgue space.*

Hypothesis 3. *The function* $u(t)$ *is such that* $_0^{\square}D_t^{\alpha}u(t) < M \ \forall t \in [0, \infty) \cap \forall \alpha \in [\alpha_1, \alpha_2],$ *where M is a constant. In other terms the fractional-order derivative is always bounded. For the limiting case where either of the order bounds tends to infinity (i.e.,* α_1 *or* $\alpha_2 \to \infty$*), the boundedness of the DO derivative requires the strength function* $\kappa(\alpha)$ *to be non-zero only over a finite range, that is,* $\kappa(\alpha)$ *must have a finite support* [45].

Pskhu [57,58] conducted early studies on the solvability of ordinary DODEs. Umarov and Gorenflo [59] extended these studies to analyze the solvability of multipoint problems. Diethelm and Ford [60,61] analyzed the existence and the uniqueness of solutions for linear DODEs, specifically for the case where Caputo-type initial conditions are available. Later, this proof was extended to the case where initial conditions are unknown [55]. It is noteworthy that these studies prove the existence and uniqueness for the fractional order $\alpha < 1$, while for $\alpha > 1$ the existence and uniqueness are still a conjecture. A similar exercise was performed on nonlinear DODEs with specific application to viscoelastic systems [62] and wave propagation [63]. The existence of solutions to hybrid DODEs was analyzed in [64], where the hybrid differential equations are quadratic perturbations to nonlinear DODEs [65,66]. Atanacković et al. also conducted similar studies on selected forms of DODEs encountered in the study of viscoelastic solids [67,68]. Note that all the aforementioned studies adopt the assumptions Hypothesis 1–3. Very recently, Fedorov studied linear DODEs that violate Hypothesis 2 resulting in an unbounded operator [69]. This study expanded the application of DODEs to initial and boundary value problems of ultra-slow diffusion.

2.3. Solution of DODEs: Analytical Methods

Concerning the analytical methods for the solution of DODEs, Caputo first proposed the use of Laplace transform to derive solutions [49]. Later, Bagley and Torvik [50,51] analyzed this approach in a systematic manner. The results obtained by the application of Laplace transform to DO derivatives are subject to minor modifications depending on the strength function and its support. Caputo derived the Laplace transform of DO derivatives with the order-distribution being an arbitrary interval $[a, b]$. Bagley and Torvik specialized this result for a restricted interval: $\alpha \in [0, 1]$, given the numerous practical examples encompassed by this choice. Diethelm and Ford extended the domain to $[0, m]$, $m \in \mathbb{N}$ [60]. The Laplace transform of a DO derivative with order distributed in $[0, m]$, based on the Caputo definition, is given as [56]

$$\mathcal{L}\left[\underbrace{\int_0^m \kappa(\alpha) {}_0^C D_t^\alpha u(t) \mathrm{d}\alpha}_{{}_{0,m}^C \mathcal{D}_{0,t}^\alpha u(t)}\right] = \int_0^m \kappa(\alpha)\left(s^\alpha \mathcal{L}[u](s) - u(0)s^{\alpha-1}\right)\mathrm{d}\alpha \tag{9}$$
$$- \sum_{j=1}^{m-1} \int_j^m \kappa(\alpha) u^{(j)}(0) s^{\alpha-j-1} \mathrm{d}\alpha$$

The Laplace transform of the DO derivative for other possible cases such as $\alpha \in [0, \infty]$ and $\alpha \in [m - 1, m]$ can be found in [45,70], respectively.

Using the Laplace transform of the DO derivative in Equation (9), Diethelm and Ford derived the analytical solution for the linear DODE: ${}_{0,m}^C \mathcal{D}_{0,t}^\alpha u(t) = f(t)$ as [60]

$$u(t) = u(0) + \mathcal{L}^{-1}\left[\frac{1}{\int_0^m \kappa(\beta)s^\beta \mathrm{d}\beta} F(s)\right] + \sum_{k=1}^{m-1} y^k(0)\mathcal{L}^{-1}\left[\frac{\int_0^m \kappa(\beta)s^{\beta-k-1}\mathrm{d}\beta}{\int_0^m \kappa(\beta)s^\beta \mathrm{d}\beta}\right] \tag{10}$$

where $\mathcal{L}^{-1}$ is the inverse Laplace transform. Note that the inverse Laplace transform in the above solution can be applied *iff* the assumptions Hypothesis 1–3, that ensure a bounded solution, are satisfied [60]. Lorenzo and Hartley derived analytical solutions for DODEs employing DO derivatives specifically for an order distributed over $\mathbb{R}^+$ [45]. Other common approaches to derive solutions of DODEs include the Fourier method [71–73], the use of Mittag–Leffler functions [74–76], the spectral representation of the fractional operator [77], and series expansion methods [78,79]. The method of Laplace transforms combined with series approximations using Laguerre polynomials was also used to solve linear and nonlinear DODEs [80]. While the work in [80] focuses on obtaining the solution for one- and two-term fractional-order relaxation equations, the method developed in [80] is highly general and may be extended to DODEs with general strength functions.

Although, in the above discussion we have primarily considered DO derivatives based on the Caputo definition, the Laplace transform of DO derivatives based on the Riemann–Liouville definition can also be derived analogously [60]. In fact, as shown in [60], the only difference appears in the terms consisting the initial conditions, similar to the CO case [4]. This difference in behavior was also highlighted by Mainardi et al. [81], who employed Laplace transforms to compare the asymptotic behaviors of fundamental solutions to time-fractional DO diffusion equations. Interestingly, different asymptotic behaviors are observed for DO derivatives based on the Riemann–Liouville and Caputo definitions. The difference in the asymptotic behaviors is primarily due to the difference in the way the initial conditions appear in the Laplace transform of the CO Riemann–Liouville and Caputo derivatives [4,82].

2.4. Solution of DODEs: Numerical Methods

Although analytical solutions are possible for special types of DODEs [45,60], the rapidly growing application of DOFC to model complex physical systems often requires the use of numerical methods. Starting from basic observations, Diethelm [83] first proposed an approximate numerical method for the solution of multi-term DODEs. Following this initial study, several other numerical methods have been developed. Note that DODEs (see, for example, Equation (5)) can be either ordinary differential equations (ODE) or partial differential equations (PDE), depending on the specific application. The numerical simulation of either a distributed-order ODE or PDE requires the numerical approximation of the DO derivative. Once the approximation of the DO derivative is obtained, the procedure to numerically simulate the DODE follows exactly from classical procedures developed for integer-order equations. In other terms, the main difference between the evaluation of classical integer-order differential equations and DODEs lies in the numerical approximation of the DO derivative. In the interest of brevity, we focus this section only on this latter aspect. In general, the procedure to numerically approximate DO derivatives can be seen as a two-step process:

1. *Step 1*: Numerical integration of the integral operator. The DO derivative consists of a continuous distribution of the fractional order α. In Step 1, a numerical integration is used to discretize the DO derivative into a multi-term CO fractional derivative.
2. *Step 2*: Approximate solution of the multi-term fractional derivative. Following the conversion of the DO derivative into a multi-term fractional derivative at step 1, different numerical methods are used to evaluate each CO fractional derivative within the multi-term derivative.

The above two steps can be more practically visualized by considering the following example of DO derivative,

$$\int_a^b \phi(\alpha) D^\alpha u(t) \mathrm{d}\alpha \overset{\text{Step 1}}{\approx} \underbrace{\sum_{i=0}^{k} W_i \phi(\alpha_i) D^{\alpha_i} u(t)}_{\text{Approximation of the integral}} \overset{\text{Step 2}}{\approx} \underbrace{\sum_{i=0}^{k} W_i \phi(\alpha_i) \Psi(\alpha_i, t)}_{\text{Incorporate approximation of } D^{\alpha_i} u(t)} \tag{11}$$

where W_i is the weight obtained from numerical integration and $\Psi(\alpha_i, t)$ is the numerical approximation of the CO derivative $D^{\alpha_i} u(t)$. In summary, at step 1, an approximation of the order integral is computed (often by quadrature rules), and at step 2, the remaining CO derivatives are approximated by employing different types of numerical methods for CO fractional derivatives. Based on this two-step approximation strategy, this section is divided into three parts: (1) a discussion of the most popular quadrature rules for the implementation of step 1, (2) a discussion of the various numerical methods for the implementation of step 2, and (3) a brief discussion on their computational aspects.

2.4.1. Numerical Integration of the Integral Operator (Step 1)

As highlighted in the previous sections, a key difference between DO derivatives and CO derivatives is the existence of an additional integration over the order. To transform the integral form into the multi-term form (first of the two-step process), two common quadrature rules are often used by researchers: (1) Gauss–Legendre quadrature rule and (2) Newton–Cotes quadrature rule. Based on the Gauss–Legendre quadrature rules [84–107], the DO derivative can be approximated using the following multi-term form,

$$\int_a^b \phi(\alpha) D^\alpha u(t) \mathrm{d}\alpha = \int_a^b g(\alpha, t) \mathrm{d}\alpha = \sum_{i=0}^{k} W_i^G g^G(\alpha_i^G, t) + R^G \tag{12}$$

where W_i^G are the weights at the Gauss points α_i^G chosen for this integration over the DO. Although the Gauss–Legendre quadrature schemes are known to achieve highly accurate results (particularly when dealing with integrands of specific type such as, for example, polynomials), an analysis of the numerical convergence and of the truncation error (including steps 1 and 2) becomes difficult when the integrand consists of fractional derivatives (like $D^\alpha u(t)$, as shown in Equation (11)). To overcome these drawbacks of the Gauss–Legendre quadrature, the Newton–Cotes scheme was considered. The Newton–Cotes quadrature scheme can be divided into closed and open approaches, depending on whether the function values at the end points are included. Following the closed approach, different quadrature rules used for DO derivatives include the trapezoid rule [56,87,106,108–117], the Simpson's rule [87,106,111,112,116–121], and the Boole's rule [122]. All these schemes are also associated with different orders of convergence. Following the open Newton–Cotes approach, the mid-point rule is widely used [107,123–143]. The truncation error at the end of step 1, when employing the Newton–Cotes approach, simply follows the classical results. More specifically, the truncation errors are $\mathcal{O}(h^2)$ for trapezoid rule and mid-point rule, $\mathcal{O}(h^4)$ for Simpson's rule, and $\mathcal{O}(h^6)$ for Boole's rule. Given the flexibility in choosing different approximations and the ease of error analysis, Newton–Cotes method is typically preferred over Gauss–Quadrature approach in step 1 approximation.

2.4.2. Approximation of the Multi-term Fractional Derivatives (Step 2)

As described in Equation (11), the second step involves the numerical approximation of the CO fractional derivatives within the multi-term fractional derivative. Strictly speaking, this approximation directly follows the techniques available for CO derivatives. The literature on numerical methods for the approximation of CO derivatives is extensive and has been the object of books [144] and papers [145–147]. Therefore, for the sake of brevity, we do not review again these methodologies.

The more interesting and challenging aspect, in the context of the DO formulation, is the combination of the step 2 approximation with the spatial and/or temporal discretization of the domain in order to develop computational models for space- and/or time-fractional DODEs. The different discretization techniques can be generally divided into (1) mesh-free approaches and (2) mesh-based approaches. The majority of mesh-free approaches are based on the spectral method, which uses basis functions to approximate the multi-term DO expression obtained in the first step. On the other hand, the mesh-based approaches involve most of the classical methods for differential equations including the finite difference method (FDM) and the finite element method (FEM). Depending on the specific implementation, that is, on the numerical technique adopted to approximate the CO fractional derivative in step 2 and the spatial and/or temporal discretization of the domain, the computational approaches differ in their accuracy and computational cost. This review focuses on this latter aspect. In this regard, we report here the accuracy of each method, wherever available. In order to unify the expressions for convergence analysis of different methods, we will use τ, h, and σ to represent the step-sizes in time, space, and order, respectively.

Mesh-Free Approaches

In this section, we briefly describe the different mesh-free approaches available in the literature to numerically simulate DODEs. The majority of these techniques adopt the common strategy of converting the DODE into a system of algebraic equations using orthogonal basis functions. This allows formulating operational matrices which approximate the CO derivatives within the step 2 approximation. Depending on the strategy adopted to develop these matrices (or, equivalently, these algebraic equations) the different mesh-free approaches can be broadly categorized as Galerkin methods, collocation methods, and tau methods. A brief discussion on these methods and some other miscellaneous techniques is provided in the following.

1. *Galerkin spectral methods* can be divided broadly into two categories depending on the specific nature of the basis functions: (1) Galerkin spectral methods based on Legendre polynomials (GLSM) and (2) Galerkin spectral methods based on Jacobi polynomials (GJSM). GLSMs were proposed very recently in [92,118,125,143,148] to solve time-fractional DODEs. These were accurate to $\mathcal{O}(\tau^{2-\beta})$ (where, $\beta \in (0,1)$). A few researchers combined the GLSM scheme with an alternating direction implicit (ADI) scheme to improve the accuracy to $\mathcal{O}(\tau^2 + \sigma^2)$ [98,139]. Numerical studies based on the GJSM approach can be found in [85,91,149]. Some interesting conclusions were presented in [150], which combined a *s*-stage implicit Runge–Kutta method in time and the GJSM/GLSM in space to solve time-space-fractional DODEs. They established that a convergence of $\mathcal{O}(s+1)$ in time could be obtained when employing an algebraically stable Runge–Kutta method with order p ($p \leq s+1$). A few researchers have compared the performance of the GLSM and GJSM techniques in [90,150,151]. The results of these studies indicate that the specific basis functions do not drastically alter the computational performance.

2. *Collocation methods* require that the approximate solution satisfies the DODE at specific locations known as the collocation points. Similar to the Galerkin spectral method, various collocation methods have been developed starting from (1) Legendre basis (LCM) [100,134] and (2) Jacobi basis (JCM) [105,152]. Zaky constructed a LCM to solve both linear and nonlinear boundary value problems [100], and later extended this method to simulate initial value DODEs [99,153]. Results indicated that the convergence error decays exponentially with an increasing number of Gauss–Legendre points. Very recently, the LCM was extended by Xu [96] to develop a higher-order Legendre–Gauss collocation method for nonlinear DODEs. JCMs were developed in [101,102,152] to solve DODEs concerning different physical applications (such as, for example, transport processes and control). A majority of the above studies achieved either first or second-order accuracy. Recently, Abdelkawy [105] proposed a fourth-order accurate scheme for time-fractional DODEs (admitting only smooth solutions) while also achieving an exponential convergence rate. Besides the popular LCM and JCM, collocation methods based on other basis functions including, for example, the Chebyshev polynomials [129,154], fractional Lagrange polynomials [92], and the wavelet method [119], were also developed. Some interesting numerical techniques were developed by combining selected aspects of the different basis functions such as, for example, the fractional-order Chelyshkov wavelets [104]. Similar to the Galerkin spectral methods, it appears that the different basis polynomials in the collocation methods, do not drastically alter computational accuracy.

3. *Tau methods* also employ different basis functions similar to the Galerkin spectral method and collocation method. Tau methods for DODEs were first developed in [155,156] using shifted Chebyshev polynomials. Building on these studies, shifted Jacobi polynomials were adopted as basis functions in [157], and shifted Legendre polynomials were adopted in [103,158]. A detailed analysis of the results from these studies suggests that the accuracy and computational cost of simulating a given DODE using the tau methods are similar to the collocation and Galerkin spectral methods.

4. *Other mesh-free methods* based on the formulation of fractional-order operational matrices have also been explored to solve DODEs. The operational matrix is based on different functions such as the block-pulse function (BPF) [89], Chebyshev polynomials [159,160], and shifted Legendre polynomials [154]. Following the same strategy, hybrid approximation methods based on the combination of different basis functions have also been developed. The specific combinations that have been explored in literature are BPFs and Bernoulli polynomials [95], BPFs and Taylor polynomials [93],

and BPFs and shifted Legendre polynomials [161]. For completeness, we mention that other numerical methods including the Laguerre spectral method [108], Legendre wavelets method [84], fractional pseudo-spectral method [162], reproducing kernel method [163], radial basis function based mesh-free methods [86,114], and element-free Galerkin method [106] have also been proposed. Further, several semi-analytical approaches including the Homotopy perturbation method [164–167], harmonic approximations [168], and the Adomian decomposition method [169–171] have also been proposed and applied to derive the solution of DODEs and multi-term fractional differential equations (FDE).

Mesh-Based Approaches

Although many mesh-free approaches can be implemented relatively easily for DO problems involving simple geometries and boundary conditions, algorithms for numerical computations on complex domains (e.g., involving irregular geometry and high-dimensional systems) still present several complexities. This also reflects from the fact that many 2D and 3D problems have been solved using mesh-based approaches, while a majority of mesh-free approaches focus primarily on 1D problems. FEM is particularly useful in exploring numerical solutions over irregular domains. Among the mesh-based approaches for DODEs, two methods have generated the most interest: finite difference methods (FDM) and finite element methods (FEM). Before proceeding to review these mesh-based approaches, it is important to note a specific challenge faced by this class of methods. More specifically, due to weak singularity of the integral kernel within the fractional derivative, numerical solutions for initial boundary-value FDEs normally have non-smooth sharp approximations near the boundary [172–174]. As the DO derivative is approximated via a weighted sum of CO derivatives (see Equation (11)), this phenomenon also occurs when solving initial boundary-value DODEs [143]. To tackle this weak singularity, the commonly used mesh-based methods need to be improved. One possible approach, commonly adopted in literature, consists in the use of a graded mesh [87,143]. Remarkably, the use of the graded mesh also helps achieving a high-order convergence [87,143].

1. Finite difference methods are one of the most widely used mesh-based approaches for the solution of DODEs because they allow easy formulation and implementation. Compared with other approaches, the convergence and accuracy of FDM are easier to analyze [175–177]. A majority of the advanced FDMs are based on the Grünwald–Letnikov method (GLM) [122,142]. Recall that GLM uses a finite number of terms from a convergent series to approximate the fractional derivative and is a widely used approach [4]. Hu [126] used a shifted GLM to simulate a time-fractional DODE with accuracy up to $\mathcal{O}(\tau^{1+\sigma/2} + h + \sigma^2)$. Second-order accurate schemes for space-fractional DODEs were developed in [136] by using a Crank–Nicolson scheme in time and a shifted GLM. Similar second-order accurate algorithms can also be found in [133,178]. The second-order accurate backward difference formula, first proposed by Diethelm [145], also appears to be popular among several researchers [124,129,138]. To further improve the numerical accuracy, more elaborate methods were developed using the weighted and shifted GLM (WSGLM). Li [179] developed a numerical scheme with high spatial accuracy ($\mathcal{O}(\tau^2 + h^{4.5} + \sigma^2)$) by combining WSGLM and the parametric quintic spline method. Another scheme capable of delivering high spatial accuracy ($\mathcal{O}(\tau^2 + h^4 + \sigma^4)$) was proposed by using the WSGLM for temporal approximation and high-order compact difference scheme for spatial approximation [117]. Yang [180] also proposed a similar composite method based on WSGLM in time and orthogonal spline collocation method in space. This scheme was shown to be unconditionally stable and accurate up to $\mathcal{O}(\tau^2 + h^{r+1} + \sigma^2)$ (here r is the polynomial degree used in the spatial domain).

FDM schemes have also been developed for high-dimensional problems, with particular attention being given to accuracy and convergence performance [141,181]. For applications requiring high accuracy, two techniques are often used: (1) compact FDM (CFDM) and (2) extrapolation method. Based on a fully discrete difference scheme [182], Ye [132] proposed a CFDM and demonstrated its convergence to be $\mathcal{O}(\tau^{1+\sigma/2} + h^4 + \sigma^2)$. Pimenov [121] constructed a linearized difference scheme for nonlinear time delay DODE. Several researchers [110,120,183] also obtained a CFDM with order $\mathcal{O}(\tau^2 + h^4 + \sigma^4)$ based on higher order temporal approximation techniques. Gao [111,116] applied two extrapolation methods in time to achieve high temporal convergence: $\mathcal{O}(\tau^2)$ and $\mathcal{O}(\tau^2|\ln\tau|^2)$. For high-dimensional problems, ADI schemes become highly popular and help achieve highly accurate (second-order in time and fourth-order in space) numerical schemes [107,184].

2. Finite element methods: Starting from the study of multi-term FDEs, Jin [185] developed a Galerkin approach, Bu [186] used a multi-grid FEM, and Zhao [187] used a spatially nonconforming FEM to solve time fractional diffusion equations. Similarly, several researchers first developed FEMs to solve multi-term FDEs and later extended them to solve DODEs [87,123,188]. Few researchers [112,189] developed the H^1-Galerkin FEM for DO sub-diffusion equations which allowed the estimation of the diffusive field variable as well as its spatial derivative. By using locally discontinuous Galerkin FEM, Aboelenen [137] and Wei [190] developed highly accurate numerical schemes with spatial convergence $\mathcal{O}(h^{k+1})$ (k is the degree of basis polynomials). Given the FEM's unique ability of handling complex geometry, several recent studies have focused on its application to irregular domains. Examples include the development of FEMs, based on unstructured meshes, to solve DO equations corresponding to different physical applications [109,191–193].

3. Other mesh-based methods: In addition to FEM and FDM, a few other mesh-based methods were also explored. Examples include the combined B-spline interpolation and the Du Fort–Frankel method [130] for time-fractional DODEs. Heris [135] and Javidi [136] introduced a fractional backward differential formulas for space DODEs and obtained a second-order accurate numerical scheme. Diethelm et al. [60,188,194] introduced a convolution quadrature method for the numerical approximation of DO operators. Based on a backward difference formula, Podlubny [195,196] proposed a matrix form to represent discrete analogs of fractional operations and extended this method to the solution of DODEs [197]. Other mesh-based techniques developed in literature to solve DODEs and multi-terms FDEs include the predictor-corrector method [56,198–201] and the finite volume method [127,128,202].

Computational Aspects of DODEs

The previously discussed numerical schemes for the approximation of fractional derivatives typically generate dense matrices; a clear consequence of the intrinsic nonlocal character of the operator. For discretizations with N number of elements (temporal or spatial), these dense matrices generally require $\mathcal{O}(N^3)$ floating point operations and $\mathcal{O}(N^2)$ memory, for each iteration. In order to reduce this high computational cost, several alternate approaches were considered. Based on the idea of relabeling employed in ADI methods, Jia [203] developed a fast FDM which stores a coefficient matrix in $\mathcal{O}(N)$ memory and performs matrix-vector multiplication in $\mathcal{O}(N\log N)$ computations. Two numerical algorithms offering comparable time and space complexity were developed by Jian [142] and Zheng [202]. By expressing the matrix of coefficients as a sum of special diagonal-Toeplitz matrices, Jian derived a fast solution technique based on the preconditioned Krylov subspace method. Zheng proposed an efficient biconjugate gradient stabilized method to solve system of equations with a Toeplitz structured coefficient matrix. More recently, a reduced-order ADI method [184] was developed to reduce the computational cost involved in the numerical solution of DODEs.

Before proceeding further, it is worth noting that the computational time for the numerical simulation of DODEs can also be reduced via parallel computation and pre-conditioning of the operational matrices used to approximate the fractional derivatives. While parallel computation has not been directly applied to DODEs, parallel solvers have been developed for CO FDEs [204–206]. Besides the parallel algorithm itself, the effect of different hardware platforms (GPU v/s CPU) [207] and different memory architectures (shared memory v/s distributed memory) [206] on the computational times for simulation of CO FDEs, have also been studied. Further, preconditioners are often designed to accelerate matrix computations in nonlinear CO FDEs involving iterative problem solving procedures. Many studies have proposed different types of preconditioners such as, for example, preconditioned biconjugate gradient method [208] and generalized minimal residual method [209], for solving nonlinear CO FDEs. Both the above described techniques, that are parallel computing and preconditioning, present possible opportunities to reduce the computational time for solving DODEs and are hence worthy of detailed investigation in the future.

3. Relevance of Distributed-Order Operators

As evident from the definitions presented in Section 2, DO operators can be interpreted as a parallel distribution of derivatives of either integer or fractional orders. It follows that one of the most immediate application of these operators is to model physical systems whose response is characterized by a superposition of different processes operating in parallel and individually described by either fractional- or integer-order operators. As an example, consider electro-rheological fluids that can change their properties following the application of an electric field. This means that, in these media, the order of a small fluid element is dependent on the local field strength. Therefore, if the applied electric field is nonuniform, a corresponding order distribution will exist throughout the material [45]. A similar example consists of modeling the response of an electrical circuit with a distributed network of capacitors exhibiting the well-known fractional-order Curie's law. According to this law, current through a capacitor varies with time t as $i(t) = V_0/Ct^\alpha$, where V_0 is a constant voltage and $\alpha \in (0,1)$ [210]. These simple examples suggest that there exists a class of physical problems that can be better described by DO operators.

Broadly speaking, the above-described class of physical problems is characterized by the presence of multifractal or equivalently multifractional systems [211]. The response of such systems is marked by the presence of multiple temporal and spatial scales, which can be accurately captured via time-fractional and space-fractional DO operators, respectively. The advantage of the DO operator in capturing the hierarchy of scales as well as anomalous scaling effects has been analyzed in detail in [44]. The occurrence of this hierarchy of scales could be better visualized by considering, for example, the modeling of turbulence via the Lévy walk approach. This approach associates a time scale with jump distances, and the multiplicity of scales is explicitly taken into account via an integral equation which contains a coupled memory kernel similar to the DO operator [212]. Other examples of such multifractional processes include the analysis of structures with simultaneous nonlocal and strain-gradient (multiscale) effects [213], diffusion of particles in microporous materials [214], analysis of financial markets where distributions of financial data usually possess fast falling power-law tails [215], and even state functions of complex quantum-mechanical systems [216,217].

From a different perspective, DO operators can also be used to retrofit models to experimental data derived from systems with an unknown fractional behavior. The fractionalization of differential equations commonly used in mathematical physics leads to the analysis of the order-parameter, say α, to be determined via experimental results. As experiments can lead to several values of the fractional order, as a result of different experimental conditions, it is convenient to introduce a DO fractional derivative. This is equivalent to integrating the product of a fractional derivative ($D_\square^\alpha(\cdot)$) of the primary response variable (say u) and a weight function (or distribution) with respect to the order of the derivative, that is, to evaluate $\int_{\text{supp }\phi} \phi(\alpha) D_\square^\alpha u \, d\alpha$. In this way, one may use several experimental results and determine a continuous function ϕ rather than focusing on a single variable that is the fractional-order α. This can be interpreted as a homogenization of the different possible fractional processes and the resulting epistemic uncertainties. In other terms, such an approach would enable a valid and accurate analysis of experimental data and allow the development of fractional-order models, without having to identify the specific underlying fractional behavior.

The above remarkable properties of DO operators have led to the development of fractional models capable of describing numerous complex physical processes. Most of the work to date has concentrated on the general areas of viscoelasticity, transport processes, and control theory. We make a few concluding remarks, before proceeding to review the most significant applications of DOFC reported to date in the different areas. Note that the application of DOFC to viscoelasticity and control theory primarily involves the use of time-fractional DO derivatives, while the application to transport processes involve both space- and time-fractional DO derivatives. This separation follows from the underlying physics being captured. In this regard, recall that, while time-fractional DO derivatives are typically used to account for memory effects and dissipation across multiple temporal scales, space-fractional derivatives are used to model nonlocal effects and spatial heterogeneity over multiple spatial scales. In the applications presented below, we do not specify if the DO model is based on a Riemann–Liouville or Caputo (or any other) definition, as it only marginally affects the overall discussion. Finally, we use the following notation in all the subsequent sections: t and x refer to the independent variables in time and space, respectively.

4. Applications to Viscoelasticity

Fractional-order derivatives are well suited to capture the dissipation in viscoelastic solids. The differ-integral definition of the fractional derivatives allows the effects of deformation history to be realized within the stress–strain constitutive models, thus combining the elastic response across different time scales. In this regard, Gemant [218,219], Caputo [46], Bagley and Torvik [5,6], and Chatterjee [7] provided seminal contributions towards the use of fractional-order models to simulate the effect of dissipation in viscoelastic solids. While an approach based on CO time-fractional derivatives is intuitive and has drawn much interest, it is not well suited for applications involving materials characterized by multiple relaxation times. In order to address this gap in modeling viscoelastic systems via the CO derivatives, DO models were proposed [48,49,220]. As mentioned in Section 3, the DO operators allow the multiple relaxation scales to be visualized as separate viscoelastic connections operating simultaneously. Thus, a superposition of multiple CO derivatives (or equivalently, multiple relaxation scales) is achieved via the definition of the DO derivative for viscoelastic solids.

4.1. Constitutive Models

As mentioned in Section 2.1, the DO derivatives were originally conceptualized to model the dissipative elastic response with several temporal relaxation scales [48]. Following this seminal work, several other models of viscoelasticity either based on DO derivatives now exist in literature. These models can be viewed as simplified versions of the following generalized DO stress–strain constitutive law, proposed by Atanacković, for viscoelastic solids [221,222]:

$$\int_0^1 \phi_\sigma(\gamma) {}_0D_t^\gamma \sigma(t)\mathrm{d}\gamma = E\int_0^1 \phi_\epsilon(\gamma) {}_0D_t^\gamma \epsilon(t)\mathrm{d}\gamma \tag{13}$$

where ϕ_σ and ϕ_ϵ represent the strength functions corresponding to stress and strain (these are constitutive functions that characterize the viscoelastic response), E is the Young's modulus, and ${}_0D_t^\gamma(\cdot)$ is the CO time-fractional derivative. The formulation in Equation (13) is referred to as the most general model because all other models, already existing in literature, can be derived from this model via suitable assumptions on the additional (fractional-order) constitutive parameters. For instance, the choice $\phi_\sigma = \delta(\gamma)$ and $\phi_\beta = \delta(\gamma - 1)$ for the for strength functions results in the standard dashpot. Additional abstractions of the DO constitutive model in Equation (13), describing different viscoelastic elements, are illustrated in the Figure 4. Further, as discussed in Equation (6), a discrete choice for the order-distribution weights in Equation (13) would result in a multi-term fractional-order expression for the DO definition given above. Employing discrete strength functions in the above equation, the stress and its temporal derivatives (of real order, not necessarily integer) can be recast in terms of strain and its (real-order) temporal derivatives as follows [223],

$$\sum_{n=0}^N a_n \left[{}_0D_t^{\alpha_n}\sigma\right] = \sum_{m=0}^M b_m \left[{}_0D_t^{\beta_m}\epsilon\right], \quad t > 0 \tag{14}$$

where the fractional-orders are assumed to satisfy: $0 \le \alpha_0 < \alpha_1... < \alpha_N < 1, 0 \le \beta_0 < \beta_1... < \beta_M < 1$. The constants $a_\square$ and $b_\square$ can be interpreted to be relaxation times for the viscoelastic solid. As demonstrated in [223], the above-presented multi-term model is effective in modeling both stress relaxation and creep response in viscoelastic structures. The integral constitutive relation given in Equation (13) can be interpreted as the continuum limit of the discrete multi-term constitutive relation given in Equation (14). This is also illustrated in Figure 4b, which depicts the DO integral model as the continuum limit of the discrete model in Figure 4a.

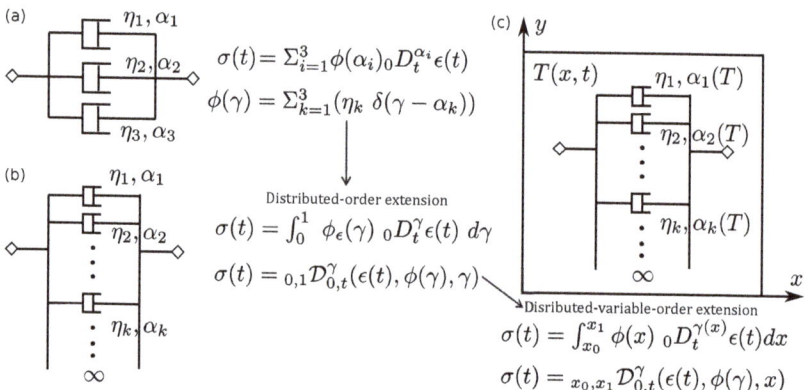

Figure 4. Examples illustrating the different DO models of viscoelasticity along with their respective constitutive relations. It appears that DO operators can model multiple viscoelastic elements within the same general formulation. Dashpots characterized by material constants η and order α indicate the individual viscoelastic elements. Schematic illustration of (**a**) the multi-term DO viscoelastic model, (**b**) the generalized DO model depicted as an infinite ensemble of elements with $\alpha_i \in (0,1]$ such that Span $\{\alpha_i\}$ is $(0,1]$, and (**c**) the generalized temperature field-dependent VO definition for the DO viscoelastic model.

4.1.1. DO Integral Models

All existing models catering to different lossy materials can be recast into the DO form in Equation (13) (or equivalently, Equation (14)) by considering different choices for order-distribution functions. In other words, each of the several distinct classifications of the viscoelastic solids proposed by Caputo and Mainardi [224] based on the creep and relaxation moduli relations, can be described by the single DO constitutive law via suitable choices of the fractional-order constitutive parameters. This highlights the relevance of DO operators and their scope in modeling viscoelastic constitutive relations when compared with other more classical integer—and fractional—(CO or VO) models available in the literature. To better illustrate this, consider the following two cases: case I: $\phi_\sigma = \delta(\gamma)$, $\phi_\epsilon = \tau_0^\alpha$, and case II: $\phi_\sigma = \tau_\sigma^\alpha$, $\phi_\epsilon = \tau_\epsilon^\alpha$, $\tau_\sigma < \tau_\epsilon$, $\tau_\square$ being a material constant. These two choices for the integral forms of the DO constitutive relation are commonly used in modeling viscoelastic solids [43,225–227]. Depending on the choice of the strength functions, Equation (13) can successfully characterize both fluid-like and solid-like viscoelastic materials. Remarkably, salient mechanical characteristics of the viscoelastic materials modeled by these choices, such as the creep and stress relaxation functions, exhibit the experimentally observed power-law attenuation [228].

4.1.2. Multi-Term Fractional Models

Compared to integral models, the discrete multi-term approach has been more widely used for the modeling of viscoelastic constitutive relations. This is a direct consequence of the simplicity with which discrete models could be modified in order to account for different lossy behaviors observed in real materials. The discrete form also facilitates a direct comparison between the viscoelastic behavior captured by DO models with respect to the more traditional and established integer-order models. This enables a better understanding of the physical relevance of DO models and it also allows a more natural approach to material characterization. The following instances of the different viscoelastic models that can be recovered from the multi-term DO law in Equation (14) further illustrate the strength of the DO approach:

1. *Kelvin-Voigt models:* The DO analogue of the Kelvin–Voigt model is obtained for the choice of $\phi_\sigma = \delta(\gamma)$, and $\phi_\epsilon = \tau^\gamma$ [229].
2. *Maxwell models:* The fractional-order Maxwell model of viscoelasticity can be obtained for $\phi_\sigma = \delta(\gamma) + \tau^\alpha \delta(\gamma - \alpha)$ and $\phi_\epsilon = E_\infty \tau^\beta \delta(\gamma - \beta)$ in Equation (13) [230]. Note that, assuming $\alpha = \beta$ in the fractional Maxwell model, allows recovering the fractional Zener model [231].
3. *Zener models:* If the material constants in Equation (14) are chosen as $a_0 = b_0 = 1$, $a_1 = a$, $b_1 = b$, and orders $\alpha_0 = \beta_0 = 0$, $\alpha_1 = \beta_1 = 1$ the classical Zener model is obtained. Similarly, $\alpha_1 = \beta_1 = \alpha$ gives the generalized Zener model [232]. Wave propagation in fractional Zener-type viscoelastic media, obtained by choosing $\phi_\sigma = \phi_\epsilon = \delta(\gamma) + \tau^\alpha \delta(\gamma - \alpha)$ in Equation (13), was studied in [233,234]. Similarly, the choice of $\phi_\sigma = \delta(\gamma) + (a/b)\delta(\gamma - (\alpha - \beta))$ and $\phi_\epsilon = a\delta(\gamma - \alpha) + c\delta(\gamma - \eta) + (ac/b)\delta(\gamma - \eta + \beta)$ in Equation (13), also results in a fractional version of the classical Zener model with springs and dashpots [223].
4. *Other models:* Viscoelastic models described for the strength functions $\phi_\sigma = \delta(\gamma) + \tau^\alpha \delta(\gamma - \alpha)$ and $\phi_\epsilon = E_0(\delta(\gamma) + \tau^\alpha \delta(\gamma - \alpha) + \tau^\beta \delta(\gamma - \beta))$ in Equation (13), were analyzed in [235]. Variations of this latter model (also referred to as the four-parameter model [236]) including the use of five-parameters [237] were studied to simulate selected types of lossy behavior in real materials. Further extensions that explored the use of additional terms were also presented [79].

In the above discussion, $\{a, b, c\}$ denote different material constants corresponding to different relaxation times and $\{\alpha, \beta, \eta\}$ are the fractional-orders associated with different lossy behaviors of the DO model (see Equation (14)). In conclusion, we note that the multi-term fractional model is highly general and offers much flexibility in modeling different types of lossy behavior in viscoelastic solids. This is unlike CO or VO approaches that require separate models to capture these different behaviors.

4.2. Material Characterization: Methods and Experiments

It is clear from the discussion in Section 4.1 that several possibilities for the viscoelastic constitutive theories exist, considering suitable choices for the DO model parameters. Before proceeding to review the application of these DO theories to the characterization of viscoelastic materials, we make an important remark. Note that the application of these DO theories to real-world viscoelastic problems requires that these models are physically as well as mathematically consistent. To ensure consistency of the DO viscoelastic theories, there exist restrictions on the choice of the fractional model parameters which are derived in accordance with the principles of (1) time invariance, (2) causality, and (3) thermodynamics (dissipation inequality given by the Clausius–Duhem inequality) [49]. The conditions over the strength distribution functions ϕ_σ and ϕ_ϵ, corresponding to the integral definition of the DO law given in Equation (13), are available in [222]. For instance, the thermodynamic law restricts the choice of DO constitutive parameters for the fluid-like viscoelastic materials, discussed in Section 4.1.1, as follows, $\tau_0 > 0$. An analogous study conducted on the discrete form of the DO constitutive law (see Equation (14)) identified the restrictions on relevant constitutive parameters [223]. The investigations conducted in the aforementioned studies were further extended in [53] which analyzed the physical as well as mathematical consistency of the generalized DO model of viscoelasticity. In this regard, note that mathematical consistency ensures the existence and uniqueness of a linear viscoelastic response corresponding to the generalized DO formulation. The framework developed in [53] provides the foundation for a rigorous and consistent application of DOFC to modeling the response of viscoelastic solids.

The discussion in Section 4.1 highlighted the ability of DO operators to capture multiple scales of relaxation time and thereby different lossy behaviors observed in real materials [220]. For this purpose, the constitutive parameters of the DO constitutive model in Equation (13) that require to be identified are the fractional-order parameters and their numerical range. Initial investigations [82,220] laid a theoretical foundation for this fractional-order system identification problem. Further experiments on the characterization of viscoelastic properties corresponding to the different class of DO models for commercial polymers are reported in [238]. Such studies were carried out by matching the experimental profiles of the loss and storage moduli for viscoelastic materials [53]. Recall from Section 4.1.2 the relevance of DO operators in modeling multiple forms of viscoelastic behavior. This feature of the DO constitutive models for viscoelastic elements presents an interesting opportunity. To better illustrate this aspect, consider the multi-term DO models depicted in Figure 4a as the sum of several independent viscoelastic connectors with their associated relaxation timescales. This type of arrangement allows incorporating multiple timescales within a single DO model in order to design an optimized fractional damper. The incorporation of multiple timescales (using the DO derivative) can also be visualized from the DO derivative of the Heaviside step function in Figure 3. The relaxation time of the viscoelastic damper can be tuned by an appropriate choice of the constituent CO derivatives and their associated weights within the definition of DO derivative. This approach presents an opportunity to identify the damper that can deliver a desired behavior in terms of overshoot, peak time, and integrated tracking error [239]. This feature is unlike the classical integer-order or CO constitutive theories that allow only a single type of lossy behavior to be captured with a given model.

4.3. Distributed-Variable-Order Models

The above discussion presented an overview of the applications that DO models, based on CO derivatives, enable in the general area of viscoelastic solids. A few studies have also explored the extension of these models to employ DO operators based on VO derivatives; here below referred to as distributed-variable-order (DVO) operators. Lorenzo and Hartley presented one of the first works exploring the combination of both VO and DO operators to the formulation of the stress–strain constitutive law of viscoelastic solids [45]. They discussed how a DVO operator defined using a spatially-dependent VO law could be used to model the response of a thermorheologically complex material subject to a spatially and temporally varying temperature field. By choosing a spatially-dependent VO law, the resulting DVO model is capable of describing the spatial variation of the viscoelastic properties. The spatial variation of viscoelastic properties can be the result of a combination of internal as well as external conditions such as, for example, varying microstructure, presence of thermal loads, and a distribution of thermal gradients. We merely note that, very recently, this concept of defining a spatially-dependent VO law was used to model nonlocal solids with spatially varying microstructure in [240]. Further, an example of the temperature-dependent DVO viscoelastic model is illustrated in Figure 4c. In this case, the DVO model is required to introduce the effect of a spatially varying temperature field $T(x, t)$ on the multiple timescales present within the DO model for viscoelasticity. This allows an accurate representation of the transient viscoelastic response [220]. It is important to mention that, unlike the DO models employing CO derivatives, the thermodynamic basis for the DVO models still remains to be ascertained.

4.4. Some Practical Applications

The DO constitutive models have been successfully applied in the analysis of viscoelastic solids. Recall that the different DO constitutive models can be classified primarily into two classes: (1) integral-models and (2) multi-term models, corresponding to the choice of DO derivative. Further, within each of these classes, further subdivisions exist depending on the specific functions chosen for (a) weights of the order-distribution functions and (b) bounds of the fractional-order α. Here, we shall present some prominent examples studied in literature that cater to a specific class of viscoelastic solids. These studies include finite solids with appropriate boundary conditions, and also the infinite solids.

Some examples of the constitutive parameters within DO integral models in Equation (13) were discussed previously in Section 4.1.1. Employing specific choices of the constitutive parameters, successful modeling of the creep response [225] and stress-relaxation [226] in finite solids is possible. Further, these integral models find relevance in modeling the vibration of fractional DO oscillators [227]. Patnaik and Semperlotti [168] demonstrated a successful application of DO viscoelastic models in the analysis of non-linear oscillators with distributed nonlinear properties. In this study, the effect of the order-distribution on the phase and frequency response was captured analytically using asymptotic techniques and some important characteristics, such as simultaneous phase and amplitude modulation (that is not seen in integer-order models) were presented. Recently, the scope of DO constitutive models is also being explored to describe viscoelasticity within complex materials like composites [43].

These studies can also be extended to modeling and analyzing the damping of the structural response. DO models can be utilized to derive moment–curvature relations of viscoelastic rods [241–243]. The DO constitutive relation between moment ($\overline{M}$) and curvature ($\overline{\kappa}$) for the viscoelastic rod is given by

$$\int_0^1 \phi_{\overline{M}}(\gamma)_0 D_t^\gamma \overline{M} \mathrm{d}\gamma = \int_0^1 \phi_{\overline{\kappa}}(\gamma)_0 D_t^\gamma \overline{\kappa} \mathrm{d}\gamma \tag{15}$$

In this equation, the choice of $\phi_{\overline{M}} = \delta(\gamma)$ and $\phi_{\overline{\kappa}} = EI\delta(\gamma)$ (EI is the bending modulus) reduces the above expression to the classical Euler–Bernoulli beam theory. The solution to the above DODE would reflect the influence of viscoelastic damping over the bending response of beams. Similar exercises can be conducted over more complex shapes with the help of advanced numerical techniques discussed in Section 2.4.

Employing the multi-term definition of the DO constitutive relations, the DO moment–curvature relations can be revisited for different classes of viscoelastic solids. For instance, DO bending relations analogous to the generalized Zener model were derived to study the dynamics of a viscoelastic rod in [243,244]. Similarly, the lateral vibration of a viscoelastic rod modeled according to the generalized Kelvin-Voigt behavior was studied in [229]. The choice of $\phi_M = \delta(\gamma) + a\delta(\gamma - \alpha)$ and $\phi_\kappa = EI(\delta(\gamma) + b\delta(\gamma - \alpha) + c\delta(\gamma - \beta))$, which is a generalization of the standard Zener model, was proposed in [235] and used in [245] to study the lateral vibrations of viscoelastic rod. DO models were also used to analyze the influence of viscoelastic foundations on the dynamic stability of local and nonlocal rods [246]. Similarly, Varghaei et al. [247] investigated the nonlinear vibration of viscoelastic beams using a generalized Kelvin–Voigt model implemented via DO derivatives. Finally, Duan and Chen [248] investigated oscillatory shear flow between two parallel plates using DO form of the constitutive law for for viscoelastic fluids. Different effects of viscoelasticity over the structural response can be realized thanks to the generality of the DO models of viscoelasticity by employing specified choices for constitutive parameters. For instance, different viscoelastic constitutive models were employed in a study over the damping influence on the propagation of an initial Dirac delta disturbance through an infinite media. This provides the necessary foundation for designing an optimized damper as in [239].

5. Applications to Transport Processes

Several experimental investigations have shown that transport processes in many classes of materials are often characterized by anomalous mechanisms exhibiting either memory effects over various temporal scales or nonlocal effects over several spatial scales [249–251]. A direct consequence of this, as an instance, is a loss of the scaling invariance (CO or VO) noted in classical transport processes. Consequently, such processes cannot be modeled by using CO (integer or fractional) or even VO differential equations, as CO and VO diffusion equations lead to self-similar probability densities with a characteristic displacement exhibiting spatio-temporal scaling. The loss of the spatio-temporal scaling is a direct result of the presence of a spectrum of temporal or spatial scales in the transport process. The presence of several temporal scales, as an example, can be the result of the presence of a mixture of delay sources of variable strength [252] while the presence of distributed spatial scales can occur in transport through multifractal materials [211,215,253] (see Figure 5). Real-world examples of such complex transport processes include applications in geophysical and atmospheric phenomena [254–257], financial markets [258], turbulence [259], and even biology and medicine [211]. As discussed in Section 3, DODEs are very well suited to model such non-scaling anomalous transport processes exhibiting effects over multiple temporal and/or spatial scales.

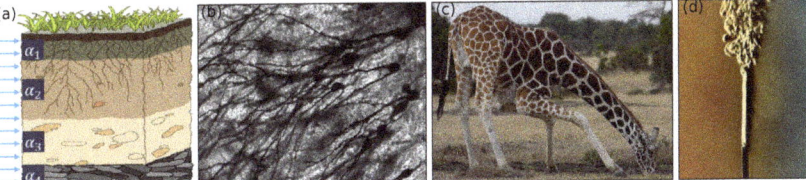

Figure 5. (**a**) Underground aquifers contain heterogenous layers of soils where each layer is characterized by a different level of porosity. The diffusion of groundwater through this multifractal media can be better described by DO operators, by replicating (mathematically) the parallel action of the different porous media in the order-distribution (see Section 5.3). Additional examples of multifractal systems where transport processes are better described via DO operators: (**b**) the diffusion of ions in neuronal dendrites [211], (**c**) the diffusion of pigments to form patterns in animals (see Section 5.2), and (**d**) turbulent flows. The subfigures (**a**–**d**) are taken from Wikipedia.

From a thorough review of the literature it appears that anomalous diffusion, among other types of anomalous transport processes, has seen the maximum applications of DOFC. Therefore, we start by reviewing the application of DO models to complex diffusive transport processes, and then move on to other processes including reaction–diffusion, advection–diffusion, and hybrid propagation. In an effort to keep this review contained and focused on the main applications of DOFC to physical modeling, we present the key aspects and mathematical characteristics of the use of DODE in the modeling of transport processes. The interested reader can find extensive mathematical details on the implementation of DO transport models in [54].

5.1. Anomalous Diffusion Processes

As highlighted previously, diffusion processes in several classes of media exhibit strong anomalies wherein the mean square displacement (MSD) is not characterized by a definite (or unique) scaling exponent, [260–263]. As an example, the MSD in several systems grows as a power of the logarithm of time (*strong anomaly*) and shares the interesting property that the probability distribution of the particle's position at long times is a double-sided exponential [261–264]. More specifically, the MSD varies as

$$\langle x^2(t)\rangle \propto \log^\nu t \tag{16}$$

where ν is a positive constant. These diffusion processes are indicated as ultraslow diffusion (or, sometimes, superslow diffusion) processes and they do not conform to self-affine random processes. The most commonly referred example of such a strong anomalous diffusion process is the Sinai diffusion ($\nu = 4$) in which the particle moves in a quenched random force field [265]. Additional examples of such ultraslow diffusion behavior include polymer physics [266], numerical experiments on an area-preserving parabolic map on a cylinder [267], motion in aperiodic environments [268], and in a family of iterated maps [269]. We highlight that, apart from ultraslow diffusion, there exist other strong anomalies including retarding subdiffusion and accelerating subdiffusion, as well as retarding superdiffusion and accelerating superdiffusion. The specific form of the DO governing equation suitable to model either phenomena depends entirely on two factors: (1) the use of time and/or space-fractional DO derivatives, and (2) support of the strength function corresponding to the time- and/or space-fractional DO derivative. In the following, we will review the different modeling possibilities arising from combinations of the aforementioned factors.

In a series of seminal studies, Chechkin et al. [261,270,271] developed a DO framework for strongly anomalous diffusion mechanisms. They considered the time-fractional DO diffusion equation:

$$\int_0^1 \tau^{\beta-1}\phi(\beta)D_t^\beta c(t,x)\,\mathrm{d}\beta = \overline{\mathbb{D}}D_x^2 c(t,x) \tag{17}$$

where $c(t,x)$ denotes the particle concentration, and $\overline{\mathbb{D}}$ denotes the diffusion coefficient. τ is a positive constant representing a characteristic time of the problem, and the strength function was chosen as $\phi(\beta) = v\beta^{v-1}$. The normalization condition for $\phi(\beta)$ on $[0,1]$, i.e., $\int_0^1 \phi(\beta)\mathrm{d}\beta = 1$ assumes $v > 0$. As established in [261], this choice of $\phi(\beta)$ leads to ultraslow kinetics. More specifically, for the above mathematical setup, the MSD is obtained as

$$\langle x^2(t)\rangle \propto \begin{cases} \frac{2\overline{\mathbb{D}}}{v}t\log(\tau/t) & t/\tau \ll 1 \\ \frac{2\overline{\mathbb{D}}}{\Gamma(1+v)}\tau\log^v(t/\tau) & t/\tau \gg 1 \end{cases} \tag{18}$$

As evident, strong diffusion anomalies are described within the above DO diffusion formalism. In fact, it appears that the DODE in Equation (17) describes a subdiffusion random process which is subordinate to the Wiener process with a diffusion exponent decreasing in time (*retarding subdiffusion*). The same behavior was further highlighted by demonstrating that the modes of the solution, obtained via separation of variables, show an ultraslow, logarithmic, decay pattern. The waiting times ($\psi(t)$) of the diffusing particles corresponding to this setup are [271]

$$\psi(t) \propto \frac{1}{t[\log(t/\tau)]^{1+v}} \tag{19}$$

and they do not have finite moments. Clearly, the DO diffusion equation can be interpreted as a limit of the continuous time random walk (CTRW) model with an extremely broad waiting-time probability density function (PDF), so that there are no finite moments [271].

We highlight that several authors have also analyzed the diffusion characteristics obtained via discrete order distributions [272–274] as well as a uniform strength distribution [261,272–274]. For the discrete time-fractional DO with $\phi(\beta) = \phi_1\delta(\beta - \beta_1) + \phi_2\delta(\beta - \beta_2)$ ($0 < \beta_1 < \beta_2 \leq 1$, $\phi_1 > 0, \phi_2 > 0$, and $\phi_1 + \phi_2 = 1$), the characteristic displacement grows initially as t^{β_2}, whereas at large times it grows as t^{β_1} indicating slow yet power-law growing diffusion. For the uniform strength function, that is $\phi(\beta) = 1$, the MSD is given as

$$\langle x^2(t)\rangle \propto \begin{cases} 2\overline{\mathbb{D}}t\log(\tau/t) & t/\tau << 1 \\ 2\overline{\mathbb{D}}\tau\log(t/\tau) & t/\tau >> 1 \end{cases} \tag{20}$$

It appears that the DODE with the uniform strength function leads to slightly anomalous superdiffusion at small times, and to ultraslow diffusion at large times.

Another example of strongly anomalous diffusion processes corresponds to accelerating superdiffusion wherein the MSD, similar to ultraslow diffusion, does not exhibit a unique spatio-temporal scaling. In this class of diffusion processes, the diffusion exponent increases with time. Such processes are characterized using the following space-fractional diffusion equation [261],

$$D_t^1 c(x,t) = \int_{0+}^2 l^{\alpha-2}\overline{\mathbb{D}}\,\Phi(\alpha)D_x^\alpha c(x,t)\,\mathrm{d}\alpha \tag{21}$$

where l is dimensional positive constant. In [261], the authors obtained the MSD behavior by considering a two-term space-fractional diffusion equation, that is by choosing the strength function to be $\Phi(\alpha) = \Phi_1\delta(\alpha - \alpha_1) + \Phi_2\delta(\alpha - \alpha_2)$ with $0 < \alpha_1 < \alpha_2 \leq 2$. For this DO diffusion equation, it was shown that at small times the characteristic displacement grows as t^{1/α_2}, whereas at large times it grows as t^{1/α_1}; clearly exhibiting superdiffusion with acceleration. The fundamental solutions for this discrete order distribution can be found in [275]. Exact solutions for a triple-order discrete distribution can be found in [276]. Random walk models corresponding to the space-fractional DO diffusion equation are presented in [275,277].

Notably, independently of the specific nature of the DODE (space-fractional or time-fractional) as well as of the strength function, the DO diffusion model no longer exhibits self-similarity or scale invariance. This is a direct result of the fact that the DO derivative modifies the constant- or even variable-order formulation, by integrating all possible orders over a certain range. The resulting solutions exhibit memory and/or nonlocal effects over several temporal and/or spatial scales leading to strong anomalies.

Building upon the time- and space-fractional DO diffusion models presented in Equations (17) and (21), several authors [278–280] developed DO diffusion models that lead to accelerating subdiffusion and retarding superdiffusion contrary to retarding subdiffusion and accelerating superdiffusion obtained via Equations (17) and (21), respectively. These DO diffusion models are given as [278–280]

$$D_t^1 c(x,t) = \int_0^1 \phi(\beta)\overline{\mathbb{D}}D_t^{1-\beta}\left[D_x^2 c(x,t)\right] \mathrm{d}\beta \tag{22a}$$

$$\int_0^2 \phi(\alpha)l^{2-\alpha}D_{|x|}^{2-\alpha}\left[D_t^1 c(x,t)\right] \mathrm{d}\alpha = \overline{\mathbb{D}}D_x^2 c(x,t) \tag{22b}$$

A direct comparison of the above equations with Equations (17) and (21) indicates an exchange in the presence of the time- and space-fractional DO derivatives, resulting in a class of mixed spatio-temporal DO derivatives. The detailed expressions of the MSD of the particles described via the above equations can be found in [278–280]. The MSD obtained via these formulations indicates that the anomalous diffusion phenomena described via Equation (22a) and Equation (22b) exhibit accelerating subdiffusion and retarding superdiffusion, respectively; that is, they become less anomalous in the course of time. Additional details on these anomalous behaviors are provided in the following. The DO time-fractional diffusion equation (Equation (22a)) describes a subdiffusion process which becomes less subdiffusive or, in other words, more classical in the course of time. The MSD demonstrates the occurrence of a transition from a growth characterized by a smaller exponent to a growth with a larger exponent. Equivalently, the probability density for a particle to remain around the origin exhibits a transition from slow to a faster decay. We highlight here that the fundamental solution for a discrete form of the Equation (22a), considering an infinite domain, can be found in [281]. The DO space fractional diffusion equation (Equation (22b)) describes power-law truncated Lévy flights, that is, a random process showing a slow convergence to a Gaussian, but exhibiting Lévy-like behavior at short times. This behavior manifests itself in the non-Gaussian Lévy scaling of the probability density to stay at the origin and in superdiffusive behavior. At short times, the central part of the PDF has a Lévy-stable shape, whereas the asymptotics decay with the power-law, faster than the decay of the Lévy-stable law. At long times, the central part of the PDF approaches the classical Gaussian shape, however, the asymptotics decay with the same power-law.

In addition to the above studies, several researchers have demonstrated the suitability of DOFC for modeling strongly anomalous diffusion behavior, particularly ultraslow diffusion, via stochastic descriptions [215,282–287]. Meerschaert et al. [282,288] developed a stochastic model based on random walks with a random waiting time between jumps. Scaling limits of these random walks are subordinated random processes whose density functions solve the DO ultraslow diffusion equation. Ultraslow diffusion has also been modeled using Langevin stochastic representations in [217,253,284,289]. As shown in [284], the solutions of DO Langevin equations have MSDs which describe retarding subdiffusion and ultraslow diffusion with logarithmic growth. Ultraslow diffusion is also obtained via the wait-first and jump-first Lévy walk models, which underlie the fractional dynamics involving DO material derivatives [289]. The approach in [289] is based on a strongly coupled CTRW, with the distribution of waiting times displaying ultraslow (logarithmic) decay of the tails. Similarly, the authors of [283,285] obtained the space-fractional DO diffusion formulation as the continuum limit of a random process which is characterized by the presence of a distribution of spatially-dependent jumping rate and the Lévy distributed jumping size. As described in [283,285], such a system is well suited to describe diffusion in multifractal systems which do not possess a unique Hurst exponent and, consequently, exhibit a lack of scaling. The lack of scaling in multifractals requires a generalization of stochastic Lévy equation by admitting a spectrum of the Lévy index. The continuum limit of this stochastic equation is the DO diffusion equation. A detailed mathematical analysis of the Lévy models is presented in [286] and a Lévy mixing based probabilistic interpretation of the DO diffusion model is presented. The characteristics of the model are exemplified by a direct application to slow diffusion, particularly the delayed Brownian motion. A similar stochastic representation, given in the form of the Brownian motion subordinated by a Lévy process was to model accelerating subdiffusion in [290]. Additionally, the authors of [290] also constructed an algorithm for computer simulations of accelerating subdiffusion paths via Monte Carlo methods.

Before proceeding further, we briefly review the contributions that several researchers made to the different mathematical aspects of the DO diffusion equations. Exact solutions corresponding to Dirichlet, Neumann, and Cauchy boundary conditions for the time-fractional DO diffusion Equation (17) can be found in [291]. The fundamental solution of the DODE corresponding to a uniform strength distribution can be found in [272–274]. Mainardi et al. [292] obtained the fundamental solution of the time-fractional DO diffusion equation based on its Mellin–Barnes integral representation. They also presented a series expansion of the fundamental solution that clearly highlights, within the solution, the presence of several time-scales related to the distribution of the fractional-orders in the DO diffusion equation. Asymptotic solutions to initial and boundary value problems based on the DO time-fractional diffusion equations can be found in [293,294]. Some additional and important mathematical aspects, such as the existence of the solution to different types of DO diffusion equations, the solvability of DO diffusion equations, subordination properties, and positivity of the solution were addressed in [59,63,263,287,295–300]. In a series of papers [71,72,301], Luchko analyzed the well-posedness of the DO formulation via maximal principles, and obtained *a priori* norm estimates for solutions to both linear and nonlinear DO diffusion equations. Luchko has also provided a survey of these maximal principles in [302]. Further, the well-posedness of the inverse problem, that is the determination of the strength distribution of the DO and its support, has been analyzed in detail in [303–307]. The analysis of the well-posedness of the inverse problem is highly essential to promote applications of DOFC since it determines whether the DO framework is suited to model a given real-world application. In other terms, given a set of experimental or real-world data, the analysis of the inverse problem determines whether DOFC is well suited to model the dataset and hence, it also indicates if the corresponding system exhibits multiscale (temporal and/or spatial) characteristics.

The remarkable properties of the DO diffusion formalism provided a strong foundation for the development of other DO transport formulations: DO reaction–diffusion, DO advection–diffusion, and DO wave propagation. Before reviewing these other applications, we briefly overview some recent, yet remarkable, real-world applications of the DO diffusion formulation (see Figure 5). Grain boundary diffusion in engineering materials at elevated temperatures, that often determines the evolution of microstructure, phase transformations, and certain regimes of plastic deformation and fracture, was modeled via a DO diffusion framework in [308]. DO diffusion equations have also been used to model the diffusion of mobile ions in different electrolytic cells [309–311]. The predictions of the DO model closely matched experimental data which indicated the presence of different diffusive regimes. A similar application was presented in [312], where DO operators were introduced into the Letokhov model of photon diffusion to model non-resonant random lasers. Very recently, the effect of disordering of nanotubes within an electrode, on the impedance of a supercapacitor, was modeled using the DO subdiffusion model in [313]. All these applications highlighted the ability of the DO diffusion formulation to accurately capture highly anomalous diffusion behavior arising out of the presence of multiple temporal and/or spatial scales.

5.2. Reaction–Diffusion Processes

An interesting application of DOFC involves the modeling of reaction–diffusion systems. Reaction–diffusion processes describe changes in the concentration of interacting chemical substances both in space and time. Reaction–diffusion processes have been linked to the formation of spots and patterns in different animals and birds [314,315], among many other real-world applications [125,316] (see Figure 5c). Distributed-order derivatives help to account for the heterogeneity and multifractal nature of the diffusing medium, typical of these applications. More importantly, the DO derivatives also account for the multiple sources of the reacting chemicals within the heterogeneous system. This allows for compact yet more comprehensive theoretical formulations of the reaction–diffusion mechanisms when compared to classical integer-order based approaches. Several authors have analyzed complex reaction–diffusion systems using DO derivatives [102,129,149,316,317]. Detailed mathematical formulations along with closed form solutions for DO reaction–diffusion equations can be found in [316,318]. The effect of different strength functions as well as the specific nature of the DO reaction–diffusion equation was analyzed numerically in [102,129,149]. Very recently, Guo et al. [148] analyzed a 3D Gordon-type reaction–diffusion model of colliding and diffusing Gordon-type solitons. The numerical results provided a deeper understanding of the complicated nonlinear behavior of the 3D Gordon-type solitons system while highlighting the remarkable capabilities of the DO derivatives in capturing the collision and diffusion of the solitons.

5.3. Advection-Diffusion Processes

The VO diffusion equation formed the basis of several interesting investigations involving strongly anomalous advection-diffusion processes in complex systems, particularly those related to hydrology such as, for example, geomigration [319], transport of solutes in heterogeneous media [257,320], the spread of contaminants in groundwater [321], as well as groundwater flow [322]. Indeed, several theoretical and experimental studies have shown that the transport of fluids and pollutants through geological aquifers exhibits the presence of multiple spatio-temporal scales arising from the multifractal nature of the aquifers. The multifractality is a direct consequence of the porous, fractured, layered, and heterogeneous nature of the aquifers (see Figure 5a). The underlying distinctive characteristics of DOFC make it a very well suited modeling approach for the aforementioned anomalous transport phenomena experienced in hydrology.

The detailed mathematical analysis of a DO advection-diffusion equation with a discrete distribution of orders was presented in [77]. Analytical solutions were obtained in [77] for a time- and space-fractional formulation and some interesting derivations including the spectral representation of the fractional Laplacian operator were presented. Later, several researchers used DOFC to model advection–diffusion in complex problems, particularly those related to hydrology. A DO advection–diffusion model was proposed in [256] to model infiltration, absorption, and water exchange in mobile and immobile zones of swelling soils. A similar formulation was adopted in [319] to model a geomigration process in a geoporous medium saturated with a salt solution that exhibits subdiffusive characteristics. Several researchers also used DOFC to model subdiffusive characteristics observed in the transportation of solutes in heterogeneous porous media [257,320,323].

Very recently, an interesting application of DOFC was proposed to simulate superdiffusion of dissolved phase contaminants in groundwater [321]. In this study, several insights including the specific impact of different geometric properties of the contaminants on their spatial distribution pattern, were derived using the DO advection-diffusion model.

5.4. Wave Propagation

Several authors investigated DO models for wave propagation by directly extending the DO diffusion approaches reviewed in Section 5.1. More specifically, this process involved altering the support of the strength function corresponding to the DO time-fractional derivative from $[0, 1]$ to an interval within $[1, 2]$. The most generalized versions of the one-dimensional DO wave equation can be obtained by modifying Equations (17) and (21) as

$$\int_1^2 \tau^{\beta-1} \phi(\beta) D_t^\beta u(t, x) \, d\beta = E_0 D_x^2 u(t, x) \tag{23a}$$

$$D_t^2 u(x, t) = \int_{0^+}^2 l^{\alpha-2} E_0 \Phi(\alpha) D_x^\alpha u(x, t) \, d\alpha \tag{23b}$$

where $u(x, t)$ denotes the particle displacement and E_0 denotes a material constant. A different set of DO wave equations can be obtained by modifying the support of the strength function and using mixed spatio-temporal DO derivatives, similar to Equations (22a) and (22b). The qualitative discussions on the application of DO models for multifractal systems, presented for other types of transport processes reviewed in this Section 5, also holds for DO wave propagation. As an example, the propagation of elastic waves through dissipative media exhibiting multifractal viscoelastic behavior (see Section 4) is described via time-fractional DO models [221,324]. Similarly, elastic wave propagation via attenuating media characterized by simultaneous microstructural and nonlocal (hence, multiscale) effects can be described via space-fractional DO models [213]. Important mathematical aspects such as the existence and uniqueness of the solution to the DO time-fractional wave equation have been outlined in detail in [63,325–327]. Additionally, the fundamental solutions of the DO wave equation have been derived in [298,325,327,328] using the technique of the Fourier and Laplace transforms. Numerical experiments highlighting the specific effects of the DO model parameters have been used to derive interesting insights into the DO wave equation in [298,325,328].

Another possible route to develop the DO wave propagation formulation consists in formulating DO stress–strain constitutive relations within the classical elastodynamic problem as proposed in [324,329]:

$$\sigma = E_0 \int_0^1 \phi(\beta) D_t^\beta \varepsilon d\beta \tag{24}$$

This approach resembles the formulation of DO viscoelastic models (see Section 4) and indeed leads to a hybrid propagation model that also captures dissipation. The DO wave propagation model was then used to simulate the interaction of compressional waves with an interface separating two dissimilar media. Further, the impact of the support and definition of the strength function were analyzed on the wave scattering at the interface.

6. Applications to Control Theory

In this section, we analyze the applications of DOFC to control theory. The foundation as well as motivation for the application of DOFC to control theory follows from a successful application of COFC to model complex control phenomena. The use of CO fractional controllers has enabled robust control and helped achieving highly desirable dynamic control characteristics. A detailed review of theory and applications of COFC in control theory can be found in [36]. In this regard, recall that a fractional derivative implicitly embeds within itself time-delays, or in other terms, it accounts for the memory of past events. Consequently, the presence of a distribution of fractional-order derivatives translates, physically, to the presence of a mixture of delay sources (similar to what is discussed in Section 5). These DO characteristics have helped achieve high performance controllers with several applications ranging from secure messaging [330], to control of motors [331,332] as well accurate frameworks to model robust stability of gene regulatory networks [332]. Broadly speaking, the applications of DOFC to control theory can be divided into two categories: (1) the development of DO controllers and (2) study of the stability and control of DO systems; the majority of the studies being focused on the latter category. In the following, we first review the DO controllers and their applications, before considering their stability. A few other studies have numerically analyzed various DO system identification techniques [220,333] and DO optimal control problems [100,334]. However, the basic DO control theory employed in the latter studies are derived from the two broad categories mentioned above.

6.1. DO Controllers and Filters

Several theoretical and experimental studies have shown that fractional-order designs can enhance both the flexibility and robustness of the controllers as a result of the additional parameters represented by the fractional-orders themselves. Tuning of the fractional-orders allows for superior control characteristics. As an example, consider the CO PID controller $\text{PI}^\lambda\text{D}^\mu$. The value of the order λ in $\text{PI}^\lambda\text{D}^\mu$ control affects the slope of the low frequency range of the system as well as the peak value of the system. On the other hand, the value of the order μ affects the accuracy of the dynamic closed-loop response, the system overshoot, and the stability. For a more detailed discussion of the roles of λ and μ, the interested reader is referred to the work in [36]. It is immediate that a distribution of several CO controllers can lead to highly accurate and robust control. In fact, DOFC allows the development of a highly generalized controller from which all other types of controllers (such as, for example, the classical integrator and differentiator, the classical PID, and the fractional $\text{PI}^\lambda\text{D}^\mu$) can be recovered.

In the most general form, the transfer function of a DO controller can be expressed as [36]

$$G(s) = \int_{\beta_1}^{\beta_2} \phi(\beta) \frac{1}{s^\beta} d\beta \tag{25}$$

where s is a complex variable. The interval $[\beta_1, \beta_2]$ dictates the specific nature of the controller. Note that a DO low-pass filter can be obtained from the DO controller via the transformation $s \to T(\beta)s + 1$ [335]. The above formulation is highly general in the sense that all the classical, CO, and DO controllers can be recovered from the same by an appropriate choice of the strength function. As an example, the classical integrator can be obtained by choosing $\phi(\beta) = \delta(\beta - 1)$, the classical differentiator can be obtained from $\phi(\beta) = \delta(\beta + 1)$, the classical PID from $\phi(\beta) = k_P\delta(\beta) + k_I\delta(\beta - 1) + k_D\delta(\beta + 1)$ (k_P, k_I and k_D are constants to be tuned), the fractional PID from $\phi(\beta) = k_P\delta(\beta) + k_I\delta(\beta - \lambda) + k_D\delta(\beta + \mu)$, and so on. It is immediate to see that a DO PID controller can be also obtained directly from the controller in Equation (25), by insisting that the support of the weight function lies within the interval $[-1, 1]$. DO PID controllers have been studied in detail in a series of papers by Jakovljević et al. [336–338]. Note that in the case of a DO controller, the strength function in Equation (25) can have infinite support. In fact, as established in [339], any DO controller can be developed by appropriate composition of the DO integrator ($0 \leq \beta_1 < \beta_2 \leq 1$), the classical integrator ($1/s$) and the classical differentiator (s). The different DO controllers have been schematically illustrated in Figure 6.

The impulse response and asymptotic behavior of the DO controllers have been derived in [335,340]. Additionally, a physical realization of the DO integrator using a series of capacitors has been developed in [210,340]. The DO controllers have been applied to control motors [338] and robots [331] among many other applications [36]. As observed in these studies, the DO controllers reduce the maximum overshoot while guaranteeing a fast dynamic response and a zero steady-state error [36,336–338]. Furthermore, the phase curves of DO PID controllers are non-constant and much wider than the corresponding CO controllers making them more robust to system uncertainties [331]. Therefore, the DO controllers exhibit unique frequency response characteristics, and provide highly robust and accurate control.

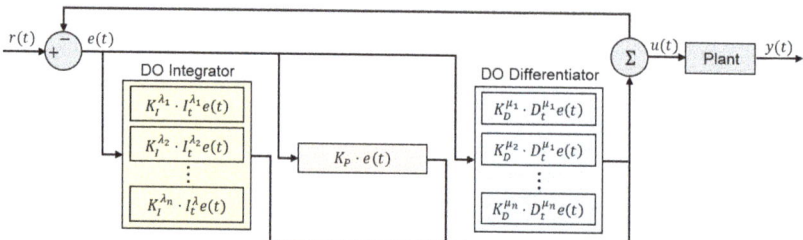

Figure 6. Block diagram illustrating the feedback DO controller based on Equation (25). The fractional-orders $\mu_k, \lambda_k \in (0,1]$. This is a highly general controller from which all classical, CO, and DO controllers, as well as the DO PID controller can be recovered by an appropriate choice of the controller constants. As an example, the DO differentiator can be obtained by setting $K_I^{\lambda_k} = 0$, $K_P = 0$, and $K_D^{\mu_k} \neq 0$. As evident, the DO differentiator consists of a network of CO differentiators. Similarly, the DO PID controller would require that $K_P \neq 0$, $K_D^{\mu_k} \neq 0$ and $K_I^{\lambda_k} \neq 0$.

6.2. Stability and Control of DO Systems

The development of robust and accurate DO controllers prompted several researchers to analyze the stability of both linear and nonlinear DO dynamical systems. Most of the studies conducted on linear systems correspond to the bounded-input bounded-output (BIBO) stability analysis of DO linear time-invariant (LTI) systems. On the other hand, the nonlinear studies have focused primarily on the Lyapunov stability of the equilibrium points of the DO system. First, we briefly review the key highlights of the DO LTI systems and their applications. Consider a DO system described via the following LTI DODE and algebraic output equation,

$$\int_0^1 \phi(\beta) D_t^\beta x(t) d\beta = Ax(t) + Bu(t)$$
$$y(t) = Cx(t) + Du(t) \tag{26}$$

where $x(t)$ is the state vector, $u(t)$ indicates the input, and $y(t)$ indicates the output of the system. A, B, C, and D are matrices of appropriate dimensions. Note that the interval of the DO derivative in the above single-input single-output (SISO) system can be converted to a more general interval $[\beta_1, \beta_2] \in [0,1]$. Applying a set of Laplace and inverse Laplace transform to the above DODE with the assumption that $x(0) = 0$ and $u(t) = \delta(t)$, the following expression can be obtained,

$$x(t) = \mathcal{L}^{-1}\left[\underbrace{\left[\left(\int_0^1 \phi(\beta) s^\beta d\beta\right) I - A\right]^{-1} B}_{G(s)}\right](t) \tag{27}$$

where I denotes the identity matrix. As established in [341–343], the DO LTI system in Equation (26) with the transfer function $H(s) = CG(s)B + D$ is BIBO stable *iff* all the roots of the secular equation corresponding to $|G(s)I - A| = 0$ have negative real parts. The contours of this stability region have been derived based on the latter principle for different definitions of the strength function in [342,344]. The stability contours are often impossible to express via elementary functions, which makes the stability tests of DO systems more complicated than their constant- and integer-order counterparts. In this regard, the Lagrange inversion theorem was utilized in [345] to obtain explicit expressions for the stability contours. Several interesting properties of these stability curves such as the slope of the tangent at very high and very low frequencies, convexity, inability to cut itself, location in the first and fourth quadrants, and shifting and enhancement of the area of the stability via multiplication of suitable functions to the strength distribution, have been presented in [346–348].The above mentioned properties of the stability boundaries were used in [347] to present a remarkable framework for the robust stability analysis of DO LTI systems with uncertain strength distributions and dynamic matrices. More specifically, these properties were used to show that the stability boundary of DO LTI systems can be accurately located in a certain region on the complex plane defined by the upper and lower bounds of the strength distribution. These results are sufficient to ensure robust stability in DO LTI systems with uncertain strength functions and uncertain dynamic matrices. The above framework presented in [347] is highly relevant for real-world applications that are commonly accompanied by uncertainties. Additional discussions on the stabilization, controllability, and passification of DO LTI systems can be found in [349–352].

The DO LTI framework discussed above has been used to analyze different systems: the solar wind-driven magnetosphere ionosphere system (a complex driven-damped dynamical system which exhibits a variety of dynamical states) [341,348], a DO Lotka–Volterra predator–prey system (a system with multiple time-delays) [353], the DO Chen system [354], and gene regulatory systems [332]. All the aforementioned applications differ primarily in the choice of the strength function which directly affects the stability and control of the system.

In nonlinear systems, researchers have focused mainly on analyzing the Lyapunov stability of systems, as also mentioned previously. The Lyapunov direct method, used for analysis of stability, was first generalized for nonlinear time-varying DO systems in [355–357] and was used to determine the stability or asymptotic stability of certain nonlinear systems including a DO analog of the Lorenz system. The theoretical framework proposed in the studies [355,356] was then used to analyze different nonlinear time-varying DO systems including a DO consensus model [358], the DO Lorenz system [359], and the DO Van der Pol oscillator [330,360]. The consensus of multi-agent systems with fixed directed graphs and described by DODE, was analyzed in [358] and sufficient conditions were obtained for robust consensus in the presence and absence of external disturbances. Recently, the stability and control of a DO Van der Pol were analyzed in [330], wherein the intervals of the different model parameters at which this oscillator exhibits periodic, chaotic, and hyperchaotic behaviors, were calculated using Lyapunov exponents. Further, a robust scheme was presented in [330] to achieve complete synchronization between two DO hyperchaotic unforced Van der Pol oscillators. This synchronization allowed the development of a secure messaging system for a text which contains alphabets, numbers, and symbols.

7. Conclusions

This paper presented an overview of the general area of Distributed-Order Fractional Calculus (DOFC) with particular focus on its applications to scientific modeling of complex systems. A branch of the broader field of fractional calculus, DOFC has rapidly emerged and captured the attention of many researchers in science and engineering. This rapid growth was mostly due to its remarkable ability to capture complex multiscale processes. Phenomena like multiple relaxation times in viscoelasticity, multiple temporal and spatial scale effects in transport processes, and mixture of time delays in control theory, just to name a few, have all illustrated the significant performance of DOFC over more traditional integer-order techniques. The main goal of this review was to provide a snapshot in time of the field of DOFC and to guide the interested reader into an introductory journey through this fascinating topic. In this regard, we highlight that the content of technical papers was only briefly addressed in order to favor a more general discussion of the evolution of the field in its different areas of application.

Despite the recent substantial growth in DOFC research, there are still many areas holding significant opportunities for further development. While some preliminary work is available on distributed-variable models, a comprehensive framework for distributed-variable-order fractional calculus (DVOFC) is still lacking. A key factor that adds to the complexity of formulating DVOFC is the existence of different definitions for VO operators that exhibit different memory characteristics. Thus, a unified definition of the different variable- and distributed-order operators and an analysis of their mathematical properties would certainly be beneficial. In these operators, the order-variation can be a function of different dependent or independent physical variables (such as, for example, temperature, space, time, and energy). The combination of the DO and VO formalisms should allow the simulation of highly complex physical systems which are both evolutionary (therefore, requiring VO operators) and multifractal (requiring DO operators) in nature. Another possible extension of currently available DO operators follows from the use of normalized self-similar strength functions within the definition of DO operators, which can be considered analogous to random-order operators. Particularly lacking is a rigorous mathematical analysis of the properties of such operators. Despite the above challenges, the extension of DOFC to these areas can have important applications in modeling random and chaotic dynamics observed, as an example, in turbulent dynamics, noise and vibration control, or even in financial systems. These models could even form the basis for the development of highly accurate risk analysis and control models.

It should be pointed out that, despite the rapidly growing number of related studies, there are still several open questions that need to be addressed before DOFC could become a mainstream modeling approach for common real-world applications. A critical step to promote the broader use of DOFC models is to establish the connection between the mathematical properties of DO operators (i.e., the strength function and its support) and the physical properties and parameters of the system to be modeled. In other terms, the identification of closed form relations linking the mathematical parameters of the DO operators to the physical parameters of the system at hand are of paramount importance to foster the use of DOFC tools in scientific modeling.

Author Contributions: W.D., S.P. and S.S. performed the literature review. All authors contributed equally to the manuscript writing. All authors have read and agreed to the published version of the manuscript.

Funding: The following work was partially supported by the National Science Foundation (NSF) under grants MOMS #1761423 and DCSD #1825837 and by the Defense Advanced Research Project Agency (DARPA) under grant #D19AP00052. The content and information presented in this manuscript do not necessarily reflect the position or the policy of the government. The material is approved for public release; distribution is unlimited.

Data Availability Statement: This article has no additional data.

Conflicts of Interest: The authors declare no conflict of interest.

References

1. Leibniz, G. Letter from Hanover, Germany, to GFA L'Hopital, 30 September 1695. *Math. Schriften* **1849**, *2*, 301–302.
2. De Oliveira, E.C.; Tenreiro Machado, J.A. A review of definitions for fractional derivatives and integral. *Math. Probl. Eng.* **2014**, *2014*. [CrossRef]
3. Erdélyi, A. On fractional integration and its application to the theory of Hankel transforms. *Q. J. Math.* **1940**, 293–303. [CrossRef]
4. Miller, K.S.; Ross, B. *An Introduction to the Fractional Calculus and Fractional Differential Equations*; Wiley: Hoboken, NJ, USA, 1993.

5. Bagley, R.; Torvik, P. A theoretical basis for the application of fractional calculus to viscoelasticity. *J. Rheol.* **1983**, *27*, 201–210. [CrossRef]

6. Torvik, P.; Bagley, R. On the appearance of the fractional derivative in the behavior of real materials. *J. Appl. Mech. Trans. ASME* **1984**, *51*, 294–298. [CrossRef]

7. Chatterjee, A. Statistical origins of fractional derivatives in viscoelasticity. *J. Sound Vib.* **2005**, *284*, 1239–1245. [CrossRef]

8. Alotta, G.; Bologna, E.; Failla, G.; Zingales, M. A fractional approach to non-Newtonian blood rheology in capillary vessels. *J. Peridyn. Nonlocal Model.* **2019**, *1*, 88–96. [CrossRef]

9. Mainardi, F. Fractional relaxation-oscillation and fractional diffusion-wave phenomena. *Chaos Solitons Fractals* **1996**, *7*, 1461–1477. [CrossRef]

10. Gritsenko, D.; Paoli, R. Theoretical Analysis of Fractional Viscoelastic Flow in Circular Pipes: General Solutions. *Appl. Sci.* **2020**, *10*, 9093. [CrossRef]

11. Gritsenko, D.; Paoli, R. Theoretical Analysis of Fractional Viscoelastic Flow in Circular Pipes: Parametric Study. *Appl. Sci.* **2020**, *10*, 9080. [CrossRef]

12. Failla, G.; Zingales, M. *Advanced Materials Modelling Via Fractional Calculus: Challenges and Perspectives*; Royal Society: Cambridge, UK, 2020.

13. Lazopoulos, K.A. Non-local continuum mechanics and fractional calculus. *Mech. Res. Commun.* **2006**, *33*, 753–757. [CrossRef]

14. Drapaca, C.S.; Sivaloganathan, S. A fractional model of continuum mechanics. *J. Elast.* **2012**, *107*, 105–123. [CrossRef]

15. Di Paola, M.; Failla, G.; Pirrotta, A.; Sofi, A.; Zingales, M. The mechanically based non-local elasticity: An overview of main results and future challenges. *Philos. Trans. R. Soc. A Math. Phys. Eng. Sci.* **2013**, *371*, 20120433. [CrossRef]

16. Sumelka, W. Thermoelasticity in the framework of the fractional continuum mechanics. *J. Therm. Stress.* **2014**, *37*, 678–706. [CrossRef]

17. Carpinteri, A.; Cornetti, P.; Sapora, A. Nonlocal elasticity: An approach based on fractional calculus. *Meccanica* **2014**, *49*, 2551–2569. [CrossRef]

18. Patnaik, S.; Sidhardh, S.; Semperlotti, F. A Ritz-based finite element method for a fractional-order boundary value problem of nonlocal elasticity. *Int. J. Solids Struct.* **2020**, *202*, 398–417. [CrossRef]

19. Sidhardh, S.; Patnaik, S.; Semperlotti, F. Geometrically nonlinear response of a fractional-order nonlocal model of elasticity. *Int. J. Nonlinear Mech.* **2020**, *125*, 103529. [CrossRef]

20. Patnaik, S.; Sidhardh, S.; Semperlotti, F. Fractional-order models for the static and dynamic analysis of nonlocal plates. *Commun. Nonlinear Sci. Numer. Simul.* **2020**, 105601. [CrossRef]

21. Patnaik, S.; Sidhardh, S.; Semperlotti, F. Geometrically nonlinear analysis of nonlocal plates using fractional calculus. *Int. J. Mech. Sci.* **2020**, *179*, 105710. [CrossRef]

22. Sidhardh, S.; Patnaik, S.; Semperlotti, F. Fractional-Order Structural Stability: Formulation and Application to the Critical Load of Slender Structures. *arXiv* **2020**, arXiv:2008.11528.

23. Sidhardh, S.; Patnaik, S.; Semperlotti, F. Analysis of the Post-Buckling Response of Nonlocal Plates via Fractional Order Continuum Theory. *J. Appl. Mech.* **2020**, 1–22. [CrossRef]

24. Patnaik, S.; Semperlotti, F. A generalized fractional-order elastodynamic theory for non-local attenuating media. *Proc. R. Soc. A* **2020**, *476*, 20200200. [CrossRef] [PubMed]

25. Sebaa, N.; Fellah, Z.E.A.; Lauriks, W.; Depollier, C. Application of fractional calculus to ultrasonic wave propagation in human cancellous bone. *Signal Process.* **2006**, *86*, 2668–2677. [CrossRef]

26. Treeby, B.E.; Cox, B. Modeling power law absorption and dispersion for acoustic propagation using the fractional Laplacian. *J. Acoust. Soc. Am.* **2010**, *127*, 2741–2748. [CrossRef]

27. Gómez-Aguilar, J.; Miranda-Hernández, M.; López-López, M.; Alvarado-Martínez, V.; Baleanu, D. Modeling and simulation of the fractional space-time diffusion equation. *Commun. Nonlinear Sci. Numer. Simul.* **2016**, *30*, 115–127. [CrossRef]

28. Saad, K.M.; Gómez-Aguilar, J. Analysis of reaction–diffusion system via a new fractional derivative with non-singular kernel. *Phys. A Stat. Mech. Its Appl.* **2018**, *509*, 703–716. [CrossRef]

29. Hollkamp, J.P.; Semperlotti, F. Application of fractional order operators to the simulation of ducts with acoustic black hole terminations. *J. Sound Vib.* **2020**, *465*, 115035. [CrossRef]

30. Buonocore, S.; Sen, M. Scattering cross sections of acoustic nonlocal inclusions: A fractional dynamic approach. *J. Appl. Phys.* **2020**, *127*, 203101. [CrossRef]

31. West, B.J.; Deering, W. Fractal physiology for physicists: Lévy statistics. *Phys. Rep.* **1994**, *246*, 1–100. [CrossRef]

32. Carpinteri, A.; Cornetti, P. A fractional calculus approach to the description of stress and strain localization in fractal media. *Chaos Solitons Fractals* **2002**, *13*, 85–94. [CrossRef]

33. Li, J.; Ostoja-Starzewski, M. Fractal solids, product measures and fractional wave equations. *Proc. R. Soc. A Math. Phys. Eng. Sci.* **2009**, *465*, 2521–2536. [CrossRef]

34. West, B.J. Fractal physiology and the fractional calculus: A perspective. *Front. Physiol.* **2010**, *1*, 12. [CrossRef] [PubMed]

35. Li, J.; Ostoja-Starzewski, M. Thermo-poromechanics of fractal media. *Philos. Trans. R. Soc. A* **2020**, *378*, 20190288. [CrossRef] [PubMed]

36. Sheng, H.; Chen, Y.; Qiu, T. *Fractional Processes and Fractional-Order Signal Processing: Techniques and Applications*; Springer Science & Business Media: New York, NY, USA, 2011.
37. Magin, R.; Ortigueira, M.D.; Podlubny, I.; Trujillo, J. On the fractional signals and systems. *Signal Process.* **2011**, *91*, 350–371. [CrossRef]
38. Lazarević, M.P.; Mandić, P.D.; Ostojić, S. Further results on advanced robust iterative learning control and modeling of robotic systems. *Proc. Inst. Mech. Eng. Part C J. Mech. Eng. Sci.* **2020**, 0954406220965996. [CrossRef]
39. Oziablo, P.; Mozyrska, D.; Wyrwas, M. Discrete-Time Fractional, Variable-Order PID Controller for a Plant with Delay. *Entropy* **2020**, *22*, 771. [CrossRef]
40. Machado, J.T.; Kiryakova, V.; Mainardi, F. Recent history of fractional calculus. *Commun. Nonlinear Sci. Numer. Simul.* **2011**, *16*, 1140–1153. [CrossRef]
41. Patnaik, S.; Hollkamp, J.P.; Semperlotti, F. Applications of variable-order fractional operators: A review. *Proc. R. Soc. A* **2020**, *476*, 20190498. [CrossRef]
42. Sun, H.; Chang, A.; Zhang, Y.; Chen, W. A review on variable-order fractional differential equations: Mathematical foundations, physical models, numerical methods and applications. *Fract. Calc. Appl. Anal.* **2019**, *22*, 27–59. [CrossRef]
43. Caputo, M.; Fabrizio, M. The kernel of the distributed order fractional derivatives with an application to complex materials. *Fractal Fract.* **2017**, *1*, 13. [CrossRef]
44. Calcagni, G. Towards multifractional calculus. *Front. Phys.* **2018**, *6*, 58. [CrossRef]
45. Lorenzo, C.F.; Hartley, T.T. Variable order and distributed order fractional operators. *Nonlinear Dyn.* **2002**, *29*, 57–98. [CrossRef]
46. Caputo, M. Linear models of dissipation whose Q is almost frequency independent. *Ann. Geophys.* **1966**, *19*, 383–393. [CrossRef]
47. Caputo, M. Linear models of dissipation whose Q is almost frequency independent—II. *Geophys. J. Int.* **1967**, *13*, 529–539. [CrossRef]
48. Caputo, M. *Elasticita e Dissipazione*; Zanichelli: Bologna, Italy, 1969.
49. Caputo, M. Mean fractional-order-derivatives differential equations and filters. *Annali dell'Universita di Ferrara* **1995**, *41*, 73–84.
50. Bagley, R.; Torvik, P. On the existence of the order domain and the solution of distributed order equations—Part I. *Int. J. Appl. Math.* **2000**, *2*, 865–882.
51. Bagley, R.; Torvik, P. On the existence of the order domain and the solution of distributed order equations—Part II. *Int. J. Appl. Math.* **2000**, *2*, 965–988.
52. Garrappa, R.; Kaslik, E.; Popolizio, M. Evaluation of fractional integrals and derivatives of elementary functions: Overview and tutorial. *Mathematics* **2019**, *7*, 407. [CrossRef]
53. Atanacković, T.M.; Pilipović, S.; Stanković, B.; Zorica, D. *Fractional Calculus with Applications in Mechanics*; Wiley Online Library: Hoboken, NJ, USA, 2014.
54. Sandev, T.; Tomovski, Ž. *Fractional Equations and Models: Theory and Applications*; Springer: Berlin/Heidelberg, Germany, 2019; Volume 61.
55. Ford, N.J.; Morgado, M.L. Distributed order equations as boundary value problems. *Comput. Math. Appl.* **2012**, *64*, 2973–2981. [CrossRef]
56. Diethelm, K.; Ford, N.J. Numerical analysis for distributed-order differential equations. *J. Comput. Appl. Math.* **2009**, *225*, 96–104. [CrossRef]
57. Pskhu, A.V. On the theory of the continual integro-differentiation operator. *Differ. Equ.* **2004**, *40*, 128–136. [CrossRef]
58. Pskhu, A. *Partial Differential Equations of Fractional Order*; Nauka: Moscow, Russia, 2005.
59. Umarov, S.; Gorenflo, R. Cauchy and nonlocal multi-point problems for distributed order pseudo-differential equations: Part one. *J. Anal. Its Appl.* **2005**, *245*, 449–466.
60. Diethelm, K.; Luchko, Y. Numerical solution of linear multi-term initial value problems of fractional order. *J. Comput. Anal. Appl* **2004**, *6*, 243–263.
61. Diethelm, K.; Ford, N.J. *Numerical Solution Methods for Distributed Order Differential Equations*; Institute of Mathematics & Informatics, Bulgarian Academy of Sciences: Sofia, Bulgaria, 2005.
62. Atanacković, T.M.; Oparnica, L.; Pilipović, S. On a nonlinear distributed order fractional differential equation. *J. Math. Anal. Appl.* **2007**, *328*, 590–608. [CrossRef]
63. Van Bockstal, K. Existence of a Unique Weak Solution to a Nonlinear Non-Autonomous Time-Fractional Wave Equation (of Distributed-Order). *Mathematics* **2020**, *8*, 1283. [CrossRef]
64. Noroozi, H.; Ansari, A.; Dahaghin, M.S. Existence results for the distributed order fractional hybrid differential equations. *Abstr. Appl. Anal.* **2012**, *2012*. [CrossRef]
65. Zhao, Y.; Sun, S.; Han, Z.; Li, Q. Theory of fractional hybrid differential equations. *Comput. Math. Appl.* **2011**, *62*, 1312–1324. [CrossRef]
66. Noroozi, H.; Ansari, A. Basic results on distributed order fractional hybrid differential equations with linear perturbations. *J. Math. Model.* **2014**, *2*, 55–73.
67. Atanackovic, T.M.; Oparnica, L.; Pilipović, S. Distributional framework for solving fractional differential equations. *Integral Transform. Spec. Funct.* **2009**, *20*, 215–222. [CrossRef]

68. Atanackovic, T.M.; Oparnica, L.; Pilipović, S. Semilinear ordinary differential equation coupled with distributed order fractional differential equation. *Nonlinear Anal. Theory Methods Appl.* **2010**, *72*, 4101–4114. [CrossRef]

69. Fedorov, V.E. Generators of analytic resolving families for distributed order equations and perturbations. *Mathematics* **2020**, *8*, 1306. [CrossRef]

70. Refahi, A.; Ansari, A.; Najafi, H.S.; Merhdoust, F. Analytic study on linear systems of distributed order fractional differential equations. *Le Matematiche* **2012**, *67*, 3–13.

71. Luchko, Y. Boundary value problems for the generalized time-fractional diffusion equation of distributed order. *Fract. Calc. Appl. Anal.* **2009**, *12*, 409–422.

72. Luchko, Y. Initial-boundary-value problems for the generalized multi-term time-fractional diffusion equation. *J. Math. Anal. Appl.* **2011**, *374*, 538–548. [CrossRef]

73. Daftardar-Gejji, V.; Bhalekar, S. Boundary value problems for multi-term fractional differential equations. *J. Math. Anal. Appl.* **2008**, *345*, 754–765. [CrossRef]

74. Li, Z.; Liu, Y.; Yamamoto, M. Initial-boundary value problems for multi-term time-fractional diffusion equations with positive constant coefficients. *Appl. Math. Comput.* **2015**, *257*, 381–397. [CrossRef]

75. Luchko, Y. Some uniqueness and existence results for the initial-boundary-value problems for the generalized time-fractional diffusion equation. *Comput. Math. Appl.* **2010**, *59*, 1766–1772. [CrossRef]

76. Luchko, Y.; Gorenflo, R. An operational method for solving fractional differential equations with the Caputo derivatives. *Acta Math. Vietnam* **1999**, *24*, 207–233.

77. Jiang, H.; Liu, F.; Turner, I.; Burrage, K. Analytical solutions for the multi-term time-space Caputo-Riesz fractional advection-diffusion equations on a finite domain. *J. Math. Anal. Appl.* **2012**, *389*, 1117–1127. [CrossRef]

78. Bazhlekova, E. Completely monotone functions and some classes of fractional evolution equations. *Integral Transform. Spec. Funct.* **2015**, *26*, 737–752. [CrossRef]

79. Bazhlekova, E.; Bazhlekov, I. Complete monotonicity of the relaxation moduli of distributed-order fractional Zener model. In *AIP Conference Proceedings*; AIP Publishing LLC: Melville, NY, USA, 2018; Volume 2048, p. 050008.

80. Stojanović, M. Fractional relaxation equations of distributed order. *Nonlinear Anal. Real World Appl.* **2012**, *13*, 939–946. [CrossRef]

81. Mainardi, F.; Mura, A.; Gorenflo, R.; Stojanović, M. The two forms of fractional relaxation of distributed order. *J. Vib. Control* **2007**, *13*, 1249–1268. [CrossRef]

82. Lorenzo, C.F.; Hartley, T.T. Initialization, conceptualization, and application in the generalized (fractional) calculus. *Crit. Rev. Biomed. Eng.* **2007**, *35*. [CrossRef]

83. Diethelm, K.; Ford, N.J. Numerical solution methods for distributed order differential equations. *Fract. Calc. Appl. Anal.* **2001**, *4*, 531–542.

84. Yuttanan, B.; Razzaghi, M. Legendre wavelets approach for numerical solutions of distributed order fractional differential equations. *Appl. Math. Model.* **2019**, *70*, 350–364. [CrossRef]

85. Abbaszadeh, M.; Dehghan, M.; Zhou, Y. Crank—Nicolson/Galerkin spectral method for solving two-dimensional time-space distributed-order weakly singular integro-partial differential equation. *J. Comput. Appl. Math.* **2020**, *374*, 112739. [CrossRef]

86. Abbaszadeh, M.; Dehghan, M. Meshless upwind local radial basis function-finite difference technique to simulate the time-fractional distributed-order advection–diffusion equation. *Eng. Comput.* **2019**, 1–17. [CrossRef]

87. Bu, W.; Ji, L.; Tang, Y.; Zhou, J. Space-time finite element method for the distributed-order time fractional reaction diffusion equations. *Appl. Numer. Math.* **2020**, *152*, 446–465. [CrossRef]

88. Dehghan, M.; Abbaszadeh, M. A Legendre spectral element method (SEM) based on the modified bases for solving neutral delay distributed-order fractional damped diffusion-wave equation. *Math. Methods Appl. Sci.* **2018**, *41*, 3476–3494. [CrossRef]

89. Duong, P.L.T.; Kwok, E.; Lee, M. Deterministic analysis of distributed order systems using operational matrix. *Appl. Math. Model.* **2016**, *40*, 1929–1940. [CrossRef]

90. Fakhar-Izadi, F. Fully Petrov—Galerkin spectral method for the distributed-order time-fractional fourth-order partial differential equation. *Eng. Comput.* **2020**, 1–10. [CrossRef]

91. Hafez, R.M.; Zaky, M.A.; Abdelkawy, M.A. Jacobi Spectral Galerkin method for Distributed-Order Fractional Rayleigh-Stokes problem for a Generalized Second Grade Fluid. *Front. Phys.* **2020**, *7*. [CrossRef]

92. Kharazmi, E.; Zayernouri, M.; Karniadakis, G.E. Petrov–Galerkin and spectral collocation methods for distributed order differential equations. *SIAM J. Sci. Comput.* **2017**, *39*, A1003–A1037. [CrossRef]

93. Jibenja, N.; Yuttanan, B.; Razzaghi, M. An Efficient Method for Numerical Solutions of Distributed-Order Fractional Differential Equations. *J. Comput. Nonlinear Dyn.* **2018**, *13*. [CrossRef]

94. Morgado, M.L.; Rebelo, M.; Ferras, L.L.; Ford, N.J. Numerical solution for diffusion equations with distributed order in time using a Chebyshev collocation method. *Appl. Numer. Math.* **2017**, *114*, 108–123. [CrossRef]

95. Mashayekhi, S.; Razzaghi, M. Numerical solution of distributed order fractional differential equations by hybrid functions. *J. Comput. Phys.* **2016**, *315*, 169–181. [CrossRef]

96. Xu, Y.; Zhang, Y.; Zhao, J. Error analysis of the Legendre-Gauss collocation methods for the nonlinear distributed-order fractional differential equation. *Appl. Numer. Math.* **2019**, *142*, 122–138. [CrossRef]

97. Pourbabaee, M.; Saadatmandi, A. Collocation method based on Chebyshev polynomials for solving distributed order fractional differential equations. *Comput. Methods Differ. Equ.* **2020**. [CrossRef]
98. Zhang, H.; Liu, F.; Jiang, X.; Zeng, F.; Turner, I. A Crank—Nicolson ADI Galerkin—Legendre spectral method for the two-dimensional Riesz space distributed-order advection-diffusion equation. *Comput. Math. Appl.* **2018**, *76*, 2460–2476. [CrossRef]
99. Zaky, M.; Doha, E.; Tenreiro Machado, J. A spectral numerical method for solving distributed-order fractional initial value problems. *J. Comput. Nonlinear Dyn.* **2018**, *13*. [CrossRef]
100. Zaky, M.A. A Legendre collocation method for distributed-order fractional optimal control problems. *Nonlinear Dyn.* **2018**, *91*, 2667–2681. [CrossRef]
101. Abdelkawy, M. A collocation method based on Jacobi and fractional order Jacobi basis functions for multi-dimensional distributed-order diffusion equations. *Int. J. Nonlinear Sci. Numer. Simul.* **2018**, *19*, 781–792. [CrossRef]
102. Abdelkawy, M.; Lopes, A.M.; Zaky, M. Shifted fractional Jacobi spectral algorithm for solving distributed order time-fractional reaction–diffusion equations. *Comput. Appl. Math.* **2019**, *38*, 81. [CrossRef]
103. Zaky, M.A.; Machado, J.T. Multi-dimensional spectral tau methods for distributed-order fractional diffusion equations. *Comput. Math. Appl.* **2020**, *79*, 476–488. [CrossRef]
104. Rahimkhani, P.; Ordokhani, Y.; Lima, P. An improved composite collocation method for distributed-order fractional differential equations based on fractional Chelyshkov wavelets. *Appl. Numer. Math.* **2019**, *145*, 1–27. [CrossRef]
105. Abdelkawy, M.; Babatin, M.M.; Lopes, A.M. Highly accurate technique for solving distributed-order time-fractional-sub-diffusion equations of fourth order. *Comput. Appl. Math.* **2020**, *39*, 1–22. [CrossRef]
106. Abbaszadeh, M.; Dehghan, M. An improved meshless method for solving two-dimensional distributed order time-fractional diffusion-wave equation with error estimate. *Numer. Algorithms* **2017**, *75*, 173–211. [CrossRef]
107. Hu, J.; Wang, J.; Nie, Y. Numerical algorithms for multidimensional time-fractional wave equation of distributed-order with a nonlinear source term. *Adv. Differ. Equ.* **2018**, *2018*, 352. [CrossRef]
108. Chen, H.; Lü, S.; Chen, W. Finite difference/spectral approximations for the distributed order time fractional reaction–diffusion equation on an unbounded domain. *J. Comput. Phys.* **2016**, *315*, 84–97. [CrossRef]
109. Gao, X.; Liu, F.; Li, H.; Liu, Y.; Turner, I.; Yin, B. A novel finite element method for the distributed-order time fractional Cable equation in two dimensions. *Comput. Math. Appl.* **2020**, *80*, 923–939. [CrossRef]
110. Gao, G.; Sun, H.; Sun, Z. Some high-order difference schemes for the distributed-order differential equations. *J. Comput. Phys.* **2015**, *298*, 337–359. [CrossRef]
111. Gao, G.; Sun, Z. Two alternating direction implicit difference schemes for two-dimensional distributed-order fractional diffusion equations. *J. Sci. Comput.* **2016**, *66*, 1281–1312. [CrossRef]
112. Li, X.; Rui, H. Two temporal second-order H^1-Galerkin mixed finite element schemes for distributed-order fractional sub-diffusion equations. *Numer. Algorithms* **2018**, *79*, 1107–1130. [CrossRef]
113. Katsikadelis, J.T. Numerical solution of distributed order fractional differential equations. *J. Comput. Phys.* **2014**, *259*, 11–22. [CrossRef]
114. Liu, Q.; Mu, S.; Liu, Q.; Liu, B.; Bi, X.; Zhuang, P.; Li, B.; Gao, J. An RBF based meshless method for the distributed order time fractional advection–diffusion equation. *Eng. Anal. Bound. Elem.* **2018**, *96*, 55–63. [CrossRef]
115. Bu, W.; Xiao, A.; Zeng, W. Finite difference/finite element methods for distributed-order time fractional diffusion equations. *J. Sci. Comput.* **2017**, *72*, 422–441. [CrossRef]
116. Gao, G.; Sun, Z. Two alternating direction implicit difference schemes with the extrapolation method for the two-dimensional distributed-order differential equations. *Comput. Math. Appl.* **2015**, *69*, 926–948. [CrossRef]
117. Gao, G.; Sun, Z. Two difference schemes for solving the one-dimensional time distributed-order fractional wave equations. *Numer. Algorithms* **2017**, *74*, 675–697. [CrossRef]
118. Fei, M.; Huang, C. Galerkin—Legendre spectral method for the distributed-order time fractional fourth-order partial differential equation. *Int. J. Comput. Math.* **2020**, *97*, 1183–1196. [CrossRef]
119. Kumar, S.; Gómez-Aguilar, J.F. Numerical Solution of Caputo-Fabrizio Time Fractional Distributed Order Reaction-diffusion Equation via Quasi Wavelet based Numerical Method. *J. Appl. Comput. Mech.* **2020**, *6*, 848–861.
120. Du, R.; Hao, Z.; Sun, Z. Lubich second-order methods for distributed-order time-fractional differential equations with smooth solutions. *East Asian J. Appl. Math.* **2016**, *6*, 131–151. [CrossRef]
121. Pimenov, V.G.; Hendy, A.S.; De Staelen, R.H. On a class of non-linear delay distributed order fractional diffusion equations. *J. Comput. Appl. Math.* **2017**, *318*, 433–443. [CrossRef]
122. Aminikhah, H.; Sheikhani, A.H.R.; Houlari, T.; Rezazadeh, H. Numerical solution of the distributed-order fractional Bagley-Torvik equation. *IEEE/CAA J. Autom. Sin.* **2017**, *6*, 760–765. [CrossRef]
123. Fan, W.; Liu, F. A numerical method for solving the two-dimensional distributed order space-fractional diffusion equation on an irregular convex domain. *Appl. Math. Lett.* **2018**, *77*, 114–121. [CrossRef]
124. Ford, N.J.; Morgado, M.L.; Rebelo, M. A numerical method for the distributed order time-fractional diffusion equation. In Proceedings of the ICFDA'14 International Conference on Fractional Differentiation and Its Applications 2014, Catania, Italy, 23–25 June 2014; IEEE: New York, NY, USA, 2014; pp. 1–6.

125. Guo, S.; Mei, L.; Zhang, Z.; Jiang, Y. Finite difference/spectral-Galerkin method for a two-dimensional distributed-order time–space fractional reaction–diffusion equation. *Appl. Math. Lett.* **2018**, *85*, 157–163. [CrossRef]

126. Hu, X.; Liu, F.; Turner, I.; Anh, V. An implicit numerical method of a new time distributed-order and two-sided space-fractional advection-dispersion equation. *Numer. Algorithms* **2016**, *72*, 393–407. [CrossRef]

127. Li, J.; Liu, F.; Feng, L.; Turner, I. A novel finite volume method for the Riesz space distributed-order diffusion equation. *Comput. Math. Appl.* **2017**, *74*, 772–783. [CrossRef]

128. Li, J.; Liu, F.; Feng, L.; Turner, I. A novel finite volume method for the Riesz space distributed-order advection–diffusion equation. *Appl. Math. Model.* **2017**, *46*, 536–553. [CrossRef]

129. Morgado, M.L.; Rebelo, M. Numerical approximation of distributed order reaction–diffusion equations. *J. Comput. Appl. Math.* **2015**, *275*, 216–227. [CrossRef]

130. Moghaddam, B.; Machado, J.T.; Morgado, M. Numerical approach for a class of distributed order time fractional partial differential equations. *Appl. Numer. Math.* **2019**, *136*, 152–162. [CrossRef]

131. Ye, H.; Liu, F.; Anh, V.; Turner, I. Numerical analysis for the time distributed-order and Riesz space fractional diffusions on bounded domains. *IMA J. Appl. Math.* **2015**, *80*, 825–838. [CrossRef]

132. Ye, H.; Liu, F.; Anh, V. Compact difference scheme for distributed-order time-fractional diffusion-wave equation on bounded domains. *J. Comput. Phys.* **2015**, *298*, 652–660. [CrossRef]

133. Wang, X.; Liu, F.; Chen, X. Novel second-order accurate implicit numerical methods for the Riesz space distributed-order advection-dispersion equations. *Adv. Math. Phys.* **2015**, *2015*. [CrossRef]

134. Zheng, R.; Liu, F.; Jiang, X.; Turner, I.W. Finite difference/spectral methods for the two-dimensional distributed-order time-fractional cable equation. *Comput. Math. Appl.* **2020**, *80*, 1523–1537. [CrossRef]

135. Heris, M.S.; Javidi, M. Fractional backward differential formulas for the distributed-order differential equation with time delay. *Bull. Iran. Math. Soc.* **2019**, *45*, 1159–1176. [CrossRef]

136. Javidi, M.; Heris, M.S. Analysis and numerical methods for the Riesz space distributed-order advection-diffusion equation with time delay. *SEMA J.* **2019**, *76*, 533–551. [CrossRef]

137. Aboelenen, T. Local discontinuous Galerkin method for distributed-order time and space-fractional convection–diffusion and Schrödinger-type equations. *Nonlinear Dyn.* **2018**, *92*, 395–413. [CrossRef]

138. Liao, H.; Lyu, P.; Vong, S.; Zhao, Y. Stability of fully discrete schemes with interpolation-type fractional formulas for distributed-order subdiffusion equations. *Numer. Algorithms* **2017**, *75*, 845–878. [CrossRef]

139. Li, X.; Rui, H.; Liu, Z. Two alternating direction implicit spectral methods for two-dimensional distributed-order differential equation. *Numer. Algorithms* **2019**, *82*, 321–347. [CrossRef]

140. Li, L.; Liu, F.; Feng, L.; Turner, I. A Galerkin finite element method for the modified distributed-order anomalous sub-diffusion equation. *J. Comput. Appl. Math.* **2020**, *368*, 112589. [CrossRef]

141. Li, X.; Rui, H. A block-centered finite difference method for the distributed-order time-fractional diffusion-wave equation. *Appl. Numer. Math.* **2018**, *131*, 123–139. [CrossRef]

142. Jian, H.; Huang, T.; Zhao, X.; Zhao, Y. A fast implicit difference scheme for a new class of time distributed-order and space fractional diffusion equations with variable coefficients. *Adv. Differ. Equ.* **2018**, *2018*, 205. [CrossRef]

143. Ren, J.; Chen, H. A numerical method for distributed order time fractional diffusion equation with weakly singular solutions. *Appl. Math. Lett.* **2019**, *96*, 159–165. [CrossRef]

144. Li, C.; Zeng, F. *Numerical Methods for Fractional Calculus*; CRC Press: Boca Raton, FL, USA, 2015; Volume 24.

145. Diethelm, K.; Ford, N.J.; Freed, A.D.; Luchko, Y. Algorithms for the fractional calculus: A selection of numerical methods. *Comput. Methods Appl. Mech. Eng.* **2005**, *194*, 743–773. [CrossRef]

146. Li, C.; Zeng, F. Finite difference methods for fractional differential equations. *Int. J. Bifurc. Chaos* **2012**, *22*, 1230014. [CrossRef]

147. Bhrawy, A.H.; Taha, T.M.; Machado, J.A.T. A review of operational matrices and spectral techniques for fractional calculus. *Nonlinear Dyn.* **2015**, *81*, 1023–1052. [CrossRef]

148. Guo, S.; Mei, L.; Zhang, Z.; Li, C.; Li, M.; Wang, Y. A linearized finite difference/spectral-Galerkin scheme for three-dimensional distributed-order time–space fractional nonlinear reaction–diffusion-wave equation: Numerical simulations of Gordon-type solitons. *Comput. Phys. Commun.* **2020**, 107144. [CrossRef]

149. Hafez, R.M.; Zaky, M.A. High-order continuous Galerkin methods for multi-dimensional advection–reaction–diffusion problems. *Eng. Comput.* **2019**, *36*, 1813–1829. [CrossRef]

150. Zhao, J.; Zhang, Y.; Xu, Y. Implicit Runge–Kutta and spectral Galerkin methods for Riesz space fractional/distributed-order diffusion equation. *Comput. Appl. Math.* **2020**, *39*, 47. [CrossRef]

151. Samiee, M.; Kharazmi, E.; Zayernouri, M.; Meerschaert, M.M. Petrov-Galerkin method for fully distributed-order fractional partial differential equations. *arXiv* **2018**, arXiv:1805.08242.

152. Bhrawy, A.; Zaky, M. Numerical simulation of multi-dimensional distributed-order generalized Schrödinger equations. *Nonlinear Dyn.* **2017**, *89*, 1415–1432. [CrossRef]

153. Zaky, M.A.; Ameen, I.G. On the rate of convergence of spectral collocation methods for nonlinear multi-order fractional initial value problems. *Comput. Appl. Math.* **2019**, *38*, 144. [CrossRef]

154. Pourbabaee, M.; Saadatmandi, A. A novel Legendre operational matrix for distributed order fractional differential equations. *Appl. Math. Comput.* **2019**, *361*, 215–231. [CrossRef]
155. Doha, E.H.; Bhrawy, A.H.; Ezz-Eldien, S.S. A Chebyshev spectral method based on operational matrix for initial and boundary value problems of fractional order. *Comput. Math. Appl.* **2011**, *62*, 2364–2373. [CrossRef]
156. Doha, E.H.; Bhrawy, A.H.; Ezz-Eldien, S. Efficient Chebyshev spectral methods for solving multi-term fractional orders differential equations. *Appl. Math. Model.* **2011**, *35*, 5662–5672. [CrossRef]
157. Bhrawy, A.H.; Zaky, M.A. A method based on the Jacobi tau approximation for solving multi-term time–space fractional partial differential equations. *J. Comput. Phys.* **2015**, *281*, 876–895. [CrossRef]
158. Zaky, M.A. A Legendre spectral quadrature tau method for the multi-term time-fractional diffusion equations. *Comput. Appl. Math.* **2018**, *37*, 3525–3538. [CrossRef]
159. Atabakzadeh, M.; Akrami, M.; Erjaee, G. Chebyshev operational matrix method for solving multi-order fractional ordinary differential equations. *Appl. Math. Model.* **2013**, *37*, 8903–8911. [CrossRef]
160. Semary, M.S.; Hassan, H.N.; Radwan, A.G. Modified methods for solving two classes of distributed order linear fractional differential equations. *Appl. Math. Comput.* **2018**, *323*, 106–119. [CrossRef]
161. Mashoof, M.; Sheikhani, A.R. Simulating the solution of the distributed order fractional differential equations by block-pulse wavelets. *UPB Sci. Bull. Ser. A Appl. Math. Phys.* **2017**, *79*, 193–206.
162. Kharazmi, E.; Zayernouri, M. Fractional pseudo-spectral methods for distributed-order fractional PDEs. *Int. J. Comput. Math.* **2018**, *95*, 1340–1361. [CrossRef]
163. Li, X.; Wu, B. A numerical method for solving distributed order diffusion equations. *Appl. Math. Lett.* **2016**, *53*, 92–99. [CrossRef]
164. Sweilam, N.; Khader, M.; Al-Bar, R. Numerical studies for a multi-order fractional differential equation. *Phys. Lett. A* **2007**, *371*, 26–33. [CrossRef]
165. Jafari, H.; Golbabai, A.; Seifi, S.; Sayevand, K. Homotopy analysis method for solving multi-term linear and nonlinear diffusion–wave equations of fractional order. *Comput. Math. Appl.* **2010**, *59*, 1337–1344. [CrossRef]
166. Jafari, M.; Aminataei, A. An algorithm for solving multi-term diffusion-wave equations of fractional order. *Comput. Math. Appl.* **2011**, *62*, 1091–1097. [CrossRef]
167. Aminikhah, H.; Sheikhani, A.H.R.; Rezazadeh, H. Approximate analytical solutions of distributed order fractional Riccati differential equation. *Ain Shams Eng. J.* **2018**, *9*, 581–588. [CrossRef]
168. Patnaik, S.; Semperlotti, F. Application of variable-and distributed-order fractional operators to the dynamic analysis of nonlinear oscillators. *Nonlinear Dyn.* **2020**, *100*, 561–580. [CrossRef]
169. Daftardar-Gejji, V.; Jafari, H. Solving a multi-order fractional differential equation using Adomian decomposition. *Appl. Math. Comput.* **2007**, *189*, 541–548. [CrossRef]
170. Daftardar-Gejji, V.; Bhalekar, S. Solving multi-term linear and non-linear diffusion–wave equations of fractional order by Adomian decomposition method. *Appl. Math. Comput.* **2008**, *202*, 113–120. [CrossRef]
171. Sadeghinia, A.; Kumar, P. One solution of multi-term fractional differential equations by Adomian decomposition method. *Int. J. Sci. Innov. Math. Res.* **2015**, *3*, 14–21.
172. Jin, B.; Li, B.; Zhou, Z. Numerical analysis of nonlinear subdiffusion equations. *SIAM J. Numer. Anal.* **2018**, *56*, 1–23. [CrossRef]
173. Jin, B.; Lazarov, R.; Zhou, Z. Numerical methods for time-fractional evolution equations with nonsmooth data: A concise overview. *Comput. Methods Appl. Mech. Eng.* **2019**, *346*, 332–358. [CrossRef]
174. Stynes, M.; O'Riordan, E.; Gracia, J.L. Error analysis of a finite difference method on graded meshes for a time-fractional diffusion equation. *SIAM J. Numer. Anal.* **2017**, *55*, 1057–1079. [CrossRef]
175. Hu, X.; Liu, F.; Anh, V.; Turner, I. A numerical investigation of the time distributed-order diffusion model. *ANZIAM J.* **2013**, *55*, C464–C478. [CrossRef]
176. Alikhanov, A.A. Numerical methods of solutions of boundary value problems for the multi-term variable-distributed order diffusion equation. *Appl. Math. Comput.* **2015**, *268*, 12–22. [CrossRef]
177. Morgado, M.; Rebelo, M. Introducing graded meshes in the numerical approximation of distributed-order diffusion equations. In *AIP Conference Proceedings*; AIP Publishing LLC: Melville, NY, USA, 2016; Volume 1776, p. 070002.
178. Abbaszadeh, M. Error estimate of second-order finite difference scheme for solving the Riesz space distributed-order diffusion equation. *Appl. Math. Lett.* **2019**, *88*, 179–185. [CrossRef]
179. Li, X.; Wong, P.J. An efficient nonpolynomial spline method for distributed order fractional subdiffusion equations. *Math. Methods Appl. Sci.* **2018**, *41*, 4906–4922. [CrossRef]
180. Yang, X.; Zhang, H.; Xu, D. WSGD-OSC scheme for two-dimensional distributed order fractional reaction–diffusion equation. *J. Sci. Comput.* **2018**, *76*, 1502–1520. [CrossRef]
181. Ran, M.; Zhang, C. New compact difference scheme for solving the fourth-order time fractional sub-diffusion equation of the distributed order. *Appl. Numer. Math.* **2018**, *129*, 58–70. [CrossRef]
182. Sun, Z.; Wu, X. A fully discrete difference scheme for a diffusion-wave system. *Appl. Numer. Math.* **2006**, *56*, 193–209. [CrossRef]
183. Qiao, H.; Liu, Z.; Cheng, A. Two unconditionally stable difference schemes for time distributed-order differential equation based on Caputo–Fabrizio fractional derivative. *Adv. Differ. Equ.* **2020**, *2020*, 36. [CrossRef]

184. Abbaszadeh, M.; Dehghan, M. A POD-based reduced-order Crank-Nicolson/fourth-order alternating direction implicit (ADI) finite difference scheme for solving the two-dimensional distributed-order Riesz space-fractional diffusion equation. *Appl. Numer. Math.* **2020**, *158*, 271–291. [CrossRef]

185. Jin, B.; Lazarov, R.; Liu, Y.; Zhou, Z. The Galerkin finite element method for a multi-term time-fractional diffusion equation. *J. Comput. Phys.* **2015**, *281*, 825–843. [CrossRef]

186. Bu, W.; Liu, X.; Tang, Y.; Yang, J. Finite element multigrid method for multi-term time fractional advection diffusion equations. *Int. J. Model. Simul. Sci. Comput.* **2015**, *6*, 1540001. [CrossRef]

187. Zhao, Y.; Zhang, Y.; Liu, F.; Turner, I.; Shi, D. Analytical solution and nonconforming finite element approximation for the 2D multi-term fractional subdiffusion equation. *Appl. Math. Model.* **2016**, *40*, 8810–8825. [CrossRef]

188. Jin, B.; Lazarov, R.; Sheen, D.; Zhou, Z. Error estimates for approximations of distributed order time fractional diffusion with nonsmooth data. *Fract. Calc. Appl. Anal.* **2016**, *19*, 69–93. [CrossRef]

189. Hou, Y.; Wen, C.; Li, H.; Liu, Y.; Fang, Z.; Yang, Y. Some Second-Order σ Schemes Combined with an H^1-Galerkin MFE Method for a Nonlinear Distributed-Order Sub-Diffusion Equation. *Mathematics* **2020**, *8*, 187. [CrossRef]

190. Wei, L.; Liu, L.; Sun, H. Stability and convergence of a local discontinuous Galerkin method for the fractional diffusion equation with distributed order. *J. Appl. Math. Comput.* **2019**, *59*, 323–341. [CrossRef]

191. Fan, W.; Jiang, X.; Liu, F.; Anh, V. The unstructured mesh finite element method for the two-dimensional multi-term time–space fractional diffusion-wave equation on an irregular convex domain. *J. Sci. Comput.* **2018**, *77*, 27–52. [CrossRef]

192. Liu, F.; Feng, L.; Anh, V.; Li, J. Unstructured-mesh Galerkin finite element method for the two-dimensional multi-term time–space fractional Bloch–Torrey equations on irregular convex domains. *Comput. Math. Appl.* **2019**, *78*, 1637–1650. [CrossRef]

193. Shi, Y.H.; Liu, F.; Zhao, Y.M.; Wang, F.L.; Turner, I. An unstructured mesh finite element method for solving the multi-term time fractional and Riesz space distributed-order wave equation on an irregular convex domain. *Appl. Math. Model.* **2019**, *73*, 615–636. [CrossRef]

194. Yin, B.; Liu, Y.; Li, H.; Zhang, Z. Approximation methods for the distributed order calculus using the convolution quadrature. *Discret. Contin. Dyn. Syst. B* **2017**, *22*. [CrossRef]

195. Podlubny, I. Matrix approach to discrete fractional calculus. *Fract. Calc. Appl. Anal.* **2000**, *3*, 359–386.

196. Podlubny, I.; Chechkin, A.; Skovranek, T.; Chen, Y.; Jara, B.M.V. Matrix approach to discrete fractional calculus II: Partial fractional differential equations. *J. Comput. Phys.* **2009**, *228*, 3137–3153. [CrossRef]

197. Podlubny, I.; Skovranek, T.; Vinagre Jara, B.M.; Petras, I.; Verbitsky, V.; Chen, Y. Matrix approach to discrete fractional calculus III: Non-equidistant grids, variable step length and distributed orders. *Philos. Trans. R. Soc. A Math. Phys. Eng. Sci.* **2013**, *371*, 20120153. [CrossRef]

198. Diethelm, K.; Ford, N.J. Multi-order fractional differential equations and their numerical solution. *Appl. Math. Comput.* **2004**, *154*, 621–640. [CrossRef]

199. Liu, F.; Meerschaert, M.; McGough, R.; Zhuang, P.; Liu, Q. Numerical methods for solving the multi-term time-fractional wave-diffusion equation. *Fract. Calc. Appl. Anal.* **2013**, *16*, 9–25. [CrossRef]

200. Ye, H.; Liu, F.; Anh, V.; Turner, I. Maximum principle and numerical method for the multi-term time–space Riesz—Caputo fractional differential equations. *Appl. Math. Comput.* **2014**, *227*, 531–540. [CrossRef]

201. Kazmi, K.; Khaliq, A.Q. An efficient split-step method for distributed-order space-fractional reaction-diffusion equations with time-dependent boundary conditions. *Appl. Numer. Math.* **2020**, *147*, 142–160. [CrossRef]

202. Zheng, X.; Liu, H.; Wang, H.; Fu, H. An efficient finite volume method for nonlinear distributed-order space-fractional diffusion equations in three space dimensions. *J. Sci. Comput.* **2019**, *80*, 1395–1418. [CrossRef]

203. Jia, J.; Wang, H. A fast finite difference method for distributed-order space-fractional partial differential equations on convex domains. *Comput. Math. Appl.* **2018**, *75*, 2031–2043. [CrossRef]

204. Gong, C.; Bao, W.; Tang, G.; Yang, B.; Liu, J. An efficient parallel solution for Caputo fractional reaction–diffusion equation. *J. Supercomput.* **2014**, *68*, 1521–1537. [CrossRef]

205. Sweilam, N.H.; Moharram, H.; Moniem, N.A.; Ahmed, S. A parallel Crank–Nicolson finite difference method for time-fractional parabolic equation. *J. Numer. Math.* **2014**, *22*, 363–382. [CrossRef]

206. Biala, T.; Khaliq, A. Parallel algorithms for nonlinear time–space fractional parabolic PDEs. *J. Comput. Phys.* **2018**, *375*, 135–154. [CrossRef]

207. Liu, J.; Gong, C.; Bao, W.; Tang, G.; Jiang, Y. Solving the Caputo fractional reaction-diffusion equation on GPU. *Discret. Dyn. Nat. Soc.* **2014**, *2014*. [CrossRef]

208. Zhao, Y.; Zhu, P.; Gu, X.; Zhao, X.; Cao, J. A limited-memory block bi-diagonal Toeplitz preconditioner for block lower triangular Toeplitz system from time–space fractional diffusion equation. *J. Comput. Appl. Math.* **2019**, *362*, 99–115. [CrossRef]

209. Zhao, Y.; Gu, X.; Li, M.; Jian, H. Preconditioners for all-at-once system from the fractional mobile/immobile advection–diffusion model. *J. Appl. Math. Comput.* **2020**, 1–23. [CrossRef]

210. Li, Y.; Chen, Y. Theory and implementation of distributed-order element networks. In Proceedings of the International Design Engineering Technical Conferences and Computers and Information in Engineering Conference, Washington, DC, USA, 28–31 August 2011; Volume 54808, pp. 361–367.

211. West, B.J.; Latka, M.; Glaubic-Latka, M.; Latka, D. Multifractality of cerebral blood flow. *Phys. A Stat. Mech. Its Appl.* **2003**, *318*, 453–460. [CrossRef]

212. Shlesinger, M.F.; West, B.J.; Klafter, J. Lévy dynamics of enhanced diffusion: Application to turbulence. *Phys. Rev. Lett.* **1987**, *58*, 1100. [CrossRef]

213. Patnaik, S.; Sidhardh, S.; Semperlotti, F. Towards a unified approach to nonlocal elasticity via fractional-order mechanics. *Int. J. Mech. Sci.* **2020**, *189*, 105992. [CrossRef]

214. Demontis, P.; Suffritti, G.B. Fractional diffusion interpretation of simulated single-file systems in microporous materials. *Phys. Rev. E* **2006**, *74*, 051112. [CrossRef] [PubMed]

215. Srokowski, T. Lévy flights in nonhomogeneous media: Distributed-order fractional equation approach. *Phys. Rev. E* **2008**, *78*, 031135. [CrossRef] [PubMed]

216. Martins, J.; Ribeiro, H.V.; Evangelista, L.R.; da Silva, L.R.; Lenzi, E.K. Fractional Schrödinger equation with noninteger dimensions. *Appl. Math. Comput.* **2012**, *219*, 2313–2319. [CrossRef]

217. Sandev, T.; Tomovski, Ž. Langevin equation for a free particle driven by power law type of noises. *Phys. Lett. A* **2014**, *378*, 1–9. [CrossRef]

218. Gemant, A. A method of analyzing experimental results obtained from elasto-viscous bodies. *Physics* **1936**, *7*, 311–317. [CrossRef]

219. Gemant, A. On fractional differentials. *Lond. Edinb. Dublin Philos. Mag. J. Sci.* **1938**, *25*, 540–549. [CrossRef]

220. Hartley, T.T.; Lorenzo, C.F. Fractional-order system identification based on continuous order-distributions. *Signal Process.* **2003**, *83*, 2287–2300. [CrossRef]

221. Atanackovic, T.M. A generalized model for the uniaxial isothermal deformation of a viscoelastic body. *Acta Mech.* **2002**, *159*, 77–86. [CrossRef]

222. Atanackovic, T.M. On a distributed derivative model of a viscoelastic body. *Comptes Rendus Mec.* **2003**, *331*, 687–692. [CrossRef]

223. Atanacković, T.M.; Konjik, S.; Oparnica, L.; Zorica, D. Thermodynamical restrictions and wave propagation for a class of fractional order viscoelastic rods. *Abstr. Appl. Anal.* **2011**, *2011*. [CrossRef]

224. Caputo, M.; Mainardi, F. Linear models of dissipation in anelastic solids. *La Rivista Del Nuovo Cimento (1971–1977)* **1971**, *1*, 161–198. [CrossRef]

225. Atanackovic, T.M.; Pilipovic, S.; Zorica, D. Distributed-order fractional wave equation on a finite domain: Creep and forced oscillations of a rod. *Contin. Mech. Thermodyn.* **2011**, *23*, 305–318. [CrossRef]

226. Atanackovic, T.M.; Pilipovic, S.; Zorica, D. Distributed-order fractional wave equation on a finite domain. Stress relaxation in a rod. *Int. J. Eng. Sci.* **2011**, *49*, 175–190. [CrossRef]

227. Atanackovic, T.M.; Budincevic, M.; Pilipovic, S. On a fractional distributed-order oscillator. *J. Phys. A Math. Gen.* **2005**, *38*, 6703. [CrossRef]

228. Duan, J.S.; Chen, Y. Mechanical response and simulation for constitutive equations with distributed order derivatives. *Int. J. Model. Simul. Sci. Comput.* **2017**, *8*, 1750040. [CrossRef]

229. Stankovic, B.; Atanackovic, T. Dynamics of a rod made of generalized Kelvin–Voigt visco-elastic material. *J. Math. Anal. Appl.* **2002**, *268*, 550–563. [CrossRef]

230. Rossikhin, Y.A.; Shitikova, M. A new method for solving dynamic problems of fractional derivative viscoelasticity. *Int. J. Eng. Sci.* **2001**, *39*, 149–176. [CrossRef]

231. Näsholm, S.P.; Holm, S. On a fractional Zener elastic wave equation. *Fract. Calc. Appl. Anal.* **2013**, *16*, 26–50. [CrossRef]

232. Atanackovic, T.M. A modified Zener model of a viscoelastic body. *Contin. Mech. Thermodyn.* **2002**, *14*, 137. [CrossRef]

233. Konjik, S.; Oparnica, L.; Zorica, D. Waves in fractional Zener type viscoelastic media. *J. Math. Anal. Appl.* **2010**, *365*, 259–268. [CrossRef]

234. Konjik, S.; Oparnica, L.; Zorica, D. Distributed-order fractional constitutive stress–strain relation in wave propagation modeling. *Zeitschrift für angewandte Mathematik und Physik* **2019**, *70*, 51. [CrossRef]

235. Rossikhin, Y.A.; Shitikova, M. Analysis of dynamic behaviour of viscoelastic rods whose rheological models contain fractional derivatives of two different orders. *ZAMM-J. Appl. Math. Mech.* **2001**, *81*, 363–376. [CrossRef]

236. Pritz, T. Analysis of four-parameter fractional derivative model of real solid materials. *J. Sound Vib.* **1996**, *195*, 103–115. [CrossRef]

237. Pritz, T. Five-parameter fractional derivative model for polymeric damping materials. *J. Sound Vib.* **2003**, *265*, 935–952. [CrossRef]

238. Petrovic, L.M.; Zorica, D.M.; Stojanac, I.L.; Krstonosic, V.S.; Hadnadjev, M.S.; Janev, M.B.; Premovic, M.T.; Atanackovic, T.M. Viscoelastic properties of uncured resin composites: Dynamic oscillatory shear test and fractional derivative model. *Dent. Mater.* **2015**, *31*, 1003–1009. [CrossRef]

239. Naranjani, Y.; Sardahi, Y.; Chen, Y.; Sun, J.Q. Multi-objective optimization of distributed-order fractional damping. *Commun. Nonlinear Sci. Numer. Simul.* **2015**, *24*, 159–168. [CrossRef]

240. Jokar, M.; Patnaik, S.; Semperlotti, F. Variable-Order Approach to Nonlocal Elasticity: Theoretical Formulation and Order Identification via Deep Learning Techniques. *arXiv* **2020**, arXiv:2008.13582.

241. Li, G.; Zhu, Z.; Cheng, C. Dynamical stability of viscoelastic column with fractional derivative constitutive relation. *Appl. Math. Mech.* **2001**, *22*, 294–303. [CrossRef]

242. Bačlić, B.; Atanacković, T. Stability and creep of a fractional derivative order viscoelastic rod. *Bulletin (Académie serbe des sciences et des arts. Classe des sciences mathématiques et naturelles. Sciences mathématiques)* **2000**, *25*, 115–131.

243. Atanackovic, T.; Stankovic, B. Dynamics of a viscoelastic rod of fractional derivative type. *ZAMM-J. Appl. Math. Mech.* **2002**, *82*, 377–386. [CrossRef]

244. Stankovic, B.; Atanackovic, T. On a model of a viscoelastic rod. *Fract. Calc. Appl. Anal.* **2001**, *4*, 501–522.

245. Stankovic, B.; Atanackovic, T. On a viscoelastic rod with constitutive equation containing fractional derivatives of two different orders. *Math. Mech. Solids* **2004**, *9*, 629–656. [CrossRef]

246. Zorica, D.; Atanacković, T.M.; Vrcelj, Z.; Novaković, B. Dynamic stability of axially loaded nonlocal rod on generalized Pasternak foundation. *J. Eng. Mech.* **2017**, *143*, D4016003. [CrossRef]

247. Varghaei, P.; Kharazmi, E.; Suzuki, J.L.; Zayernouri, M. Vibration analysis of geometrically nonlinear and fractional viscoelastic cantilever beams. *arXiv* **2019**, arXiv:1909.02142.

248. Duan, J.; Chen, L. Oscillatory shear flow between two parallel plates for viscoelastic constitutive model of distributed-order derivative. *Int. J. Numer. Methods Heat Fluid Flow* **2019**. [CrossRef]

249. Metzler, R.; Klafter, J. The restaurant at the end of the random walk: Recent developments in the description of anomalous transport by fractional dynamics. *J. Phys. A Math. Gen.* **2004**, *37*, R161. [CrossRef]

250. Hanyga, A. Anomalous diffusion without scale invariance. *J. Phys. A Math. Theor.* **2007**, *40*, 5551. [CrossRef]

251. Tateishi, A.A.; Ribeiro, H.V.; Lenzi, E.K. The role of fractional time-derivative operators on anomalous diffusion. *Front. Phys.* **2017**, *5*, 52. [CrossRef]

252. Meerschaert, M.M.; Nane, E.; Vellaisamy, P. Distributed-order fractional diffusions on bounded domains. *J. Math. Anal. Appl.* **2011**, *379*, 216–228. [CrossRef]

253. Sandev, T.; Chechkin, A.V.; Korabel, N.; Kantz, H.; Sokolov, I.M.; Metzler, R. Distributed-order diffusion equations and multifractality: Models and solutions. *Phys. Rev. E* **2015**, *92*, 042117. [CrossRef]

254. Koscielny-Bunde, E.; Bunde, A.; Havlin, S.; Roman, H.E.; Goldreich, Y.; Schellnhuber, H.J. Indication of a universal persistence law governing atmospheric variability. *Phys. Rev. Lett.* **1998**, *81*, 729. [CrossRef]

255. Ashkenazy, Y.; Baker, D.R.; Gildor, H.; Havlin, S. Nonlinearity and multifractality of climate change in the past 420,000 years. *Geophys. Res. Lett.* **2003**, *30*, 2146. [CrossRef]

256. Su, N. Distributed-order infiltration, absorption and water exchange in mobile and immobile zones of swelling soils. *J. Hydrol.* **2012**, *468*, 1–10. [CrossRef]

257. Yang, Z.; Zheng, X.; Wang, H. A variably distributed-order time-fractional diffusion equation: Analysis and approximation. *Comput. Methods Appl. Mech. Eng.* **2020**, *367*, 113118. [CrossRef]

258. Oświecimka, P.; Kwapień, J.; Drożdż, S. Wavelet versus detrended fluctuation analysis of multifractal structures. *Phys. Rev. E* **2006**, *74*, 016103. [CrossRef] [PubMed]

259. Benzi, R.; Biferale, L.; Toschi, F. Multiscale velocity correlations in turbulence. *Phys. Rev. Lett.* **1998**, *80*, 3244. [CrossRef]

260. Beghin, L. Random-time processes governed by differential equations of fractional distributed order. *Chaos Solitons Fractals* **2012**, *45*, 1314–1327. [CrossRef]

261. Chechkin, A.V.; Gorenflo, R.; Sokolov, I.M. Retarding subdiffusion and accelerating superdiffusion governed by distributed-order fractional diffusion equations. *Phys. Rev. E* **2002**, *66*, 046129. [CrossRef]

262. Sokolov, I.M.; Chechkin, A.V.; Klafter, J. Fractional diffusion equation for a power-law-truncated Lévy process. *Phys. A Stat. Mech. Its Appl.* **2004**, *336*, 245–251. [CrossRef]

263. Sandev, T.; Sokolov, I.M.; Metzler, R.; Chechkin, A. Beyond monofractional kinetics. *Chaos Solitons Fractals* **2017**, *102*, 210–217. [CrossRef]

264. Sandev, T.; Iomin, A.; Kantz, H.; Metzler, R.; Chechkin, A. Comb model with slow and ultraslow diffusion. *Math. Model. Nat. Phenom.* **2016**, *11*, 18–33. [CrossRef]

265. Sinai, Y.G. The limiting behavior of a one-dimensional random walk in a random medium. *Theory Probab. Its Appl.* **1983**, *27*, 256–268. [CrossRef]

266. Schiessel, H.; Sokolov, I.M.; Blumen, A. Dynamics of a polyampholyte hooked around an obstacle. *Phys. Rev. E* **1997**, *56*, R2390. [CrossRef]

267. Prosen, T.; Žnidarič, M. Anomalous diffusion and dynamical localization in polygonal billiards. *Phys. Rev. Lett.* **2001**, *87*, 114101. [CrossRef]

268. Iglói, F.; Turban, L.; Rieger, H. Anomalous diffusion in aperiodic environments. *Phys. Rev. E* **1999**, *59*, 1465. [CrossRef]

269. Dräger, J.; Klafter, J. Strong anomaly in diffusion generated by iterated maps. *Phys. Rev. Lett.* **2000**, *84*, 5998. [CrossRef] [PubMed]

270. Chechkin, A.V.; Gorenflo, R.; Sokolov, I.M.; Gonchar, V.Y. Distributed order time fractional diffusion equation. *Fract. Calc. Appl. Anal.* **2003**, *6*, 259–280.

271. Chechkin, A.V.; Klafter, J.; Sokolov, I.M. Fractional Fokker-Planck equation for ultraslow kinetics. *EPL (Europhys. Lett.)* **2003**, *63*, 326. [CrossRef]

272. Mainardi, F.; Pagnini, G.; Gorenflo, R. Some aspects of fractional diffusion equations of single and distributed order. *Appl. Math. Comput.* **2007**, *187*, 295–305. [CrossRef]

273. Mainardi, F.; Mura, A.; Pagnini, G.; Gorenflo, R. Sub-diffusion equations of fractional order and their fundamental solutions. In *Mathematical Methods in Engineering*; Springer: Berlin/Heidelberg, Germany, 2007; pp. 23–55.

274. Mainardi, F.; Mura, A.; Pagnini, G.; Gorenflo, R. Time-fractional diffusion of distributed order. *J. Vib. Control* **2008**, *14*, 1267–1290. [CrossRef]

275. Shen, S.; Liu, F.; Anh, V. Fundamental solution and discrete random walk model for a time-space fractional diffusion equation of distributed order. *J. Appl. Math. Comput.* **2008**, *28*, 147. [CrossRef]

276. Saxena, R.K.; Pagnini, G. Exact solutions of triple-order time-fractional differential equations for anomalous relaxation and diffusion I: The accelerating case. *Phys. A Stat. Mech. Its Appl.* **2011**, *390*, 602–613. [CrossRef]

277. Umarov, S.; Steinberg, S. Random walk models associated with distributed fractional order differential equations. In *High Dimensional Probability*; Giné, E., Koltchinskii, V., Li, W., Zinn, J., Eds.; Institute of Mathematical: Statistics: Hayward, CA, USA, 2006; pp. 117–127.

278. Sokolov, I.M.; Chechkin, A.V.; Klafter, J. Distributed-order fractional kinetics. *arXiv* **2004**, arXiv:cond-mat/0401146.

279. Chechkin, A.V.; Gonchar, V.Y.; Gorenflo, R.; Korabel, N.; Sokolov, I.M. Generalized fractional diffusion equations for accelerating subdiffusion and truncated Lévy flights. *Phys. Rev. E* **2008**, *78*, 021111. [CrossRef]

280. Chechkin, A.; Sokolov, I.M.; Klafter, J. Natural and modified forms of distributed-order fractional diffusion equations. In *Fractional Dynamics: Recent Advances*; Klafter, J., Lim, S.C., Metzler, R., Eds.; World Scientific: Singapore, 2012; pp. 107–127.

281. Langlands, T.A.M. Solution of a modified fractional diffusion equation. *Phys. A Stat. Mech. Its Appl.* **2006**, *367*, 136–144. [CrossRef]

282. Meerschaert, M.M.; Scheffler, H.P. Stochastic model for ultraslow diffusion. *Stoch. Process. Their Appl.* **2006**, *116*, 1215–1235. [CrossRef]

283. Hahn, M.; Umarov, S. Fractional Fokker-Planck-Kolmogorov type equations and their associated stochastic differential equations. *Fract. Calc. Appl. Anal.* **2011**, *14*, 56–79. [CrossRef]

284. Eab, C.H.; Lim, S.C. Fractional Langevin equations of distributed order. *Phys. Rev. E* **2011**, *83*, 031136. [CrossRef]

285. Hahn, M.; Kobayashi, K.; Umarov, S. SDEs driven by a time-changed Lévy process and their associated time-fractional order pseudo-differential equations. *J. Theor. Probab.* **2012**, *25*, 262–279. [CrossRef]

286. Toaldo, B. Lévy mixing related to distributed order calculus, subordinators and slow diffusions. *J. Math. Anal. Appl.* **2015**, *430*, 1009–1036. [CrossRef]

287. Awad, E. On the time-fractional Cattaneo equation of distributed order. *Phys. A Stat. Mech. Its Appl.* **2019**, *518*, 210–233.

288. Meerschaert, M.; Nane, E.; Vellaisamy, P. The fractional Poisson process and the inverse stable subordinator. *Electron. J. Probab.* **2011**, *16*, 1600–1620.

289. Magdziarz, M.; Teuerle, M. Fractional diffusion equation with distributed-order material derivative. Stochastic foundations. *J. Phys. A Math. Theor.* **2017**, *50*, 184005.

290. Orzeł, S.; Mydlarczyk, W.; Jurlewicz, A. Accelerating subdiffusions governed by multiple-order time-fractional diffusion equations: Stochastic representation by a subordinated Brownian motion and computer simulations. *Phys. Rev. E* **2013**, *87*, 032110.

291. Naber, M. Distributed order fractional sub-diffusion. *Fractals* **2004**, *12*, 23–32.

292. Mainardi, F.; Pagnini, G. The role of the Fox–Wright functions in fractional sub-diffusion of distributed order. *J. Comput. Appl. Math.* **2007**, *207*, 245–257.

293. Li, Z.; Luchko, Y.; Yamamoto, M. Asymptotic estimates of solutions to initial-boundary-value problems for distributed order time-fractional diffusion equations. *Fract. Calc. Appl. Anal.* **2014**, *17*, 1114–1136.

294. Tomovski, Ž.; Sandev, T. Distributed-order wave equations with composite time fractional derivative. *Int. J. Comput. Math.* **2018**, *95*, 1100–1113.

295. Kochubei, A.N. Distributed order calculus and equations of ultraslow diffusion. *J. Math. Anal. Appl.* **2008**, *340*, 252–281.

296. Atanackovic, T.M.; Pilipovic, S.; Zorica, D. Existence and calculation of the solution to the time distributed order diffusion equation. *Phys. Scr.* **2009**, *2009*, 014012.

297. Li, Z.; Yamamoto, M. Initial-boundary value problems for linear diffusion equation with multiple time-fractional derivatives. *arXiv* **2013**, arXiv:1306.2778.

298. Gorenflo, R.; Luchko, Y.; Stojanović, M. Fundamental solution of a distributed order time-fractional diffusion-wave equation as probability density. *Fract. Calc. Appl. Anal.* **2013**, *16*, 297–316. [CrossRef]

299. Sandev, T.; Tomovski, Z.; Crnkovic, B. Generalized distributed order diffusion equations with composite time fractional derivative. *Comput. Math. Appl.* **2017**, *73*, 1028–1040. [CrossRef]

300. Li, Z.; Kian, Y.; Soccorsi, E. Initial-boundary value problem for distributed order time-fractional diffusion equations. *Asymptot. Anal.* **2019**, *115*, 95–126. [CrossRef]

301. Al-Refai, M.; Luchko, Y. Analysis of fractional diffusion equations of distributed order: Maximum principles and their applications. *Analysis* **2016**, *36*, 123–133. [CrossRef]

302. Luchko, Y. Maximum principle and its application for the time-fractional diffusion equations. *Fract. Calc. Appl. Anal.* **2011**, *14*, 110–124. [CrossRef]

303. Rundell, W.; Zhang, Z. Fractional diffusion: Recovering the distributed fractional derivative from overposed data. *Inverse Probl.* **2017**, *33*, 035008. [CrossRef]
304. Li, Z.; Luchko, Y.; Yamamoto, M. Analyticity of solutions to a distributed order time-fractional diffusion equation and its application to an inverse problem. *Comput. Math. Appl.* **2017**, *73*, 1041–1052. [CrossRef]
305. Ruan, Z.; Wang, Z. A backward problem for distributed order diffusion equation: Uniqueness and numerical solution. *Inverse Probl. Sci. Eng.* **2020**. [CrossRef]
306. Li, Z.; Fujishiro, K.; Li, G. Uniqueness in the inversion of distributed orders in ultraslow diffusion equations. *J. Comput. Appl. Math.* **2020**, *369*, 112564. [CrossRef]
307. Sun, C.; Liu, J. An inverse source problem for distributed order time-fractional diffusion equation. *Inverse Probl.* **2020**, *36*, 055008. [CrossRef]
308. Sibatov, R.T. Anomalous grain boundary diffusion: Fractional calculus approach. *Adv. Math. Phys.* **2019**, *2019*. [CrossRef]
309. Santoro, P.A.; De Paula, J.L.; Lenzi, E.K.; Evangelista, L.R. Anomalous diffusion governed by a fractional diffusion equation and the electrical response of an electrolytic cell. *J. Chem. Phys.* **2011**, *135*, 114704. [CrossRef] [PubMed]
310. Ciuchi, F.; Mazzulla, A.; Scaramuzza, N.; Lenzi, E.K.; Evangelista, L.R. Fractional diffusion equation and the electrical impedance: Experimental evidence in liquid-crystalline cells. *J. Phys. Chem. C* **2012**, *116*, 8773–8777. [CrossRef]
311. Lenzi, E.K.; Fernandes, P.R.G.; Petrucci, T.; Mukai, H.; Ribeiro, H.V.; Lenzi, M.K.; Gonçalves, G. Anomalous diffusion and electrical response of ionic solutions. *Int. J. Electrochem. Sci.* **2013**, *8*, 2849–2862.
312. Chen, Y.; Fiorentino, A.; Dal Negro, L. A fractional diffusion random laser. *Sci. Rep.* **2019**, *9*, 1–14. [CrossRef]
313. Kitsyuk, E.P.; Sibatov, R.T.; Svetukhin, V.V. Memory Effect and Fractional Differential Dynamics in Planar Microsupercapacitors Based on Multiwalled Carbon Nanotube Arrays. *Energies* **2020**, *13*, 213. [CrossRef]
314. Bard, J.B.L. A model for generating aspects of zebra and other mammalian coat patterns. *J. Theor. Biol.* **1981**, *93*, 363–385. [CrossRef]
315. Murray, J.D. On pattern formation mechanisms for Lepidopteran wing patterns and mammalian coat markings. *Philos. Trans. R. Soc. Lond. B Biol. Sci.* **1981**, *295*, 473–496.
316. Saxena, R.K.; Mathai, A.M.; Haubold, H.J. Distributed order reaction-diffusion systems associated with Caputo derivatives. *J. Math. Phys.* **2014**, *55*, 083519. [CrossRef]
317. Lenzi, E.K.; Neto, R.M.; Tateishi, A.A.; Lenzi, M.K.; Ribeiro, H.V. Fractional diffusion equations coupled by reaction terms. *Phys. A Stat. Mech. Its Appl.* **2016**, *458*, 9–16. [CrossRef]
318. Saxena, R.K.; Mathai, A.M.; Haubold, H.J. Computational solutions of distributed order reaction-diffusion systems associated with Riemann-Liouville derivatives. *Axioms* **2015**, *4*, 120–133. [CrossRef]
319. Bulavatsky, V.M.; Krivonos, Y.G. Mathematical modeling of the dynamics of anomalous migration fields within the framework of the model of distributed order. *Cybern. Syst. Anal.* **2013**, *49*, 390–396. [CrossRef]
320. Yin, M.; Ma, R.; Zhang, Y.; Wei, S.; Tick, G.R.; Wang, J.; Sun, Z.; Sun, H.; Zheng, C. A distributed-order time fractional derivative model for simulating bimodal sub-diffusion in heterogeneous media. *J. Hydrol.* **2020**, 125504. [CrossRef]
321. Yin, M.; Zhang, Y.; Ma, R.; Tick, G.R.; Bianchi, M.; Zheng, C.; Wei, W.; Wei, S.; Liu, X. Super-diffusion affected by hydrofacies mean length and source geometry in alluvial settings. *J. Hydrol.* **2020**, *582*, 124515. [CrossRef]
322. Su, N.; Nelson, P.N.; Connor, S. The distributed-order fractional diffusion-wave equation of groundwater flow: Theory and application to pumping and slug tests. *J. Hydrol.* **2015**, *529*, 1262–1273. [CrossRef]
323. Liang, Y.; Chen, W.; Xu, W.; Sun, H. Distributed order Hausdorff derivative diffusion model to characterize non-Fickian diffusion in porous media. *Commun. Nonlinear Sci. Numer. Simul.* **2019**, *70*, 384–393. [CrossRef]
324. Caputo, M. Diffusion with space memory modelled with distributed order space fractional differential equations. *Ann. Geophys.* **2003**. [CrossRef]
325. Atanackovic, T.M.; Pilipovic, S.; Zorica, D. A diffusion wave equation with two fractional derivatives of different order. *J. Phys. A Math. Theor.* **2007**, *40*, 5319. [CrossRef]
326. Atanackovic, T.M.; Pilipovic, S.; Zorica, D. Time distributed-order diffusion-wave equation. I. Volterra-type equation. *Proc. R. Soc. A Math. Phys. Eng. Sci.* **2009**, *465*, 1869–1891. [CrossRef]
327. Sandev, T.; Tomovski, Z.; Dubbeldam, J.L.; Chechkin, A. Generalized diffusion-wave equation with memory kernel. *J. Phys. A Math. Theor.* **2018**, *52*, 015201. [CrossRef]
328. Atanackovic, T.M.; Pilipovic, S.; Zorica, D. Time distributed-order diffusion-wave equation. II. Applications of Laplace and Fourier transformations. *Proc. R. Soc. A Math. Phys. Eng. Sci.* **2009**, *465*, 1893–1917. [CrossRef]
329. Caputo, M.; Carcione, J.M. Wave simulation in dissipative media described by distributed-order fractional time derivatives. *J. Vib. Control* **2011**, *17*, 1121–1130. [CrossRef]
330. Mahmoud, G.M.; Farghaly, A.A.; Abed-Elhameed, T.M.; Aly, S.A.; Arafa, A.A. Dynamics of distributed-order hyperchaotic complex van der Pol oscillators and their synchronization and control. *Eur. Phys. J. Plus* **2020**, *135*, 32. [CrossRef]
331. Zhou, F.; Zhao, Y.; Li, Y.; Chen, Y. Design, implementation and application of distributed order PI control. *ISA Trans.* **2013**, *52*, 429–437. [CrossRef] [PubMed]

332. Wang, C.; Guo, Y.; Zheng, S.; Chen, Y. Robust stability analysis of LTI systems with fractional degree generalized frequency variables. *Fract. Calc. Appl. Anal.* **2019**, *22*, 1655–1674. [CrossRef]

333. Adams, J.L.; Hartley, T.T.; Lorenzo, C.F. Fractional-order system identification using complex order-distributions. *IFAC Proc. Vol.* **2006**, *39*, 200–205. [CrossRef]

334. Zaky, M.; Machado, J.T. On the formulation and numerical simulation of distributed-order fractional optimal control problems. *Commun. Nonlinear Sci. Numer. Simul.* **2017**, *52*, 177–189. [CrossRef]

335. Li, Y.; Sheng, H.; Chen, Y. On distributed order low-pass filter. In Proceedings of the 2010 IEEE/ASME International Conference on Mechatronic and Embedded Systems and Applications, Qingdao, China, 15–17 July 2010; IEEE: New York, NY, USA, 2010; pp. 588–592.

336. Jakovljević, B.B.; Rapaić, M.R.; Jelicić, Z.D.; Sekara, T.B. Optimization of distributed order fractional PID controller under constraints on robustness and sensitivity to measurement noise. In Proceedings of the ICFDA'14 International Conference on Fractional Differentiation and Its Applications 2014, Catania, Italy, 23–25 June 2014; IEEE: New York, NY, USA, 2014; pp. 1–6.

337. Jakovljević, B.B.; Šekara, T.B.; Rapaić, M.R.; Jeličić, Z.D. On the distributed order PID controller. *AEU-Int. J. Electron. Commun.* **2017**, *79*, 94–101. [CrossRef]

338. Jakovljević, B.; Lino, P.; Maione, G. Fractional and Distributed Order PID Controllers for PMSM Drives. In Proceedings of the 2019 18th European Control Conference (ECC), Naples, Italy, 25–28 June 2019; IEEE: New York, NY, USA, 2019; pp. 4100–4105.

339. Li, Y.; Sheng, H.; Chen, Y.Q. On distributed order integrator/differentiator. *Signal Process.* **2011**, *91*, 1079–1084. [CrossRef]

340. Li, Y.; Chen, Y.Q. Theory and implementation of weighted distributed order integrator. In Proceedings of the 2012 IEEE/ASME 8th IEEE/ASME International Conference on Mechatronic and Embedded Systems and Applications, Suzhou, China, 8–10 July 2012; IEEE: New York, NY, USA, 2012; pp. 119–124.

341. Najafi, H.S.; Sheikhani, A.R.; Ansari, A. Stability analysis of distributed order fractional differential equations. *Abstr. Appl. Anal.* **2011**, *2011*. [CrossRef]

342. Jiao, Z.; Chen, Y.; Podlubny, I. *Distributed Order Dynamic Systems, Modeling, Analysis and Simulation*; Springer: New York, NY, USA, 2012.

343. Jiao, Z.; Chen, Y.; Zhong, Y. Stability analysis of linear time-invariant distributed-order systems. *Asian J. Control* **2013**, *15*, 640–647. [CrossRef]

344. Rivero, M.; Rogosin, S.V.; Tenreiro Machado, J.A.; Trujillo, J.J. Stability of fractional order systems. *Math. Probl. Eng.* **2013**, *2013*. [CrossRef]

345. Taghavian, H.; Tavazoei, M.S. Algebraic conditions for stability analysis of linear time-invariant distributed order dynamic systems: A Lagrange inversion theorem approach. *Asian J. Control* **2019**, *21*, 879–890. [CrossRef]

346. Tavazoei, M.S. Fractional/distributed-order systems and irrational transfer functions with monotonic step responses. *J. Vib. Control* **2014**, *20*, 1697–1706. [CrossRef]

347. Taghavian, H.; Saleh Tavazoei, M. Robust Stability Analysis of Distributed-Order Linear Time-Invariant Systems With Uncertain Order Weight Functions and Uncertain Dynamic Matrices. *J. Dyn. Syst. Meas. Control* **2017**, *139*, 121010. [CrossRef]

348. Majma, E.; Tavazoei, M.S. Properties of the stability boundary in linear distributed-order systems. *Int. J. Syst. Sci.* **2020**, *51*, 1733–1743. [CrossRef]

349. Fernández-Anaya, G.; Flores-Godoy, J.; Lugo-Peñaloza, A.; Muñoz-Vega, R. Stabilization and passification of distributed-order fractional linear systems using methods of preservation. *J. Frankl. Inst.* **2013**, *350*, 2881–2900. [CrossRef]

350. Li, Y.; Chen, Y. Lyapunov stability of fractional-order nonlinear systems: A distributed-order approach. In Proceedings of the ICFDA'14 International Conference on Fractional Differentiation and Its Applications 2014, Catania, Italy, 23–25 June 2014; IEEE: New York, NY, USA, 2014; pp. 1–6.

351. Boyadzhiev, D.; Kiskinov, H.; Veselinova, M.; Zahariev, A. Stability analysis of linear distributed order fractional systems with distributed delays. *Fract. Calc. Appl. Anal.* **2017**, *20*, 914. [CrossRef]

352. He, B.; Chen, Y.; Kou, C. On the Controllability of Distributed-Order Fractional Systems With Distributed Delays. In Proceedings of the ASME 2017 International Design Engineering Technical Conferences and Computers and Information in Engineering Conference, Cleveland, OH, USA, 6–9 August 2017; American Society of Mechanical Engineers Digital Collection: New York, NY, USA, 2017.

353. Aminikhah, H.; Refahi Sheikhani, A.; Rezazadeh, H. Stability analysis of distributed order fractional Chen system. *Sci. World J.* **2013**, *2013*. [CrossRef]

354. Aminikhah, H.; Sheikhani, A.R.; Rezazadeh, H. Stability analysis of linear distributed order system with multiple time delays. *UPB Sci. Bull.* **2015**, *77*, 207–218.

355. Fernández-Anaya, G.; Nava-Antonio, G.; Jamous-Galante, J.; Muñoz-Vega, R.; Hernández-Martínez, E. Asymptotic stability of distributed order nonlinear dynamical systems. *Commun. Nonlinear Sci. Numer. Simul.* **2017**, *48*, 541–549.

356. Taghavian, H.; Tavazoei, M.S. Stability analysis of distributed-order nonlinear dynamic systems. *Int. J. Syst. Sci.* **2018**, *49*, 523–536.

357. Fernández-Anaya, G.; Quezada-Téllez, L.; Franco-Pérez, L. Stability analysis of distributed order of Hilfer nonlinear systems. *Math. Methods Appl. Sci.* **2020**. [CrossRef]

358. Nava-Antonio, G.; Fernandez-Anaya, G.; Hernandez-Martinez, E.; Jamous-Galante, J.; Ferreira-Vazquez, E.; Flores-Godoy, J. Consensus of multi-agent systems with distributed fractional order dynamics. In Proceedings of the 2017 International Workshop on Complex Systems and Networks (IWCSN), Doha, Qatar, 8–10 December 2017; IEEE: New York, NY, USA, 2017; pp. 190–197.
359. Mahmoud, G.M.; Aboelenen, T.; Abed-Elhameed, T.M.; Farghaly, A.A. Generalized Wright stability for distributed fractional-order nonlinear dynamical systems and their synchronization. *Nonlinear Dyn.* **2019**, *97*, 413–429.
360. Al Themairi, A.; Farghaly, A. The Dynamics Behavior of Coupled Generalized van der Pol Oscillator with Distributed Order. *Math. Probl. Eng.* **2020**, *2020*. [CrossRef]

MDPI

St. Alban-Anlage 66

4052 Basel

Switzerland

Tel. +41 61 683 77 34

Fax +41 61 302 89 18

www.mdpi.com

Entropy Editorial Office

E-mail: entropy@mdpi.com

www.mdpi.com/journal/entropy

www.ingramcontent.com/pod-product-compliance
Lightning Source LLC
LaVergne TN
LVHW070201100526
838202LV00015B/1979